McClure

"A study of Th Time value of money"

GRADIENT COLLECTS UP 1 YR. BEFORE POINT OF Δ

← COLLECTS UP 1 YR. BEFORE 1ST PAYMENT
→ COLLECTS UP AT TIME OF 1ST PAYMENT

ENGINEERING ECONOMY

ENGINEERING ECONOMY

Second Edition

Leland T. Blank, P.E.
Department of Industrial Engineering
Texas A&M University

Anthony J. Tarquin, P.E.
Department of Civil Engineering
University of Texas at El Paso

McGraw-Hill Book Company

New York St. Louis San Francisco Auckland Bogotá Hamburg
Johannesburg London Madrid Mexico Montreal New Delhi
Panama Paris São Paulo Singapore Sydney Tokyo Toronto

This book was set in Press Roman by The Total Book (ACS).
The editor was Julienne V. Brown;
the production supervisor was Diane Renda.
Halliday Lithograph Corporation was printer and binder.

ENGINEERING ECONOMY

1234567890 HALHAL 89876543

ISBN 0-07-062961-7

Library of Congress Cataloging in Publication Data

Blank, Leland T.
Engineering economy.

Rev. ed. of: Engineering economy–a behavioral approach / Anthony J. Tarquin, Leland T. Blank. 1st ed. 1976.
Bibliography: p.
Includes index.
1. Engineering economy. I. Tarquin, Anthony J. II. Tarquin, Anthony J. Engineering economy–a behavioral approach. III. Title.
TA177.4.B58 1983 658.1'52 82-15340
ISBN 0-07-062961-7 AACR2

To our parents
Edna
Alvin
Virginia
Ercole

for all their encouragement

CONTENTS

*Optional section or chapter.

PREFACE

The first edition of this text presented, in a clearly written fashion, the basic principles of economic analysis for application in the decision making process. Our objective was to present the material in the clearest, most concise method possible without sacrificing coverage or true understanding on the part of the reader. In this second edition, we have attempted to do the same thing while retaining the basic structure of the text developed in the first edition.

MATERIAL AND ORGANIZATIONAL CHANGES

New chapters or sections have been added for *rate of return analysis* of a project by inclusion of discussions on internal, multiple and external rates. The new depreciation and tax laws introduced by the Economic Recovery Tax Act of 1981 are covered. In particular, the *Accelerated Cost Recovery System (ACRS)* is presented and discussed. The effects of this new capital recovery system are examined in examples and problems. Optional material has been prepared for the topics of inflation considerations, escalating (gradient) series, continuous compounding of interest (with tables), and the correct approach to solving capital budgeting problems.

Reorganization and condensation of material has taken place in the areas of cash flow diagrams, uniform gradients and mutually exclusive alternatives. In all the overriding consideration has been the preservation of the free flowing, easy to understand format which characterized edition one. We are confident that this text represents an up-to-date, well-balanced presentation on economic analysis with coverage that is particularly relevant to engineers and other decision makers.

USE OF TEXT

This text has been prepared in an easy to read fashion for use in teaching and as a reference book for the basic computations used in an engineering economic analysis.

It is best suited for use in a one semester or one quarter undergraduate course in engineering economic analysis, project analysis or engineering cost analysis. The students should have at least a junior standing. A background in calculus is not necessary to understand the material contained herein, but a basic understanding of economics and accounting (especially from the cost viewpoint) is very helpful. However, the building block approach used in the text's design allows a practitioner unacquainted with economics and engineering principles to use the text to learn, understand and correctly apply the techniques in the process of decision making.

Computer programs for microprocessors which assist in the analysis of certain problems are discussed briefly throughout the chapters and in Appendix E. These programs are detailed and presented in the Instructor's Guide, as are the solutions to all problems at the end of each chapter.

COMPOSITION OF TEXT

Each chapter contains an overall and several specific behavioral objectives followed by the study material with section headings that correspond to the specific objectives. Thus, section 5.1, for example, contains the material which discusses objective 5.1. Virtually all sections contain one or more illustrative solved examples which are separated from the textual material and include comments about the solution and pertinent relations to other topics in the book. Each section has reference to advanced solved problems at the end of the chapter and unsolved problems which the reader should now be able to understand and solve. The final answer to these problems is contained in Appendix F. This approach allows the opportunity to apply material on a section-by-section basis or wait until an entire chapter is completed.

Some material is considered optional and is indicated by an asterisk (*) in the table of contents, in the chapter and by each problem applicable to a starred section. These sections may be omitted with no loss of understanding of how to correctly apply the technique. Also, it is not necessary in a subsequent chapter and section to have covered the material in earlier, optional sections, unless possibly the section itself is optional.

For special topics discussed in other texts which may be of interest to the reader a Further Information section is included in some chapters. Reference by subject area is given to citations detailed in the Bibliography.

TEXT OVERVIEW

The text is composed of 19 chapters in five levels as shown on the flowchart on the facing page. Coverage of the material should approximate the flow in this chart to ensure understanding of the material. The material in Level One emphasizes basic computational skills. Level Two discusses the three most commonly used techniques to evaluate alternatives and Level Three extends these analysis techniques to other methods commonly used by the engineering economist. Level Four presents the important

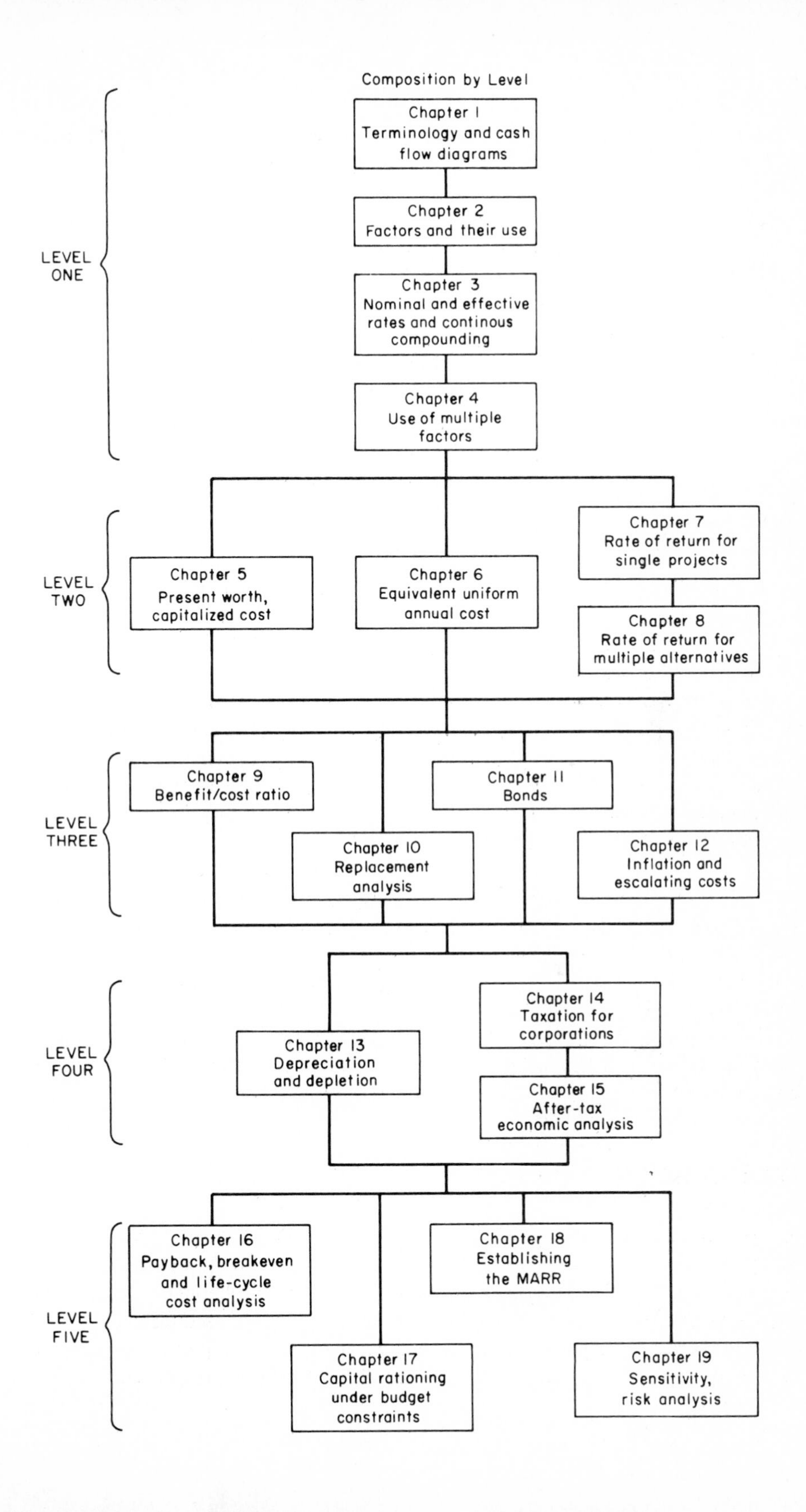

Composition by Level
LEVEL ONE
Chapter 1
Terminology and cash flow diagrams
Chapter 2
Factors and their use
Chapter 3
Nominal and effective rates and continous compounding
Chapter 4
Use of multiple factors
LEVEL TWO
Chapter 5
Present worth, capitalized cost
Chapter 6
Equivalent uniform annual cost
Chapter 7
Rate of return for single projects
Chapter 8
Rate of return for multiple alternatives
LEVEL THREE
Chapter 9
Benefit/cost ratio
Chapter 10
Replacement analysis
Chapter 11
Bonds
Chapter 12
Inflation and escalating costs
LEVEL FOUR
Chapter 13
Depreciation and depletion
Chapter 14
Taxation for corporations
Chapter 15
After-tax economic analysis
LEVEL FIVE
Chapter 16
Payback, breakeven and life-cycle cost analysis
Chapter 17
Capital rationing under budget constraints
Chapter 18
Establishing the MARR
Chapter 19
Sensitivity, risk analysis

topics of depreciation and corporation taxation while Level Five presents supplementary and advanced analysis procedures. The entire material in chapters 12 (inflation), 17 (capital rationing), 18 (setting the MARR) and 19 (sensitivity and risk) may be considered optional if a shortened version of the text is of interest.

In the preparation of this text, we have been assisted by many students and teachers whose comments and suggestions have been of inestimable value. From the typists to the editor, the cooperation we have received has been outstanding. We would take this opportunity to thank everyone who contributed either directly or indirectly to the successful completion of this book. A special thanks is given to our spouses for their support and encouragement during manuscript preparation.

Leland T. Blank
Anthony J. Tarquin

INTRODUCTION

In most cases when a specific task is to be performed there are several alternative ways to accomplish the task. In a business setting (or in personal life) much of the information for each alternative can be expressed quantitatively in terms of dollar incomes and disbursements. When capital investments for equipment, materials, and labor are needed to fund the alternatives and engineering activity of some kind is involved, the techniques of engineering economy may be used to assist in determining which is the best alternative. Usually the monetary values are future estimates that would result if each alternative were selected for implementation. These estimates are based on facts, experience, judgment, and comparison with other similar projects.

In many cases an engineer rather than an economist, accountant, financier, banker, or tax expert performs the analysis, because the technical details are already known by the engineer, so it is easier for the engineer to learn and perform the analysis procedures than it is for people in other fields to learn the technical details. Therefore, this subject matter is extremely important for the engineer in his or her profession as well as personal life for evaluating alternatives with respect to investments, automobile purchases, and the like.

Some typical industrially based questions addressed by the techniques covered in this text may be generally stated as follows:

Given a buy or lease plan for a computer, which one should be selected?

Should a particular processing line be fully automated or should automation be developed at each station?

How long should a presently owned asset remain in service before it pays for itself and returns 20% on the investment?

Which of several separate investment projects should be selected given that only a fixed amount of investment capital is available?

What rate of return can be realized on an investment in a particular project if the yearly revenues and disbursements are reliably estimated?

Can tax dollar be saved for additional investment if a different method of depreciation (capital recovery) is considered?

In short, engineering economy will allow you to take into account the fact that money makes money. The engineering design to accomplish a specific task may be the best possible; but if it is not economically competitive, the design should not be implemented.

Some of the features you will understand and learn to include in your own economic analyses are:

Time value of money
Break even
Depletion
Bonds
Cash flow
Effective interest rate
Minimum cost life
Inflation
Depreciation
Income taxes
Cost of capital
Payback period
Replacement
Sensitivity analysis

It is important to remember throughout your work in engineering economic analysis that the numerical data you use in the evaluation are only estimates of what is expected to take place. Therefore, the more accurate the estimates at the time the analysis is made, the better the decision when one of the alternatives is chosen.

LEVEL ONE

In the first four chapters of this book, you learn how to account for the fact that money has different values over different time intervals. This involves the derivation and use of engineering-economy factors, which greatly simplify computations that would otherwise be rather laborious.

Chapter	Subject
1	Terminology and cash-flow diagrams
2	Economy factors
3	Nominal and effective rates
4	Multiple factor use

CHAPTER

ONE

TERMINOLOGY AND CASH-FLOW DIAGRAMS

The purpose of this chapter is to provide you with the basic terminology of engineering economy and the fundamental concepts that form the basis for economic analysis.

This chapter will also teach you the meaning of the symbols used in engineering economy and how to construct a cash-flow diagram. The material you learn here will be used throughout the remainder of this book. In particular, you will find the cash-flow diagram exceptionally useful in simplifying complicated descriptive problems.

SECTION OBJECTIVES

In order to complete this chapter, you must be able to do the following:

1.1 Define the terms *alternative*, *evaluation criterion*, *intangible factors*, *time value of money*, *interest*, and *principal*.

1.2 Define *interest rate* and *interest period* and calculate the interest that has been accrued in one interest period, given the principal and the interest rate or total accrued amount.

1.3 Define *equivalence* and (*a*) calculate the amount of money equivalent to a present sum, given the present sum, the future or past date of equivalence, and the interest rate; and (*b*) calculate the interest rate per year at which different sums separated by one year are equivalent. Explain how different loan-repayment schemes can be equivalent yet allow different repayment schedules for a given principal, interest rate, and time period.

1.4 Define *simple interest* and *compound interest* and calculate the total amount of money accrued after one or more years using simple- and compound-interest methods, given the annual interest rate and the original principal.
1.5 Define and recognize in a problem statement the economy symbols P, F, A, n, and i.
1.6 Define *cash flow*, state what is meant by *end-of-period convention*, and construct a cash-flow diagram, given a statement describing the amount and the times at which the cash flows take place.

STUDY GUIDE

1.1 Basic Terminology

Before we begin to develop the terminology and fundamental concepts upon which engineering economy are based, it would be appropriate to define what is meant by *engineering economy*. In the simplest terms, engineering economy is a collection of mathematical techniques which simplify economic comparisons. With these techniques, a rational, meaningful approach to evaluating the economic aspects of different methods of accomplishing a given objective can be developed. Engineering economy is, therefore, a decision assistance tool by which one method will be chosen as the most economical one.

In order for you to be able to apply the techniques, however, it is necessary for you to understand the basic terminology and fundamental concepts that form the foundation for engineering-economy studies. Some of these terms and concepts are described below.

An *alternative* is a stand-alone solution for a given situation. We are faced with alternatives in virtually everything we do, from selecting the method of transportation we use to get to work every day to deciding between buying a house or renting one. Similarly, in engineering practice, there are always several ways of accomplishing a given task, and it is necessary to be able to compare them in a rational manner so that the most economical alternative can be selected. The alternatives in engineering considerations usually involve such items as asset purchase cost (first cost), the anticipated life of the asset, the yearly costs of maintaining the asset (annual maintenance and operating cost), the anticipated asset resale value (salvage value), and the interest rate. After the facts and all the relevant estimates have been collected, an engineering-economy analysis can be conducted to determine which is best from an economic point of view. It should be clear that in most instances, if there were not alternative ways of accomplishing a particular task, there would be no need for an engineering-economy analysis. The procedures developed in this book will enable you to make accurate economic decisions when one or more alternatives are being considered. What these procedures will not do, however, is help you identify what the alternatives are. It should be understood that unless the alternative which is actually the most economical one is *recognized as an alternative*, there is no way that it can be selected, no matter what analytical techniques are used. Thus, the importance of alternative identification in the decision-making process cannot be overemphasized, because it is only when this aspect of the process has been thoroughly completed that the analysis techniques presented in this book can be of greatest value.

In order to be able to compare different methods of accomplishing a given objective, it is necessary to have an *evaluation criterion* that can be used as a basis for judging the alternatives. In engineering economy, *dollars* are used as the basis for comparison. Thus, when there are several ways of accomplishing a given objective, the method that has the lowest overall cost is *usually* selected. However in most cases, the alternatives involve *intangible factors*, such as the effect of a process change on employee morale, which cannot be expressed in terms of dollars. When the alternatives available have approximately the same equivalent cost, the nonquantifiable, or intangible, factors may be used as the basis for selecting the best alternative.

For items of an alternative which can be quantified in terms of dollars, it is important to recognize the concept of the time value of money. It is often said that money makes money. The statement is indeed true, for if we elect to invest money today (for example, in a bank or savings and loan association), by tomorrow we will have accumulated more money than we had originally invested. This change in the amount of money over a given time period is called the *time value of money*; it is the most important concept in engineering economy. You should also realize that if a person or company finds it necessary to borrow money today, by tomorrow more money than the original loan will be owed. This fact is also explained by the time value of money.

The manifestation of the time value of money is termed *interest*, which is a measure of the increase between the original sum borrowed or invested and the final amount owed or accrued. Thus, if you invested money at some time in the past, the interest would be

$$\text{Interest} = \text{total amount accumulated} - \text{original investment} \qquad (1.1)$$

On the other hand, if you borrowed money at some time in the past, the interest would be

$$\text{Interest} = \text{present amount owed} - \text{original loan} \qquad (1.2)$$

In either case, there is an increase in the amount of money that was originally invested or borrowed, and the increase over the original amount is the interest. The original investment or loan is referred to as *principal*.

Probs. P1.1 to P1.3

1.2 Interest Calculations

When interest is expressed as a percentage of the original amount per unit time, the result is an *interest rate*. This rate is calculated as follows:

$$\text{Percent interest rate} = \frac{\text{interest accrued per unit time}}{\text{original amount}} \times 100\% \qquad (1.3)$$

By far the most common time period used for expressing interest rates is one year. However, since interest rates are often expressed over periods of time shorter than one year (i.e., 1% per month), the time unit used in expressing an interest rate is termed an *interest period*. The following two examples illustrate the computation of interest rate.

Example 1.1 The Get-Rich-Quick (GRQ) Company invested $100,000 on May 1 and withdrew a total of $106,000 exactly one year later. Compute (*a*) the interest gained from the original investment and (*b*) the interest rate from the investment.

SOLUTION (*a*) Using Eq. (1.1),

$$\text{Interest} = 106{,}000 - 100{,}000 = \$6000$$

(*b*) Equation (1.3) is used to obtain

$$\text{Percent interest rate} = \frac{6000/\text{year}}{100{,}000} \times 100\% = 6\% \text{ per year}$$

COMMENT For borrowed money, computations are similar to those shown above except that interest is computed by Eq. (1.2). For example, if GRQ borrowed $100,000 now and repaid $110,000 in one year, using Eq. (1.2), we find that interest is $10,000 and the interest rate from Eq. (1.3) is 10% per year.

Example 1.2 Joe Bilder plans to borrow $20,000 for one year at 15% interest. Compute (*a*) the interest and (*b*) the total amount due after one year.

SOLUTION (*a*) Equation (1.3) may be solved for the interest accrued to obtain

$$\text{Interest} = 20{,}000\ (0.15) = \$3000$$

(*b*) Total amount due is the sum of principal and interest or

$$\text{Total due} = 20{,}000 + 3000 = \$23{,}000$$

COMMENT Note that in *b* above, the total amount due may also be computed as

$$\text{Total due} = \text{principal}\ (1 + \text{interest rate}) = 20{,}000\ (1.15) = \$23{,}000$$

In each example the interest period was one year and the interest was calculated as of the end of one period. When more than one yearly interest period is involved (for example, if we had wanted to know the amount of interest GRQ would owe on the above loan after 3 years), it becomes necessary to determine whether the interest is payable on a *simple* or *compound* basis. The concepts of simple and compound interest are discussed in Sec. 1.4.

Solved Problems 1.12 and 1.13
Probs. P1.4 to P1.6

1.3 Equivalence

The time value of money and interest rate utilized together generate the concept of *equivalence*, which means that different sums of money at different times can be equal in economic value. For example, if the interest rate is 12% per year, \$100 today (i.e., at present) would be equivalent to \$112 one year from today since

$$\text{Amount accrued} = 100 + 100(0.12) = 100(1 + 0.12) = 100(1.12)$$
$$= \$112$$

Thus, if someone offered you a gift of \$100 today or \$112 one year from today, it would make no difference which offer you accepted, since in either case you would have \$112 one year from today. The two sums of money are therefore equivalent to each other when the interest rate is 12% per year. At either a higher or a lower interest rate, however, \$100 today is not equivalent to \$112 one year from today. In addition to considering future equivalence, one can apply the same concepts for determining equivalence in previous years. Thus, \$100 now would be equivalent to 100/1.12 = \$89.29 one year ago if the interest rate is 12% per year. From these examples, it should be clear that \$89.29 last year, \$100 now, and \$112 one year from now are equivalent when the interest rate is 12% per year. The fact that these sums are equivalent can be established by computing the interest rate as follows:

$$\frac{112}{100} = 1.12, \text{ or } 12\% \text{ per year}$$

and

$$\frac{100}{89.29} = 1.12, \text{ or } 12\% \text{ per year}$$

The concept of equivalence can be further illustrated by considering different loan-repayment schemes. Each scheme represents repayment of a \$5000 loan in 5 years at 15%-per-year interest. Table 1.1 presents the details for the four repayment methods described below. (The methods for determining the amount of the payments are presented in Chaps. 2 and 3.)

Plan 1 No interest or principal is recovered until the fifth year. Interest accumulates each year on the total of principal and all accumulated interest.

Plan 2 The accrued interest is paid each year and the principal is recovered at the end of 5 years.

Plan 3 The accrued interest and 20% of the principal, that is, \$1000, is paid each year. Since the remaining loan balance decreases each year, the accrued interest decreases each year.

Plan 4 Equal payments are made each year with a portion going toward principal recovery and the remainder covering the accrued interest. Since the loan balance decreases at a rate which is slower than in plan 3 due to the equal end-of-year payments, the interest decreases, but at a rate slower than in plan 3.

Table 1.1 Different repayment schedules of $5,000 at 15% for 5 years

(1) End of year	(2) = 0.15(5) Interest for year	(3) = (2) + (5) Total owed at end of year	(4) Payment per plan	(5) = (3) − (4) Balance after payment
Plan 1				
0				$5,000.00
1	$ 750.00	$ 5,750.00	$0	5,750.00
2	862.50	6,612.50	0	6,612.50
3	991.88	7,604.38	0	7,604.38
4	1,140.66	8,745.04	0	8,745.04
5	1,311.76	10,056.80	10,056.80	0
			$10,056.80	
Plan 2				
0				$5,000.00
1	$750.00	$5,750.00	$ 750.00	5,000.00
2	750.00	5,750.00	750.00	5,000.00
3	750.00	5,750.00	750.00	5,000.00
4	750.00	5,750.00	750.00	5,000.00
5	750.00	5,750.00	5,750.00	0.00
			$8,750.00	
Plan 3				
0				$5,000.00
1	$750.00	$5,750.00	$1,750.00	4,000.00
2	600.00	4,600.00	1,600.00	3,000.00
3	450.00	3,450.00	1,450.00	2,000.00
4	300.00	2,300.00	1,300.00	1,000.00
5	150.00	1,150.00	1,150.00	0
			$7,250.00	
Plan 4				
0				$5,000.00
1	$750.00	$5,750.00	$1,491.58	4,258.42
2	638.76	4,897.18	1,491.58	3,405.60
3	510.84	3,916.44	1,491.58	2,424.86
4	363.73	2,788.59	1,491.58	1,297.01
5	194.57	1,491.58	1,491.58	0
			$7,457.90	

Note that the total amount repaid in each case would be different, even though each repayment scheme would require exactly 5 years to repay the loan. The difference in the total amounts repaid can of course be explained by the time value of money, since the amount of the payments is different for each plan. With respect to equivalence, the table shows that when the interest rate is 15% per year, $5000 at time 0 is equivalent to $10,056.80 at the end of year 5 (plan 1), or $750 per year for 4 years and $5750 at the end of year 5 (plan 2), or the decreasing amounts shown in years 1 through 5 (plan 3), or $1,491.58 per year for 5 years (plan 4). Using the formulas developed in

Chaps. 2 and 3, we can easily show that if the payments in each plan (column 4) are reinvested at 15% per year when received, the total amount of money available at the end of year 5 will be \$10,056.80 from each repayment plan.

Solved Problems 1.14 and 1.15
Probs. P1.7 and P1.8

1.4 Simple and Compound Interest

The concepts of interest and interest rate were introduced in Secs. 1.1 and 1.2 and used in Sec. 1.3 to calculate for one interest period past and future sums of money equivalent to a present sum (principal). When more than one interest period is involved, the terms *simple* and *compound* interest must be considered.

Simple interest is calculated using the principal only, ignoring any interest that was accrued in preceding interest periods. The total interest can be computed using the relation

$$\text{Interest} = (\text{principal})(\text{number of periods})(\text{interest rate}) = Pni \tag{1.4}$$

Example 1.3 If you borrow \$1000 for 3 years at 6%-per-year simple interest, how much money will you owe at the end of 3 years?

SOLUTION The interest for each of the 3 years is

$$\text{Interest per year} = 1000(0.06) = \$60$$

Total interest for 3 years from Eq. (1.4) is

$$\text{Total interest} = 1000(3)(0.06) = \$180$$

Finally, the amount due after 3 years is

$$1000 + 180 = \$1180$$

COMMENT The \$60 interest accrued in the first year and the \$60 accrued in the second year did not earn interest. The interest due was calculated on the principal only. The results of this loan are tabulated in Table 1.2. The end-of-year figure of zero represents the present, that is, when the money is borrowed. Note that no

Table 1.2 Simple-interest computation

(1) End of year	(2) Amount borrowed	(3) Interest	(4) = (2) + (3) Amount owed	(5) Amount paid
0	\$1000			
1	...	\$60	\$1060	\$ 0
2	...	60	1120	0
3	...	60	1180	1180

payment is made by the borrower until the end of year 3. Thus, the amount owed each year increases uniformly by $60, since interest is figured only on the principal of $1000.

In calculations of *compound interest*, the interest for an interest period is calculated on the principal *plus the total amount of interest accumulated in previous periods*. Thus, compound interest means "interest on top of interest" (i.e., it reflects the effect of the time value of money).

Example 1.4 If you borrow $1000 at 6%-per-year *compound* interest, as in the preceding example, compute the total amount due after a 3-year period.

SOLUTION The interest and total amount due for each year is computed as follows:

Interest, year 1 = 1000(0.06) = $60
Total amount due after year 1 = 1000 + 60 = $1060
Interest, year 2 = 1060(0.06) = $63.60
Total amount due after year 2 = 1060 + 63.60 = $1123.60
Interest, year 3 = 1123.60(0.06) = $67.42
Total amount due after year 3 = 1123.60 + 67.42 = $1191.02

COMMENT The details are shown in Table 1.3. The repayment scheme is the same as that for the simple-interest example; that is, no amount is repaid until the principal, plus all interest, is due at the end of year 3. The time value of money is recognized in compound interest. Thus, with compound interest, the original $1000 would accumulate an extra $1191.02 - 1180.00 = $11.02 compared with simple interest in the 3-year period.

In Chap. 2, formulas are developed which simplify compound-interest calculations. The same concepts are involved when the interest period is less than a year. A discussion of this case is deferred until Chap. 3, however. Since real-world calculations almost always involve compound interest, the interest rates specified hereinafter refer to compound interest unless specified otherwise.

Solved Problem 1.16
Probs. P1.9 to P1.21

Table 1.3 Compound-interest computation

(1) End of year	(2) Amount borrowed	(3) Interest	(4) = (2) + (3) Amount owed	(5) Amount paid
0	$1000			
1	...	$60.00	$1060.00	$ 0
2	...	63.60	1123.60	0
3	...	67.42	1191.02	1191.02

1.5 Symbols and Their Meaning

The mathematical relations used in engineering economy employ the following symbols:

P = value or sum of money at a time denoted as the present; dollars
F = value or sum of money at some future time; dollars
A = a series of consecutive, equal, end-of-period amounts of money; dollars per month, dollars per year
n = number of interest periods; months, years
i = interest rate per interest period; percent per month, percent per year

watch units

The symbols P and F represent single-time occurrence values: A occurs each interest period for a specified number of periods with the same dollar value. The units of the symbols aid in clarifying their meaning. The present sum P and future sum F are expressed in dollars; A is referred to in dollars per interest period. It is important to note here that in order for a series to be represented by the symbol A, it must be uniform (i.e., the dollar value must be the same for each period) and the uniform dollar amounts must extend through *consecutive* periods. Both conditions must exist before the dollar value can be represented by A. Since n is commonly expressed in years, A is usually expressed in units of dollars per year. The compound interest rate i is expressed in percent per interest period. For example, 5% per year. Except where noted otherwise, this rate applies throughout the entire n years or n interest periods. The i value is often the minimum attractive rate of return (Sec. 18.2). The most common engineering-economy problems involve the use of n and i and at least two of the three terms, P, F, and A. The following four examples illustrate the use of the symbols.

Example 1.5 If you borrow \$2000 now and must repay the loan plus interest at a rate of 12% per year in 5 years, what is the total amount you must pay? List the values of P, F, n, and i.

SOLUTION In this situation P and F, but not A, are involved, since all transactions are single payments. The values are as follows:

P = \$2000 F = ? i = 12% per year n = 5 years

Example 1.6 If you borrow \$2000 now at 17% per year for 5 years and must repay the loan in equal yearly payments, what will you be required to pay? Determine the value of the symbols involved.

SOLUTION

P = 2000
A = ? per year for 5 years
i = 17% per year
n = 5 years

There is no F value involved.

In both examples, the P value of \$2000 is a receipt and F or A is a disbursement. It is equally correct to use these symbols in reserve roles, as in the examples below.

Example 1.7 If you deposit \$500 in a savings account on May 1, 1984, which pays 7% per year, what annual amount can you withdraw for the following 10 years? List the symbol values.

SOLUTION

P = \$500
A = ? per year
i = 7% per year
n = 10 years

COMMENT The value for the \$500 disbursement P and receipt A are given the same symbol names as before, but they are considered in a different context. Thus, a P value may be a receipt (Examples 1.5 and 1.6) or a disbursement (this example).

Example 1.8 If you deposit \$100 in a savings account each year for 7 years at an interest rate of 6% per year, what single amount will you be able to withdraw after 7 years? Define the symbols and their roles.

SOLUTION In this example, the equal annual deposits are in a series A and the withdrawal is a future sum, or F value. There is no P value here, thus

A = \$100 per year for 7 years
F = ?
i = 6% per year
n = 7 years

Solved Problem 1.17
Probs. P1.22 to P1.24

1.6 Cash-Flow Diagrams

Every person or company has cash receipts (income) and cash disbursements (costs) which occur over a particular time span. These receipts and disbursements in a given time interval are referred to as *cash flow*, with positive cash flows usually representing receipts and negative cash flows representing disbursements. At any point in time, the net cash flow would be represented as

$$\text{Net cash flow} = \text{receipts} - \text{disbursements} \qquad (1.5)$$

Since cash flow normally takes place at frequent and varying time intervals within an interest period, a simplifying assumption is made that all cash flow occurs at the end

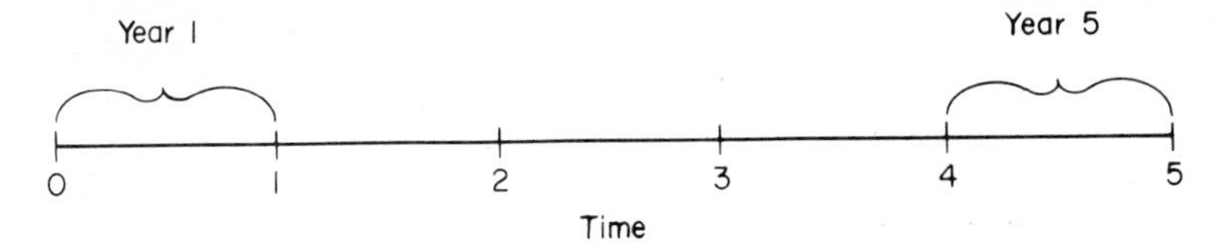

Figure 1.1 A typical cash-flow time scale.

of the interest period. This is known as the *end-of-period convention*. Thus, when several receipts and disbursements occur within a given interest period, the net cash flow is assumed to occur at the end of the interest period. However, it should be understood that although the dollar amounts of F or A are always considered to occur at the *end of the interest period*, this does not mean that the end of the period is December 31. In the situation of Example 1.7, since investment took place on May 1, 1984, the withdrawals will take place on May 1, 1985, and each succeeding May 1 for 10 years (the last withdrawal will be on May 1, 1994, not 1995). Thus, *end of the period* means one time period from the date of the transaction (whether it be receipt or disbursement). In the next chapter you will learn how to determine the equivalent relations between P, F, and A values at different times.

A *cash-flow diagram* is simply a graphical representation of cash flows drawn on a time scale. The diagram should represent the statement of the problem and should include what is given and what is to be found. That is, after the cash-flow diagram has been drawn, an outside observer should be able to work the problem by looking at only the diagram. Time 0 is considered to be the present and time 1 the end of time period 1. (We will assume that the periods are in years until Chap. 3.) The time scale of Fig. 1.1 is set up for 5 years. Since it is assumed that cash flows occur only at the end of the year, we will be concerned only with the times marked 0, 1, 2, . . . , 5.

The direction of the arrows on the cash-flow diagram is important to problem solution. Therefore, in this text, a vertical arrow pointing up will indicate a positive cash flow. Conversely, an arrow pointing down will indicate a negative cash flow. The cash-flow diagram in Fig. 1.2 illustrates a receipt (income) at the end of year 1 and a disbursement at the end of year 2.

It is important that you thoroughly understand the meaning and construction of the cash-flow diagram, since it is a valuable tool in problem solution. The three examples below illustrate the construction of cash-flow diagrams.

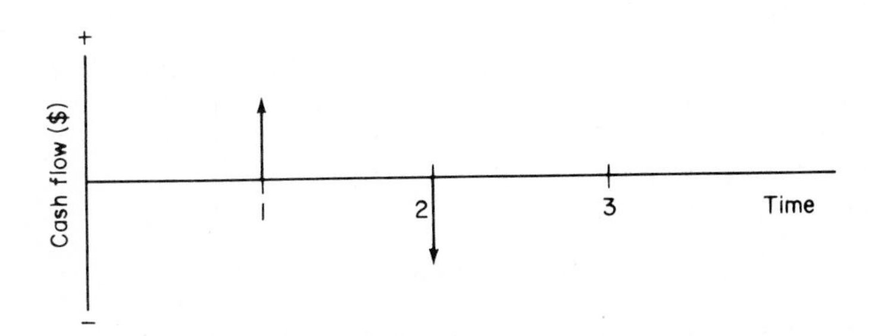

Figure 1.2 Example of positive and negative cash flows.

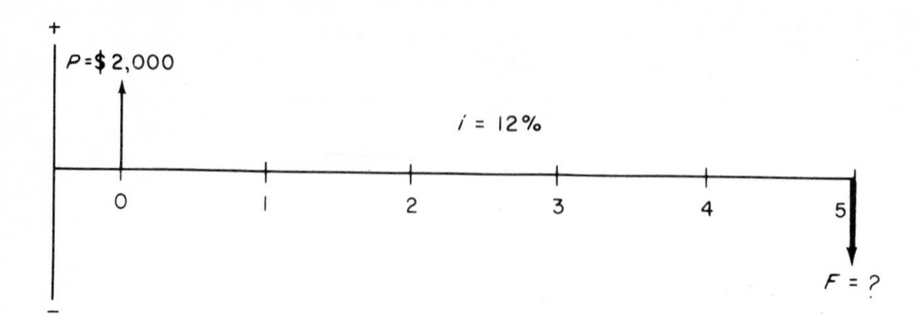

Figure 1.3 Cash-flow diagram for Example 1.9.

Example 1.9 Consider the situation presented in Example 1.5, where \$2000 (P) is borrowed and F is to be found after 5 years. Construct the cash-flow diagram for this case, assuming an interest rate of 12% per year.

SOLUTION Figure 1.3 presents the cash-flow diagram.

COMMENT While it is not necessary to use an exact scale on the cash-flow axes, you will probably avoid errors later on if you make a neat diagram. Note also that the present sum P is a *receipt* at year 0 and the future sum F is a *disbursement* at the end of year 5.

Example 1.10 If you start now and make five deposits of \$1000 per year (A) in a 7%-per-year account, how much money will be accumulated immediately after you have made the last deposit? Construct the cash-flow diagram.

SOLUTION The cash flows are shown in Fig. 1.4.

COMMENT Since you have decided to start now, the first deposit is at year 0 and the fifth deposit and withdrawal occur at the end of year 4. Note that in this ex-

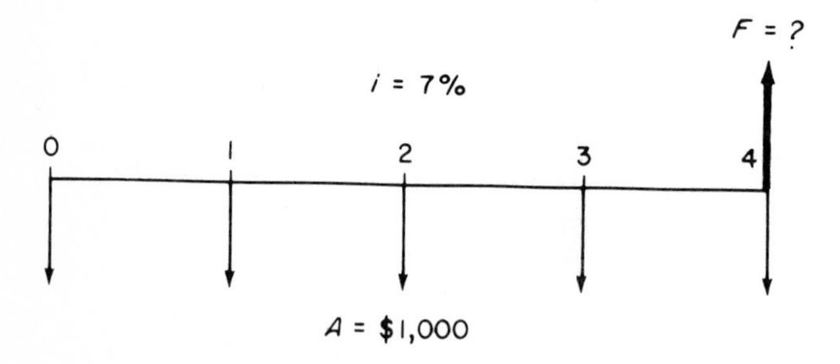

Figure 1.4 Cash-flow diagram for Example 1.10.

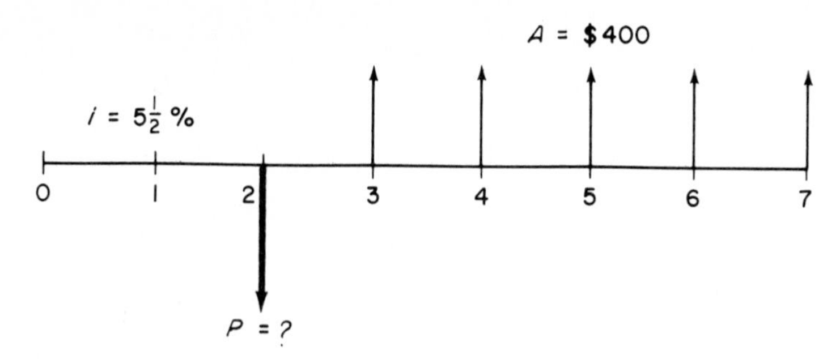

Figure 1.5 Cash-flow diagram for Example 1.11.

ample, the amount accumulated after the fifth deposit is to be computed; thus, the future amount is represented by a question mark (that is, $F = ?$)

Example 1.11 Assume that you want to deposit an amount P into an account 2 years from now in order to be able to withdraw \$400 per year for 5 years starting 3 years from now. Assume that the interest rate is $5\frac{1}{2}\%$ per year. Construct the cash-flow diagram.

SOLUTION Figure 1.5 presents the cash flows, where P is to be found. Note that the diagram shows what was given and what is to be found.

Solved Problems 1.18 to 1.20
Probs. P1.25 to P1.41

SOLVED PROBLEMS

You should refer to the section noted after each example if you don't understand the solution.

Example 1.12 Calculate the interest and total amount accrued after one year if \$2000 is invested at an interest rate of 15% per year.

SOLUTION

$$\text{Interest earned} = 2000(0.15) = \$300$$

$$\text{Total amount accrued} = 2000 + 2000(0.15) = 2000(1 + 0.15) = \$2300$$

Sec. 1.2

Example 1.13 (*a*) Calculate the amount of money that must have been deposited one year ago for you to have \$1000 now at an interest rate of 5% per year. (*b*) Calculate the interest that was earned in the same time period.

SOLUTION

(*a*) Total amount accrued = original deposit + (original deposit)(interest rate)

Let X = original deposit, then

$$1000 = X + X(0.05) = X(1 + 0.05)$$

$$1000 = 1.05X$$

$$X = \frac{1000}{1.05} = 952.38$$

Original deposit = \$952.38

(*b*) By using Eq. (1.1), we have

$$\text{Interest} = \$1000 - \$952.38 = \$47.62$$

Sec. 1.2

Example 1.14 Calculate the amount of money that must have been deposited one year ago for the investment to earn \$100 in interest in one year, if the interest rate is 6% per year.

SOLUTION Let a = total amount accrued and b = original deposit.

$$\text{Interest} = a - b$$

$$a = b + b\,(\text{interest rate})$$

$$\text{Interest} = b + b\,(\text{interest rate}) - b$$

$$\text{Interest} = b\,(\text{interest rate})$$

$$\$100 = b(0.06)$$

$$b = \frac{100}{0.06} = \$1666.67$$

Sec. 1.3

Example 1.15 Make the calculations necessary to show which of the statements below are true and which are false, if the interest rate is 5% per year:

(*a*) \$98 now is equivalent to \$105.60 one year from now.
(*b*) \$200 one year past is equivalent to \$205 now.
(*c*) \$3000 now is equivalent to \$3150 one year from now.
(*d*) \$3000 now is equivalent to \$2887.14 one year ago.
(*e*) Interest accumulated in one year on an investment of \$2000 is \$100.

SOLUTION

(*a*) Total amount accrued = 98(1.05) = \$102.90 $\neq$ \$105.60; therefore false. Another way to solve this is as follows: Required investment = 105.60/1.05 = \$100.57 $\neq$ \$98. Therefore false.

Sec. 1.3

(*b*) Required investment = 205.00/1.05 = \$195.24 ≠ \$200; therefore false.

Secs. 1.2 and 1.3

(*c*) Total amount accrued = 3000(1.05) = \$3150; therefore true.

Secs. 1.2 and 1.3

(*d*) Total amount accrued = 2887.14(1.05) = \$3031.50 ≠ \$3000; therefore false.

Secs. 1.2 and 1.3

(*e*) Interest = 2000(0.05) = \$100; therefore true.

Sec. 1.2

Example 1.16 Calculate the total amount due after 2 years if \$2500 is borrowed now and the compound-interest rate is 8% per year.

SOLUTION The results may be presented as in the table to obtain a total amount due of \$2916.

(1) End of year	(2) Amount borrowed	(3) Interest	(4) = (2) + (3) Amount owed	(5) Amount paid
0	\$2500			
1	. . .	\$200	\$2700	\$0
2	. . .	216	2916	2916

Sec. 1.4

Example 1.17 Assume that you plan to make a lump-sum deposit of \$5000 now into an account that pays 6% per year, and you plan to withdraw an equal end-of-year amount of \$1000 for 5 years starting next year. At the end of the sixth year, you plan to close your account by withdrawing the remaining money. Define the engineering-economy symbols involved.

SOLUTION

P = \$5000
A = \$1000 per year for 5 years
F = ? at end of year 6
i = 6% per year
n = 5 years for A

Sec. 1.5

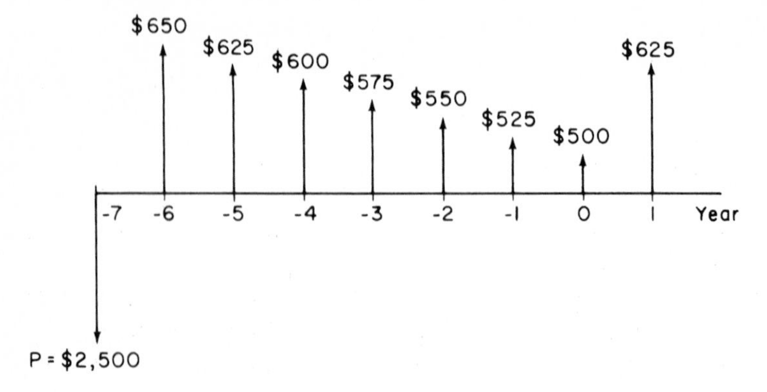

Figure 1.6 Cash-flow diagram for Example 1.18.

Example 1.18 The Hot-Air Company invested $2500 in a new air compressor 7 years ago. Annual income from the compressor was $750. During the first year, $100 was spent on maintenance, a cost that increased each year by $25. The company plans to sell the compressor for salvage at the end of next year for $150. Construct the cash-flow diagram for the piece of equipment.

SOLUTION The income and cost for years −7 through 1 (next year) are tabulated below with net cash flow computed using Eq. (1.5). The cash flows are diagramed in Fig. 1.6.

End of year	Income	Cost	Net cash flow
−7	$ 0	$2500	$−2500
−6	750	100	650
−5	750	125	625
−4	750	150	600
−3	750	175	575
−2	750	200	550
−1	750	225	525
0	750	250	500
1	750 + 150	275	625

Example 1.19 Suppose that you want to make a deposit into your account now such that you can withdraw an equal annual amount of A_1 = $200 per year for the first 5 years starting one year after your deposit and a different annual amount of A_2 = $300 per year for the following 3 years. How would the cash-flow diagram appear if i is $4\frac{1}{2}\%$ per year?

SOLUTION The cash flows would appear as shown in Fig. 1.7.

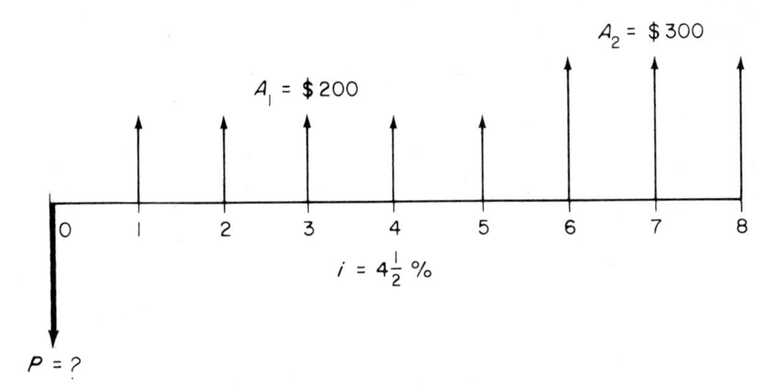

Figure 1.7 Cash-flow diagram for two different A values, Example 1.19.

COMMENT The first withdrawal (positive cash flow) occurs at the end of year 1, exactly one year after P is deposited.

Sec. 1.6

Example 1.20 If you buy a new television set in 1985 for $300, maintain it for 3 years at a cost of $20 per year, and then sell it for $50, diagram your cash flows and label each arrow as P, F, or A with its respective dollar value so that you can find the single amount in 1984 that would be equivalent to all the cash flows shown. Assume an interest rate of 12%.

SOLUTION Figure 1.8 presents the cash-flow diagram.

COMMENT The two $20 negative cash flows form a series of two equal end-of-year values. As long as the dollar values are equal and in two or more consecutive periods, they can be represented by A, regardless of where they begin or end.

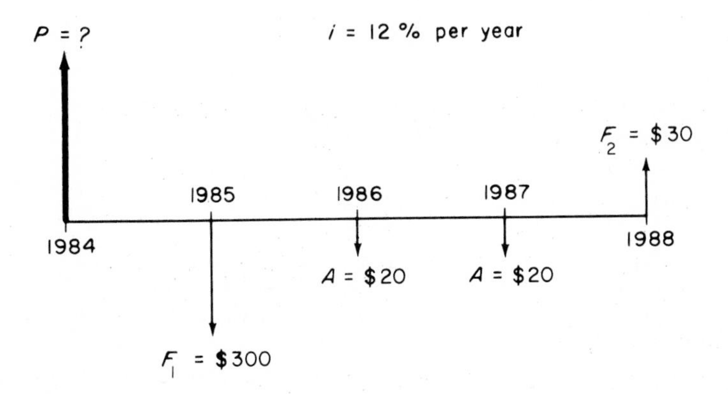

Figure 1.8 A cash-flow diagram for Example 1.20.

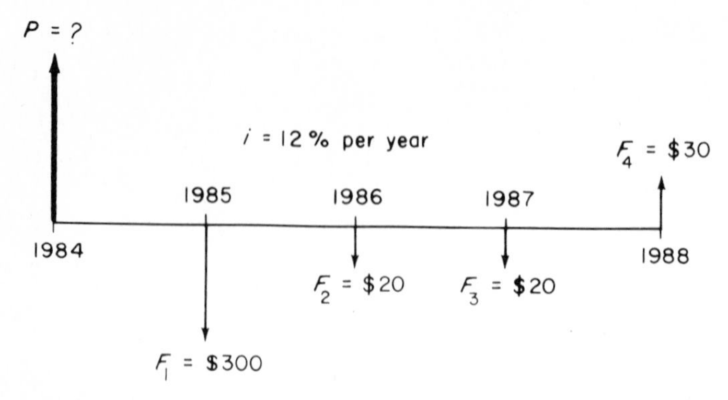

Figure 1.9 A cash flow for Example 1.20 considering all values as future sums.

However, the $30 positive cash flow in 1988 is a single-occurrence value in the future and is therefore labeled an F value. It is possible, however, to view all the individual cash flows as F values. The diagram could be drawn as shown in Fig. 1.9. In general, however, if two or more equal end-of-period amounts occur consecutively, by the definition in Sec. 1.5 they should be labeled A values because, as is described in Chap. 2, the use of A values when possible simplifies calculations considerably. Thus, the interpretation pictured by the diagram of Fig. 1.9 is discouraged and will not be used further in this text.

Sec. 1.6

PROBLEMS

P1.1 What is meant by *the time value of money*?

P1.2 What is the difference between the terms *principal* and *present amount owed*?

P1.3 List three items that might be regarded as intangible factors.

P1.4 If a bank advertises 14%-per-year interest compounded semiannually, what is the interest period?

P1.5 Find the interest due after one year on a loan of $5000 if interest is 12% per year.

P1.6 Find the original amount of a loan if interest is $1\frac{1}{2}$% per month payable monthly and the borrower just made the first monthly payment of $25 in interest.

P1.7 At what interest rate are $450 a year ago and $550 one year from now equivalent?

P1.8 How do you explain the fact that two different amounts of money can be equivalent to each other?

P1.9 If the Get-Rich-Quik Company invested $50,000 in a new process one year ago and has just now realized a profit of $7500, what was its rate of return based on this investment?

P1.10 Why is the minimum attractive rate of return of a business organization greater than the interest rate obtainable from a bank or savings and loan association?

P1.11 Assume that you have been offered an investment opportunity in which you may invest $1000 at 7%-per-year simple interest for 3 years or you may invest the same $1000 at 6%-per-year compound interest for 3 years. Which investment offer would you accept?

P1.12 (*a*) How much interest would you pay if you borrowed \$600 at $1\frac{1}{2}\%$ per month compounded monthly for 3 months?

(*b*) What percent of the principal is this interest amount?

P1.13 Work the two parts of Prob. P1.12 for $1\frac{1}{2}\%$-per-month simple interest.

P1.14 How much money will your friend owe after 4 years if she borrows \$1000 now at 7%-per-year simple interest?

P1.15 How much money will be owed after 2 years if a person borrows \$500 at 1%-per-month simple interest?

P1.16 How much money can you borrow now if you repay the lender \$850 in 2 years and the interest rate is 6% per year compounded yearly?

P1.17 If you borrow \$1500 now and must repay \$1800 two years from now, what is the interest rate on your loan? Assume that interest is compounded yearly.

P1.18 If you invest \$10,000 now in a business venture that promises to return \$14,641, how soon must you receive the \$14,641 in order to make at least 10% per year compounded yearly on your investment?

P1.19 Your friend tells you that he has just repaid a loan he got 3 years ago at 10%-per-year simple interest. If, upon questioning, you learn that his payment was \$195, how much did he borrow?

P1.20 If you invest \$3500 now in return for a guaranteed \$5000 income at a later date, when must you receive your money in order to earn at least 8%-per-year simple interest?

P1.21 Section 1.4 shows that \$1000 at 6%-per-year simple interest is equivalent to \$1180 in 3 years. Find the compound-interest rate per year for which the equivalence is also correct.

P1.22 Five equal deposits of \$1000 will be made every 2 years starting next year at 10% per year and the total accrued amount withdrawn when the last deposit is made. List the economy symbols and values involved in this problem.

P1.23 The GRQ Company plans to deposit \$709.90 now at 6% per year and withdraw \$100 per year for the next 5 years and \$200 per year for the following 2 years. What are the economy symbols and their respective values?

P1.24 How many years would it take for \$1400 to triple in value at an interest rate of 10% per year? Define the economy symbols.

P1.25 What is meant by *end-of-period convention*?

P1.26 Diagram the yearly net cash flows for Example 1.5.

P1.27 Diagram the yearly net cash flows for Example 1.7.

P1.28 Assume that you have developed an investment plan that is carried out as follows: Invest \$500 now and every other year through year 10 and withdraw \$300 every year starting 5 years from now and continuing for 8 more years. Diagram the yearly net cash flows.

P1.29 Construct cash-flow diagrams for the loan-repayment schedules in Table 1.1 (column 4).

P1.30 If you plan to make a deposit now such that you will have \$3000 in your account 5 years from now, how much must you deposit if the interest rate is 8%? Draw the cash-flow diagram.

P1.31 Your uncle has agreed to make five \$700-per-year deposits into a savings account for you starting now. You have in turn agreed not to withdraw any money until the end of year 9, at which time you plan to remove \$3000 from the account. Further, you plan to withdraw the remaining amount in three equal year-end installments after the initial withdrawal. Diagram the cash flows for your uncle and yourself.

P1.32 The president of a company wants to make two equal lump-sum deposits, one 2 years and the second 4 years from now, so he can make five \$100-per-year withdrawals starting when the second deposit is made. Further, he plans to withdraw an additional \$500 the year after the withdrawal series ends. Draw his cash-flow diagram.

P1.33 You want to invest money at 8% per year so that 6 years from now you can withdraw an amount F in a lump sum. The investment consultant at the bank has developed the following two plans for you: (*1*) Deposit $351.80 now and $351.80 three years from now. (*2*) Deposit $136.32 per year starting next year and ending in year 6. Draw the cash-flow diagram for each plan if F is to be found.

P1.34 How much could you spend now in order to avoid spending $580 eight years from now if the interest rate is 6%? Draw the cash-flow diagram.

P1.35 If you deposit $100 per year for 5 years starting one year from now, how much will you have in your account 15 years from now if the interest rate is 10% per year? Draw the cash-flow diagram.

P1.36 What is the present worth of an expenditure of $1200 five years from now and $2200 eight years from now if the interest rate is 10% per year? Construct the cash-flow diagram.

P1.37 Calculate the present worth of an expenditure of $85 per year for 6 years that starts 3 years from now if the interest rate is 20%. Construct the cash-flow diagram.

P1.38 If you invest $10,000 now in a real estate venture, how much must you sell your property for 10 years from now if you want to make a 12% rate of return on your investment? Define the economy symbols and draw the cash-flow diagram.

P1.39 If you invest $4100 now and receive $7500 five years from now, what is the rate of return on your investment? Define the economy symbols and construct the cash-flow diagram.

P1.40 How much money would be accumulated in 6 years if a person deposited $500 now and amounts increasing by $50 per year for the next 6 years? Assume i is 16% per year and draw the cash-flow diagram.

P1.41 What uniform payment for 8 years beginning one year from now would be equivalent to spending $4500 now, $3300 three years from now, and $6800 five years from now if the interest rate is 8% per year? Define the economy symbols and draw the cash flow diagram.

CHAPTER

TWO

FACTORS AND THEIR USE

The objective of this chapter is to teach you the derivation of the engineering-economy factors and the use of these basic factors in economy computations. This chapter is one of the most important in the book, since the concepts presented here will be used throughout the remainder of the text.

SECTION OBJECTIVES

To complete this chapter you must be able to define and derive the formulas for the following:

2.1. Single-payment compound-amount factor and single-payment present-worth factor
2.2. Uniform-series present-worth factor and capital-recovery factor using the single-payment present-worth factor
2.3. Uniform-series compound-amount factor and sinking-fund factor using the single-payment compound-amount factor and the capital-recovery factor

You must also be able to do the following:

2.4. Find the correct numerical value of a factor in a table, given the standard factor notation
2.5. Define and develop the uniform-gradient present-worth and annual-series factors using the single-payment present-worth factor

2.6. Linearly interpolate to find a correct factor value, given an interest rate and/or year value not listed in the tables.
2.7. Calculate the present worth P, future worth F, or equivalent uniform annual series A of an investment, given the interest rate i, the number of years n, and the monetary value of one of the terms P, F, or A.
2.8. Calculate the present worth P and equivalent uniform annual series A of alternatives involving a *conventional* uniform gradient, given the interest rate and statement of the problem.
2.9. Calculate the interest rate (rate of return) of a sequence of cash flows, given the number of years and two of the following: present worth P, future worth F, or uniform series A starting at the end of year 1 and ending at the end of year n.
2.10. Determine the number of years n for a sequence of cash flows, given the interest rate i and two of the following: present worth P, future worth F, uniform series A starting at the end of year 1 and ending at the end of year n, or uniform gradient G starting at the end of year 2 and continuing through year n.

STUDY GUIDE

2.1 Derivation of Single-Payment Formulas

In this section, a formula is developed which allows determination of the amount of money that is accumulated (F) after n years from a *single* investment (P) when interest is compounded one time per year (or period).

You will recall from Chap. 1 that compound interest refers to interest paid on top of interest. Therefore, if an amount of money P is invested at some time $t = 0$, the amount of money F_1 that will be accumulated one year hence will be

$$F_1 = P + Pi$$

$$F_1 = P(1 + i)$$

At the end of the second year, the amount of money accumulated (F_2) will be equal to the amount that accumulated after year 1 plus the interest from the end of year 1 to the end of year 2. Thus,

$$\begin{aligned} F_2 &= F_1 + F_1 i \\ &= P(1 + i) + P(1 + i)i \end{aligned} \tag{2.1}$$

or

$$\begin{aligned} F_2 &= P(1 + i + i + i^2) \\ &= P(1 + 2i + i^2) \\ &= P(1 + i)^2 \end{aligned}$$

Similarly, the amount of money accumulated at the end of year 3, using Eq. (2.1), will be

$$F_3 = F_2 + F_2 i$$
$$= [P(1+i) + P(1+i)i] + [P(1+i) + P(1+i)i]i$$
$$= P(1+i) + 2P(1+i)i + P(1+i)i^2$$

Factoring out $P(1+i)$, we have

$$F_3 = P(1+i)(1+2i+i^2)$$
$$= P(1+i)(1+i)^2$$
$$= P(1+i)^3$$

From the preceding values, it is evident by mathematical induction that the formula can be generalized for n years as

$$F = P(1+i)^n \tag{2.2}$$

The expression $(1+i)^n$, called the *single-payment compound-amount factor* (SPCAF), will yield the future amount F of an initial investment P after n years at interest rate i.

Expressing P from Eq. (2.2) in terms of F results in the expression

$$P = F\left[\frac{1}{(1+i)^n}\right] \tag{2.3}$$

The expression in brackets is known as the *single-payment present-worth factor* (SPPWF). This expression will allow determination of the present worth P of a given future amount F after n years at interest rate i. The cash-flow diagram for this formula is shown in Fig. 2.1. Conversely, if you used the SPCAF for the diagram in Fig. 2.1, you could find F, given P.

It is important to note that the two formulas derived here are *single-payment* formulas; that is, they are used to find the present or future amount when only one payment or receipt is involved. In the next two sections, formulas are developed for

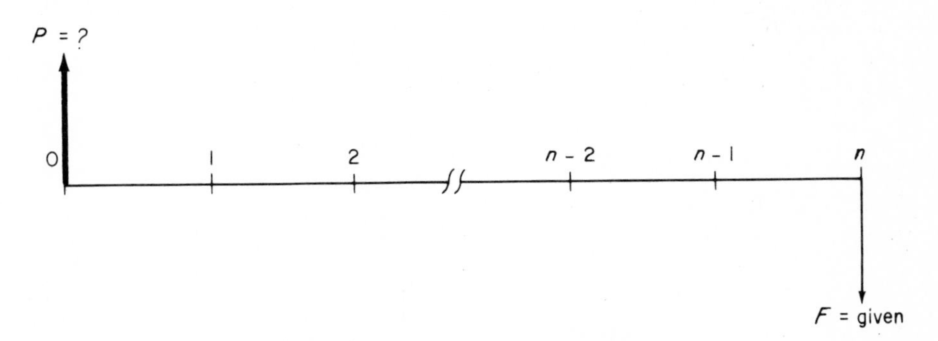

Figure 2.1 Cash-flow diagram to find P given F using the SPPWF.

calculating the present or future worth when several payments or receipts must be considered.

2.2 Derivation of the Uniform-Series Present-Worth Factor and the Capital-Recovery Factor

The present worth of the uniform series shown in Fig. 2.2 can be determined by considering each A value as a future worth F in the single-payment present-worth factor and then summing the present-worth values. The general formula is

$$P = A\left[\frac{1}{(1+i)^1}\right] + A\left[\frac{1}{(1+i)^2}\right] + A\left[\frac{1}{(1+i)^3}\right] + \cdots + A\left[\frac{1}{(1+i)^{n-1}}\right] + A\left[\frac{1}{(1+i)^n}\right]$$

where the terms in brackets represent the SPPWF for years 1 through n, respectively. Factoring out A,

$$P = A\left[\frac{1}{(1+i)^1} + \frac{1}{(1+i)^2} + \frac{1}{(1+i)^3} + \cdots + \frac{1}{(1+i)^{n-1}} + \frac{1}{(1+i)^n}\right] \tag{2.4}$$

Equation (2.4) may be simplified by multiplying both sides by $1/(1+i)$ to yield

$$\frac{P}{1+i} = A\left[\frac{1}{(1+i)^2} + \frac{1}{(1+i)^3} + \frac{1}{(1+i)^4} + \cdots + \frac{1}{(1+i)^n} + \frac{1}{(1+i)^{n+1}}\right] \tag{2.5}$$

Subtracting Eq. (2.4) from Eq. (2.5) yields

$$\frac{P}{1+i} - P = A\left[-\frac{1}{(1+i)^1} + \frac{1}{(1+i)^{n+1}}\right]$$

Factoring out P and rearranging, we have

$$P\left(\frac{1}{1+i} - 1\right) = A\left[\frac{1}{(1+i)^{n+1}} - \frac{1}{1+i}\right]$$

Simplifying both sides of the equation yields

$$P\left(\frac{-i}{1+i}\right) = A\,\frac{1}{1+i}\left[\frac{1}{(1+i)^n} - 1\right]$$

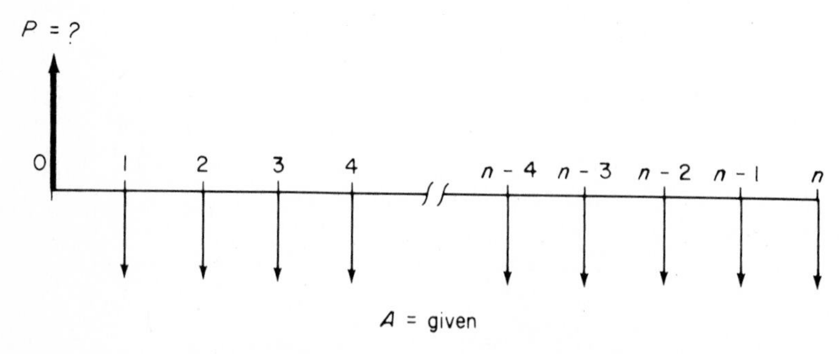

Figure 2.2 Diagram used to determine the present worth of a uniform series.

Dividing by $-i/(1+i)$ yields the following for $i \neq 0$.

$$
\begin{aligned}
P &= A\left(\frac{1}{1+i}\right)\frac{[1/(1+i)^n]-1}{-i/(1+i)} \\
&= A\left(\frac{1}{-i}\right)\left[\frac{1-(1+i)^n}{(1+i)^n}\right] \\
&= A\left[\frac{(1+i)^n-1}{i(1+i)^n}\right] \qquad i \neq 0 \qquad (2.6)
\end{aligned}
$$

The term in brackets is called the *uniform-series present-worth factor* (USPWF). This factor will give the present worth P of an equivalent uniform annual series A which begins *at the end of year 1* and extends for n years at an interest rate i.

By rearranging Eq. (2.6), we can express A in terms of P:

$$A = P\left[\frac{i(1+i)^n}{(1+i)^n-1}\right] \qquad (2.7)$$

The term in brackets, called the *capital-recovery factor* (CRF), yields the equivalent uniform annual cost A over n years of a given investment P when the interest rate is i.

It is very important to commit to memory the fact that these formulas were derived with the present worth P and the first uniform annual-cost value A *one year (period) apart*. That is, the present sum P *must always* be located one period *prior* to the first A. The correct use of these factors is illustrated in Sec. 2.7.

2.3 Derivation of the Uniform-Series Compound-Amount Factor and the Sinking-Fund Factor

While the *sinking-fund factor* (SFF) and the *uniform-series compound-amount factor* (USCAF) could be derived using the SPCAF, the simplest way to derive the formulas is to substitute into those already developed. Thus, if P from Eq. (2.3), which uses the SPPWF, is substituted into Eq. (2.7), the following formula results:

$$A = F\left[\frac{1}{(1+i)^n}\right]\left[\frac{i(1+i)^n}{(1+i)^n-1}\right] = F\left[\frac{i}{(1+i)^n-1}\right] \qquad (2.8)$$

The expression in brackets in Eq. (2.8) is the SFF. Equation (2.8) is used to determine the uniform annual series that would be equivalent to a given future worth F. This is shown graphically in Fig. 2.3. Note that the uniform series A begins at the end of period 1 and continues through the period of the given F. This is unlike the uniform-series present-worth formulas in the preceding section, where the P and the first A were always one period apart.

Equation (2.8) can be rearranged to express F in terms of A:

$$F = A\left[\frac{(1+i)^n-1}{i}\right] \qquad (2.9)$$

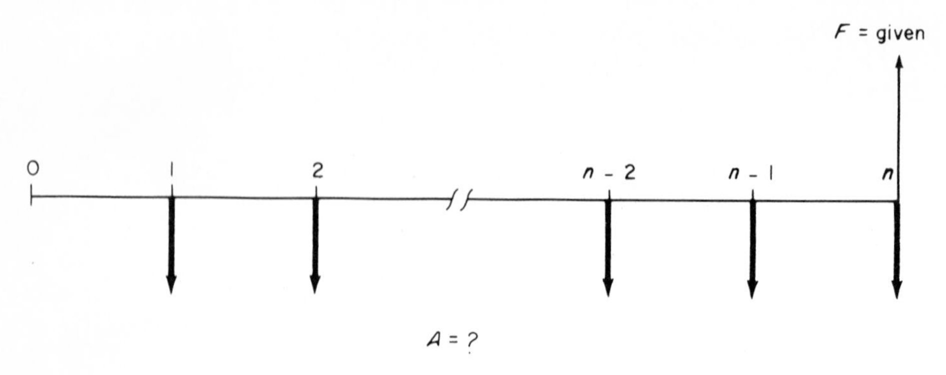

Figure 2.3 Transformation of a given F value into an equivalent A series.

The term in brackets is called the *uniform-series compound-amount factor* (USCAF) and when multiplied by the given uniform annual amount A yields the future worth of the uniform series. The cash-flow diagram for this case would be just the opposite of that shown in Fig. 2.3. Again, it is important to remember that the future amount F occurs in the same period as the last A.

Prob. P2.1

2.4 Standard Factor Notation and Use of Interest Tables

To avoid the cumbersome task of writing out the formulas each time one of the factors is used, a standard notation has been adopted which represents the various factors. This standard notation, which also includes the interest rate and the number of periods, will always be of the general form $(X/Y, i\%, n)$. The first letter inside the parentheses represents what you want to find, while the second letter Y represents what is given. For example, F/P means "find F when given P." The i is the interest rate in percent and the n represents the number of periods involved. Thus, $(F/P, 6\%, 20)$ means obtain the factor which when multiplied by a given P allows you to find the future amount of money F that will be accumulated in 20 periods if the interest rate is 6%.

Table 2.1 Standard factor notations

Factor name	Standard notation
Single-payment present-worth (SPPWF)	$(P/F, i\%, n)$
Single-payment compound-amount (SPCAF)	$(F/P, i\%, n)$
Uniform-series present-worth (USPWF)	$(P/A, i\%, n)$
Capital-recovery (CRF)	$(A/P, i\%, n)$
Sinking-fund (SFF)	$(A/F, i\%, n)$
Uniform-series compound-amount (USCAF)	$(F/A, i\%, n)$

Table 2.2 Computations using standard notation

To find	Given	Factor	Formula
P	F	$(P/F, i\%, n)$	$P = F(P/F, i\%, n)$
F	P	$(F/P, i\%, n)$	$F = P(F/P, i\%, n)$
P	A	$(P/A, i\%, n)$	$P = A(P/A, i\%, n)$
A	P	$(A/P, i\%, n)$	$A = P(A/P, i\%, n)$
A	F	$(A/F, i\%, n)$	$A = F(A/F, i\%, n)$
F	A	$(F/A, i\%, n)$	$F = A(F/A, i\%, n)$

The standard notation is simpler than factor names for identifying factors and will be used exclusively hereinafter. Table 2.1 shows the standard notation for the formulas derived thus far in this chapter. For ready reference the formulas used in computations are collected in Table 2.2.

In order to simplify the routine engineering-economy calculations involving the factors above, tables of factor values have been prepared for interest rates from 0.5 to 50% and time periods from 1 to 100 years. These tables, found in Appendix A and identified as Tables A-1 through A-30, are arranged with various factors across the top and the number of years n down the left and right column. The word "discrete" in the title of each table is printed to emphasize that these tables are for factors which utilize the end-of-period convention (Sec. 1.6) and that interest is compounded once each interest period. For a given factor, interest rate, and time, the correct factor value would be found in the respective interest-rate table at the intersection of the given factor and n. For example, the value of the factor $(P/A, 5\%, 10)$ is found in the P/A column of Table A-7 at year 10 as 7.7217. The value 7.7217 could, of course, have been computed using the mathematical expression for the USPWF in Eq. (2.6).

$$(P/A, 5\%, 10) = \frac{(1+i)^n - 1}{i(1+i)^n} = \frac{1.05^{10} - 1}{0.05(1.05^{10})} = 7.7217$$

Table 2.3 presents several examples of the use of the interest tables in Appendix A.

Prob. P2.2

Table 2.3 Use of interest tables

Standard notation	i, %	n	Table	Factor value
$(F/A, 10\%, 3)$	10	3	A-12	3.310
$(A/P, 7\%, 20)$	7	20	A-9	0.09439
$(P/F, 25\%, 35)$	25	35	A-25	0.0004

2.5 Definition and Derivation of Gradient Formulas

A *uniform gradient* is a cash-flow *series* which either increases or decreases *uniformly*. That is, the cash flow, whether income or disbursements, changes by the same amount each year. The *amount* of the increase or decrease is the *gradient*. For example, if a clothing manufacturer predicts that the cost of maintaining a cutting machine will increase by $500 per year until the machine is retired, a gradient series is involved and the amount of the gradient is $500. Similarly, if the company expects income to decrease by $3000 per year for the next 5 years, the decreasing income represents a gradient in the amount of $3000 per year.

The formulas previously developed for uniform-series cash flows were generated on the basis of year-end payments of equal value. In the case of a gradient, each year-end cash flow is different, so a new formula must be derived. In developing a formula which can be used for uniform gradients, it is convenient to assume that the payment that occurs at the end of year 1 does not involve a gradient but is rather a *base payment*. In actual applications, the base payment is usually larger or smaller than the gradient increase or decrease. For example, if you purchase a new car with a 12,000-mile complete guarantee, you might reasonably expect to have to pay for only the gasoline during the first year of operation. Let us assume that this cost is $400; that is, $400 is the base amount. After the first year, however, you would have to absorb the cost of repair or replacement yourself, and these costs could reasonably be expected to increase each year that you own the car. So if you estimate your operation costs to increase by $25 each year, the amount you would pay after the second year would be $425, after the third $450, and so on to year n, when the total cost would be $400 + (n - 1)25$. The cash-flow diagram for this is shown in Fig. 2.4. Note that the gradient is first observed between year 1 and year 2, and the first (base) payment ($400) is not equal to the gradient ($25). We will now define a new symbol for gradients:

G = annual arithmetic change in the magnitude of receipts or disbursements

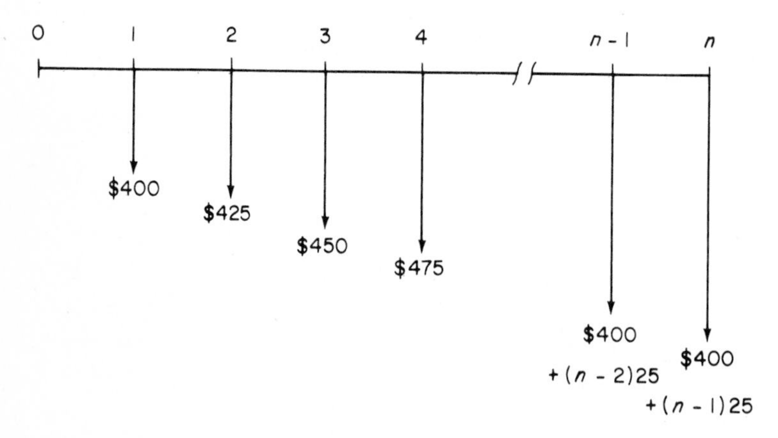

Figure 2.4 Diagram of a uniform-gradient series with a gradient of $25.

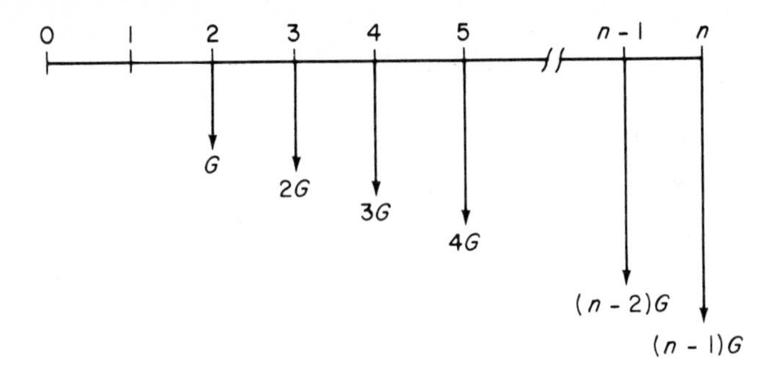

Figure 2.5 Uniform-gradient series ignoring the base amount.

The value of G may be positive or negative. If we ignore the base payment, we can construct a generalized uniformly increasing-gradient cash-flow diagram as shown in Fig. 2.5. Note that the gradient begins in year 2. This is called a *conventional gradient*.

Example 2.1 The Free Spirit Company expects to realize a revenue of \$100,000 next year from the sale of a new product. However, sales are expected to decrease uniformly with new competition to a level of \$47,500 in 8 years. Determine the gradient and construct the cash-flow diagram.

SOLUTION

$$\text{Base amount} = \$100{,}000$$

$$\text{Revenue loss by year 8} = 100{,}000 - 47{,}500 = \$52{,}500$$

$$\text{Gradient} = \text{loss}/(n-1)$$

$$= 52{,}500/(8-1) = \$7500 \text{ per year}$$

The cash-flow diagram is shown in Fig. 2.6.

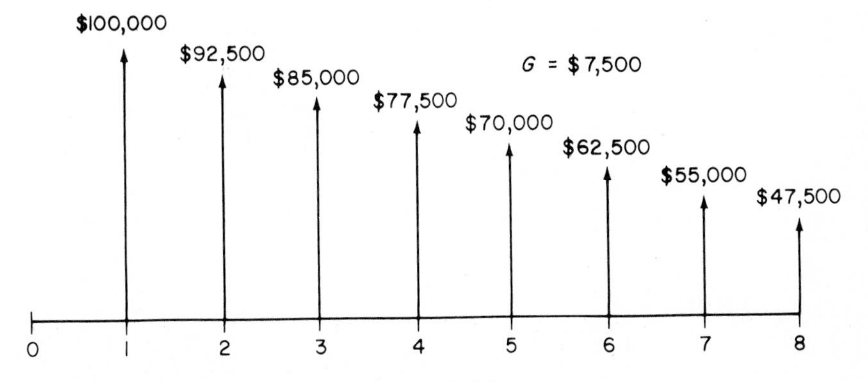

Figure 2.6 Diagram for gradient series, Example 2.1.

There are several ways by which the uniform-gradient factors can be derived. We will use the single-payment present-worth factor $(P/F, i\%, n)$ but the same result can be obtained using single-payment compound-amount factors, uniform-series compound-amount factors, or uniform-series present-worth factors.

Referring to Fig. 2.5, we find that the present worth at year 0 of the gradient payments would be equal to the sum of the present worths of the individual payments. Thus,

$$P = G(P/F, i\%, 2) + 2G(P/F, i\%, 3) + 3G(P/F, i\%, 4) + \cdots + [(n-2)G](P/F, i\%, n-1) + [(n-1)G](P/F, i\%, n)$$

Factoring out G yields

$$P = G[(P/F, i\%, 2) + 2(P/F, i\%, 3) + 3(P/F, i\%, 4) + \cdots + (n-2)(P/F, i\%, n-1) + (n-1)(P/F, i\%, n)]$$

Replacing the symbols with the appropriate single-payment present-worth factor expression in Eq. (2.3) yields

$$P = G\left[\frac{1}{(1+i)^2} + \frac{2}{(1+i)^3} + \frac{3}{(1+i)^4} + \cdots + \frac{n-2}{(1+i)^{n-1}} + \frac{n-1}{(1+i)^n}\right] \qquad (2.10)$$

Multiplying both sides of Eq. 2.10 by $(1+i)^1$ to simplify yields

$$P(1+i)^1 = G\left[\frac{1}{(1+i)^1} + \frac{2}{(1+i)^2} + \frac{3}{(1+i)^3} + \cdots + \frac{n-2}{(1+i)^{n-2}} + \frac{n-1}{(1+i)^{n-1}}\right] \qquad (2.11)$$

Subtracting Eq. (2.10) from Eq. (2.11), noting that the first term of Eq. (2.11) and the last term of Eq. (2.10) have no matching terms, yields the following relations.

$$P(1+i)^1 - P = G\left[\frac{1}{(1+i)^1} + \frac{(2-1)}{(1+i)^2} + \frac{(3-2)}{(1+i)^3} + \cdots + \frac{(n-1)-(n-2)}{(1+i)^{n-1}} - \frac{n-1}{(1+i)^n}\right]$$

$$P(1+i)^1 - P = G\left[\frac{1}{(1+i)^1} + \frac{1}{(1+i)^2} + \frac{1}{(1+i)^3} + \cdots + \frac{1}{(1+i)^{n-1}} + \frac{1-n}{(1+i)^n}\right]$$

If we write the left side of this equation as $P + Pi - P$, factor out the n in the last term, and divide by i, we have

$$P = \frac{G}{i}\left[\frac{1}{(1+i)^1} + \frac{1}{(1+i)^2} + \frac{1}{(1+i)^3} + \cdots + \frac{1}{(1+i)^{n-1}} + \frac{1}{(1+i)^n}\right] - \frac{Gn}{i(1+i)^n}$$

Since the expression in the brackets is the present worth of a uniform series of 1 for n years, we substitute the expression for the P/A factor from Eq. (2.6).

$$P = \frac{G}{i}\left[\frac{(1+i)^n - 1}{i(1+i)^n}\right] - \frac{Gn}{i(1+i)^n}$$

$$= \frac{G}{i}\left[\frac{(1+i)^n - 1}{i(1+i)^n} - \frac{n}{(1+i)^n}\right] \tag{2.12}$$

Equation (2.12) is the general relation to convert a uniform gradient G for n years into a present worth at year 0; that is, Fig. 2.7*a* is converted into the equivalent cash flow in Fig. 2.7*b*. The *uniform-gradient present-worth* and standard factor notation is

$$(P/G, i\%, n) = \frac{1}{i}\left[\frac{(1+i)^n - 1}{i(1+i)^n} - \frac{n}{(1+i)^n}\right]$$

Note that the gradient starts in year 2 in Fig. 2.7*a* and P is found in year 0.

The equivalent uniform annual series of the gradient G is found by multiplying the present worth in Eq. (2.12) by the $(A/P, i\%, n)$ factor expression from Eq. (2.7).

$$A = P(A/P, i\%, n)$$

$$= \frac{G}{i}\left[\frac{(1+i)^n - 1}{i(1+i)^n} - \frac{n}{(1+i)^n}\right]\left[\frac{i(1+i)^n}{(1+i)^n - 1}\right]$$

$$= G\left[\frac{1}{i} - \frac{n}{(1+i)^n - 1}\right] \tag{2.13}$$

The expression in brackets in Eq. (2.13) is called the *uniform-gradient annual-series* factor and is identified by $(A/G, i\%, n)$. This factor converts Fig. 2.8*a* into Fig. 2.8*b*. You should realize that the annual series is nothing but an A value equivalent to the gradient. Note from Fig. 2.8 that the gradient starts in year 2 and the A values occur from year 1 to year n inclusive.

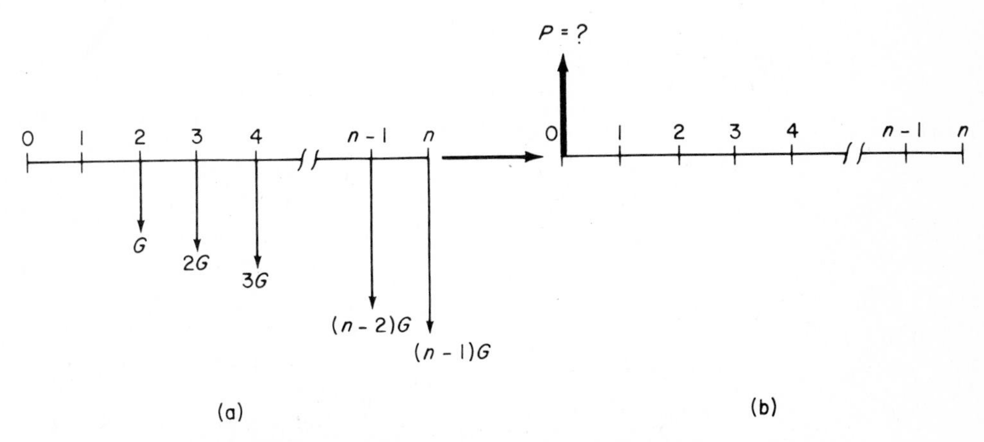

Figure 2.7 Conversion diagram from a uniform gradient to a present worth.

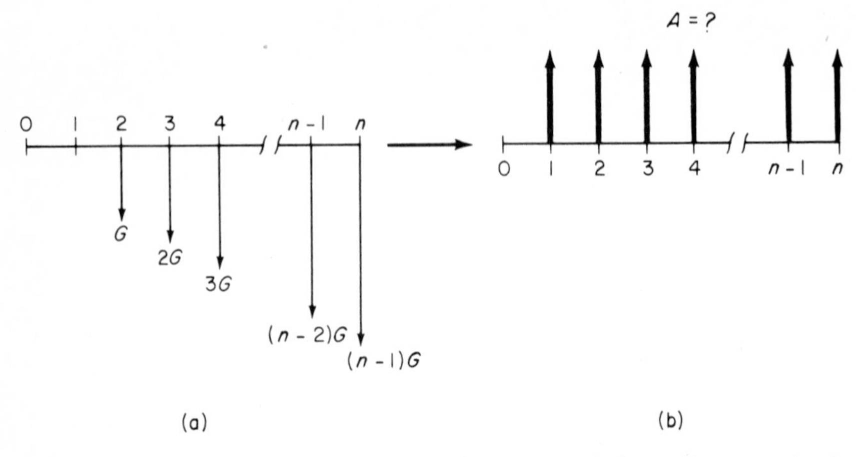

Figure 2.8 Conversion diagram of a uniform gradient to an equivalent uniform annual series.

The formulas used to compute P and A in terms of standard notation factors are

$$P = G(P/G, i\%, n) \tag{2.14}$$

$$A = G(A/G, i\%, n) \tag{2.15}$$

Even though P in Fig. 2.7*b* is drawn as in income, it is actually the present worth of the disbursements shown in the gradient of Fig. 2.7*a*. This convention is used to avoid drawing equivalent cash flows over actual cash flows in more complicated situations. In fact, this confusion would be present in Fig. 2.8 if the annual series were superimposed on the gradient.

Tables A-31 through A-38 present the P/G and A/G factors for interest rates up to 50%. Both factors are arranged and can be used in the same manner as the compound-interest tables; that is, for a specified i and n. Table 2.4 lists several examples of gradient factors taken from the tables. Throughout this chapter we will assume the n value is given in years. In Chap. 3 you will learn how to use the tables in Appendix A for interest periods other than a year.

Probs. P2.3 to P2.5

Table 2.4 Examples of gradient factors

Value to be computed	Standard notation	i, %	n	Table	Factor
P	$(P/G, 5\%, 10)$	5	10	A-31	31.652
P	$(P/G, 30\%, 24)$	30	24	A-34	10.943
A	$(A/G, 6\%, 19)$	6	19	A-36	7.287
A	$(A/G, 35\%, 8)$	35	8	A-38	2.060

Table 2.5 Linear interpolation setup

	i or *n*	Factor	
b [*a*]	→ tabulated	value 1	[*c*] *d*
	→ desired	unlisted	
	→ tabulated	value 2	

2.6 Interpolation in Interest Tables

Sometimes it is necessary to locate a factor value for an interest rate *i* or *n* that is not in the interest tables. When this occurs, the desired factor value can be obtained in one of two ways: (1) by using the formulas that were derived in Secs. 2.1 to 2.3 and 2.5 or (2) by interpolating between the tabulated values on both sides of the unlisted desired value. It is generally easier and faster to use the formulas rather than to interpolate for determining the factor value which corresponds to the unlisted *i* or *n* value. However, *linear* interpolation is acceptable and is considered sufficient as long as the values of *i* or *n* are not too distant from each other.

The first step in linear interpolation is to set up the known and the unknown values as shown in Table 2.5. A ratio equation is then set up and solved for *c*, as follows (refer to Table 2.5):

$$\frac{a}{b} = \frac{c}{d} \quad \text{or} \quad c = \frac{a}{b}d \tag{2.16}$$

where *a*, *b*, *c*, and *d* represent the differences between the numbers shown in the interest tables. The value of *c* from Eq. (2.16) is added to or subtracted from value 1, depending on whether the factor is increasing or decreasing in value, respectively. The following examples illustrate the procedure just described.

Example 2.2 Determine the value of the *A/P* factor for an interest rate of 7.3% and *n* of 10 years, that is, (*A/P*, 7.3%, 10).

SOLUTION The values of the *A/P* factor for interest rates of 7 and 8% are listed in Tables A-9 and A-10, respectively. Thus we have the following situation:

b [*a*]	→ 7%	0.14238	[*c*] *d*
	→ 7.3%	*X*	
	→ 8%	0.14903	

The unknown *X* is the desired factor value. From Eq. (2.16),

$$c = \frac{7.3 - 7}{8 - 7}(0.14903 - 0.14238)$$

$$= \frac{0.3}{1} 0.00665$$

$$= 0.00199$$

Since the factor is increasing in value as the interest rate increases from 7 to 8%, the value of c must be *added* to the value of the 7% factor. Thus,

$$X = 0.14238 + 0.00199 = 0.14437$$

COMMENT It is good practice to check the "reasonableness" of your final answer by verifying that X lies between the values of the known factors used in the interpolation in approximately the correct proportions. In this case, since 0.14437 is less than 0.5 of the distance between 0.14238 and 0.14903, the answer seems reasonable. Rather than interpolating, a simpler procedure in some cases may be to use the formula to compute the factor value directly.

Example 2.3 Find the value of the $(P/F, 4\%, 48)$ factor.

SOLUTION From Table A-6 for 4% interest, the values of the P/F factor for 45 and 50 years can be found as follows:

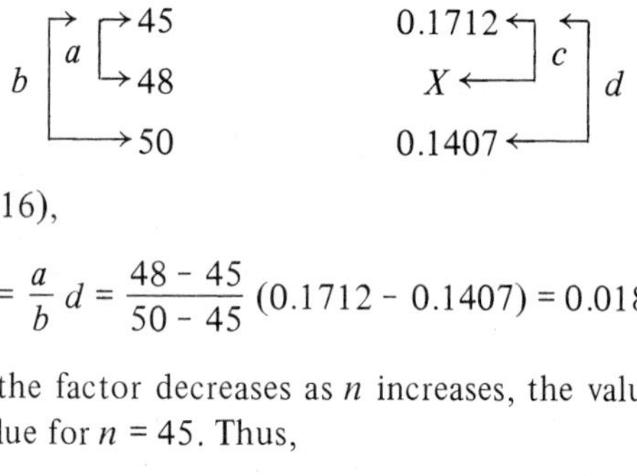

b	a	45	0.1712	c	d
		48	X		
		50	0.1407		

Again, from Eq. (2.16),

$$c = \frac{a}{b} d = \frac{48 - 45}{50 - 45}(0.1712 - 0.1407) = 0.0183$$

Since the value of the factor decreases as n increases, the value of c must be *subtracted* from the value for $n = 45$. Thus,

$$X = 0.1712 - 0.0183 = 0.1529$$

Solved Problem 2.14
Probs. P2.6 to P2.9

2.7 Present-Worth, Future-Worth, and Equivalent Uniform Annual-Series Calculations

The first and probably most important step in solving engineering-economy problems is construction of a cash-flow diagram. In addition to more clearly illustrating "the problem," the cash-flow diagram immediately shows which formulas should be used and whether the conditions of cash flow presented allow straightforward application of the formulas as derived in the preceding sections. Obviously, the formulas can be used only when the cash flow of the problem conforms exactly to the cash-flow diagram for the formulas. For example, the uniform-series factors could not be used if payments or receipts occurred *every other year* instead of every year. It is very important, therefore, to remember the conditions for which the formulas apply. The correct use of the formulas for finding P, F, or A is illustrated in the examples that follow. All equations used are taken from Table 2.2. See the Solved Problems for cases in which some of these formulas cannot be applied.

Example 2.4 If a woman deposits \$600 now, \$300 two years from now, and \$400 five years from now, how much will she have in her account ten years from now if the interest rate is 5%?

SOLUTION The first step is to draw the cash-flow diagram (Fig. 2.9), which indicates that an F value is to be computed. Since each value is different and they do not take place each year, the future worth F is equal to the sum of the individual single payments at year 10. Thus,

$$F = 600(F/P, 5\%, 10) + 300(F/P, 5\%, 8) + 400(F/P, 5\%, 5)$$

$$= 600(1.6289) + 300(1.4775) + 400(1.2763)$$

$$= \$1931.11$$

COMMENT The problem could also be solved by finding the present worth in year 0 of the \$300 and \$400 deposits using the P/F factors and then finding the future worth of the total.

$$P = 600 + 300(P/F, 5\%, 2) + 400(P/F, 5\%, 5)$$

$$= 600 + 300(0.9070) + 400(0.7835)$$

$$= \$1185.50$$

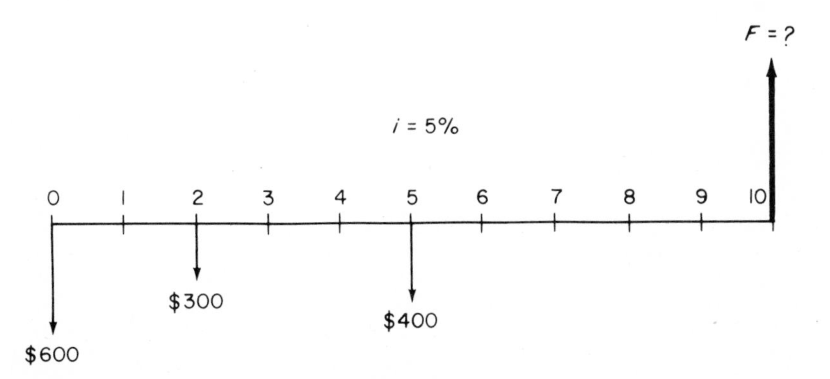

Figure 2.9 Diagram for a future value, Example 2.4.

Then,

$$\begin{aligned} F &= 1185.50(F/P, 5\%, 10) \\ &= 1185.50(1.6289) \\ &= \$1931.06 \end{aligned}$$

It should be obvious that there are a number of ways the problem could be worked, since any year could be used to find the equivalent total of the deposits before finding the value at year 10. As an exercise, you should work the problem using year 5 for finding the equivalent total before determining the final amount in year 10. All answers should be the same, except for round-off error.

Example 2.5 How much money would a man have in his account after 8 years if he deposited \$100 per year for 8 years at 4% starting one year from now?

SOLUTION The cash-flow diagram is shown in Fig. 2.10. Since the payments start at the end of year 1 and end in the year the future worth is desired, the F/A formula can be used. Thus,

$$F = 100(F/A, 4\%, 8) = 100(9.214) = \$921.40$$

Example 2.6 How much money would you be willing to spend now in order to avoid spending \$500 seven years from now if the interest rate is 18%?

SOLUTION The cash-flow diagram appears in Fig. 2.11. The problem might be easier if it were stated in another manner, such as, what is the present worth of \$500 seven years from now if the interest rate is 18%; or, what present amount would be equivalent to \$500 seven years hence if the interest is 18%; or, what initial investment is equivalent to spending \$500 seven years from now at an interest rate of 18%? In all cases F is given and P is to be computed.

$$P = 500(P/F, 18\%, 7) = 500(0.3139) = \$156.95$$

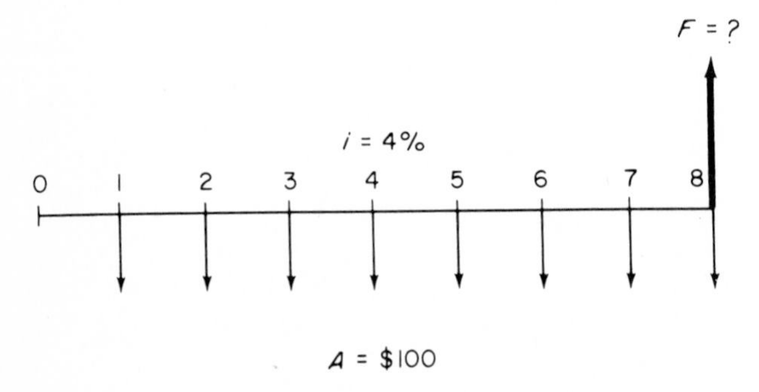

Figure 2.10 Diagram to find F for a uniform series, Example 2.5.

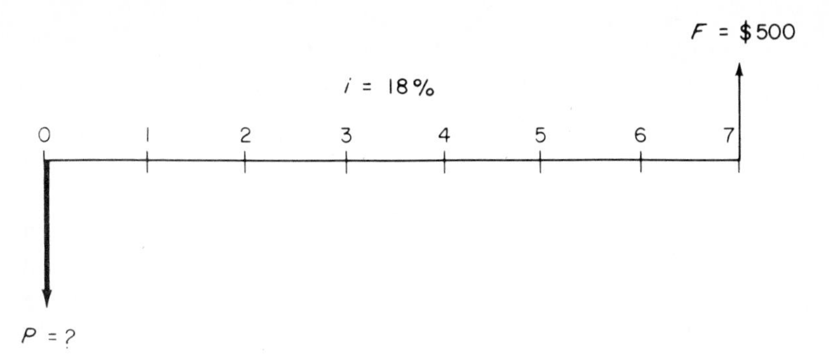

Figure 2.11 Diagram for Example 2.6.

COMMENT Although there are several ways to state the same problem, the cash-flow diagram remains the same in each case.

Example 2.7 How much money would you be willing to pay now for a note that will yield \$600 per year for 9 years starting next year if the interest rate is 7%?

SOLUTION The cash-flow diagram is shown in Fig. 2.12. Since the cash-flow diagram fits the P/A uniform-series formula, the problem can be solved directly.

$$P = 600(P/A, 7\%, 9) = 600(6.5152) = \$3909.12$$

COMMENT You should recognize that P/F factors could be used for each of the 9 years and the resulting present worths added to get the correct answer. Another way would be to find the future worth F of the \$600 payments and then find the present worth of the F value. There are many ways to solve an engineering-economy problem. Only the most direct method will be presented here, but you should work the problems in at least one other way to become more familiar with the use of the formulas.

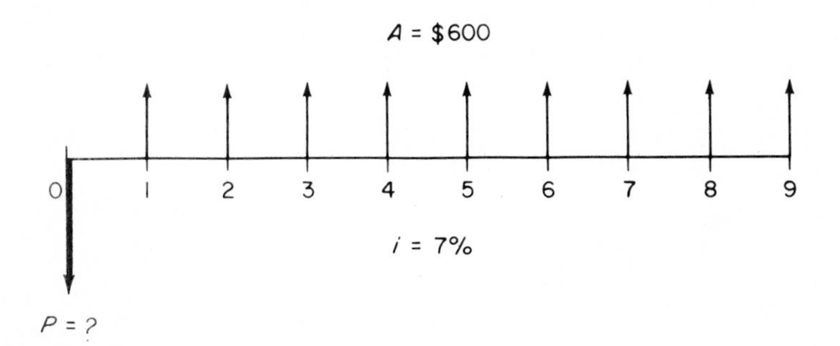

Figure 2.12 Diagram for Example 2.7.

Example 2.8 How much money must a person deposit every year starting one year from now at $5\frac{1}{2}\%$ per year in order to accumulate \$6000 seven years from now?

SOLUTION The cash-flow diagram is shown in Fig. 2.13. The cash-flow diagram fits the A/F formula as derived. Thus,

$$A = 6000(A/F, 5.5\%, 7) = 6000(0.12096) = \$725.76 \text{ per year}$$

Solved Problem 2.15
Probs. P2.10 to P2.38

2.8 Present Worth and Equivalent Uniform Annual Series of Conventional Gradients

If the gradient begins in year 2, year 0 of the gradient and year 0 of the entire cash-flow diagram coincide, and the gradient is referred to as *conventional.* In this case the present worth P or equivalent uniform annual series A of the *gradient only* can be determined by using Eq. (2.14) or (2.15), respectively. The cash flow that forms the base amount of the gradient must be considered separately, as illustrated below.

Example 2.9 A couple plan to start saving money by depositing \$500 into their savings account one year from now. They estimate that the deposits will increase by \$100 per year for 9 years thereafter. What would be the present worth of the investments if the interest rate is 5% per year?

SOLUTION The cash-flow diagram is shown in Fig. 2.14. Two computations must be made: the first to compute the present worth of the base amount (P_A), and a second to compute the present worth of the gradient (P_G). Then the total present

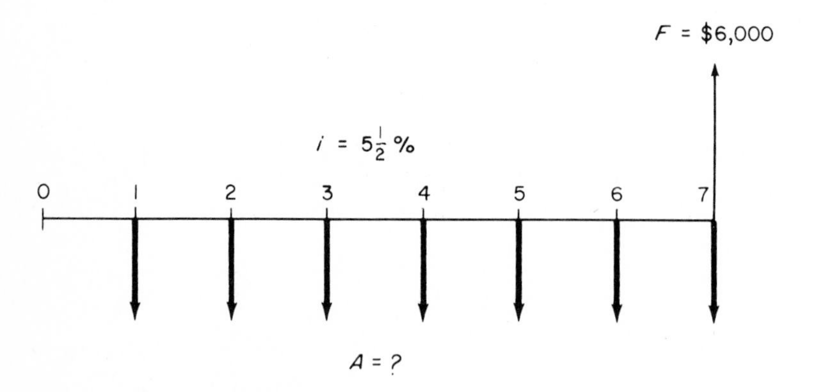

Figure 2.13 Diagram for Example 2.8.

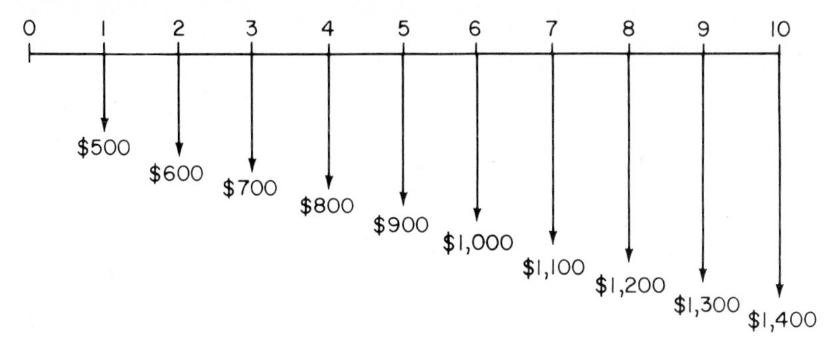

Figure 2.14 Cash flow, Example 2.9.

worth P_T equals P_A plus P_G, since P_A and P_G occur in year 0. This is clearly illustrated by converting the cash-flow diagram as in Fig. 2.15. The present worth would be calculated as follows:

$$
\begin{aligned}
P_T &= P_A + P_G \\
&= 500(P/A, 5\%, 10) + 100(P/G, 5\%, 10) \\
&= 500(7.7217) + 100(31.652) \\
&= \$7026.05
\end{aligned}
$$

COMMENT It is important to emphasize again that the gradient factor represents the present worth of the *gradient only*. Any other cash flow involved must be considered separately.

Example 2.10 Work Example 2.9 solving for the equivalent uniform annual series.

SOLUTION Here too it is necessary to consider the gradient and the other costs involved in the cash flow separately. From the diagrams shown in Fig. 2.15*b* and *c*, the annual series would be

$$A = A_1 + A_G$$

where A_1 = equivalent annual series of the base amount $500
A_G = equivalent annual series of the gradient = $100(A/G, 5\%, 10)$

Then,

$$
\begin{aligned}
A &= 500 + 100(A/G, 5\%, 10) = 500 + 100(4.099) \\
&= \$909.90 \text{ per year from years 1 to 10}
\end{aligned}
$$

COMMENT It is often helpful to remember that the present worth of the base amount and gradient can simply be multiplied by the appropriate A/P factor to

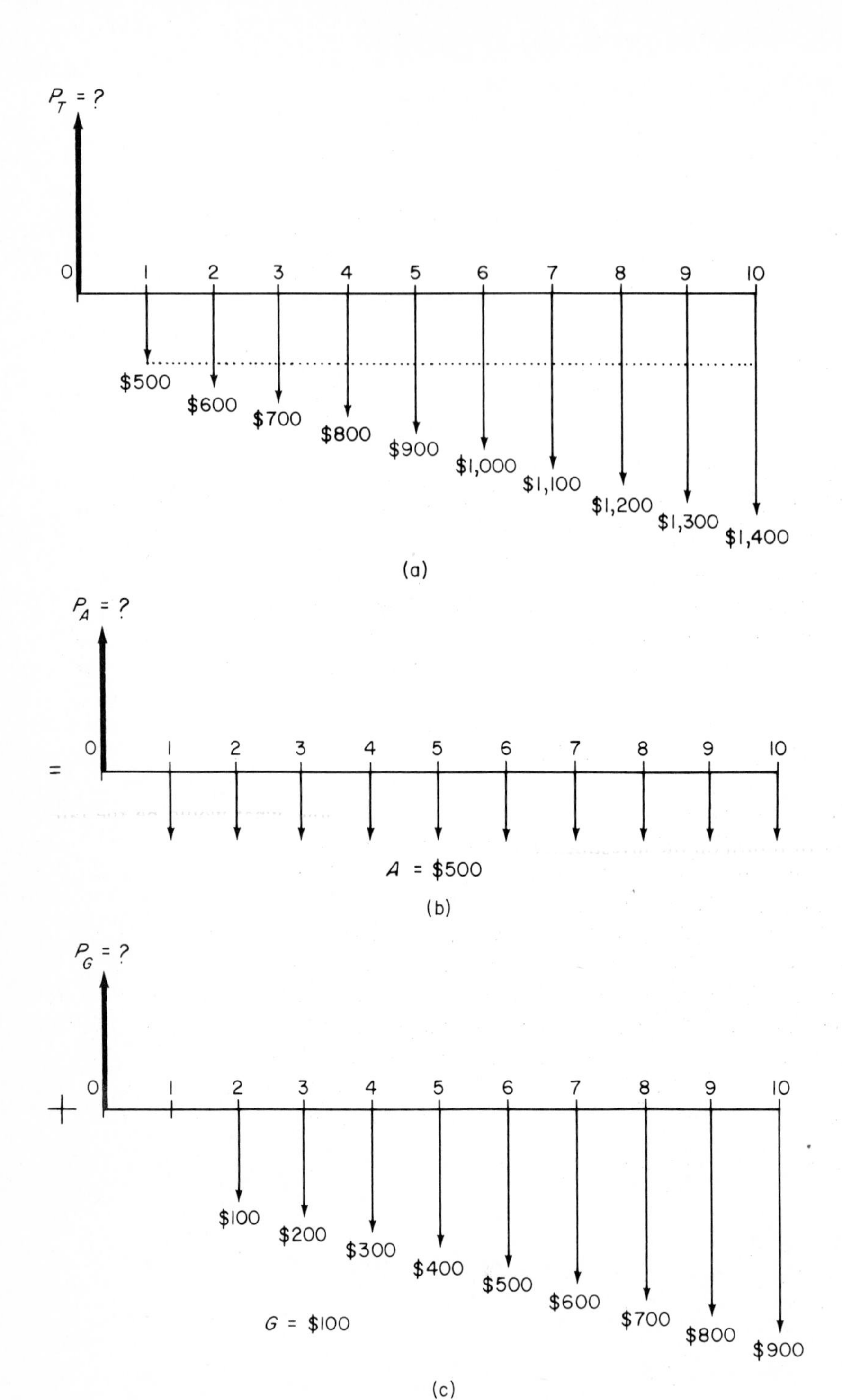

Figure 2.15 Partitioned diagram, Example 2.9, $a = b + c$.

get A. Here,

$$A = P_T(A/P, 5\%, 10) = 7026.05(0.12950)$$
$$= \$909.87$$

Probs. P2.39 to P2.50

2.9 Calculation of Unknown Interest Rates

In some cases, the amount of money invested and the amount of money received after a specified number of years are known, and it is desired to determine the interest rate or rate of return. When only a single payment and a single receipt, or a uniform series of payments or receipts are involved, the unknown interest rate can be determined by direct solution of the economy equation. When nonuniform payments or several factors are involved, however, the problem must be solved by the trial-and-error method. In this section, only single-payment or uniform-series cash-flow problems are considered. The more complicated trial-and-error problems are deferred until Chap. 7, which deals with rate-of-return analysis.

Although the single-payment and uniform-series formulas can be rearranged and expressed in terms of i, it is generally simpler to *solve for the value of the factor* and then look up the interest rate in the interest tables. This method is illustrated in the examples that follow.

Example 2.11

(*a*) If a person can make a business investment requiring an expenditure of \$3000 now in order to receive \$5000 five years from now, what would be the rate of return on the investment?

(*b*) If the same person can receive 9% interest from certificates of deposit, which investment should be made?

SOLUTION

(*a*) The cash-flow diagram is shown in Fig. 2.16. The interest rate can be found by setting up the P/F or F/P equations and solving directly for the factor

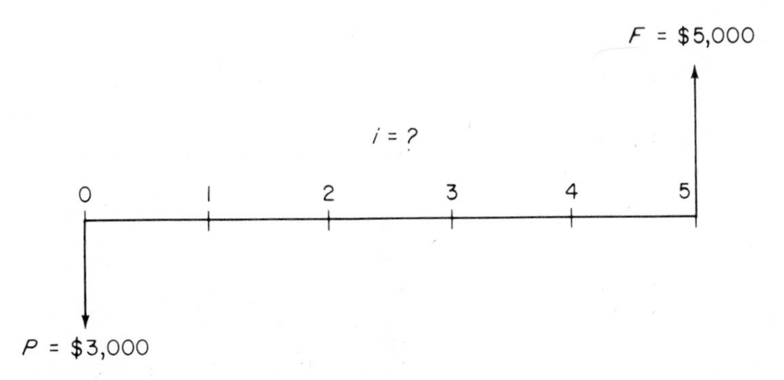

Figure 2.16 Diagram used to determine the rate of return, Example 2.11*a*.

value. Using P/F,

$$P = F(P/F, i\%, n)$$

$$3000 = 5000(P/F, i\%, 5)$$

$$(P/F, i\%, 5) = \frac{3000}{5000} = 0.6000$$

From the interest tables, a P/F factor of 0.6000 for n equal to 5 lies between 10 and 11%. Interpolating between these two values using Eq. (2.16), we have

$$\begin{array}{ccc|ccc} & & 0.6209 & 10\% & & \\ b & a & 0.6000 & i\% & c & d \\ & & 0.5935 & 11\% & & \end{array}$$

$$c = \left(\frac{0.6209 - 0.6000}{0.6209 - 0.5935}\right)(11 - 10) = \frac{0.0209}{0.0274}(1) = 0.7628$$

Therefore,

$$i = 10 + 0.76 = 10.76\%$$

(*b*) Since 10.76% is greater than the 9% available in certificates of deposit, the person should make the business investment.

COMMENT Since the higher rate of return would be received on the business investment, the investor would probably select this option instead of the certificates of deposit. However, the degree of risk associated with the business investment was not specified. Obviously the amount of risk associated with a particular investment is an important parameter and oftentimes causes selection of the lower-rate-of-return investment. Unless specified to the contrary, the problems in this text will assume equal risks for all alternatives.

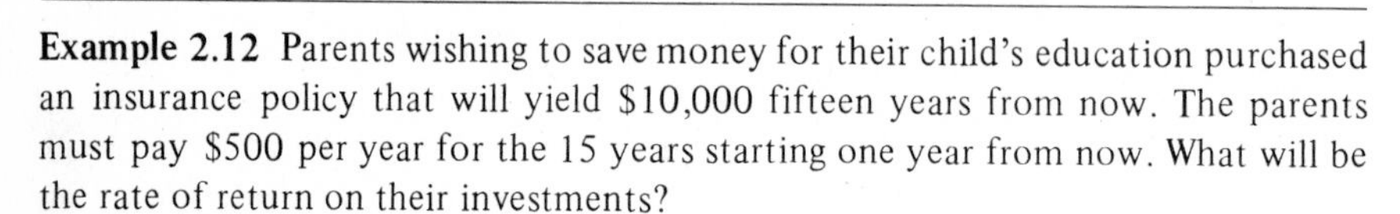

Example 2.12 Parents wishing to save money for their child's education purchased an insurance policy that will yield $10,000 fifteen years from now. The parents must pay $500 per year for the 15 years starting one year from now. What will be the rate of return on their investments?

SOLUTION The cash-flow diagram is shown in Fig. 2.17. Either the A/F or F/A factor could be used. Using A/F,

$$A = F(A/F, i\%, n)$$

$$500 = 10{,}000(A/F, i\%, 15)$$

$$(A/F, i\%, 15) = 0.0500$$

From the interest tables under the A/F column for 15 years, the value 0.0500 is found to lie between 3 and 4%. By interpolation, $i = 3.98\%$.

Probs. P2.51 to P2.58

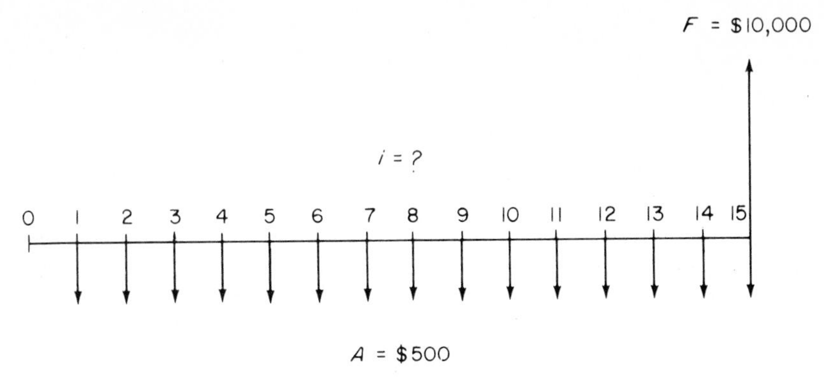

Figure 2.17 Diagram to determine the rate of return, Example 2.12.

2.10 Calculation of Unknown Years

In break-even economic analysis, it is sometimes necessary to determine the number of years required before an investment pays off. Other times it is desirable to be able to determine when given amounts of money will be available from a proposed investment. In these cases, the unknown value is n, and techniques similar to those of the preceding section on unknown interest rates can be used to find n.

Though these problems can be solved directly for n by proper manipulation of the single-payment and uniform-series formulas, it is generally easier to solve for the factor value and interpolate in the interest tables, as illustrated below.

Example 2.13 How long will it take for $1000 to double if the interest rate is 5%?

SOLUTION The cash-flow diagram is shown in Fig. 2.18. The problem can be solved using either the F/P or P/F factor. Using the P/F factor,

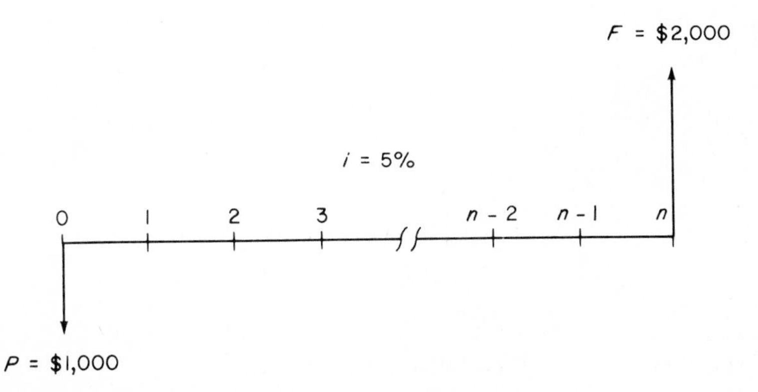

Figure 2.18 Diagram to determine an n value, Example 2.13.

$$P = F(P/F, i\%, n)$$

$$1000 = 2000(P/F, 5\%, n)$$

$$(P/F, 5\%, n) = 0.500$$

From the 5% interest table, the value 0.500 under the P/F column lies between 14 and 15 years. By interpolation, $n = 14.2$ years.

COMMENT Problems of this type become more complicated when two or more nonuniform payments are involved. See the Solved Problems for an illustration using trial and error.

Solved Problem 2.16
Probs. P2.59 to P2.67

SOLVED PROBLEMS

Example 2.14 A new building has been purchased by Waldorf Concession Stands, Inc. The present worth of future maintenance costs is to be calculated with a P/A factor. If $i = 13\%$ per year and the life is expected to be 42 years, estimate what factor value is correct by using two-way interpolation in the tables at $i = 12$ and 15%.

SOLUTION The P/A factor requires two-way interpolation for i and n. First, we will find the P/A factor for $i = 13\%$ at $n = 40$ and $n = 45$ using 12 and 15%.

	i	$n = 40$			$n = 45$		
1	12%	8.2438	c_{40}	1.6020	8.2825	c_{45}	1.6282
3	13%	X_{40}			X_{45}		
	15%	6.6418			6.6543		

The subscripts correspond to the n value for which the factor is computed.

$$c_{40} = \tfrac{1}{3}(1.6020) = 0.5340 \qquad X_{40} = 8.2438 - 0.5340 = 7.7098$$

$$c_{45} = \tfrac{1}{3}(1.6282) = 0.5427 \qquad X_{45} = 8.2825 - 0.5427 = 7.7398$$

Now estimate the P/A factor for $n = 42$.

$$X_{42} = 7.7098 + \tfrac{2}{5}(0.0300) = 7.7218$$

Thus, we have

$$(P/A, 13\%, 42) = 7.7218$$

The correct value using Eq. (2.6) is 7.6469.

Sec. 2.6

Example 2.15 Explain why the uniform-series factors *cannot* be used to compute P or F *directly* for the cash flows of Fig. 2.19.

SOLUTION

(*a*) The P/A factor cannot be used to compute P since the $100-per-year receipt does not occur each year from year 1 through year 5.

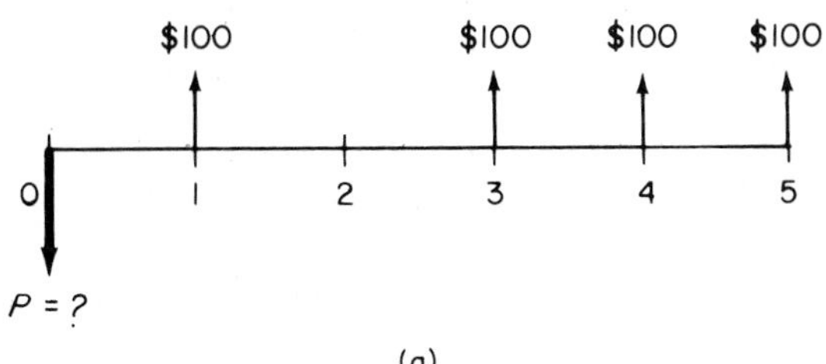

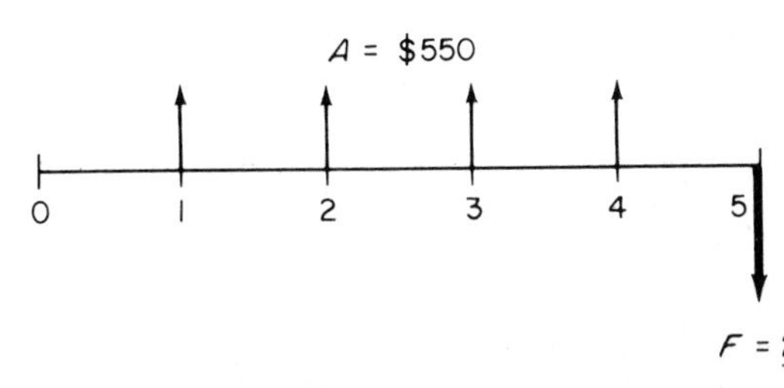

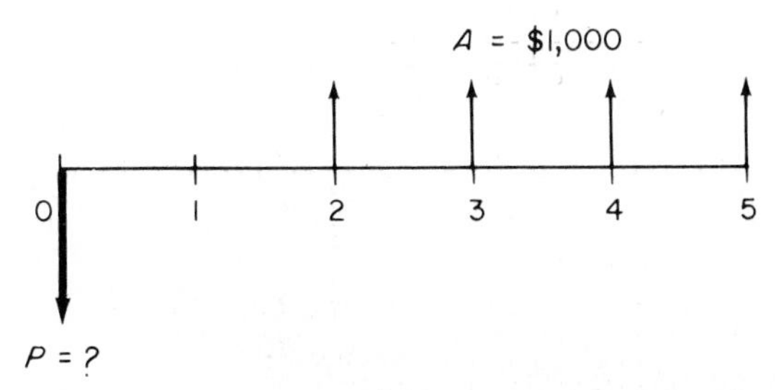

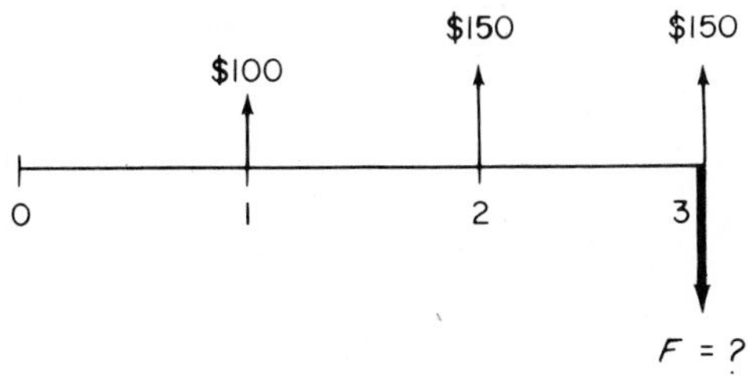

Figure 2.19 Cash-flow diagrams, Example 2.15.

(*b*) Since there is no $A = \$550$ in year 5, the F/A factor cannot be used. The relation $F = 550(F/A, i\%, 4)$ would furnish the future worth in year 4, not year 5 as desired.
(*c*) The first $A = \$1000$ value occurs in year 2. Use of the relation $P = 1000(P/A, i\%, 4)$ will compute P in year 1, not year 0.
(*d*) The receipt values are unequal; thus the relation $F = A(F/A, i\%, 3)$ cannot be used to compute F.

COMMENT Naturally, there are ways to compute P or F without resorting to only P/F and F/P factors; these methods are discussed in Chap. 4.

Sec. 2.7

Example 2.16 If an investor deposits \$2000 now, \$500 three years from now, and \$1000 five years from now, how many years will it take from now for his total investment to amount to \$10,000 if the interest rate is 6%?

SOLUTION The cash-flow diagram (Fig. 2.20) requires that the following equation be satisfied:

$$F = P_1(F/P, i\%, n) + P_2(F/P, i\%, n-3) + P_3(F/P, i\%, n-5)$$

$$10{,}000 = 2000(F/P, 6\%, n) + 500(F/P, 6\%, n-3) + 1000(F/P, 6\%, n-5)$$

This relation must be solved by selecting various values of n and solving until the equation is satisfied. Interpolation for n will be necessary to obtain an exact equality. The procedure shown in Table 2.6 indicates that 20 years is too long and 15 years is too short. Therefore, we interpolate between 15 and 20 years.

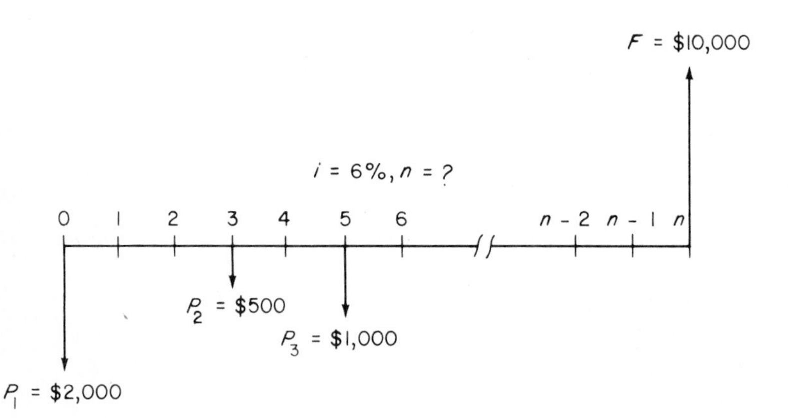

Figure 2.20 Diagram to determine n for a nonuniform series, Example 2.16.

Table 2.6 Trial-and-error solution for *n*

n	$2000(F/P, 6\%, n)$	$500(F/P, 6\%, n-3)$	$1000(F/P, 6\%, n-5)$	F	Comment
5	$2676.40	$ 561.80	$1000.00	$ 4,238.20	Too small
15	4793.20	1006.10	1790.80	7,590.10	Too small
20	6414.20	1346.40	2396.60	10,157.20	Too large

$$c = \frac{10{,}000 - 7590.10}{10{,}157.20 - 7590.10}(20 - 15) = 4.69$$

$$n = 15 + c$$

$$= 19.69 \text{ years}$$

COMMENT The final answer is close to 20 years, as it should be, since the F value calculated for $n = 20$ in Table 2.6 is close to the desired value of $10,000.

Sec. 2.10

PROBLEMS

P2.1 Construct the cash-flow diagrams and derive the SPPWF, USPWF, and USCAF formulas for beginning-of-year amounts rather than the end-of-year convention. The P value should take place at the same time as for the end-of-year convention.

P2.2 Find the correct numerical value for the following factors from the interest tables:

1. $(A/F, 6\%, 20)$
2. $(F/P, 8\%, 5)$
3. $(P/A, 20\%, 8)$
4. $(A/P, 10\%, 25)$
5. $(P/G, 15\%, 20)$

P2.3 Construct the cash-flow diagram for the following deposits (year = k):

Year, k	1	2	3	4-7
Deposit, $	60	60	60	$100 + 10(k - 4)$

P2.4 Use the factor formulas to show that the following expressions are correct for the uniform-gradient factors:

(*a*) $(A/G, i\%, n) = (P/G, i\%, n)(A/P, i\%, n)$
(*b*) $(P/G, i\%, n) = (A/G, i\%, n)(P/A, i\%, n)$

P2.5 Find the value of the factor to convert a 12-year gradient to an equivalent uniform annual series if interest is at 20% per year.

P2.6 Find the numerical value of the following factors by (*a*) interpolation and (*b*) use of the appropriate formula:

1. $(P/A, 8.5\%, 13)$
2. $(F/A, 37\%, 24)$
3. $(P/F, 7.7\%, 9)$
4. $(A/F, 49\%, 28)$

P2.7 Find the numerical value of the following factors by (*a*) interpolation and (*b*) use of the appropriate formula:

1. $(F/P, 3\%, 39)$
2. $(A/P, 10\%, 9.8)$
3. $(A/F, 6\%, 52)$
4. $(P/F, 18\%, 37)$

P2.8 Find the numerical value of the following factors by (*a*) interpolation and (*b*) use of the appropriate formula:

1. $(P/F, 3.8\%, 7.7)$
2. $(P/A, 9.7\%, 68)$
3. $(F/A, 23\%, 11.6)$
4. $(A/F, 17\%, 23)$

P2.9 Find the correct factor value for the following from the interest tables and, when necessary, interpolation:

1. $(P/G, 10\%, 8)$
2. $(A/G, 15\%, 5)$
3. $(A/G, 17\%, 23)$
4. $(P/G, 28\%, 41)$

P2.10 How much money will Mr. Jones have in his bank account in 12 years if he deposits $3500 now at an interest rate of 12% per year?

P2.11 If Ms. James wants to have $8000 in her account 8 years from now to buy a new sports car, how much money will she have to deposit every year starting one year from now if the interest rate is 9% per year?

P2.12 What is the present worth of $700 now, $1500 four years from now, and $900 six years from now at an interest rate of 8% per year?

P2.13 How much money would be accumulated in 14 years if $1290 were deposited each year starting one year from now at an interest rate of 15% per year?

P2.14 If Mr. Savum borrowed $4500 with a promise to make 10 equal annual payments starting one year from now, how much would his payments be if the interest rate were 20% per year?

P2.15 How much money must be deposited in a lump sum 4 years from now in order to accumulate $20,000 eighteen years from now if the interest rate is 8% per year?

P2.16 Ms. Lendup would like to know the present worth of a 35-year $600-per-year annuity beginning one year from now at an interest rate of $6\frac{1}{2}\%$ per year?

P2.17 How much money can you borrow now if you promise to pay $600 per year beginning one year from now for 7 years at an interest rate of 17% per year?

P2.18 How much money now will be equivalent to $5000 six years from now at an interest rate of 7% per year?

P2.19 What uniform annual amount must you deposit for 5 years to have an equivalent present-investment sum of $9000 at an interest rate of 10% per year?

P2.20 How much money would be accumulated in 25 years if $800 were deposited one year from now, $2400 six years from now, and $3300 eight years from now, all at an interest rate of 18% per year?

P2.21 What is the future worth of a uniform annual series of $1000 for 10 years at an interest rate of $8\frac{3}{4}$% per year?

P2.22 What is the present worth of $600 per year for 52 years beginning one year from now at an interest rate of 10% per year?

P2.23 How much money would be accumulated in 43 years from an annual deposit of $1200 per year starting one year from now if the interest rate were $19\frac{1}{4}$% per year?

P2.24 I plan to buy some property which my uncle has generously offered to me. The payment scheme is $700 every other year through year 8 starting 2 years from now. What is the present worth of this generous offer if the interest rate is 17% per year?

P2.25 If a college student can save $600 per year from her part-time job, how long will it take her to save enough money to purchase a $2500 dune buggy if she can get 10% per year interest on her money?

P2.26 What single amount of money will have to be deposited 4 years from now if you want to have $8000 in your account 11 years from now? Use an interest rate of 10% per year.

P2.27 For the cash-flow diagram shown below, calculate the amount of money in year 3 that would be equivalent to all the cash flows shown, using an interest rate of 11% per year.

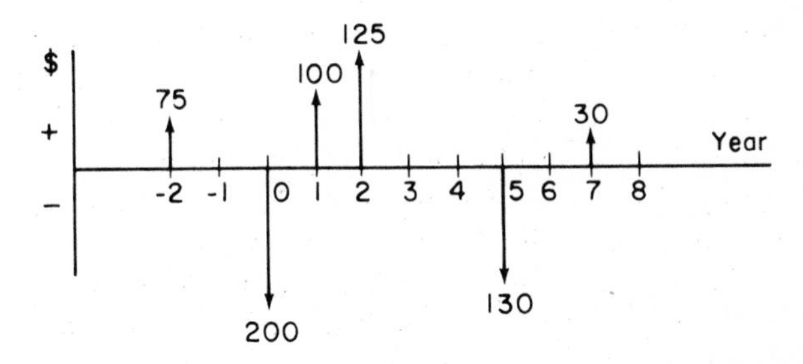

P2.28 What year-end payment will be required to pay off a $25,000 debt in 9 years if the first payment is to be made one year from now and the interest rate is 15% per year?

P2.29 How much money must be deposited into a savings account each year, starting in 1985, if you want to have $150,000 when you retire in the year 2020? The interest rate is 16% per year.

P2.30 If annual deposits of $1000 are made into a savings account for 30 years beginning one year from now, how much will be in the fund immediately after the last deposit if the fund pays interest at a rate of 10% per year?

P2.31 What single payment 12 years from now would be equivalent to a payment of $6200 five years from now for an interest rate of 13% per year?

P2.32 How much money would have to be invested at the end of each year for 25 years starting next year into a fund which is to amount to $180,000 at the end of the 25 years? Assume that the interest rate is 14% per year.

P2.33 The 4-Sight Steel Fabricating Co. is planning to make two equal deposits such that 10 years from now the company will have $49,000 to replace a small machine. If the first deposit is to be made 2 years from now and the second is to be made 8 years from now, how much must be deposited each time if the interest rate is 15% per year?

P2.34 Rework Prob. P2.33 for deposits made in years 1 and 9.

P2.35 The GRQ Company is planning to borrow $58,000 at 15% per year. The company expects to repay the loan with six equal annual payments at the end of each year, beginning one year after the loan is received. Determine the amount of interest that would be charged in the first and second year's payment.

P2.36 The Lee-Key Roof Company has offered a small company two methods by which to pay for some needed roof repairs. Method 1 involves a payment of \$2500 as soon as the job is done, i.e., now. Method 2 allows the company to defer payment for 5 years, at which time a payment of \$5000 would be required. If the interest rate is 16% per year, compute the P value for each method and select the one with the smaller P value.

P2.37 If a company has an opportunity to invest \$33,000 now for 14 years at 15% per year simple interest or 13% per year compound interest, which investment should be made?

P2.38 A plant manager is trying to decide whether to buy a new machine now or wait and purchase a similar one 3 years from now. The machine at the present time would cost \$25,000, but 3 years from now it is expected to cost \$39,000. If the interest rate the company uses is 20% per year, should the plant manager buy now or should she buy 3 years from now?

P2.39 A cash-flow sequence starts in year 1 at \$200 and increases to \$354 in year 8. Do the following: (*a*) construct the cash-flow diagram; (*b*) determine the amount of the annual gradient; (*c*) locate the gradient present worth on the diagram; and (*d*) determine the value of n for the gradient factor.

P2.40 For the cash-flow shown below, calculate (*a*) the equivalent uniform annual cost in years 1 through 5 and (*b*) the present worth of the cash flow. Assume that the interest rate is 12% per year.

Year	1	2	3	4	5
Cash flow, \$	5000	5400	5800	6200	6600

P2.41 For the diagram shown below, find the value of x that will make the negative cash flows equal to the positive cash flow of \$800 at time 0. Assume that i = 15% per year.

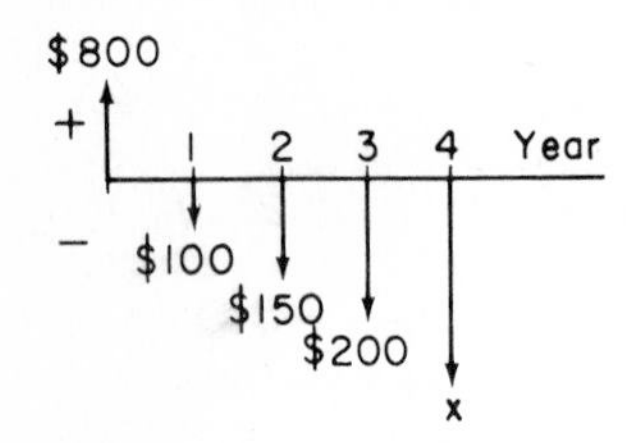

P2.42 Find the present worth of an income series wherein the cash flow in year 1 is \$1200 and it increases by \$300 per year through year 11. Use an interest rate of 15% per year.

P2.43 Determine the equivalent uniform annual cost of a machine which has costs of \$8500 at the end of year 1 and costs increasing by \$500 per year through year 8. Assume that the interest rate is 25% per year.

P2.44 Determine the present worth of a machine which has an initial cost of \$10,000 and operating costs of \$1200 the first year and \$1350 the second year and amounts increasing by \$150 per year through year 10. Use an interest rate of 18% per year.

P2.45 Determine the equivalent uniform annual cost of a process which will involve an initial outlay of \$70,000 followed by costs of \$8000 in year 1, \$9000 in year 2, and amounts increasing by \$1000 per year through its 13-year life. Assume that the interest rate is 14% per year.

P2.46 A company borrows \$15,000 at an interest rate of 15% per year with the agreement that the loan will be repaid over an 8-year period. The repayment scheme will be such that each payment will be \$250 larger than the preceding one, with the first payment to be made one year after the loan is negotiated. Determine the amount of the third payment.

P2.47 Assume that the GRQ Company wants to have $500,000 available for investment 10 years from now. The company plans to invest $4000 the first year and amounts increasing by a uniform gradient thereafter. If the company's interest rate is 20% per year, what must be the size of the gradient in order for GRQ to meet its objective?

P2.48 Find the present worth of the cash flow shown below, using an interest rate of 18% per year.

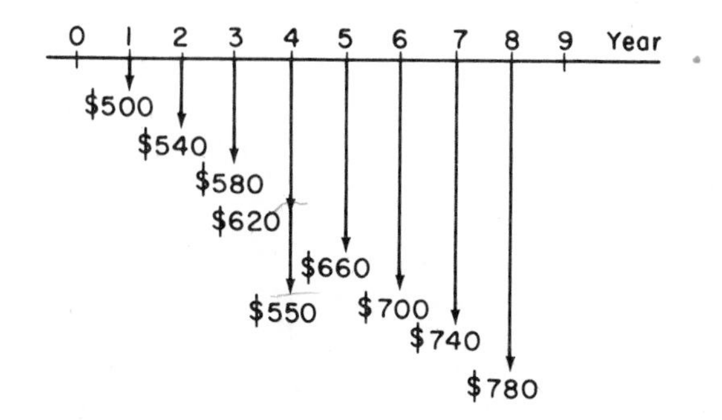

P2.49 Calculate the value of G in the cash-flow series shown below such that the present worth of the cash flow will be $28,500 when the interest rate is 15% per year.

Year	1	2	3	4	5	6	7
Cash flow, $	4000	$4000 + G$	$4000 + 2G$	$4000 + 3G$	$4000 + 4G$	$4000 + 5G$	$4000 + 6G$

P2.50 The Cimeron Cinnamon Co. is undertaking a program to reduce operating costs. The vice president in charge of operations has established a goal of saving an equivalent total present amount of $90,000 in the next 4 years through reduced finished-product losses. He estimates that the company will be able to save $40,000 the first year, but cost reductions will become more difficult each year. If the reductions are expected to follow a uniformly decreasing gradient, what must the reductions be in years 2, 3, and 4 in order for the company to meet its established goal? The company's interest rate is 15% per year.

P2.51 If a company invests $3000 now and will receive $5000 twelve years from now, what is the rate of return on the investment?

P2.52 The Playmore Company, a group of investors, is considering the attractiveness of purchasing a piece of property for $18,000. The group anticipates that the value of the property will increase to $21,500 in 5 years. What is the rate of return on this investment?

P2.53 The Juicer Utility Company has a retirement program in which employees invest $1200 every year for 25 years, starting one year after their initial employment. If the company guarantees at least $50,000 at the time of retirement, what is the rate of return on the investment?

P2.54 If a person borrows $6000 for a new car with an agreement to pay the loan company $2500 per year for 3 years, what is the interest rate on the loan?

P2.55 At what interest rate would $1500 accumulate to $2500 in 5 years?

P2.56 If a person purchased stock for $8000 twelve years ago and received dividends of $1000 per year, what is the rate of return on the investment?

P2.57 At what rate of return would $900 per year for 9 years accumulate to (a) $8100 and (b) $10,000?

P2.58 The Gotcha Loan Company offers qualified borrowers the opportunity to borrow $1000 now and repay the loan in 10 "easy" yearly installments of $155, the first installment due one year from now. At what interest rate are their customers borrowing their money?

P2.59 How many years would it take for $1750 to triple in value if the interest rate were 12%?

P2.60 If a person deposits $5000 now at 8% interest and plans to withdraw $500 per year every year starting one year from now, how long can the full withdrawals be made?

P2.61 What is the minimum number of years a person must deposit $400 per year in order to have at least $10,000 on the date of the last deposit? Use an interest rate of 8% and round off to the higher integer year.

P2.62 What is the minimum number of year-end deposits that have to be made before the total value of the deposits is at least ten times greater than the value of a single year-end deposit if the interest rate is $12\frac{1}{2}$%?

P2.63 How many years would it take for a $800 deposit now and a $1600 deposit 3 years from now to accumulate to $3500 at an interest rate of 8%?

P2.64 If you start saving money by depositing $1000 per year in a bank account which pays 11%-per-year interest, how many years will it take to accumulate $10,000 if the first deposit is made one year from now?

P2.65 How long would it take to recover an investment of $10,000 which pays 15% per year interest if $1000 were saved the first year and $1100 the second year and the amounts continued to increase by $100 per year?

P2.66 If a company invests $15,000 in energy-saving devices, how long will it take to recover the investment if the annual savings are $2000 the first year and they increase by $500 per year thereafter? Use an interest rate of 25% per year.

P2.67 The cash flow associated with a particular project is expected to be $2500 in year 1 and $2800 in year 2 and amounts increasing by $300 per year. How many years will it take to recover an initial investment of $20,000 if the interest rate is 18% per year?

CHAPTER

THREE

NOMINAL AND EFFECTIVE INTEREST RATES AND CONTINUOUS COMPOUNDING

This chapter teaches you how to make engineering-economy computations using interest periods other than one year. The material of this chapter is often helpful for handling personal financial matters. Optional sections on continuous compounding are included in this chapter.

SECTION OBJECTIVES

To complete this chapter you must be able to:

3.1. Define *compounding period*, *payment period*, *nominal interest rate*, and *effective interest rate*.
3.2. Write the formula for computing the effective interest rate and define each term in the formula.
3.3. Compute the effective interest rate and find the numerical value of any specific engineering-economy factor for that rate, given the nominal interest rate and number of compounding periods.
3.4. Calculate the present worth or future worth of a specified cash flow when the

payment period is *longer* than the compounding period, given the amount and times of the payments, the compounding period, and the nominal interest rate.

*3.5. Calculate the present worth or future worth of a specified cash flow when the payment period is *shorter* than the compounding period, given the amount and times of the payments, the compounding period, and the nominal interest rate.

*3.6. Derive and use the effective interest-rate formula for *continuous compounding*, given the nominal interest rate.

*3.7. Derive and state the interest factors for discrete cash flows and continuous compounding. Use the factors to compute the present worth, future worth, or equivalent uniform annual series, given the nominal rate and the cash-flow values and times.

*3.8. Do the same as in objective 3.7 except for continuous cash flows and continuous compounding of interest.

STUDY GUIDE

3.1 Nominal and Effective Rates

In Chap. 1, the concepts of simple- and compound-interest rates were introduced. The basic difference between the two is that compound interest includes interest on the interest earned in the previous year. In essence, nominal and effective interest rates have the same relation as do simple and compound interest; the difference is that nominal and effective interest rates are used when the *compounding period* (or interest period) is less than one year. Thus, when an interest rate is expressed over a period of time shorter than a year, such as 1% per month, the terms *nominal* and *effective* interest rates must be considered. Specifically, the *nominal interest rate per year* is defined as the period interest rate multiplied by the number of periods per year. A period interest rate listed as 1.5% per month could thus also be expressed as a *nominal* 18% per year (that is, 1.5% per month × 12 months per year). The calculation for the nominal interest rate obviously ignores the time value of money, similar to the calculation of simple interest. When the time value of money is taken into consideration in calculating annual interest rates from period interest rates, the annual rate is called the *effective interest rate*.

In addition to considering the interest or compounding period, it is also necessary to consider the frequency of the payments or receipts within the one-year interval. For simplicity, the frequency of the payments or receipts is known as the *payment period*. It is important to distinguish between the compounding period and the payment period because in many instances the two do not coincide. For example, if a company deposited money each month into an account that pays a nominal interest rate of 14% per year compounded semiannually, the payment period would be one month while the compounding period would be 6 months. Similarly, if a person deposits money each year into a savings account which compounds interest quarterly, the payment period is one year, while the compounding period is 3 months. If the payment and compounding periods are the same, the rate is expressed as in preceding chapters (for example, 1% per month, where the compounding period is a month and payments are to be made at the end of each month).

Probs. P3.1 and P3.2

3.2 Effective Interest-Rate Formulation

To illustrate the difference between nominal and effective interest rates, the future worth of $100 after one year is determined using both rates. If a bank pays 12% interest compounded annually, the future worth of $100 using a nominal interest rate of 12% per year is

$$F = P(1 + i)^n = 100(1.12^1) = \$112.00 \tag{3.1}$$

On the other hand, if the interest is compounded semiannually, the future worth must include the *interest on the interest earned in the first period*. An interest rate of 12% per year compounded semiannually means that the bank will pay 6% interest two times per year (i.e., every 6 months). Figure 3.1 is the cash-flow diagram for semiannual compounding for a nominal interest rate of 12% per year. Equation (3.1) obviously ignores the interest earned in the first period. Taking this into consideration, therefore, the future worth of the $100 would actually be

$$\begin{aligned} F &= 100(1 + 0.06)^2 = 100(1.06^2) \qquad (3.2) \\ &= 100(1.1236) \\ &= \$112.36 \end{aligned}$$

where 6% is the *effective semiannual interest rate* found by taking 12%/2 = 6%, since there are two compounding periods per payment period. The effective annual interest rate, therefore, would be 12.36%, instead of 12%, since $12.36 interest would be earned. The equation for acquiring the effective interest rate from the nominal interest may be obtained by recognizing that the first term in parentheses in Eq. (3.2) is equal to $(1 + i)$ when the interest period is one year. Setting this term equal to the generalized form of the term in parentheses in Eq. (3.1) yields

$$(1 + i) = \left(1 + \frac{r}{t}\right)^t$$

$$i = \left(1 + \frac{r}{t}\right)^t - 1 \tag{3.3}$$

where i = effective interest rate per period
r = nominal interest rate
t = number of compounding periods

Equation (3.3) is referred to as the effective-interest-rate equation. As the number of compounding periods increases, t approaches infinity, in which case the equation represents the interest rate for *continuous compounding*. See optional Secs. 3.6 to 3.8 for a detailed discussion of this subject.

Prob. P3.3

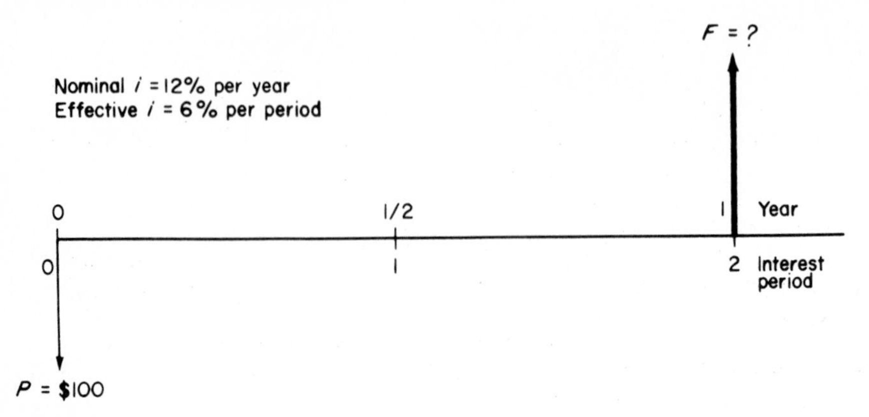

Figure 3.1 Cash-flow diagram for semiannual compounding periods.

3.3 Calculation of Effective Interest Rates

For an annual interest rate that is compounded over a specified number of interest periods, the effective interest rate can be calculated by substituting into Eq. (3.3). After the effective rate has been determined, the factors in the interest tables of Appendix A can be used in the normal manner (Chap. 2). As is shown in the next section, this is only one of two methods that may be used when the interest period is shorter than the payment period. Since the annual effective interest rate is usually found to be a rate for which no tables are prepared, it is necessary to interpolate between interest rates to determine the factor value or use the actual equation to obtain the factor.

Example 3.1 A university credit union advertises that its interest rate on loans is 1% per month. Calculate the annual effective interest rate and use the tables in Appendix A to find the corresponding P/F factor for $n = 8$.

SOLUTION Substituting $r/t = 0.01$ and $t = 12$ into Eq. (3.3) yields

$$\begin{aligned} i &= (1 + 0.01)^{12} - 1 \\ &= 1.1268 - 1 \\ &= 0.1268 \qquad (12.68\%) \end{aligned}$$

In order to find the P/F factor, it is necessary to interpolate between $i = 12\%$ and $i = 13\%$.

$$b\left[\begin{array}{l} a\left[\begin{array}{l} 12\% \\ 12.68\% \end{array}\right. \\ 13\% \end{array}\right. \qquad \left.\begin{array}{r} \left.\begin{array}{r} 0.4039 \\ P/F \end{array}\right] c \\ 0.3762 \end{array}\right] d$$

$$c = \frac{0.68}{1}(0.0277) = 0.0188$$

Then,

$$(P/F, 12.68\%, 8) = 0.4039 - 0.0188 = 0.3851$$

If the values $i = 12.68\%$ and $n = 8$ are substituted into the P/F factor relation, Eq. (2.3), the result is

$$(P/F, 12.68\%, 8) = \frac{1}{(1 + 0.1268)^8} = 0.3848$$

COMMENT The term r/t is equal to 0.01 in the effective-interest-rate equation because the interest rate stated in the problem is already expressed as a period interest rate rather than a nominal rate. You might find it easier to think of the interest rate of 1% per month as being the same as an annual nominal rate of 12% per year compounded monthly, in which case Eq. (3.3) would be

$$i = \left(1 + \frac{0.12}{12}\right)^{12} - 1 = 0.1268 \qquad (12.68\%)$$

It is important for you to remember that 1% per month is already an effective interest rate, if you consider the interest period as one month. In this case, the P/F factor desired above would be represented as $(P/F, 1\%, 96)$. This is discussed more thoroughly in the next section.

Table 3.1 presents the effective interest rate i for various nominal interest rates r using Eq. (3.3) and compounding periods of 6 months, 3 months, 1 month, 1 week, and 1 day. The continuous compounding column is discussed in Sec. 3.6.

Solved Problem 3.11
Probs. P3.4 to P3.9

3.4 Calculations for Payment Periods Longer Than Compounding Periods

When the compounding period of an investment (or loan) does not coincide with the payment period, it becomes necessary to manipulate the interest rate and/or payment period in order to determine the correct amount of money accumulated or paid at various times. Remember that if the payment and compound periods do not agree, the interest tables cannot be used until appropriate corrections are made. In this section, we consider the situation where the payment period (for example, year) is equal to or longer than the compounding period (for example, month). The two conditions that occur are

1. The cash flows require the use of the single-payment factors $(P/F, F/P)$.
2. The cash flows require the use of the uniform-series factors.

3.4.1 Single-payment factors There are two correct procedures that can be used when only single factors are involved. The first is the procedure that was illustrated in

Table 3.1 Tabulation of effective interest rates for nominal annual rates

Nominal rate, r%	Semiannually (t = 2)	Quarterly (t = 4)	Monthly (t = 12)	Weekly (t = 52)	Daily (t = 365)	Continuously ($t = \infty; e^r - 1$)
0.25	0.250	0.250	0.250	0.250	0.250	0.250
0.50	0.501	0.501	0.501	0.501	0.501	0.501
0.75	0.751	0.752	0.753	0.753	0.753	0.753
1.00	1.003	1.004	1.005	1.005	1.005	1.005
1.50	1.506	1.508	1.510	1.511	1.511	1.511
2	2.010	2.015	2.018	2.020	2.020	2.020
3	3.023	3.034	3.042	3.044	3.045	3.046
4	4.040	4.060	4.074	4.079	4.081	4.081
5	5.063	5.095	5.116	5.124	5.126	5.127
6	6.090	6.136	6.168	6.180	6.180	6.184
7	7.123	7.186	7.229	7.246	7.247	7.251
8	8.160	8.243	8.300	8.322	8.328	8.329
9	9.203	9.308	9.381	9.409	9.417	9.417
10	10.250	10.381	10.471	10.506	10.516	10.517
11	11.303	11.462	11.572	11.614	11.623	11.628
12	12.360	12.551	12.683	12.734	12.745	12.750
13	13.423	13.648	13.803	13.864	13.878	13.883
14	14.490	14.752	14.934	15.006	15.022	15.027
15	15.563	15.865	16.076	16.158	16.177	16.183
16	16.640	16.986	17.227	17.322	17.345	17.351
17	17.723	18.115	18.389	18.497	18.524	18.530
18	18.810	19.252	19.562	19.684	19.714	19.722
19	19.903	20.397	20.745	20.883	20.917	20.925
20	21.000	21.551	21.939	22.093	22.132	22.140
21	22.103	22.712	23.144	23.315	23.358	23.368
22	23.210	23.883	24.359	24.549	24.598	24.608
23	24.323	25.061	25.586	25.796	25.849	25.860
24	25.440	26.248	26.824	27.054	27.113	27.125
25	26.563	27.443	28.073	28.325	28.390	28.403
26	27.690	28.646	29.333	29.609	29.680	29.693
27	28.823	29.859	30.605	30.905	30.982	30.996
28	29.960	31.079	31.888	32.213	32.298	32.313
29	31.103	32.309	33.183	33.535	33.626	33.643
30	32.250	33.547	34.489	34.869	34.968	34.986
31	33.403	34.794	35.807	36.217	36.327	36.343
32	34.560	36.049	37.137	37.578	37.693	37.713
33	35.723	37.313	38.478	38.952	39.076	39.097
34	36.890	38.586	39.832	40.339	40.472	40.495
35	38.063	39.868	41.198	41.740	41.883	41.907
40	44.000	46.410	48.213	48.954	49.150	49.182
45	50.063	53.179	55.545	56.528	56.788	56.831
50	56.250	60.181	63.209	64.479	64.816	64.872

Example 3.1, where the effective interest rate coinciding with the payment period or one-year period (whichever comes first) is found from Eq. (3.3) or Table 3.1, and the interest tables are used for the appropriate number of periods or years. The second procedure is to *divide* the nominal interest rate r by the number of compounding periods per year t and *multiply* the number of years by the number of compounding periods per year. Using standard notation, the single-payment equations may then be written

$$P = F\left(P/F, \frac{r}{t}\%, tn\right) \tag{3.4}$$

$$F = P\left(F/P, \frac{r}{t}\%, tn\right) \tag{3.5}$$

where n is the number of years and r and t are as defined in Sec. 3.2. Example 3.2 demonstrates both these procedures.

Example 3.2 If a woman deposits $1000 now, $3000 four years from now, and $1500 six years from now at an interest rate of 6% per year compounded semi-annually, how much money will she have in her account 10 years from now?

SOLUTION The cash-flow diagram is shown in Fig. 3.2: According to the first procedure specified above, the effective interest per year should be calculated and then used to find F in year 10. From Table 3.1, with $r = 12\%$ and semiannual compounding, effective $i = 12.36\%$; or by Eq. (3.3),

$$i = \left(1 + \frac{0.12}{2}\right)^2 - 1 = (1.06^2) - 1 = 0.1236 \qquad (12.36\%)$$

Then,

$$F = 1000(F/P, 12.36\%, 10) + 3000(F/P, 12.36\%, 6) + 1500(F/P, 12.36\%, 4)$$

$$= \$11{,}634.50$$

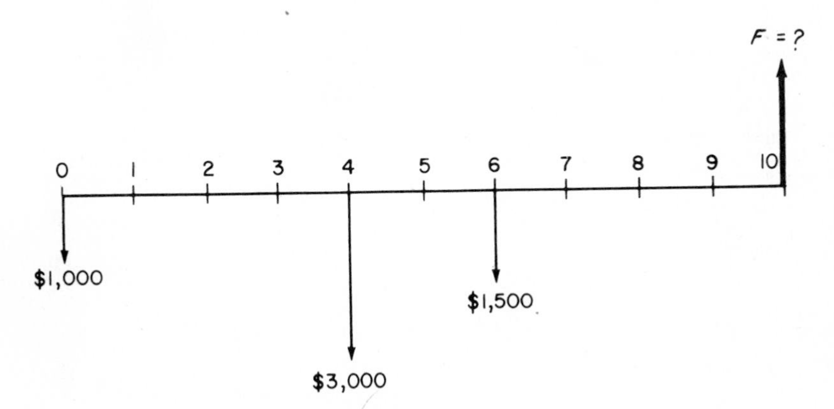

Figure 3.2 Cash-flow diagram, Example 3.2.

By the second procedure Eq. (3.5) is used with $t = 2$.

$$F = 1000[F/P, \tfrac{12}{2}\%, 2(10)] + 3000[F/P, \tfrac{12}{2}\%, 2(6)] + 1500[F/P, \tfrac{12}{2}\%, 2(4)]$$

$$F = 1000(F/P, 6\%, 20) + 3000(F/P, 6\%, 12) + 1500(F/P, 6\%, 8)$$

$$= \$11{,}634.50$$

COMMENT The second procedure is the easier of the two, since it usually does not require interpolation or use of the factor formulas.

3.4.2 Uniform-series factors When the cash flow of the problem dictates the use of one or more of the uniform-series factors, one procedure used is to *express the effective interest rate over the same time period as the payments*. When a nominal interest rate is given and the *compounding period is equal to the payment period* (for example, *semiannual* payments at an interest rate of 8% per year compounded *semiannually*), then, as for the single-payment factors, simply divide the interest rate by t, the number of compounding periods per year, and set n equal to the total number of payments. For example, the P/A factor would appear as $(P/A, r/t\%, tn)$. The procedure is illustrated in Table 3.2, where the standard notation for the P/A factor is shown for three payment schemes and interest rates.

When compounding occurs more frequently than the payments (such as *semiannual* payments at an interest rate of 8% per year compounded *quarterly*), the effective interest rate per Eq. (3.3) must be used. This procedure is illustrated in Example 3.3.

Example 3.3 If a woman deposits \$500 every 6 months for 7 years, how much money will she have in her account after she makes her last deposit if the interest rate is 8% per year compounded quarterly?

SOLUTION The cash-flow diagram is shown in Fig. 3.3. Since interest is compounded quarterly, the effective interest rate per payment period (i.e., the effective semiannual rate) must be determined. The effective semiannual rate can be obtained by finding the quarterly interest rate and plugging it into the effective interest rate, Eq. (3.3). Here $r/t = 0.08/4 = 0.02$; and since there are two quarters

Table 3.2 Finding *i* and *n* for compounding period equal to payment period

Payment scheme (A value)	Nominal interest rate r	$A(P/A, i\%, n)$
\$500 semiannually for 5 years	6% per year compounded semiannually	$500(P/A, 3\%, 10)$
\$75 monthly for 3 years	12% per year compounded monthly	$75(P/A, 1\%, 36)$
\$180 quarterly for 15 years	8% per year compounded quarterly	$180(P/A, 2\%, 60)$

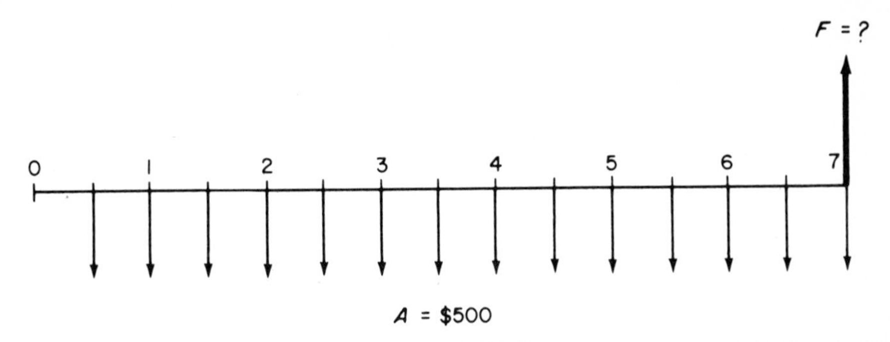

Figure 3.3 Diagram of semiannual deposits used to determine F, Example 3.3.

in one semiannual period,

$$i = (1 + 0.02)^2 - 1 = 0.0404 \qquad (4.04\%)$$

Alternatively, the effective interest rate could have been obtained from Table 3.1 by using the r value of 4% with "semiannual" compounding to get $i = 4.04\%$.

The value $i = 4.04\%$ seems reasonable, since we expect the effective rate to be slightly higher than the nominal rate of 4% per 6-month period. The effective rate can now be used in the F/A formula to find the future worth of the semiannual deposits, where $n = 2(7) = 14$ periods. Thus,

$$\begin{aligned} F &= A(F/A, 4.04\%, 14) \\ &= 500(18.344) \\ &= \$9172.12 \end{aligned} \tag{3.6}$$

COMMENT It is important to note that the effective interest *per payment period* (6 months) was used for i and that the *number of payment periods* was used for n.

It is possible to rewrite the uniform-series factor formulas to accommodate different compounding-period and payment-period values. This allows direct computation of the P/A or F/A factor values without first determining the effective interest rate. First, alter the effective interest rate, Eq. (3.3), by including p, the number of payment periods per year. Now

$$i = \left(1 + \frac{r}{t}\right)^{t/p} - 1 \tag{3.7}$$

Substitute Eq. (3.7) into the factor formulas in brackets in Eqs. (2.6) and (2.9), respectively, to obtain

$$(P/A, i\%, n) = \frac{1 - (1 + r/t)^{-ts/p}}{(1 + r/t)^{t/p} - 1} = \frac{1 - (1 + r/t)^{-tn}}{i}. \tag{3.8}$$

$$(F/A, i\%, n) = \frac{(1 + r/t)^{ts/p} - 1}{(1 + r/t)^{t/p} - 1} = \frac{(1 + r/t)^{tn} - 1}{i}. \tag{3.9}$$

where r = nominal interest rate per year
t = number of compounding periods per year
p = number of payments (deposits) per year
s = total number of payments (deposits)
n = number of years = s/p
i = value from Eq. (3.7)

As an illustration, the exact F/A factor value in Example 3.3 may be computed using Eq. (3.9) with $r = 0.08$, $t = 4$ (quarterly compounding), $p = 2$ (semiannual deposits), and $n = 7$ years. (Note that $s = 14$ payments, so $n = s/p = 7$.)

$$(F/A, 4.04\%, 7) = \frac{(1 + 0.08/4)^{4(7)} - 1}{(1 + 0.08/4)^{4/2} - 1} = \frac{0.7410}{0.0404} = 18.342$$

The interpolated factor value from the tables was 18.344.

It is also possible to use Eqs. (3.8) and (3.9) when $t = p$, as discussed in the first paragraph of Sec. 3.4.2; however, it may be more work to obtain the factor. As an illustration, consider the first entry in Table 3.2, where $r = 0.06$, $t = p = 2$, and $n = 5$ years. By Eq. (3.8),

$$(P/A, i\%, n) = \frac{1 - (1 + 0.06/2)^{-2(5)}}{(1 + 0.06/2)^{2/2} - 1} = 8.5302$$

This is the same as the value $(P/A, 3\%, 10) = 8.5302$ obtained from Table A-5. Either method can therefore be used to obtain the uniform-series factor values.

The general procedure to be followed when the payment period is equal to or less frequent (i.e., longer) than the compounding period can be summarized as follows:

1. Determine whether the single-payment formulas or the uniform-series formulas must be used.
2. If the single-payment formulas are required, use one of the methods specified in Sec. 3.4.1 for single-payment factors.
3. If uniform-series formulas are required, use one of the methods specified in Sec. 3.4.2 for uniform payments.
4. Use the factor obtained in step 2 or 3 to find the desired P, F, or A.

Solved Problem 3.12
Probs. P3.10 to P3.24

*3.5 Calculations for Payment Periods Shorter Than Compounding Periods

When the compounding period occurs less frequently than the payment period, there are several ways to calculate the future amount or present worth depending on the conditions specified (or assumed) regarding the interperiod compounding. *Interperiod compounding*, as used here, refers to the handling of the payments made *between* compounding periods. The two cases discussed below are as follows:

1. There is no interest paid on the money deposited (or withdrawn) between compounding periods.
2. The money deposited (or withdrawn) between compounding periods earns simple interest. That is, interest is not paid on the interest earned in the preceding interperiod.

In the first case any amount of money that is deposited or withdrawn between compounding periods is regarded as having been *deposited* at the *beginning of the next compounding period* or *withdrawn* at the *end of the previous compounding period*. This is the usual mode of operation of banks and other lending institutions. Thus, if the compounding period were a *quarter*, the actual transactions shown in Fig. 3.4*a* would be treated as shown in Fig. 3.4*b*. To find the present worth of the cash flow represented by Fig. 3.4*b*, the nominal yearly interest rate is divided by 4 (since interest is compounded quarterly) and the appropriate P/F or F/P factor is used.

For the second case, any amount of money that is deposited between compounding periods earns simple interest; in order to obtain the interest earned in the inter-

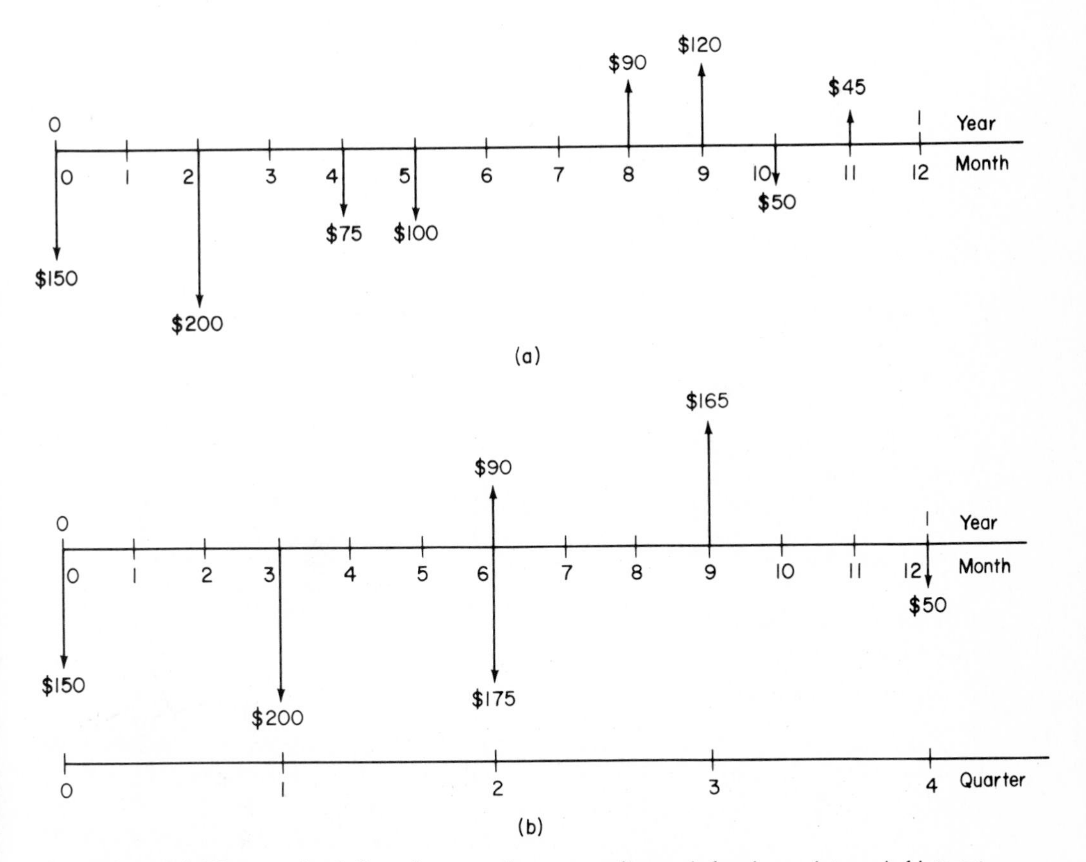

Figure 3.4 Diagram of cash flows for quarterly compounding periods using no interperiod interest.

period *each interperiod deposit* must then be *multiplied* by

$$\left(\frac{m}{N}\right)i \tag{3.10}$$

where N = number of periods in a compounding period
m = number of periods prior to the end of a compounding period
i = interest rate per compounding period

Note that the value obtained by Eq. (3.10) yields only the interest that has accumulated and is not the total end-of-period amount. Example 3.4 illustrates the calculations described here.

Example 3.4 Calculate the amount of money that would be in a person's savings account after 12 months if deposits were made as shown in Fig. 3.5. Assume that the bank pays 6% per year compounded semiannually and pays simple interest on the interperiod deposits.

SOLUTION The first step is to find the amount of money that will be accumulated at each compounding period (i.e., every 6 months) using the effective rate of 3% semiannually. The future or present worth of such deposits can then be calculated with the regular interest formulas. Thus, for the deposits made within the first compounding period, the total value F_6 after 6 months is

$$\begin{aligned} F_6 &= [100 + 100(\tfrac{5}{6})0.03] + [90 + 90(\tfrac{3}{6})0.03] + 80 \\ &= (100 + 2.50) + (90 + 1.35) + 80 \\ &= \$273.85 \end{aligned}$$

Similarly, the amount F_{12} accumulated in the second compounding period is

$$\begin{aligned} F_{12} &= [75 + 75(\tfrac{5}{6})0.03] + [85 + 85(\tfrac{4}{6})0.03] + [70 + 70(\tfrac{1}{6})0.03] \\ &= (75 + 1.88) + (85 + 1.70) + (70 + 0.35) \\ &= \$233.93 \end{aligned}$$

The cash-flow diagram has now been reduced to that shown in Fig. 3.6. Thus, the future worth F at the end of the year is

$$F = 273.85(F/P, 3\%, 1) + 233.93 = \$516$$

A lot of trouble, right?

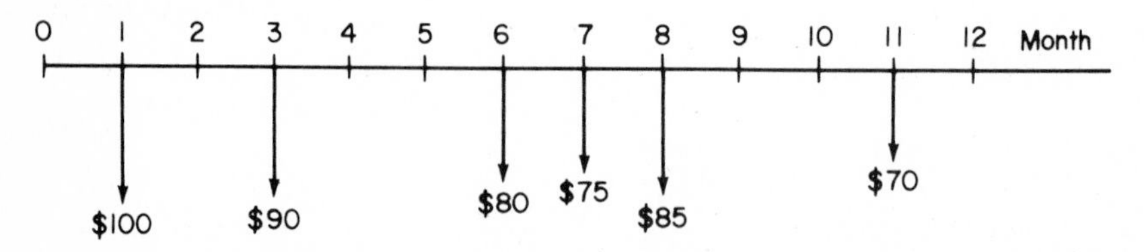

Figure 3.5 Actual deposits made with simple interperiod interest paid, Example 3.4.

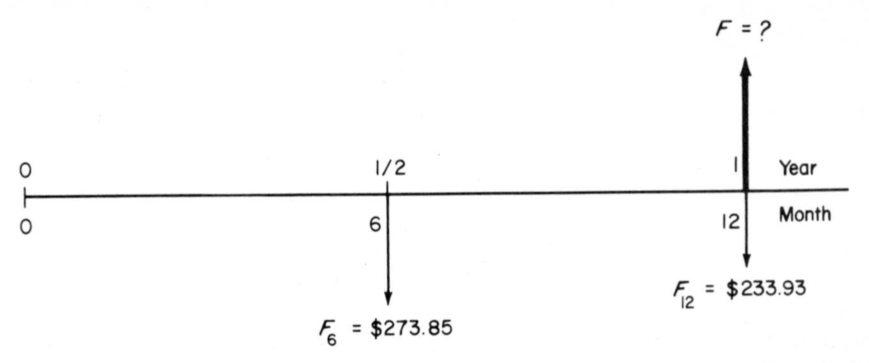

Figure 3.6 Equivalent deposits after simple interest is computed for interperiods, Example 3.4.

COMMENT In calculating the amounts of money accumulated after 6 months (F_6) and 12 months (F_{12}), note that the amount of the deposit was added to each interest term, since Eq. (3.10) represents only the *interest accumulated*, not the total amount.

Determination of the compound-period amount (i.e., end-of-period amount) is simplified when the interperiod payments are regular and uniform. That is, if the same amount of money is deposited in *each* interperiod, the equivalent compound-period amount can be calculated from the following equation.

$$A = ND + \frac{N-1}{2}(iD) \tag{3.11}$$

where A = equivalent uniform compound-period payment
D = value of the deposit made each interperiod
N = number of interperiods per compounding period

See the Solved Problems for an illustration.

Solved Problem 3.13
Probs. P3.25 to P3.33

*3.6 Effective Interest Rates for Continuous Compounding

As the compounding period becomes shorter and shorter, the value of t, number of compounding periods per interest period, increases. In the situation where interest is compounded continuously, t approaches infinity and the effective-interest-rate formula in Eq. (3.3) may be written in a new form. First recall the definition of the natural logarithm base e.

$$\lim_{h \to \infty} \left(1 + \frac{1}{h}\right)^h = e = 2.71828+ \tag{3.12}$$

The limit of Eq. (3.3) as t approaches infinity is found using $r/t = 1/h$, which makes $t = hr$.

$$\lim_{t\to\infty} i = \lim_{t\to\infty} \left(1 + \frac{r}{t}\right)^t - 1$$

$$= \lim_{h\to\infty} \left(1 + \frac{1}{h}\right)^{hr} - 1 = \lim_{h\to\infty} \left[\left(1 + \frac{1}{h}\right)^h\right]^r - 1$$

$$i = e^r - 1 \tag{3.13}$$

Equation (3.13) is used to compute the effective continuous-interest rate. For example, if $r = 15\%$ per year, the effective continuous rate is

$$i = e^{0.15} - 1 = 0.16183$$

or 16.183%. For convenience, Table 3.1 includes the effective continuous rate for many nominal rates as computed by Eq. (3.13).

Example 3.5

(*a*) Compute the effective interest rates for a nominal annual rate of 10% for all the compounding periods used in Table 3.1. Compare the answers with the tabulated values.

(*b*) If an investor requires at least an effective return of 15% on his money, what is the minimum annual nominal rate that is acceptable if continuous compounding takes place?

SOLUTION

(*a*) Use $r = 0.10$ and Eq. (3.3) to compute the effective i value as shown in Table 3.3. For example, daily compounding results in $t = 365$ and

$$i = \left(1 + \frac{0.10}{365}\right)^{365} - 1 = 1.10516 - 1$$

$$= 0.10516 \qquad (10.516\%)$$

For continuous compounding use Eq. (3.13).

$$i = e^{0.10} - 1 = 0.10517 \qquad (10.517\%)$$

Comparison shows that the Table 3.3 values and Table 3.1 entries are identical.

Table 3.3 Effective rates for $r = 0.10$ per year

Compounding frequency	t	$[1 + 0.10/t)^t - 1]100\%$
Annually	1	10.000
Semiannually	2	10.250
Quarterly	4	10.381
Monthly	12	10.471
Weekly	52	10.506
Daily	365	10.516
Continuously	∞	10.517

(*b*) Use Eq. (3.13) with $i = 15\%$ to solve for r by taking the natural logarithm.

$$e^r - 1 = 0.15$$

$$e^r = 1.15$$

$$\ln e^r = \ln 1.15$$

$$r = 0.13976 \qquad (13.976\%)$$

Therefore, a rate of 13.976% per year compounded continuously will generate an effective 15% return.

COMMENTS The general formula to find the nominal rate given the effective, continuous rate i is

$$r = \ln(1 + i) \tag{3.14}$$

Probs. P3.34 to P3.36

*3.7 Interest Factors and Calculations Using Discrete Cash Flows and Continuous Compounding

Before discussing computations involving continuous compounding, it is important to understand assumptions about the flow of cash through time and how the flow is related to the compounding of interest. The three most commonly used situations are:

Discrete cash flows and discrete compounding (Chap. 2)
Discrete cash flows and continuous compounding (this section)
Continuous cash flows and continuous compounding (next section)

All analysis performed thus far has assumed that the cash flow occurs at the end of an interest period, which is commonly one year. However, it is often more realistic to allow the flow of money to take place throughout the interest period. When cash flow is assumed to occur at all time points, it is called continuous cash flow. We discuss this situation in Sec. 3.8.

The first case listed above—discrete cash flows and discrete compounding—has been the assumption in all previous work. Now we will derive the economy factors for discrete cash flows with continuous compounding of interest. As the length of the compounding period approaches zero, $t \to \infty$, as in the previous section, and the base amount of money used for interest computation is increased every moment by the interest earned in the last moment. The continuous-compounding-interest factors are obtained by taking the limit as t increases toward infinity of each compound factor as written in the effective-rate form. We have previously used the equation $F = P(F/P, i\%, n)$, where $(F/P, i\%, n) = (1 + i)^n$. For a nominal rate r compounded t times per period, we can write the F/P factor as

$$(F/P, i\%, n) = \left(1 + \frac{r}{t}\right)^{tn}$$

Using the definition of e in Eq. (3.12), we have

$$\lim_{t \to \infty} (F/P, i\%, n) = \lim_{t \to \infty} \left(1 + \frac{r}{t}\right)^{tn} = e^{rn}$$

The term e^{rn} is the single-payment compound-amount factor for continuous compounding and discrete cash flows. The reciprocal e^{-rn} is the continuous-compounding P/F factor. The uniform-series compound-amount factor is derived from Eq. (2.9).

$$\lim_{t \to \infty} (F/A, i\%, n) = \lim_{t \to \infty} \frac{(1+i)^n - 1}{i}$$

$$= \lim_{t \to \infty} \frac{(1 + r/t)^{tn} - 1}{(1 + r/t)^t - 1} = \frac{e^{rn} - 1}{e^r - 1}$$

where $i = (1 + r/t)^t - 1$ is the effective interest-rate equation.

The capital-recovery factor is derived using Eq. (2.7) and the limit of the effective interest-rate formula, Eq. (3.13).

$$\lim_{t \to \infty} (A/P, i\%, n) = \lim_{t \to \infty} \frac{i(1+i)^n}{(1+i)^n - 1}$$

$$= \lim_{t \to \infty} \left\{ \frac{[(1 + r/t)^t - 1](1 + r/t)^{tn}}{(1 + r/t)^{tn} - 1} \right\}$$

$$= \frac{(e^r - 1)e^{rn}}{e^{rn} - 1} = \frac{e^r - 1}{1 - e^{-rn}}$$

Table 3.4 presents the relations for all discrete cash flow, continuous-compounding factors, including the gradient factors $(A/G, r\%, n)$ and $(P/G, r\%, n)$. The nominal

Table 3.4 Factor formulas for discrete cash flow, continuous compounding

Factor	Formula	Equation
$(F/P, r\%, n)$	e^{rn}	$F = P(F/P, r\%, n)$
$(P/F, r\%, n)$	e^{-rn}	$P = F(P/F, r\%, n)$
$(A/F, r\%, n)$	$\dfrac{e^r - 1}{e^{rn} - 1}$	$A = F(A/F, r\%, n)$
$(F/A, r\%, n)$	$\dfrac{e^{rn} - 1}{e^r - 1}$	$F = A(F/A, r\%, n)$
$(A/P, r\%, n)$	$\dfrac{e^r - 1}{1 - e^{-rn}}$	$A = P(A/P, r\%, n)$
$(P/A, r\%, n)$	$\dfrac{1 - e^{-rn}}{e^r - 1}$	$P = A(P/A, r\%, n)$
$(A/G, r\%, n)$	$\dfrac{1}{e^r - 1} - \dfrac{n}{e^{rn} - 1}$	$A = G(A/G, r\%, n)$
$(P/G, r\%, n)$	$\dfrac{1 - e^{-rn}}{(e^r - 1)^2} - \dfrac{n}{e^{rn}(e^r - 1)}$	$P = G(P/G, r\%, n)$

interest-rate symbol $r\%$ replaces $i\%$ in the factor notation to distinguish these from the discrete compounding factors. Values for the factors in Table 3.4 are given in Appendix B for interest rates from 0.50 to 50%. To assist you in turning to the correct table in Appendix B, note that the nominal interest rate is followed by the wording "continuous-compound interest factors." Additionally, the effective interest rate i as computed by Eq. (3.13) is shown in parentheses below the nominal rate. Example 3.6 illustrates the use of these tables.

Example 3.6 Use the appropriate formulas and appendix tables to determine the following.

(*a*) The effective interest rate and P/A factor value of 15 years and an interest rate of 10% per year compounded continuously.

(*b*) A comparison of the $(F/P, i\%, 10)$ and the $(F/P, r\%, 10)$ factor values for $i = r = 15\%$.

(*c*) The value of the $(A/F, r\%, 5)$ factor where the effective continuous-interest rate is 12%.

SOLUTION

(*a*) From Appendix B using $r = 10\%$ the effective rate is $i = 10.517\%$ and

$$(P/A, 10\%, 15) = 7.3867$$

(*b*) From Appendix A for $i = 15\%$ per year compounded annually $(F/P, 15\%, 10) = 4.0456$, which is computed using the equation

$$(F/P, 15\%, 10) = (1 + i)^n = 1.15^{10} = 4.0456$$

From Appendix B for $r = 15\%$ per year compounded continuously $(F/P, 15\%, 10) = 4.4817$, which is computed using the equation

$$(F/P, 15\%, 10) = e^{rn} = e^{1.5} = 4.4817$$

The factor value for $r = 15\%$ and therefore the F value will be larger due to the continuous compounding of interest.

(*c*) First determine the nominal rate using Eq. (3.14).

$$r = \ln(1 + 0.12) = 0.1133 \qquad (11.33\%)$$

By interpolating between 10 and 12% in the Appendix B tables,

$$(A/F, 11.33\%, 5) = 0.16212 - \frac{11.33 - 10}{12 - 10}(0.16212 - 0.15508)$$

$$= 0.16212 - 0.00458 = 0.15754$$

If the A/F formula from Table 3.4 is used, the mathematically correct value is 0.15742, which is very close to the interpolated value.

The equations in Table 3.4 are used to compute the P, F, and A values for a sequence of cash flows and a continuously compounded interest rate. All computations are parallel to those for annual compounding, as shown in the following examples.

Example 3.7 Mr. Blunder and Mr. Smart both plan to invest \$5000 for 10 years at 10% per year. Compute the future worth for both men if Mr. Blunder figures interest compounded annually and Mr. Smart assumes continuous compounding.

SOLUTION

Mr. Blunder For annual compounding the future worth is

$$F = P(F/P, 10\%, 10) = 5000(2.5937) = \$12{,}969$$

Mr. Smart Using continuous compounding and Appendix B,

$$F = P(F/P, 10\%, 10) = 5000(2.7183) = \$13{,}591$$

COMMENT Continuous compounding represents a \$622, or 4.8%, increase in earnings. This is not a tremendous difference, as might have been expected. Just for comparison, note that a savings and loan association might compound daily, which yields an effective rate of 10.516% (F = \$13,590), whereas 10% continuous compounding offers only a very slight increase to 10.517%.

Example 3.8 Compare the present worth of \$2000 a year for 10 years at 10% per year (*a*) compounded annually and (*b*) compounded continuously.

SOLUTION

(*a*) For annual compounding,

$$P = 2000(P/A, 10\%, 10) = 2000(6.1446) = \$12{,}289$$

(*b*) For continuous compounding, r = 10% and the P/A factor is from Appendix B.

$$\begin{aligned} P = A(P/A, r\%, n) &= 2000(P/A, 10\%, 10) \\ &= 2000(6.0104) \\ &= \$12{,}021 \end{aligned}$$

As expected, the present worth for continuous compounding is less since the accumulation of interest requires a smaller investment to accrue the same amount as annual compounding at a later date.

Probs. P3.37 to P3.42

*3.8 Interest Factors and Calculations Using Continuous Cash Flows and Continuous Compounding

If money is assumed to flow throughout each period, the continuous cash-flow assumption is made. This approach is commonly called *funds flow* and interest is continuously compounded, as in the previous section. Cash may flow in several manners, as shown in Fig. 3.7, uniform flow being the most commonly made assumption. In all cases the

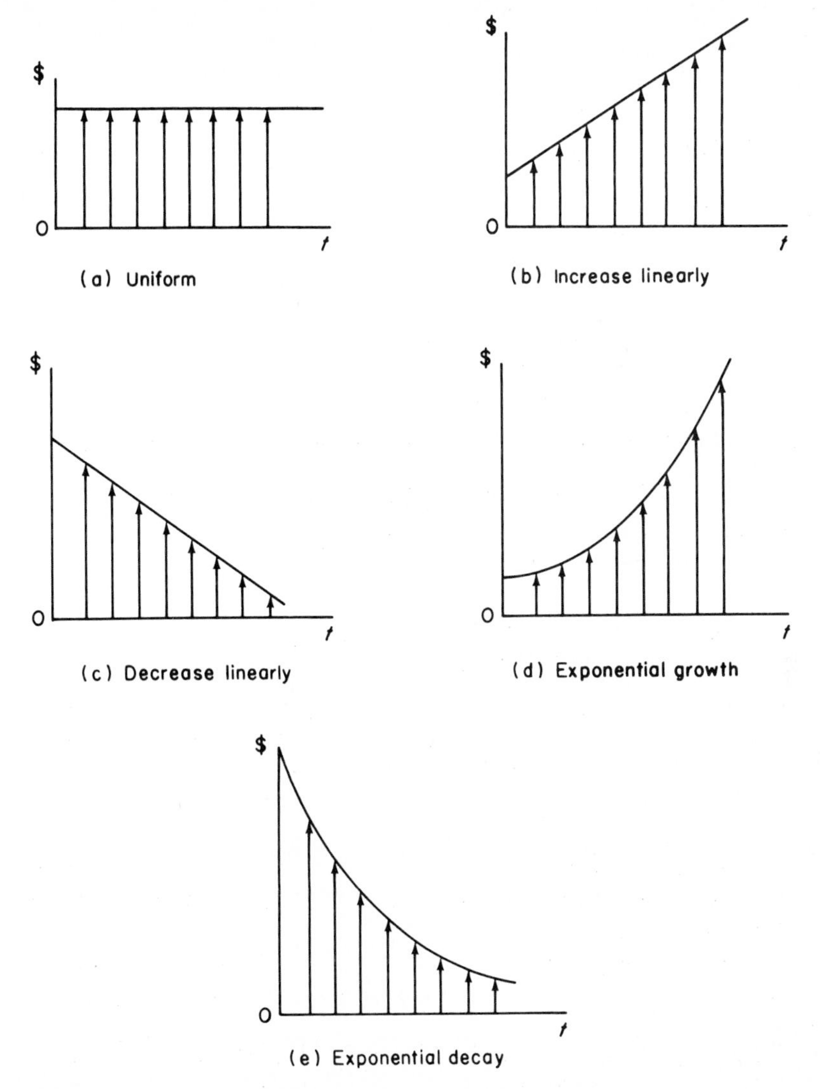

Figure 3.7 Different assumptions for the continuous distribution of cash flow over a period of time.

total flow amount symbolized by $\bar{A}$ is accumulated over k increments in one period. For the uniform flow the amount is $\bar{A}/k$ for k times per period (Fig. 3.8).

The interest factors for the uniform cash-flow case are derived by taking the limit as k approaches infinity, which allows interest to be compounded continuously while requiring that the time between each cash flow approaches zero. The future-worth value for the uniform cash flow for one year may be written

$$F = \frac{\bar{A}}{k}\,(F/A, r\%, 1) \tag{3.15}$$

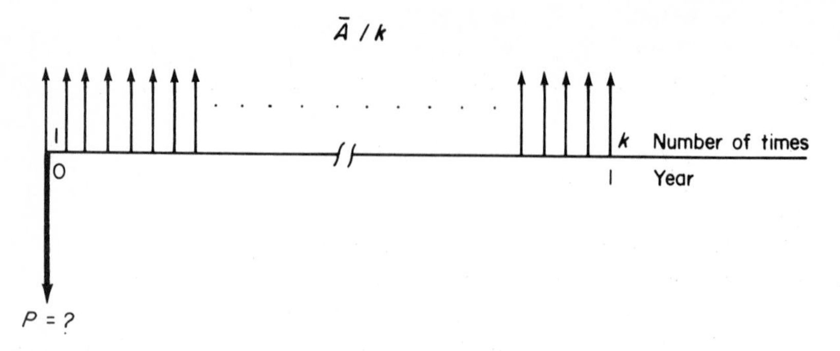

Figure 3.8 Continuous uniform flow of $\bar{A}$ through one year.

where $r\%$ is used to indicate a nominal interest rate. The form of Eq. (2.9) is used to write Eq. (3.15) as

$$F = \frac{\bar{A}}{k} \frac{(1 + r/k)^k - 1}{r/k}$$

Using the definition of e in Eq. (3.12),

$$\lim_{k \to \infty} F = \lim_{k \to \infty} \bar{A} \frac{(1 + r/k)^k - 1}{r}$$

$$= \bar{A} \left[\frac{e^r - 1}{r} \right]$$

The expression in parentheses is used to convert from discrete to continuous cash flows when interest is continuously compounded. The factor formulas in Table 3.4 may therefore be converted from discrete to continuous cash flow by simply multiplying by $(e^r - 1)/r$. The *funds-flow compound-amount* factor can now be written as

$$(F/\bar{A}, r\%, n) = (F/A, r\%, n) \left(\frac{e^r - 1}{r} \right) = \left(\frac{e^{rn} - 1}{e^r - 1} \right) \left(\frac{e^r - 1}{r} \right)$$

$$= \frac{e^{rn} - 1}{r}$$

where $\bar{A}$ indicates that this is a continuous cash-flow factor.

The *funds-flow present-worth* factor is the multiplication of the conversion factor and the $(P/A, r\%, n)$ factor in Table 3.4.

$$(P/\bar{A}, r\%, n) = \left(\frac{1 - e^{-rn}}{e^r - 1} \right) \left(\frac{e^r - 1}{r} \right)$$

$$= \frac{1 - e^{-rn}}{r} = \frac{e^{rn} - 1}{re^{rn}}$$

The F/P factor for continuous cash flow is e^{rn}, the same formula derived in Sec. 3.7, because only one cash-flow amount is involved. Relations for the factors and computa-

Table 3.5 Factor formulas for continuous cash flow, continuous compounding

Factor	Formula	Equation
$(F/P, r\%, n)$	e^{rn}	$F = P(F/P, r\%, n)$
$(P/F, r\%, n)$	e^{-rn}	$P = F(P/F, r\%, n)$
$(\bar{A}/F, r\%, n)$	$\dfrac{r}{e^{rn} - 1}$	$\bar{A} = F(\bar{A}/F, r\%, n)$
$(F/\bar{A}, r\%, n)$	$\dfrac{e^{rn} - 1}{r}$	$F = \bar{A}(F/\bar{A}, r\%, n)$
$(\bar{A}/P, r\%, n)$	$\dfrac{re^{rn}}{e^{rn} - 1}$	$\bar{A} = P(\bar{A}/P, r\%, n)$
$(P/\bar{A}, r\%, n)$	$\dfrac{e^{rn} - 1}{re^{rn}}$	$P = \bar{A}(P/\bar{A}, r\%, n)$

tions are summarized in Table 3.5; and Appendix C tabulates the six factor values for selected nominal interest rates. The nominal and effective rates are given in parentheses in all the interest tables. Computations for P, F, and A parallel those for the discrete case, as shown in the equations in Table 3.5 and the examples below.

Example 3.9 Jack plans to invest \$5000 now and \$2000 per year for 10 years in a real estate franchise. Determine the future worth of the investment under the following conditions and compare the results:

1. Discrete cash flow, 15% per year compounded annually
2. Discrete cash flow, 15% per year compounded continuously
3. Continuous cash flow, 15% per year compounded continuously

SOLUTION

1. For $i = 15\%$ using Appendix A,

$$F = 5000(F/P, 15\%, 10) + 2000(F/A, 15\%, 10)$$
$$= 5000(4.0455) + 2000(20.304) = \$60{,}836.00$$

2. For $r = 15\%$ using Appendix B,

$$F = 5000(F/P, 15\%, 10) + 2000(F/A, 15\%, 10)$$
$$= 5000(4.4817) + 2000(21.514) = \$65{,}436.50$$

3. For $r = 15\%$ using Appendix C,

$$F = 5000(F/P, 15\%, 10) + 2000(F/\bar{A}, 15\%, 10)$$
$$= 5000(4.4817) + 2000(23.211) = \$68{,}830.50$$

The future worth increases when interest is compounded continuously and again when the funds-flow (continuous-cash-flow) assumption is made.

Example 3.10 Two investment plans have been proposed for the Buy Now–See Later Corporation. Plan D requires an investment of \$10,000 with an expected annual positive cash flow of \$2500 for 10 years. Plan Z requires \$5000 and promises an annual cash flow of \$10,000 for 30 years. Use A/P-factor analysis to determine the nominal rate of return for each plan using continuous compounding with (*a*) discrete cash flows and (*b*) continuous cash flows.

SOLUTION

(*a*) For discrete cash flows compute the A value for each plan using Appendix B tables.

$$A_D = -10{,}000(A/P, r\%, 10) + 2500 = 0$$

$$(A/P, r\%, 10) = 2500/10{,}000 = 0.2500$$

Interpolation gives a nominal rate of $r_D = 19.36\%$ for plan D.

$$A_Z = -50{,}000\ (A/P, r\%, 30) + 10{,}000 = 0$$

$$(A/P, r\%, 10) = 10{,}000/50{,}000 = 0.2000$$

Interpolation yields $r_Z = 18.11\%$ for plan Z, which indicates that plan D is a slightly better investment.

(*b*) For continuous cash flow the $\bar{A}/P$ factors have the same values as in part *a*, but interpolation in Appendix C tables results in the nominal rates of

$$r_D = 22.28\% \qquad \text{and} \qquad r_Z = 19.95\%$$

Again, plan D is preferred, but the rates are somewhat higher.

Probs. P3.43 to P3.46

SOLVED PROBLEMS

Example 3.11 Mr. Jones plans to place money in a certificate of deposit that pays 18% per year compounded daily. What effective rate will he receive (*a*) yearly and (*b*) semiannually?

SOLUTION

(*a*) Using Eq. (3.3), with $t = 365$,

$$i = \left(1 + \frac{0.18}{365}\right)^{365} - 1$$

$$= 0.19716 \qquad (19.716\%)$$

That is, Mr. Jones will get an effective 19.716% per year on his deposit.

(*b*) Here $r = 0.09$ per 6 months and $t = 182$ days:

$$i = \left(1 + \frac{0.09}{182}\right)^{182} - 1$$

$$= 0.09415 \qquad (9.415\%)$$

Sec. 3.3

Example 3.12 Ms. Warren wants to purchase a new compact car for \$8500. She plans to borrow the money from her credit union and to repay it monthly over a period of 4 years. If the nominal interest rate is 12% per year compounded monthly, what will her monthly installments be?

Solution Since the compounding period equals the payment period, the effective monthly rate is $i = 1\%$ per month and there are $n = 12(4) = 48$ payments. The monthly payments are

$$A = 8500(A/P, 1\%, 48) = 8500(0.02633) = \$233.81$$

Comment It is equally correct in this example to use the reciprocal of Eq. (3.8) to find the A/P factor value of 0.02633 with $r = 0.12, t = p = 12$, and $n = 4$ years.

Sec. 3.4

Example 3.13 Calculate the equivalent compound-period amount if \$200 is deposited every month at an interest rate of 6% per year compounded semiannually with simple interest paid on interperiod deposits.

Solution From Eq. (3.11) and a semiannual interest rate of 3%,

$$A = 6(200) + \tfrac{5}{2}(0.03)(200) = \$1215$$

Thus, \$200 per month is equivalent to \$1215 every 6 months, since the compounding period is semiannual.

Comment The present worth or future worth is calculated using the P/A or F/A factors for $i = 3\%$ and $n =$ number of compounding periods (for example, 3 years is six compounding periods).

Sec. 3.5

FURTHER INFORMATION

(Numbers in brackets are references in the bibliography.)

Continuous compounding (Secs. 3.6 to 3.8): Barish and Kaplan [2], pp. 72-76; Canada and White [4], pp. 39-43; DeGarmo, Canada, and Sullivan [5], pp. 97-102; Grant, Ireson, and Leavenworth [7], pp. 579-590; Reisman [12], pp. 10-39; Thuesen, Fabrycky, and Thuesen [17], pp. 83-94.

PROBLEMS

P3.1 What is the nominal interest rate per year if interest is (*a*) 0.50% every 2 weeks and (*b*) 2% every semiannual period?

P3.2 What time period is usually used in expressing a nominal interest rate?

P3.3 What is the dimensional unit of the term r/t in the effective interest-rate equation?

P3.4 Calculate the nominal and effective interest rates per year for a finance charge of $1\frac{1}{2}$% per month.

P3.5 What are the nominal and effective interest rates per year for an interest charge of 4% every 6 months?

P3.6 What effective interest rate per year is equivalent to a nominal rate of 12% per year compounded semiannually?

P3.7 What effective interest rate per year is equivalent to a nominal rate of 16% per year compounded quarterly?

P3.8 Calculate the nominal and effective interest rates per year for a finance charge of $1\frac{3}{4}$% per month and find the value of the P/A factor for 10 years for each rate.

P3.9 What quarterly interest rate is equivalent to an effective annual rate of 6%?

P3.10 How much money would be accumulated in 8 years if an investor deposits $2500 now at a nominal interest rate of 8% per year compounded semiannually?

P3.11 If a person buys a car for $5500 and must make monthly payments of $200 for 36 months, what are the nominal and effective interest rates per year for this transaction?

P3.12 How much should you be willing to pay for an annuity that will provide $300 every 3 months for 6 years starting 3 months from now if you want to make a nominal 12% per year compounded quarterly?

P3.13 If a person deposits $75 into a savings account every month, how much money will be accumulated after 10 years if the interest rate is a nominal 12% per year compounded monthly?

P3.14 If a person borrows $3000 and must repay the loan in 2 years with equal monthly installments, how much is the monthly payment if the interest rate is 1% per month?

P3.15 If a person made a lump-sum deposit 12 years ago which has accumulated to $9500 now, how much was the original deposit if the interest rate received was a nominal 4% per year compounded semiannually?

P3.16 What is the present worth of $10,000 now, $6000 eight years from now, and $9000 twelve years from now if the interest rate is a nominal 6% per year compounded quarterly?

P3.17 What would be the value of the deposits in Prob. 3.16 25 years from the date of the original deposit?

P3.18 A man has been presented with the opportunity to buy a second mortgage note valued at $1500 for $1300. The note is due 4 months from now. If he purchases the note, what nominal and effective rate of return will he make? Assume interest is compounded monthly.

P3.19 What monthly interest rate is equivalent to an effective semiannual rate of 4%?

P3.20 If a pants manufacturing company spends $14,000 in order to improve the efficiency of a sewing operation, how much must it save each month in reduced manpower costs in order to recover its investment in $2\frac{1}{2}$ years if the effective interest rate is 12.68% per year compounded monthly?

P3.21 A woman who has just won $45,00 in a lottery wants to deposit enough of her winnings into a savings account so that she will have $10,000 for her son's college education. Assume that her son just turned 3 years old and will begin college when he is 18 years of age. How much must the woman deposit to earn 7% per year compounded quarterly on her investment?

P3.22 How much money can be withdrawn semiannually for 20 years from a retirement fund which earns 5% per year interest compounded semiannually and has a present amount of $36,000 in it?

P3.23 What year-end payment is equivalent to the monthly payments that would be paid on a \$2000 loan that must be repaid in one year at an interest rate of 1% per month?

P3.24 Compute the monthly deposit required to accumulate \$5000 in 5 years at a nominal 6% per year compounded daily.

***P3.25** Draw two cash-flow diagrams illustrating the timing of cash flow if banks paid interest on interperiod deposits but not on withdrawals. Assume that interest is payable quarterly and represent deposits by X's and withdrawals by Y's, with at least one deposit and one withdrawal per interest period. Draw one diagram for actual deposits and withdrawals and another illustrating deposits and withdrawals from the bank's point of view.

***P3.26** How much money would be in a savings account in which a person had deposited \$100 every month for 5 years at an interest rate of 5% per year compounded quarterly? Assume simple interperiod interest.

***P3.27** How much money would the person in Prob. P3.26 have if interest were compounded (*a*) monthly? (*b*) daily?

***P3.28** Calculate the future worth of the transactions shown in Fig. 3.4*a* if interest is 6% per year compounded semiannually and interperiod interest is not paid.

***P3.29** Rework Prob. P3.28, except assume that simple interest is paid on interperiod deposits but not withdrawals.

***P3.30** A tool-and-die company expects to have to replace one of its lathes in 5 years at a cost of \$18,000. How much would the company have to deposit every month in order to accumulate \$18,000 in 5 years if the interest rate is 6% per year compounded semiannually? Assume simple interperiod interest.

***P3.31** What monthly deposit would be equivalent to a deposit of \$600 every 3 months for 2 years if the interest rate is 6% compounded semiannually? Assume simple interperiod interest on the \$600 deposits only.

***P3.32** How much money would be in the savings account of a person who had deposited \$150 every month and withdrawn \$300 every 6 months for 2 years? Use an interest rate of 4% per year compounded semiannually and assume that interperiod interest is not paid.

***P3.33** How many monthly deposits of \$75 would a person have to make in order to accumulate \$15,000 if the interest rate is 6% per year compounded semiannually? Assume that simple interperiod interest is paid.

***P3.34** Jerri had just invested money at 14.5% per year compounded continuously. What is the effective rate to be expected?

***P3.35** Is it true that the effective rate yielded by a nominal rate of 15% per year compounded quarterly is just about the same as the effective rate of 14.726% per year compounded continuously? Why or why not?

***P3.36** What is the annual rate necessary to yield the following annual effective rates if continuous compounding is in effect: (*a*) 22%, (*b*) 13.75%.

***P3.37** Verify that the $(P/A, r\%, n)$ factor may be determined from the relation $(P/A, r\%, n) = (F/A, r\%, n)(P/F, r\%, n)$. Perform this verification by using formulas and table values.

***P3.38** What uniform annual amount would you have to deposit for 5 years to have an equivalent present-investment sum of \$10,000 at an interest rate of 10% per year compounded continuously?

***P3.39** Determine the difference in future worth of the following annual cash flows if interest is 12% per year compounded annually and 12% per year compounded continuously.

Year	0	1	2	3	4
Discrete cash flow, \$	-15,000	5,000	5,000	5,000	12,000

***P3.40** If Ms. Watson has agreed to repay a loan of $4500 in 10 equal annual payments starting one year from now, determine the size of each payment if interest is 15% per year compounded continuously.

***P3.41** Deposits of $900 per year for 9 years are to be made. If $10,000 is to be accumulated, determine (*a*) the continuously compounded nominal annual rate (*b*) the continuously compounded effective annual rate.

***P3.42** Mr. Adams plans to borrow $5000 now and repay the loan in eight equal payments of $875 per year starting one year from now. Compute the interest rate if interest is (*a*) compounded annually and (*b*) compounded continuously.

***P3.43** J. Smith plans to purchase a new computer for $200,000 and sell data-processing services for the next 10 years. What is the total annual revenue requirement to make 20% per year on the investment if revenues are assumed to be received (*a*) continuously throughout each year and (*b*) at the end of each year and interest is compounded annually.

***P3.44** Find the present worth of $5000 per year continuous cash flow for 10 years if the effective continous rate is 20%. Use the correct factor formula to obtain the factor value.

***P3.45** S. Johnson wants to accumulate $10,000 in 5 years. If the nominal annual rate is 10%, determine the total annual deposit under the following conditions and compare the amounts for each plan to determine which requires the smallest annual deposit.

(*a*) End-of-year deposits, semiannual compounding
(*b*) 6-month deposits, semiannual compounding
(*c*) End-of-year deposits, continuous compounding
(*d*) Continuous deposits, continuous compounding

***P3.46** Compute the (*a*) nominal and (*b*) effective annual rate of return of an investment of $18,000 now which will return $25,500 in 5 years if the continuous-cash-flow assumption is made.

CHAPTER

FOUR

USE OF MULTIPLE FACTORS

The purpose of this chapter is to teach you how to solve problems which involve the multiplication of several engineering-economy factors and values of P, F, A, and G.

SECTION OBJECTIVES

To complete this chapter you must be able to:

4.1. Determine the year in which the present worth or future worth is located, given a statement describing a randomly placed uniform series.
4.2. Determine the present worth, equivalent uniform annual series, or future worth of a uniform series of disbursements or receipts which begin at a time other than year 1, given the amount and time of each payment and the interest rate.
4.3. Calculate the present worth or future worth of randomly distributed single amounts and uniform-series amounts, given the times and amounts of the payments and the interest rate.
4.4. Calculate the equivalent uniform annual series of randomly distributed single amounts and uniform-series amounts, given the times, amounts, and the interest rate.
4.5. Calculate the present-worth and equivalent uniform annual series of cash flows involving shifted gradients, given the interest rate and statement of the problem.
4.6. Calculate the present-worth or equivalent uniform annual series of cash flows which include a decreasing gradient, given the interest rate and statement of the problem.

STUDY GUIDE

4.1 Location of Present Worth and Future Worth

When a uniform series of payments begins at a time other than at the end of year (period) 1, several methods can be used to find the present worth. For example, the present worth of the uniform series of disbursements shown in Fig. 4.1 could be determined by any of the following methods:

1. Use the single-payment present-worth factor $(P/F, i\%, n)$ to find the present worth of each disbursement at year 0 and add them.
2. Use the single-payment compound-amount factor $(F/P, i\%, n)$ to find the future worth of each disbursement in year 13, add them, and then find the present worth of the total using $P = F(P/F, i\%, 13)$.
3. Use the uniform-series compound-amount factor $(F/A, i\%, n)$ to find the future amount by $F = A(F/A, i\%, 10)$ and then find the present worth using $P = F(P/F, i\%, 13)$.
4. Use the uniform-series present-worth factor $(P/A, i\%, n)$ to compute the "present worth" (which will *not* be located in year 0!) and then find the present worth in year 0 by using the $(P/F, i\%, n)$ factor. (Present worth is enclosed in quotation marks to represent the present worth as determined by the uniform-series present-worth factor and to differentiate it from the present worth in year 0.)

This and the next section illustrate the fourth method for calculating the present worth of a uniform series that does not begin at the end of year 1.

For the cash-flow diagram shown in Fig. 4.1, the "present worth" that would be obtained by using the $(P/A, i\%, n)$ factor would be located in *year 3*, not year 4. This

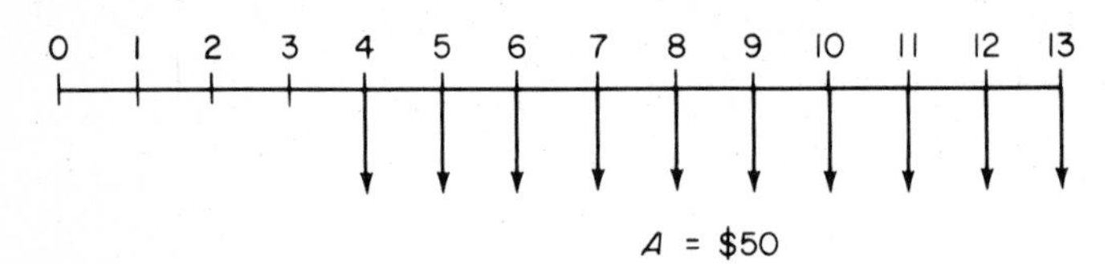

Figure 4.1 A randomly placed uniform series.

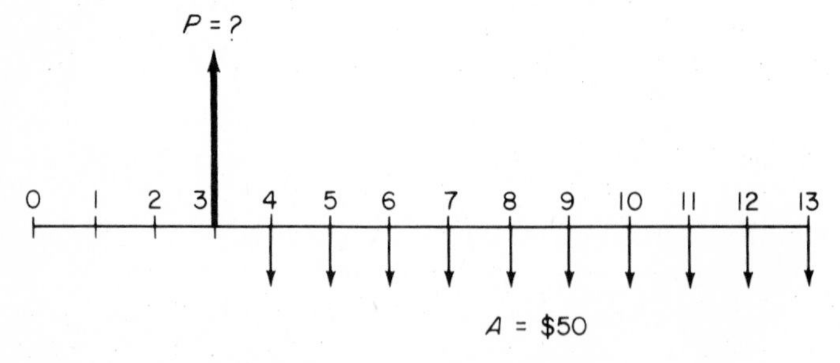

Figure 4.2 Location of P for the randomly placed uniform series in Fig. 4.1.

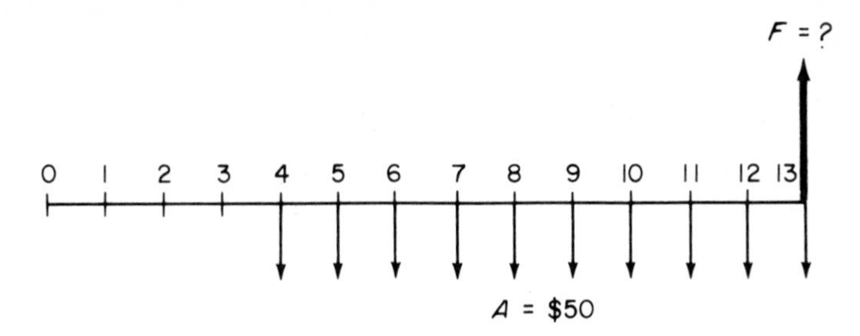

Figure 4.3 Placement of F for the uniform series of Fig. 4.1.

is shown in Fig. 4.2, with the present-worth arrow pointing upward to represent an equivalent amount in year 3. Note that P is located *one year prior* to the beginning of the first annual disbursement. Why? Because the P/A factor was derived with the P in year 0 and the A beginning at the end of year 1; that is, the P must always be *one year ahead* of the first A value (Sec. 2.2). The most common mistake made in working problems of this type is improper placement of P. Therefore, it is extremely important that you remember the following rule. *The present worth is always located one year prior to the first annual payment when using the uniform-series present-worth factor, $(P/A, i\%, n)$.*

On the other hand, the uniform-series compound-amount factor, $(F/A, i\%, n)$, was derived (Sec. 2.3) with the future worth F located in the *same year* as the last payment. Figure 4.3 shows the location of the future worth when F/A is used for the cash flow shown in Fig. 4.1. Thus, *the future worth is always located in the same year as the last annual payment when using the uniform-series compound-amount factor $(F/A, i\%, n)$.*

It is also important to remember that the number of years n that should be used with the P/A or F/A factors is equal to the number of payments. It is generally helpful to *renumber* the cash-flow diagram to avoid counting errors. Figure 4.4 shows the cash-flow diagram of Fig. 4.1 renumbered for determination of n. Note that in this example $n = 10$.

Solved Problem 4.12
Prob. P4.1

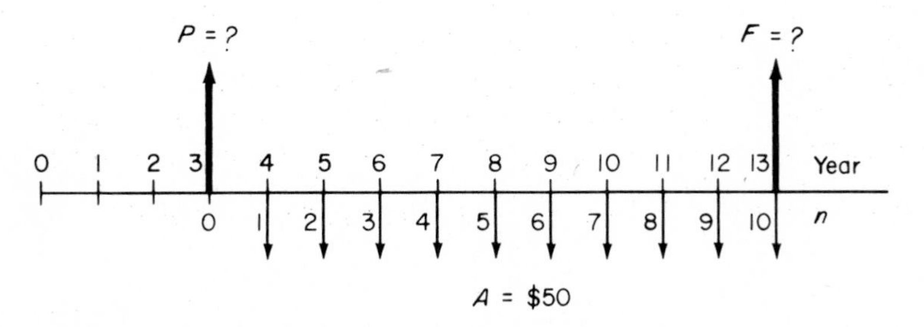

Figure 4.4 Renumbering of payments in Fig. 4.1 to show that $n = 10$ in the P/A or F/A factors.

4.2 Calculations for a Uniform Series that Begins after Year 1

As stated in Sec. 4.1, there are many methods that can be used to solve problems having a uniform series that begins at a time other than the end of year 1. However, it is generally much more convenient to use the uniform-series formulas than it is to use the single-payment formulas.

There are specific steps which should be followed in solving problems of this type in order to avoid unnecessary errors:

1. Draw a cash-flow diagram of the receipts and disbursements of the problem.
2. Locate the present worth or future worth on the cash-flow diagram.
3. Determine n by renumbering the cash-flow diagram.
4. Draw the cash-flow diagram representing the desired equivalent cash flow.
5. Set up and solve the equations.

These steps are illustrated in the following two examples.

Example 4.1 A person buys a piece of property for \$5000 down and deferred annual payments of \$500 a year for 6 years starting 3 years from now. What is the present worth of the investment if the interest rate is 8%?

SOLUTION The cash-flow diagram is shown in Fig. 4.5. The nomenclature P_A is used throughout this book to represent the present worth of a uniform annual series A and P'_A represents the present worth at a time other than year 0. Similarly, P_T represents the total present worth at time 0. The correct placement of P'_A and diagram renumbering to obtain n are also indicated in Fig. 4.5. Note that P'_A is located in year 2, not year 3, and $n = 6$, not 8. To solve this problem first find the value of P'_A:

$$P'_A = \$500(P/A, 8\%, 6)$$

Since P'_A is located in year 2, it is necessary to find P_A in year 0:

$$P_A = P'_A(P/F, 8\%, 2)$$

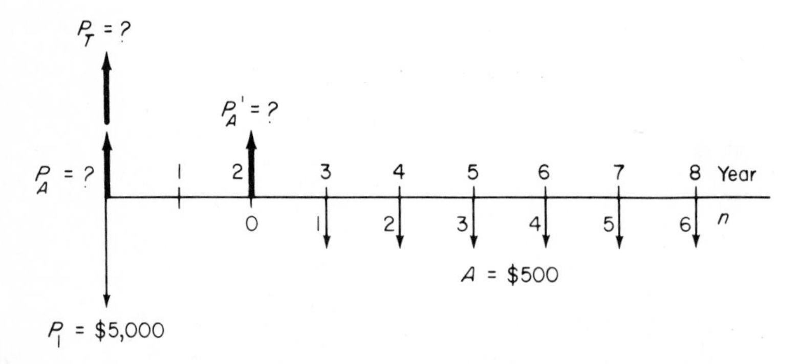

Figure 4.5 Placement of present-worth values, Example 4.1.

The total present worth can now be determined by adding P_A and P_1 (the initial investment is labeled P_1):

$$\begin{aligned} P_T = P_1 + P_A &= 5000 + 500(P/A, 8\%, 6)(P/F, 8\%, 2) \\ &= 5000 + 500(4.6229)(0.8573) \\ &= \$6981.60 \end{aligned}$$

Example 4.2 Calculate the 8-year equivalent uniform annual series at 6% interest for the uniform disbursements shown in Fig. 4.6.

SOLUTION Figure 4.7 shows the original cash-flow diagram and the desired equivalent diagram. In order to convert uniform cash flows that begin sometime after year 1 into an equivalent uniform annual cost over *all* the years, the first step is to convert the cash flow into a present worth or future worth. Then either the conventional capital-recovery factor $(A/P, i\%, n)$ or the sinking-fund factor $(A/F, i\%, n)$ can be used to determine the equivalent uniform annual cost. Both these methods are illustrated below.

1. Present-worth method (refer to Fig. 4.7).

$$\begin{aligned} P'_A &= 800(P/A, 6\%, 6) \\ P_T &= P'_A(P/F, 6\%, 2) = 800(P/A, 6\%, 6)(P/F, 6\%, 2) \\ &= \$3501.12 \end{aligned}$$

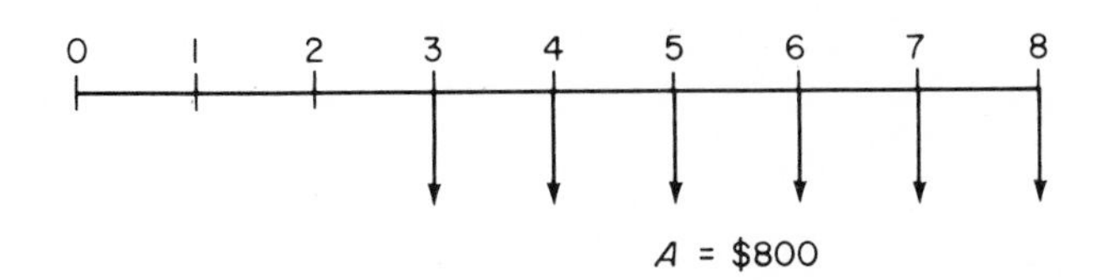

Figure 4.6 Series of uniform disbursements, Example 4.2.

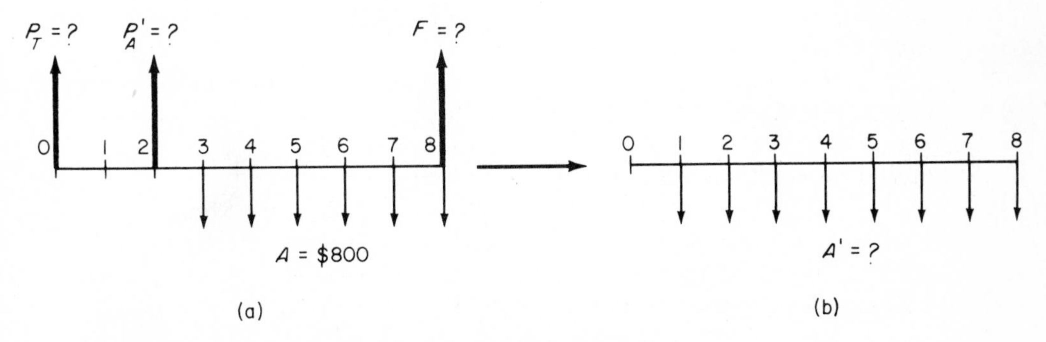

Figure 4.7 Desired equivalent diagram for the series of Fig. 4.6.

where P_T is the total present worth of the cash flow. The equivalent series A' can now be determined with the A/P factor

$$A' = P_T(A/P, 6\%, 8) = \$563.82$$

for 8 years as shown in Fig. 4.7*b*.

2. Future-worth method (Fig. 4.7): The first step is to calculate the future worth F:

$$F = 800(F/A, 6\%, 6) = \$5580$$

The sinking-fund factor $(A/F, i\%, n)$ can now be used to obtain A':

$$A' = F(A/F, 6\%, 8) = \$563.80$$

COMMENT In the present-worth method, note that P'_A was located in year 2, not year 3. After the present worth was determined, the equivalent series was calculated using $n = 8$. In the future-worth method, $n = 6$ was used to find F, and $n = 8$ for finding the equivalent series, since the cost must be spread uniformly over *all* the years.

Solved Problem 4.13
Probs. P4.2 to P4.17

4.3 Calculations Involving Uniform-Series and Randomly Distributed Amounts

When a uniform series of payments is included in a cash flow that also contains randomly distributed single amounts, the procedures learned in Sec. 4.2 should be applied to the uniform-series amounts and the single-payment formulas applied to the single-payment amounts. This type of problem, illustrated below, is merely a combination of previous types.

Example 4.3 A couple owning 50 hectares of valuable land decided to sell the mineral rights on their property to a mining company. Their primary objective was to obtain long-term investment income and sufficient money to finance the college education of their two children. Since the children were 12 and 2 years of age at the time they were negotiating the contract, they knew that the children would be in college 6 and 16 years from the present. They therefore made a proposal to the company that it pay them \$20,000 per year for 20 years beginning one year hence plus \$10,000 six years from now and \$15,000 sixteen years from now. If the company wanted to pay off its lease immediately, how much would it have to pay now if the interest rate is 6%?

SOLUTION The cash-flow diagram for this problem is shown in Fig. 4.8. This problem is solved by finding the present worth of the uniform series and adding it to the present worths of the two individual payments. Thus,

$$\begin{aligned} P &= 20{,}000(P/A, 6\%, 20) + 10{,}000(P/F, 6\%, 6) + 15{,}000(P/F, 6\%, 16) \\ &= \$242{,}352 \end{aligned}$$

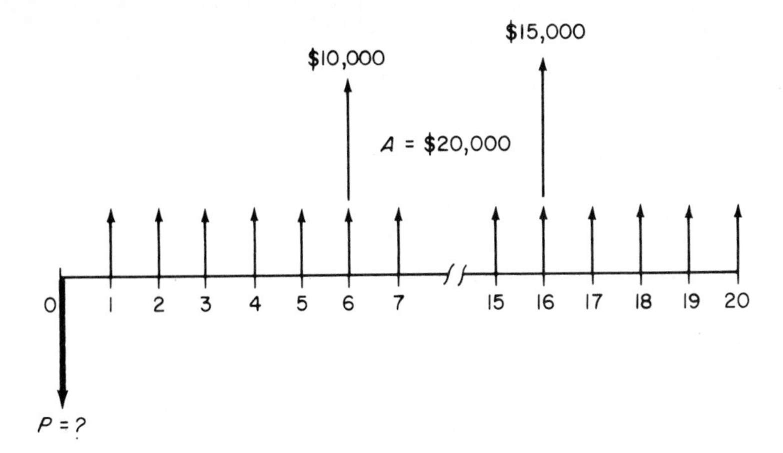

Figure 4.8 Diagram including a uniform series and single amounts, Example 4.3.

COMMENT In this example, note that the uniform series started at the end of year 1 so that the present worth obtained with the P/A factor represented the present worth at year 0. It was not necessary to use the P/F factor on the uniform series.

Example 4.4 If the uniform payments described in Example 4.3 did not begin until 3 years from the time the contract was signed, what would be the present worth of the receipts?

SOLUTION The cash-flow diagram is shown in Fig. 4.9, with the n scale shown above the time axis. The number of years for the uniform series is still 20.

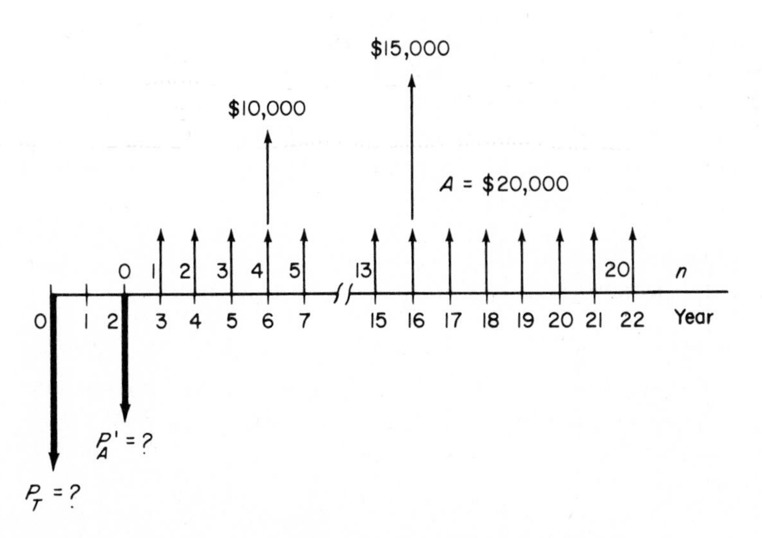

Figure 4.9 Diagram from Fig. 4.8 with the A series shifted 2 years.

$$P'_A = 20{,}000(P/A, 6\%, 20)$$

$$\begin{aligned} P_T &= P'_A(P/F, 6\%, 2) + 10{,}000(P/F, 6\%, 6) + 15{,}000(P/F, 6\%, 16) \\ &= 20{,}000(P/A, 6\%, 20)(P/F, 6\%, 2) + 10{,}000(P/F, 6\%, 6) \\ &\quad + 15{,}000(P/F, 6\%, 16) \\ &= \$217{,}118 \end{aligned}$$

COMMENT Displacement of the annual series by 2 years has decreased the present worth of the cash flows by $25,234.

Example 4.5 Calculate the future worth of the receipts shown in Fig. 4.9 using $i = 6\%$.

SOLUTION The future worth of the uniform series and the single, random receipts can be calculated as follows:

$$\begin{aligned} F &= 20{,}000(F/A, 6\%, 20) + 10{,}000(F/P, 6\%, 16) + 15{,}000(F/P, 6\%, 6) \\ &= \$782{,}381 \end{aligned}$$

COMMENT Although the determination of n is straightforward, make sure that you fully understand how the values 20, 16, and 6 were obtained for calculating the future worth.

Solved Problem 4.14
Probs. P4.18 to P4.25

4.4 Equivalent Uniform Annual Series of Both Uniform and Single Payments

Whenever it is desired to calculate the equivalent uniform annual series of randomly distributed single payments and/or uniform amounts, the most important fact to remember is that the payments *must first be converted to a present worth or future worth*. Then the equivalent uniform annual series can be obtained with the appropriate A/P or A/F factor.

Example 4.6 Calculate the 20-year equivalent uniform annual series for the receipts described in Example 4.3 (Fig. 4.8).

SOLUTION The desired equivalent cash-flow diagram is shown in Fig. 4.10. From the cash-flow diagram shown in Fig. 4.8, it is evident that the uniform-series receipts (that is, $20,000) are already distributed through all 20 years of the diagram. It is therefore necessary to convert only the single amounts to an equivalent uniform annual series and add the value obtained to the $20,000. This can be done by either the present-worth or future-worth method.

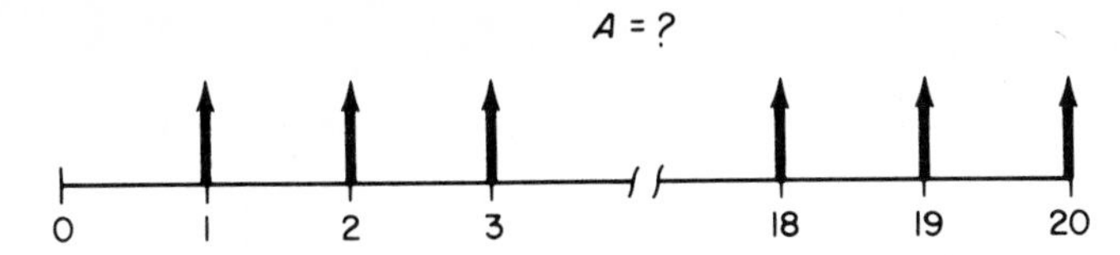

Figure 4.10 Desired equivalent series, Example 4.6.

(*a*) Present-worth method:

$$A = 20{,}000 + 10{,}000(P/F, 6\%, 6)(A/P, 6\%, 20)$$
$$+ 15{,}000(P/F, 6\%, 16)(A/P, 6\%, 20)$$
$$= 20{,}000 + [10{,}000(P/F, 6\%, 6)$$
$$+ 15{,}000(P/F, 6\%, 16)](A/P, 6\%, 20)$$
$$= \$21{,}129 \text{ per year}$$

(*b*) Future-worth method:

$$A = 20{,}000 + 10{,}000(F/P, 6\%, 14)(A/F, 6\%, 20)$$
$$+ 15{,}000(F/P, 6\%, 4)(A/F, 6\%, 20)$$
$$= 20{,}000 + [10{,}000(F/P, 6\%, 14)$$
$$+ 15{,}000(F/P, 6\%, 4)](A/F, 6\%, 20)$$
$$= \$21{,}129 \text{ per year}$$

COMMENT Note that it was necessary to take the single payments to either end of the time scale before annualizing. Failure to do so would result in unequal receipts in some years.

Example 4.7 Convert the cash flow shown in Fig. 4.9 to an equivalent uniform annual series over 22 years. Use $i = 6\%$.

SOLUTION Since the uniform-series receipts are not distributed through all 20 years of the time scale, it is first necessary to find the present worth or future worth of the series. This was done in Examples 4.4 and 4.5, respectively. The equivalent uniform annual series can now be obtained by multiplying the values previously obtained by the (*A*/*P*, 6%, 22) factor or the (*A*/*F*, 6%, 22) factor as follows:

$$A = P_T(A/P, 6\%, 22) = 217{,}118(A/P, 6\%, 22) = \$18{,}032$$

or

$$A = F(A/F, 6\%, 22) = 782{,}381(A/F, 6\%, 22) = \$18{,}034$$

COMMENT When a uniform series begins at a time other than at the end of year 1, or when intermediate single amounts are involved, it is most important to remember that the equivalent present or future worth of the uniform series must be determined before the equivalent uniform annual series can be obtained.

Probs. P4.26 to P4.34

4.5 Present Worth and Equivalent Annual Series of Shifted Gradients

In Sec. 2.5, Eq. (2.12), used for calculating the present worth of a uniform gradient, was derived. You will recall that the equation was derived for a present worth in year 0 with the gradient starting in year 2 (see Fig. 2.5). Therefore, the present worth of a uniform gradient will always be located *2 years before the gradient starts*. The examples that follow illustrate where the present worth of the gradient is located.

Example 4.8 For the cash-flow diagram shown in Fig. 4.11, locate the gradient present worth.

SOLUTION The present worth of the gradient, P_G, is shown in Fig. 4.12. In the derivation of the present worth of a gradient series, the present worth was located 2 years before the start of the gradient. Therefore, for the gradient in Fig. 4.11, the present worth would be located at the end of year 2. It is usually advantageous to renumber the cash-flow diagram so that the gradient year 0 and the number of

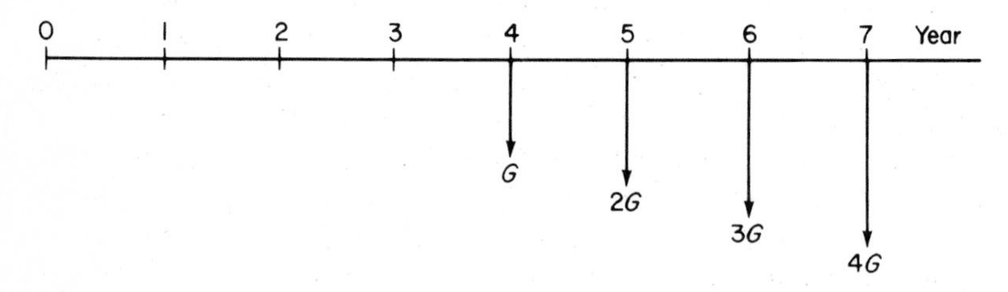

Figure 4.11 Diagram of gradient, Example 4.8.

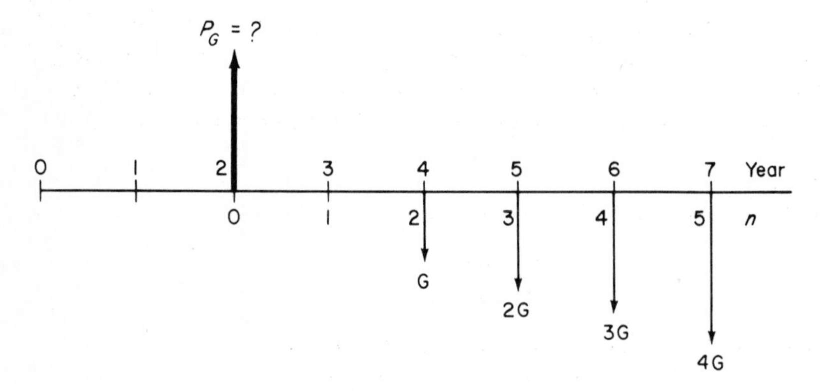

Figure 4.12 Diagram locating the present worth of gradient in Fig. 4.11.

years n of the gradient can be determined. The best method for accomplishing this is to determine where the gradient begins and label that time as year 2 and then work backward and forward. In this example, since the gradient started in year 4, gradient year 2 was placed below year 4 on the original diagram (Fig. 4.12). Year 0 for the gradient was then located by moving back 2 years.

Example 4.9 For the cash-flow diagram shown in Fig. 4.13, explain why the present worth of the gradient is located in year 3.

SOLUTION The gradient is \$50 and begins between years 4 and 5 of the original cash-flow diagram. Therefore, year 5 represents year 2 of the gradient; the present worth of the gradient would then be located in year 3. If Fig. 4.13 is divided into two cash-flow diagrams, the location of the gradient becomes quite clear, as in Fig. 4.14.

When a gradient of a cash-flow sequence starts in year 2, it is called a conventional gradient, as discussed in Sec. 2.5. When a gradient begins at a time before or after year 2, it is referred to as a *shifted gradient*. To determine n, the same renumbering procedure used to determine where the present worth of the gradient is located is necessary. For the cash-flow diagrams shown in Fig. 4.15a to c, the gradients G, number of years n, and gradient factors used to calculate the present worth and annual series of the gradients are shown on each cash-flow diagram, assuming an interest rate of 6% per year.

It is important to note that the A/G factor *cannot* be used to find an equivalent A value for cash flows involving a shifted gradient. Consider the cash-flow diagram of Fig. 4.16. To find the equivalent annual disbursement, you must first find the present worth of the gradient, take this present worth back to year 0, then annualize the pres-

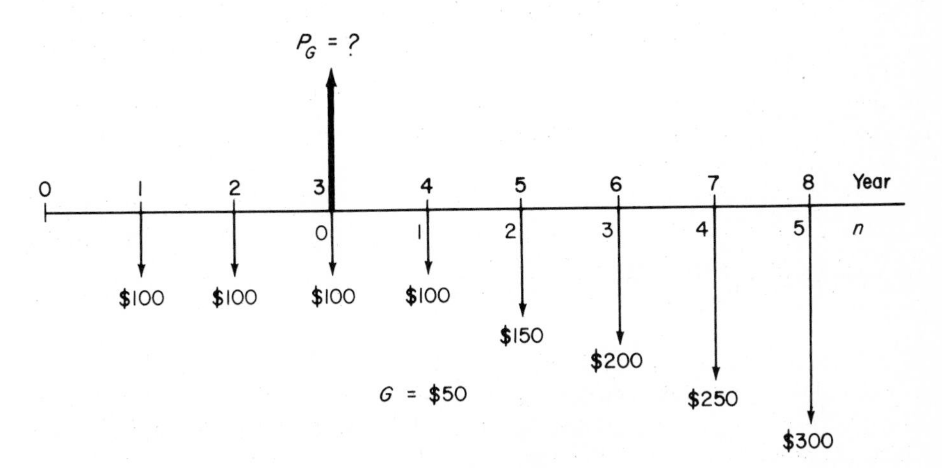

Figure 4.13 Location of the present worth of a gradient.

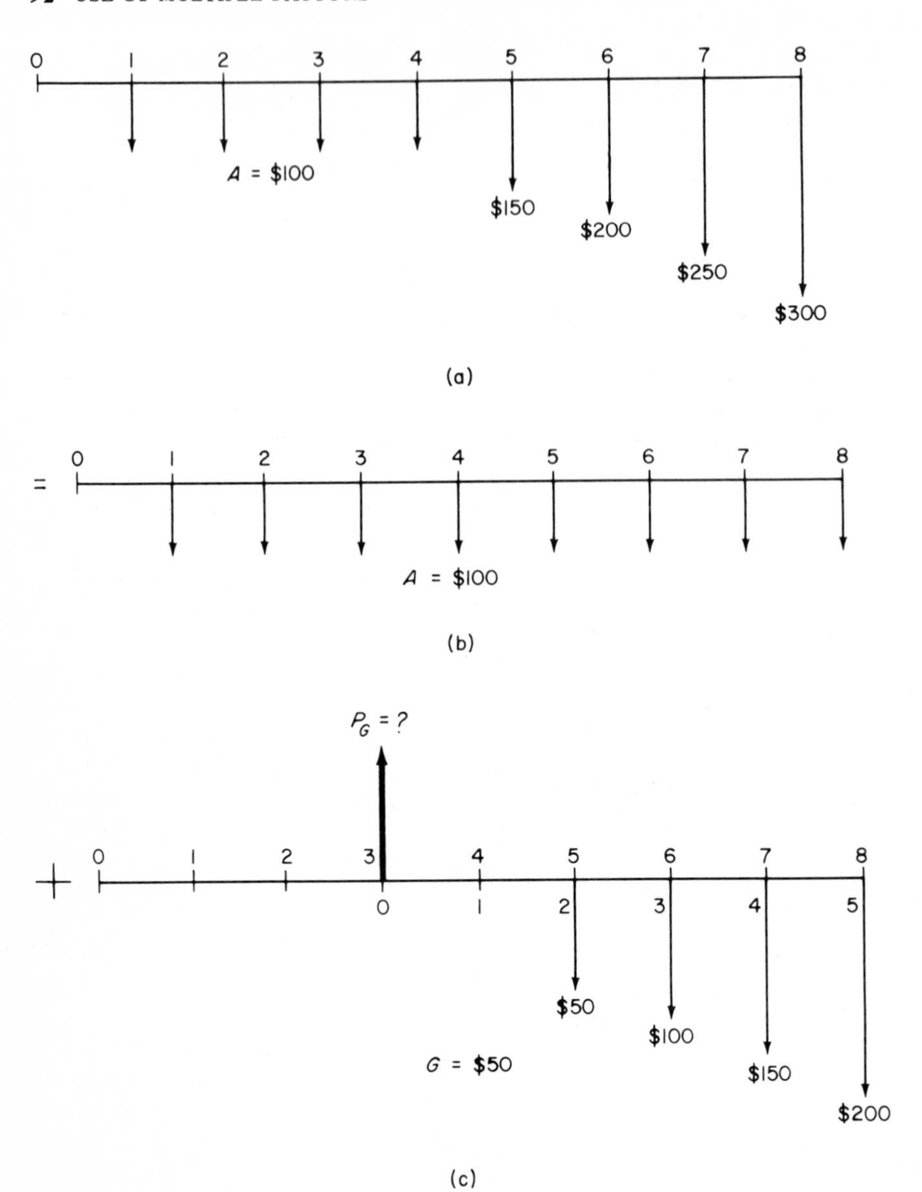

Figure 4.14 Partitioned cash flow of Fig. 4.13: (*a*) = (*b*) + (*c*).

ent worth from year 0 with the A/P factor. If you used the annual-series gradient factor $(A/G, i\%, n)$, the gradient would be converted to an annual cost over years 3 to 7 only. For this reason, the first step in this type of problem is always to find the present worth of the gradient at actual year 0. The steps involved in handling problems of this type are illustrated in Example 4.10.

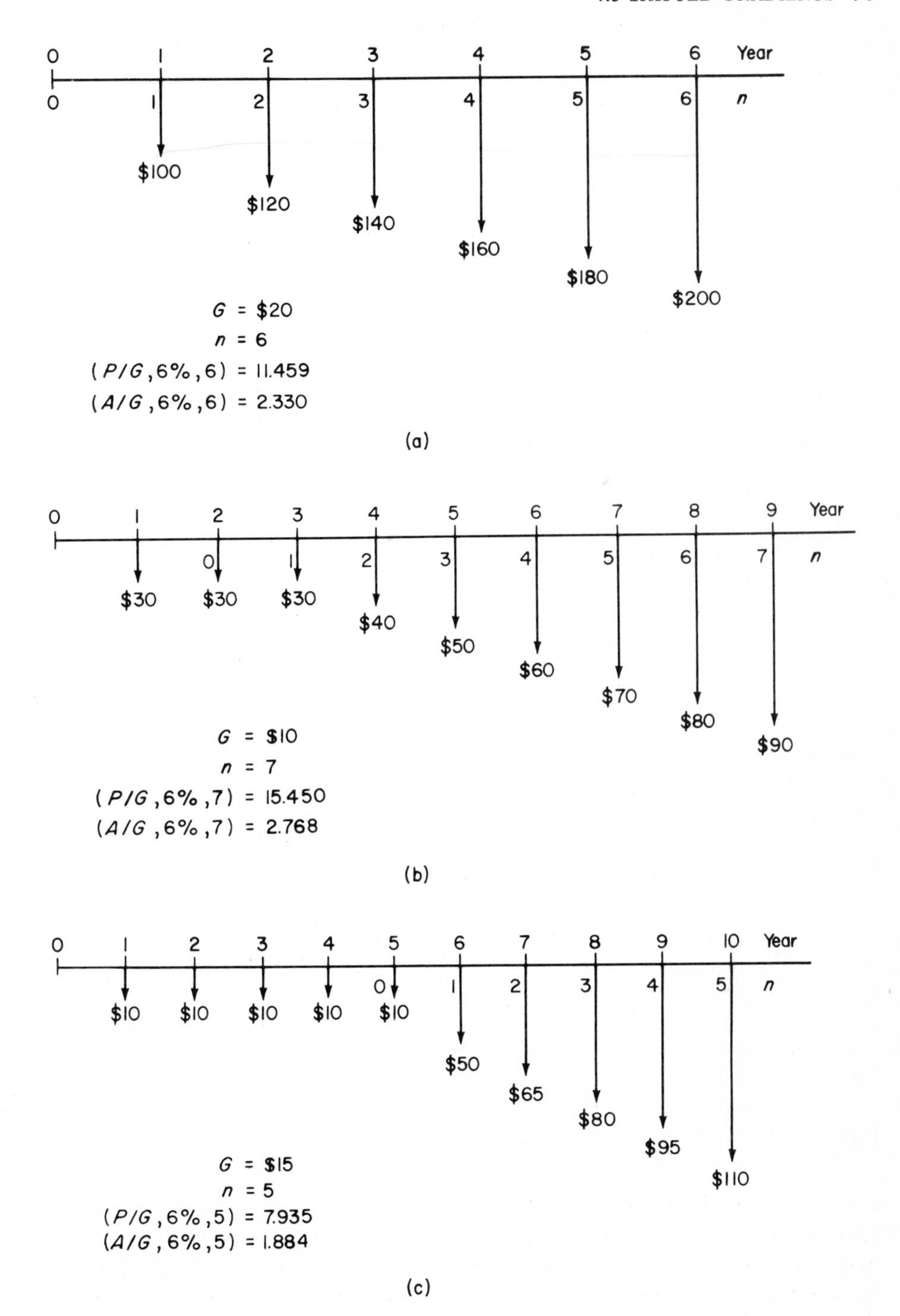

Figure 4.15 Determination of G and n values used in the gradient factors.

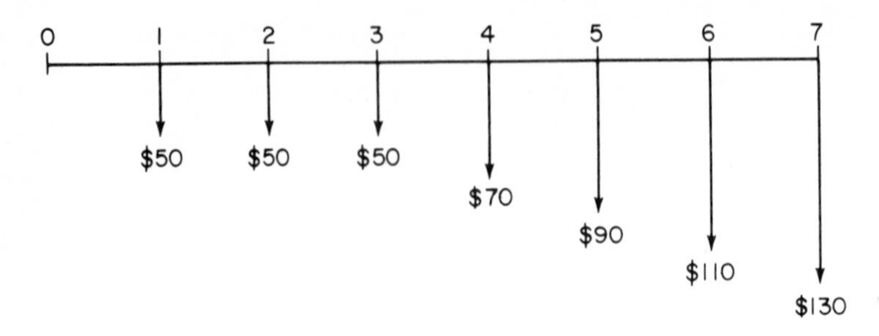

Figure 4.16 Diagram illustrating a shifted gradient.

Example 4.10 Compute the equivalent annual series for the payments of Fig. 4.16.

SOLUTION The solution steps are:

1. Consider the $50 base amount as an annual cost for all 7 years (Fig. 4.17).
2. Find the present worth of the gradient P_G that occurs in year 2 as shown by the gradient-year time scale.

$$P_G = 20(P/G, i\%, 5)$$

3. Bring the gradient present worth back to actual year 0.

$$P_1 = P_G(P/F, i\%, 2)$$

4. Annualize the gradient present worth from year 0 through year n.

$$A = P_1(A/P, i\%, 7)$$

5. Finally, add the remaining annual costs to the gradient annual cost. Here the base amount is $50 for all 7 years and the annual series is

$$A = 20(P/G, i\%, 5)(P/F, i\%, 2)(A/P, i\%, 7) + 50$$

Solved Problems 4.15 and 4.16
Probs. P4.35 to P4.49

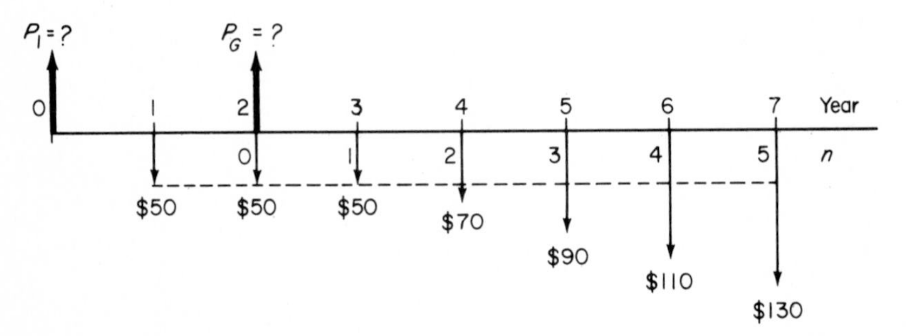

Figure 4.17 Completed diagram, Example 4.10.

4.6 Decreasing Gradients

Example 2.1 involved the use of a decreasing uniform gradient, $G = \$7500$ per year. The use of the gradient factors is the same for increasing and decreasing gradients, except that in the case of decreasing gradients the following are true:

1. The base amount is equal to the *largest* amount attained in the gradient series.
2. The gradient has a negative value; thus, the term $-G(A/G, i\%, n)$ or $-G(P/G, i\%, n)$ must be used in the computations.

The present worth of the gradient will still take place 2 years before the gradient starts and the A value will start at year 1 and continue through year n.

Example 4.11 Find the (*a*) present worth and (*b*) annual series of the receipts shown in Fig. 4.18 for $i = 7\%$ per year.

SOLUTION

(*a*) The cash flows of Fig. 4.18 may be separated as in Fig. 4.19. The dashed line in Fig. 4.19*a* indicates that the gradient is subtracted from an annual receipt of \$900. The present worth is computed as

$$
\begin{aligned}
P_T = P_A - P_G &= 900(P/A, 7\%, 6) - 100(P/G, 7\%, 6) \\
&= 900(4.7665) - 100(10.978) \\
&= \$3192.05
\end{aligned}
$$

(*b*) The annual series is made up of two components: the base amount and the equivalent uniform-gradient amount. The annual-receipt series ($A_1 = \$900$) is the base amount, and the annual series A_G, which is equivalent to the gradient, is subtracted from A_1.

$$
\begin{aligned}
A = A_1 - A_G &= 900 - 100(A/G, 7\%, 6) \\
&= 900 - 100(2.303) \\
&= \$669.70 \text{ per year for years 1 to 6}
\end{aligned}
$$

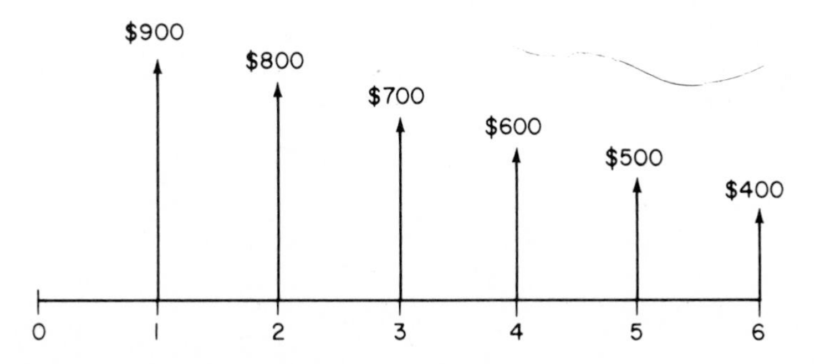

Figure 4.18 Diagram including a decreasing gradient.

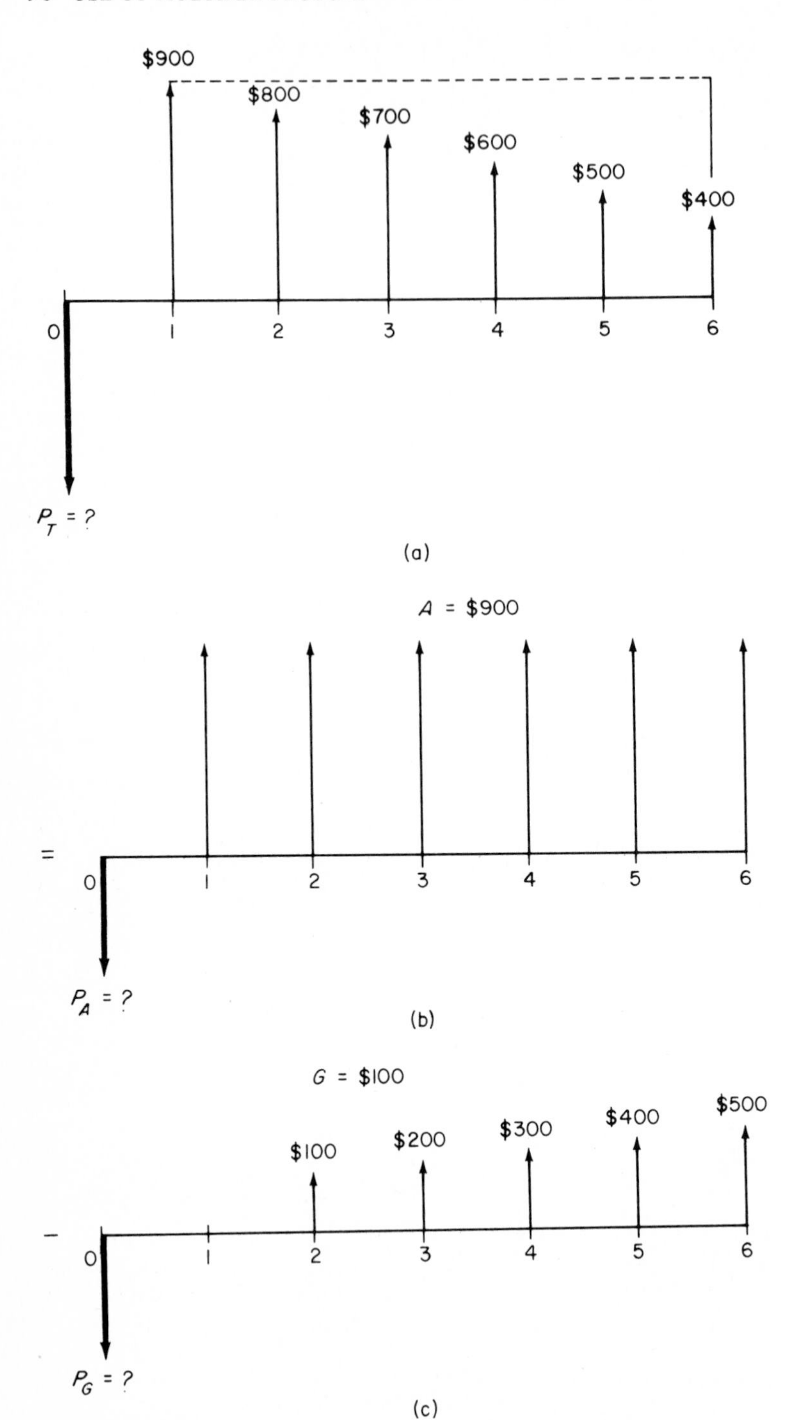

Figure 4.19 Partitioned cash flow of Fig. 4.18: (*a*) = (*b*) − (*c*).

COMMENT Shifted decreasing gradients are handled in a fashion similar to shifted increasing gradients. For an example that combines conventional increasing and shifted decreasing gradients, see the Solved Problems.

Solved Problem 4.17
Probs. P4.50 to P4.56

SOLVED PROBLEMS

Example 4.12 A family decides to buy a new refrigerator on credit. The payment scheme calls for a \$100 down payment now (the month is March) and \$55 a month from June to November with interest at $1\frac{1}{2}\%$ per month compounded monthly. Construct the cash-flow diagram and indicate P in the month in which you can compute an equivalent value using one P/A and one F/P factor. Give the n values for all computations.

SOLUTION Since the payment period, months, equals the compounding period, the interest tables of Appendix A can be used. Figure 4.20 solves the problem by placing P in May. The relation using only the two factors is $P = 100(F/P, 1.5\%, 2) + 55(P/A, 1.5\%, 6)$, where $n = 2$ for the F/P and $n = 6$ for the P/A factor.

COMMENT The control for placement of P must be retained by a uniform series, since the P/A factor is inflexible in the computational procedure for P.

Sec. 4.1

Example 4.13 Consider the two uniform series shown in Fig. 4.21. Compute the present worth at 15% using three different methods.

SOLUTION There are numerous ways to find the present worth. The two simplest are probably the future-worth and present-worth methods. For a third method, the use of the *intermediate-year method* at year 7 is demonstrated.

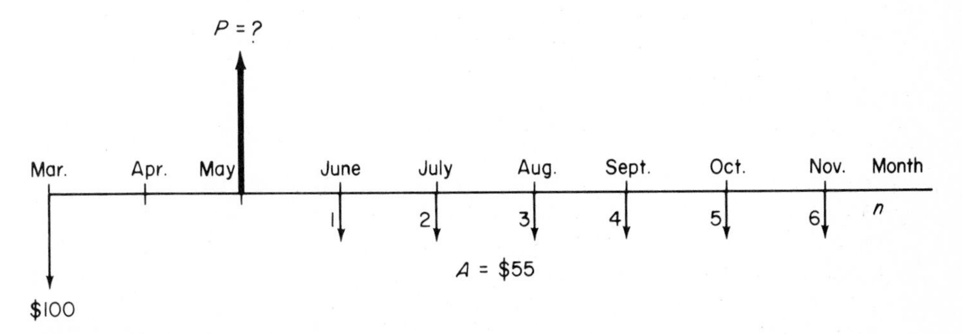

Figure 4.20 Placement of an equivalent amount using only P/A and F/P factors, Example 4.12.

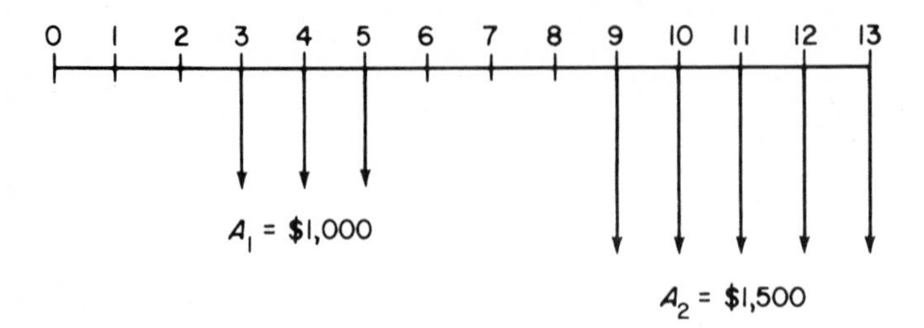

Figure 4.21 Uniform series used to compute a present worth by several methods, Example 4.13.

(*a*) Present-worth method (see Fig. 4.22*a*): The use of P/A factors for the uniform series and P/F factors to obtain the actual present worths allows us to find P_T.

$$P_T = P_{A1} + P_{A2}$$

where
$$\begin{aligned} P_{A1} &= P'_{A1}(P/F, 15\%, 2) = A_1(P/A, 15\%, 3)(P/F, 15\%, 2) \\ &= 1000(2.2832)(0.7561) \\ &= \$1726 \\ P_{A2} &= P'_{A2}(P/F, 15\%, 8) = A_2(P/A, 15\%, 5)(P/F, 15\%, 8) \\ &= 1500(3.3522)(0.3269) \\ &= \$1644 \end{aligned}$$

$$P_T = 1726 + 1644 = \$3370$$

(*b*) Future-worth method (Fig. 4.22*b*): Using the F/A, F/P, and P/F factors, we have

$$P_T = (F_{A1} + F_{A2})(P/F, 15\%, 13)$$

where
$$\begin{aligned} F_{A1} &= F'_{A1}(F/P, 15\%, 8) = A_1(F/A, 15\%, 3)(F/P, 15\%, 8) \\ &= 1000(3.472)(3.0590) \\ &= \$10{,}621 \\ F_{A2} &= A_2(F/A, 15\%, 5) = 1500(6.742) \\ &= \$10{,}113 \end{aligned}$$

Then
$$\begin{aligned} P_T &= 20{,}734(0.1625) \\ &= \$3369 \end{aligned}$$

(*c*) Intermediate-year method (Fig. 4.22*c*): If we find the present worth of both series at year 7 and use the P/F factor, we have

$$P_T = (F_{A1} + P_{A2})(P/F, 15\%, 7)$$

The P_{A2} value is computed as a present worth; but to find the value at year 0, it must be treated as an F value. Thus,

$$\begin{aligned} F_{A1} &= F'_{A1}(F/P, 15\%, 2) = A_1(F/A, 15\%, 3)(F/P, 15\%, 2) \\ &= 1000(3.472)(1.3225) \\ &= \$4592 \end{aligned}$$

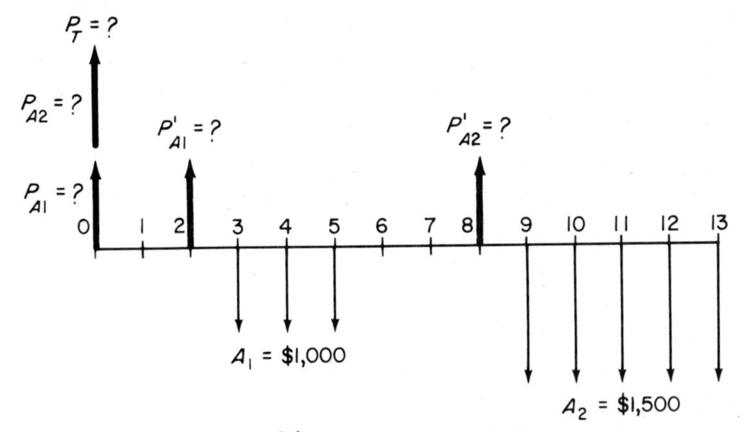

(a) Present worth method

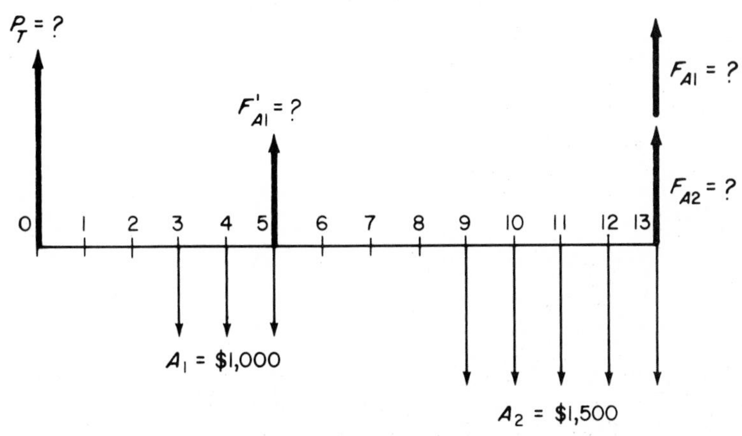

(b) Future worth method

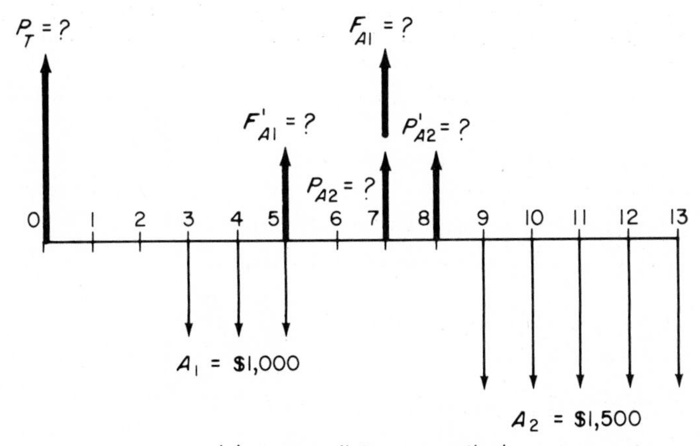

(c) Intermediate year method

Figure 4.22 Computation of the present worth of Fig. 4.21 by three methods.

$$P_{A2} = P'_{A2}(P/F, 15\%, 1) = A_2(P/A, 15\%, 5)(P/F, 15\%, 1)$$
$$= 1500(3.3522)(0.8696)$$
$$= \$4373$$

Then
$$P_T = (F_{A1} + P_{A2})(P/F, 15\%, 7)$$
$$= 8965(0.3759)$$
$$= \$3370$$

Sec. 4.2

Example 4.14 Calculate the present worth of the following series of cash flows if $i = 8\%$.

Year	0	1	2	3	4	5	6	7
Cash flow, $	+460	+460	+460	+460	+460	+460	+460	−5000

SOLUTION The cash-flow diagram is shown in Fig. 4.23. Since the disbursement in year 0 is equal to the disbursements of the A series, the P/A factor can be used for either 6 or 7 years. The problem is worked both ways below.

1. Using P/A and $n = 6$: For this case, the disbursement in year 0 (P_1) is added to the present worth of the remaining payments, since the P/A factor for $n = 6$ will place P_A in year 0. Thus,

$$P = P_1 + P_A - P_F$$
$$= 460 + 460(P/A, 8\%, 6) - 5000(P/F, 8\%, 7)$$
$$= \$-331$$

   Note that the present worth of the $-5000 cash flow ($P_F$) is negative, since it is a negative cash flow.

2. Using P/A and $n = 7$: By using the P/A factor for $n = 7$, the "present worth" is located in year − 1, not year 0, because the P must always be one year ahead

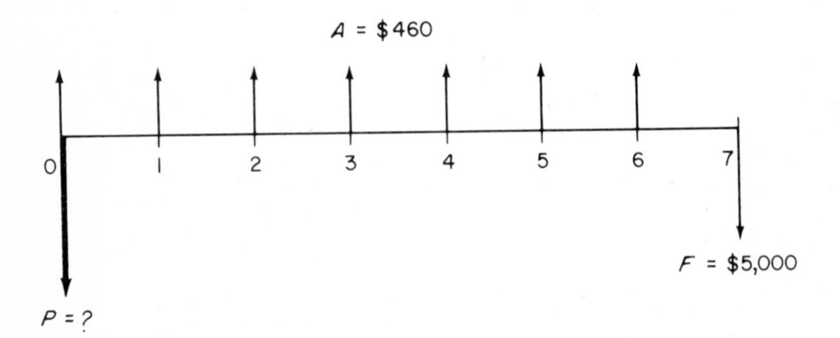

Figure 4.23 Cash-flow diagram, Example 4.14.

of the first A when the P/A factor is used. It is therefore necessary to move the P_A value one year forward with the F/P factor. Thus,

$$P = 460(P/A, 8\%, 7)(F/P, 8\%, 1) - 5000(P/F, 8\%, 7)$$
$$= \$-331$$

COMMENT Rework the problem by first finding the future worth of the series and then solving for P. You should obtain the same answer as above, if you work it correctly.

Sec. 4.3

Example 4.15 Determine the amount of the gradients, the location of the present worth of the gradients, and the n values of the cash flow of Fig. 4.24.

SOLUTION You should construct your own cash-flow diagram upon which you locate the gradient present worths and determine the n values. If we call the series from year 1 to year 4 G_1, the base amount is \$25, G_1 is \$15, n_1 equals 4 years, and P_{G1} occurs in year 0. For the second series, the base amount is also \$25, but G_2 is \$5, n_2 equals 7 years, and P_{G2} takes place in year 5.

COMMENT Even though Fig. 4.24 shows a series of disbursements and receipts, both gradients are increasing. Decreasing gradients are discussed in Sec. 4.6.

Sec. 4.5

Example 4.16 Using $i = 8\%$ for the cash flows of Fig. 4.25, compute (*a*) the equivalent annual series and (*b*) present worth.

SOLUTION

(*a*) The dashed lines of Fig. 4.25 should help you in the solution for present worth and equivalent annual series. For the annual series, use the steps outlined in Example 4.10.

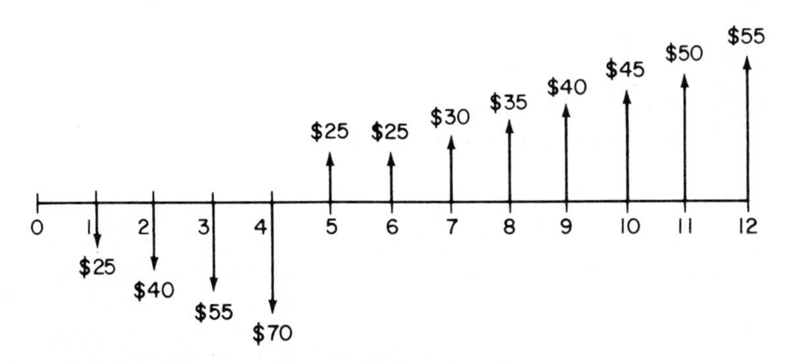

Figure 4.24 Cash flow of two gradients, Example 4.15.

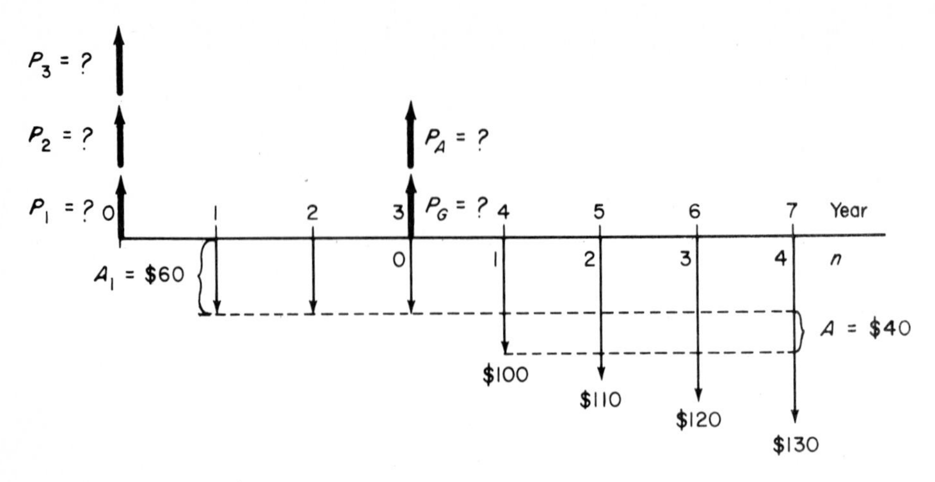

Figure 4.25 Shifted gradient, Example 4.16.

1. A_1 = \$60 for 7 years
   A = \$40 base amount of gradient for 4 years
   A_3 = equivalent series of base amount for 7 years
   $= P_2(A/P, 8\%, 7)$

   where P_2 = present worth of A = \$40 series
   $= 40(P/A, 8\%, 4)(P/F, 8\%, 3)$
   $= \$105.17$

   Then, $A_3 = 105.17(A/P, 8\%, 7)$
   $= \$20.20$
2. P_G = present worth of gradient in year 3
   $= G(P/G, 8\%, 4) = 10(4.650)$
   $= \$46.50$
3. P_1 = present worth of P_G in year 0
   $= P_G(P/F, 8\%, 3) = 46.50(0.7938)$
   $= \$36.91$
4. A_2 = equivalent 7 year A value of gradient
   $= P_1(A/P, 8\%, 7) = 36.91(0.19207)$
   $= \$7.09$
5. The equivalent annual series is

$$A = A_1 + A_2 + A_3 = 60.00 + 7.09 + 20.20$$
$$= \$87.29$$

(*b*) To find the present worth of the cash flows shown in Fig. 4.25, note that the present worth P_1 of the gradient is the same as that calculated in step 3 above. Then the \$40 series has a present worth P_2 of

$$P_2 = 40(P/A, 8\%, 4)(P/F, 8\%, 3) = 40(3.3121)(0.7938)$$
$$= \$105.17$$

The \$60 annual series has a present worth P_3 of

$$P_3 = 60(P/A, 8\%, 7) = 60(5.2064)$$
$$= \$312.38$$

The total present worth P_T is

$$P_T = P_1 + P_2 + P_3 = 36.91 + 105.17 + 312.38$$
$$= \$454.46$$

which is equivalent to \$87.29 per year, as in part *a*.

Sec. 4.5

Example 4.17 Assume that you are planning to invest money at 7% per year shown by the increasing gradient of Fig. 4.26. In addition, you plan to withdraw according to the decreasing gradient shown. Find the present worth and equivalent annual series for the investment and withdrawal sequence.

SOLUTION For the investment sequence, G is \$500, the base amount is \$2000, and n equals 5; for the withdrawal sequence, G is \$-1000, the base amount is \$5000, and n equals 5; there is also a 2-year annual series with A equal to \$1000 in years 11 and 12. For the investment sequence,

$$P_I = \text{present worth of investments}$$
$$= 2000(P/A, 7\%, 5) + 500(P/G, 7\%, 5)$$
$$= 2000(4.1002) + 500(7.646)$$
$$= \$12{,}023.40$$

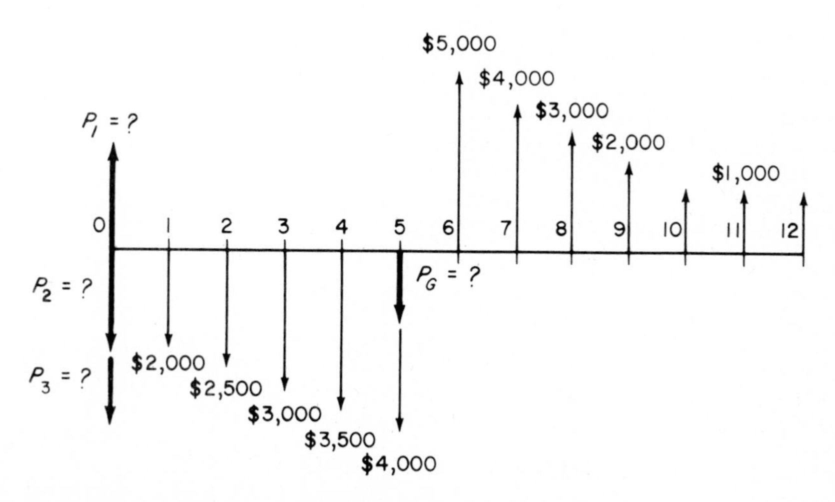

Figure 4.26 An investment and withdrawal sequence, Example 4.17.

For the withdrawal sequence,

$$P_W = \text{present worth of withdrawal gradient}$$
$$+ \text{present worth of withdrawals in years 11 and 12}$$
$$= P_2 + P_3 = P_G(P/F, 7\%, 5) + P_3$$
$$= [5000(P/A, 7\%, 5) - 1000(P/G, 7\%, 5)](P/F, 7\%, 5)$$
$$+ 1000(P/A, 7\%, 2)(P/F, 7\%, 10)$$
$$= [5000(4.1002) - 1000(7.646)](0.7130) + 1000(1.8080)(0.5084)$$
$$= \$10{,}084.80$$

Since P_I is actually a negative cash flow and P_W positive, the resultant present worth P is

$$P = P_W - P_I = 10{,}084.80 - 12{,}023.40$$
$$= \$-1938.60$$

The A value may be computed using

$$A = P(A/P, 7\%, 12)$$
$$= \$-244.07$$

COMMENT Thus, in present-worth equivalent, you will invest \$1938.60 more than you plan to withdraw. This is equivalent to an annual savings of \$244.07 per year for the 12 years.

Sec. 4.6

PROBLEMS

P4.1 Construct a net cash-flow diagram for each of the following and locate the present worth and future worth of each uniform series *separately*.

(*a*) A \$600 deposit for 7 years starting 4 years from now.

(*b*) A \$2000-per-month deposit for 6 months starting next month and a \$1500-per-month withdrawal for 4 months starting 3 months from now.

(*c*) The following transaction in your Christmas club account for the past 12 months:

Month	Deposit	Withdrawal	Month	Deposit	Withdrawal
January	\$20	\$ 0	July	\$75	\$ 25
February	20	0	August	75	25
March	20	10	September	25	10
April	20	10	October	25	10
May	75	10	November	25	10
June	75	25	December	0	250

P4.2 Determine the amount of money a person must deposit now in order to be able to make ten \$3600-per-year withdrawals starting 20 years from now if the interest rate is 14% per year.

P4.3 What is the equivalent uniform annual cost in years 1 through 16 of \$250 per year for 12 years, with the first payment starting 5 years from now, if the interest rate is 20% per year?

P4.4 A couple purchases an insurance policy which they plan to use to finance their child's college education. If the policy provides \$10,000 twelve years from now, how much can be withdrawn every year for 5 years if the child starts college 15 years from now? Assume $i = 16\%$ per year.

P4.5 A woman deposited \$700 per year for 8 years. Starting in the ninth year she increased her deposits to \$1200 per year for 5 more years. How much money did she have in her account immediately after she made her last deposit if the interest rate was 15% per year?

P4.6 How much money will the woman in Prob. P4.5 have in her account 30 years from the present time if she makes no deposits after the one in year 13?

P4.7 What is the present worth one year prior to the first deposit for the investment specified in Prob. P4.5?

P4.8 A man plans to begin saving for his retirement such that he will be able to withdraw money every year for 30 years starting 25 years from now. He estimates that he will be able to start saving money one year from now and plans to deposit \$500 per year. What uniform annual amount will he be able to withdraw when he retires, if the interest rate is 12% per year?

P4.9 A businessman purchased a used building and found that the ceiling was poorly insulated. He estimated that with 6 inches of foam insulation, he could cut the heating bill by \$25 per month and the air conditioning cost by \$20 per month. Assuming that the winter season is the first 6 months of the year and the summer season is the next 6 months, how much can he afford to spend on insulation if he expects to keep the building for only 2 years? Assume $i = 1\frac{1}{2}\%$ per month.

P4.10 A couple plan to make an investment now in order to finance their child's college education. If the child should be able to withdraw \$1000 per year in years 15 through 20 such that the last withdrawal will close the account, how much must be invested now if interest is computed at 9% per year?

P4.11 If the couple in Prob. P4.10 wanted to make a uniform deposit for 14 years instead of the lump-sum investment, how much would they have to deposit every year starting one year from now?

P4.12 How much money would the heirs of the man in Prob. P4.8 receive if he died 4 years after his retirement?

P4.13 How much money would the woman in Prob. P4.5 have if the interest rate increased from 15 to 16% after 5 years?

P4.14 Giovanni's Pizza Palace has a 10-year lease on 200 square meters of space in an enclosed shopping center. The rent is paid yearly at a rate of \$100 per square meter. At the end of the fourth year of the lease, the owner of the pizza shop decides to purchase a building and relocate the business. How much must the owner of the shopping center be paid for the remainder of the lease if the interest rate is 12% per year?

P4.15 In order to compare leasing versus purchasing a microprocessor, an engineer was told to convert beginning-of-period payments (for leasing) to end-of-period payments. The beginning-of-period amount was \$1000 per month, and the computer was to be leased for 3 years. If the company's nominal interest rate is 18% per year compounded monthly, what is the equivalent end-of-period amount?

P4.16 Determine the beginning-of-year payments which would be equivalent to the cash-flow shown below. Use an interest rate of 15% per year.

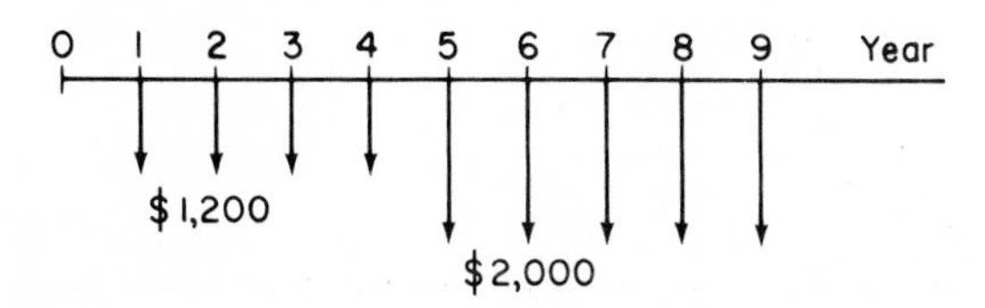

P4.17 Calculate the value of x in the cash-flow shown below such that the equivalent total value in month 7 is \$8000, using an interest rate of $1\frac{1}{2}\%$ per month.

Month	Cash flow
0	200
1	200
2	200
3	200
4	200
5	x
6	x
7	x
8	x
9	500
10	500
11	500

P4.18 If a couple opens a savings account by depositing \$1500 now and deposits \$1500 every year for 14 years, how much will be in the account after the last deposit if the interest rate is 12% per year?

P4.19 How much will be in the account in Prob. P4.18 if the interest rate changes to 14% per year after the first 5 years?

P4.20 Find the value of x such that the positive cash flows will be exactly equivalent to the negative cash flows if the interest rate is 15% per year.

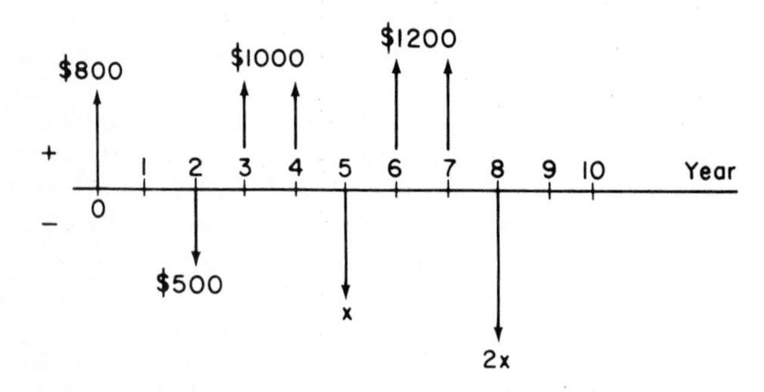

P4.21 Find the value of x in the diagram below that would make the equivalent present worth of the cash flow equal to \$22,000, if the interest rate is 15% per year.

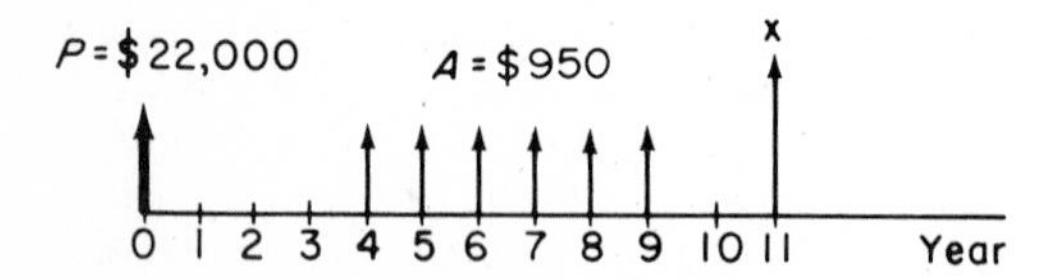

P4.22 Calculate the amount of money in year 7 that would be equivalent to the following cash flows if the interest rate is a nominal 13% per year compounded semiannually.

Year	0	1	2	3	4	5	6	7	8	9
Amount, $	900	900	900	900	1300	1300	1300	0	900	900

P4.23 The Gro Company purchased a machine for $12,000 with an expected salvage value of $2000. Operating expenses for the machine will be $1800 per year. In addition, a major overhaul will be required every 5 years at a cost of $2800. What is the equivalent present cost of the machine if it will have an 18-year life and the interest rate is 15% per year?

P4.24 A petroleum company is planning to sell a number of existing oil wells. The wells are expected to produce 100,000 barrels of oil per year for 11 more years. If the selling price per barrel of oil is currently $35, how much would you be willing to pay for the wells if the price of oil is expected to increase by $3 per barrel every 3 years, with the first increase to occur 2 years from now? Assume that the interest rate is 12% per year for the first 4 years and 15% per year thereafter, and that oil sales are made at the end of each year.

P4.25 Calculate the (*a*) present worth and (*b*) future worth in year 10 of the following series of disbursements:

Year	Disbursement	Year	Disbursement
0	$3500	6	$ 5000
1	3500	7	5000
2	3500	8	5000
3	3500	9	5000
4	5000	10	15,000
5	5000		

Assume that i = 16% per year compounded semiannually.

P4.26 A building contractor purchased a dirt scraper for $35,000. He maintained the scraper at a cost of $2500 per year. He overhauled the machine 4 years after the purchase at a cost of $4000. Two years after the overhaul, he sold the scraper for $18,000. What was his equivalent uniform annual cost if the interest rate was 10% per year?

P4.27 How much money would you have to deposit for 6 consecutive years starting one year from now if you want to be able to withdraw $45,000 eleven years from now? Assume that the interest rate is 15% per year.

P4.28 What is the equivalent uniform annual cost of the disbursements shown in Prob. P4.25?

P4.29 A large manufacturing company purchased a semiautomatic machine for $13,000. Its annual maintenance and operation cost was $1700. Five years after the initial purchase, the company decided to purchase an additional unit for the machine which would make it fully automatic. The additional unit had a first cost of $7100. The cost for operating the machine in the fully automatic condition was $900 per year. If the company used the machine for a total of 16 years and then sold the automatic addition for $1800, what was the equivalent uniform annual cost of the machine at an interest rate of 9%?

P4.30 Calculate the number of $15,000 payments that would be required in the cash-flow diagram shown below in order for the annual payments to be equivalent to the initial $37,000 saving. Use an interest rate of an effective 17% per year compounded monthly.

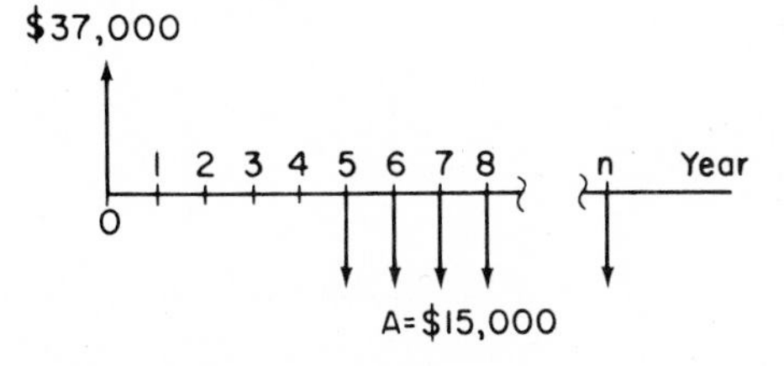

P4.31 A company borrows $8000 at an interest rate of a nominal 12% per year compounded monthly. The company desires to repay the loan in 14 equal monthly payments, with the first payment starting one month from now.

(*a*) What should be the size of each payment?

(*b*) If after making eight payments the company decides to pay off the balance of the loan in the ninth month, how much must the company pay?

P4.32 If you want to have $125,000 for your son's college education 21 years from now, how much will you have to deposit each year if your first deposit is now and the last deposit is 18 years from now? Assume that the interest rate is 15% per year.

P4.33 A woman plans to make a total of eight deposits, with the first deposit now and succeeding deposits at one-year intervals, so that she will be able to withdraw $4000 per year for 10 years, the first withdrawal starting 16 years from now. How much must she deposit each year if the interest rate is a nominal 12% per year compounded quarterly?

P4.34 An elderly man starts a retirement account by depositing $10,000 now and $300 each month for 10 years. How much money can he withdraw per month for 5 years if he makes his first withdrawal 3 months after his last deposit? Assume that the interest rate is a nominal 18% per year compounded monthly.

P4.35 A couple expect to borrow $500 each year for the next 2 years to cover Christmas expenses. Due to increasing costs, the couple expects to have to borrow $550 three years from now, $600 the next year, and $650 the following year. However, due to their children's ages, they hope to have to borrow only $300 per year after that. Calculate (*a*) the present worth and (*b*) the equivalent uniform annual cost of the disbursements for a total of 15 years using an interest rate of 13% per year.

P4.36 The UR-OK Company is considering two types of machines, both of which will do the same job. The net cash flows for each are tabulated below. Calculate the present worth of each machine using an interest rate of 18% per year.

	Cash flows			Cash flows	
Year	Machine *A*	Machine *B*	Year	Machine *A*	Machine *B*
0	$+2000	$+2000	7	$+3000	$+2500
1	2000	2000	8	3500	2500
2	2000	2500	9	4000	2500
3	2500	3000	10	4500	3000
4	2500	3500	11	3000	3500
5	2500	4000	12	3000	4000
6	2500	2500			

P4.37 A person borrows $8000 at a nominal 7% per year compounded quarterly. It is desired to repay the loan in 12 semiannual payments, with the first payment to start 3 months from now. If the payments are to increase by $50 each time, determine the size of the first payment.

P4.38 Find the value of G in the diagram below that would make the income stream equivalent to the disbursement stream, using an interest rate of 20% per year.

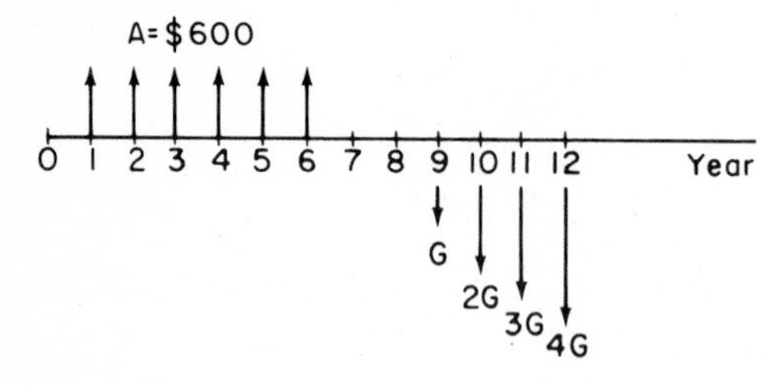

P4.39 For the diagram shown below, find the value of the last receipt in the income stream that would make the receipts equivalent to the $500 initial investment at time 0. Use an interest rate of 15% per year.

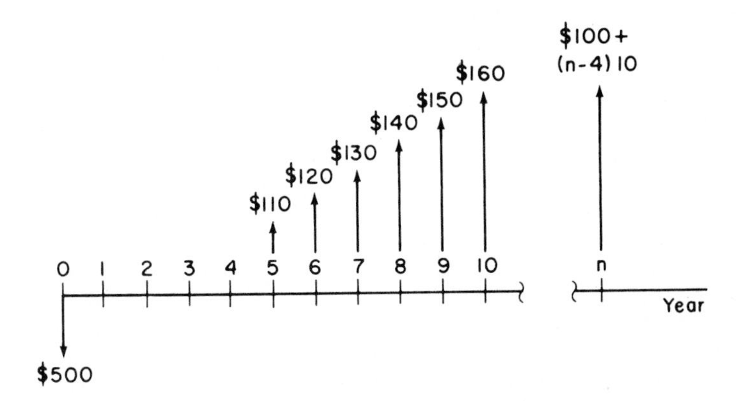

P4.40 Calculate the value of x for the cash-flow series shown below such that the equivalent total value in month 5 is $25,000, using an interest rate of a nominal 12% per year compounded monthly.

Month	Cash flow
0	100
1	$100 + x$
2	$100 + 2x$
3	$100 + 3x$
4	$100 + 4x$
5	$100 + 5x$
6	$100 + 6x$
7	$100 + 7x$
8	$100 + 8x$
9	$100 + 9x$
10	$100 + 10x$
11	$100 + 11x$
12	$100 + 12x$
13	$100 + 13x$
14	$100 + 14x$

P4.41 Solve for the value of G such that the left cash-flow diagram is equivalent to the one on the right. Use an interest rate of 13% per year.

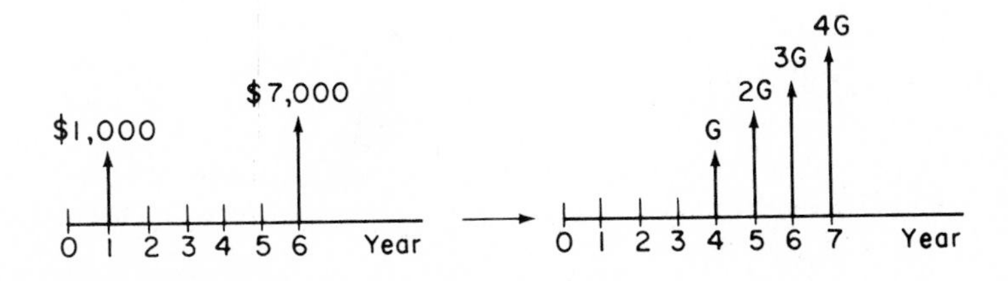

P4.42 Solve for the value of x, using an interest rate of 12% per year:

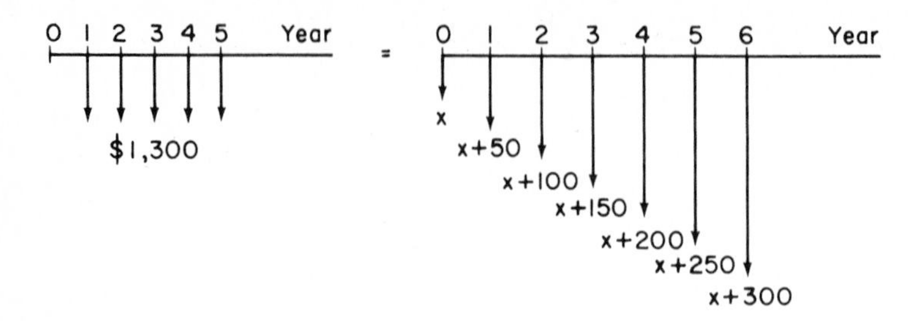

P4.43 If a machine costs \$15,000 to purchase and the operating costs are \$1000 at the end of the first year, \$1200 at the end of the second, and amounts increasing by \$200 per year through year 12, what is the present worth of the machine if the interest rate is a nominal 15% per year compounded semiannually?

P4.44 If the operating costs in Prob. P4.43 start now instead of at the end of the first year, what will the present worth be at time 0 for the 12 years (i.e., 13 payments)?

P4.45 For the diagram shown below, find the value of x that will make the negative cash flows equal to the positive cash flow of \$800 at time 0. Assume $i = 15\%$ per year.

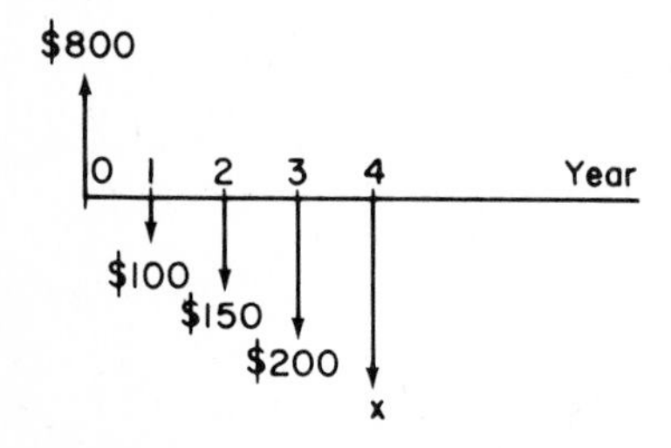

P4.46 Rework Prob. P4.45 above, using an interest rate of a nominal 15% per year compounded quarterly.

P4.47 Find the future worth (in month 9) of the cash flow shown in the figure below, using an interest rate of 1% per month.

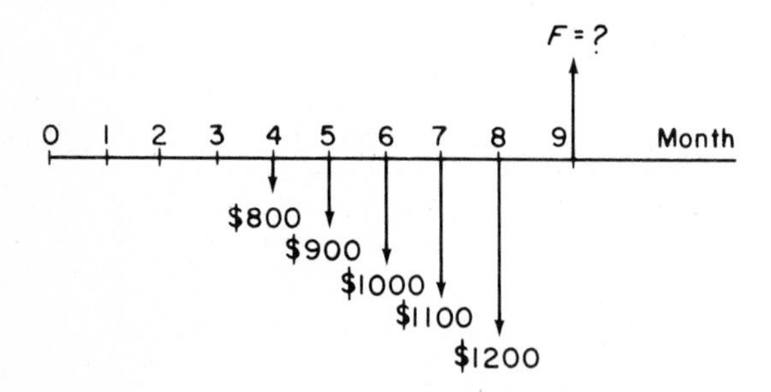

P4.48 Mr. Alum Nye is planning to make a contribution to the junior college from which he graduated. He would like to donate an amount of money now so that the college can support students.

Specifically, he would like to provide financial support for tuition for five students per year for a total of 20 years (i.e., 21 grants), with the first tuition grant to be made immediately and continuing at one-year intervals. The cost of tuition at the school is $3800 per year and is expected to stay at that amount for 4 more years. After that time, however, the tuition cost will increase by $90 per year. If the school can deposit the donation and earn interest at a rate of a nominal 8% per year compounded semiannually, how much must Mr. Alum Nye donate?

P4.49 Find the present worth of the cash flow shown below using an interest rate of 16% per year.

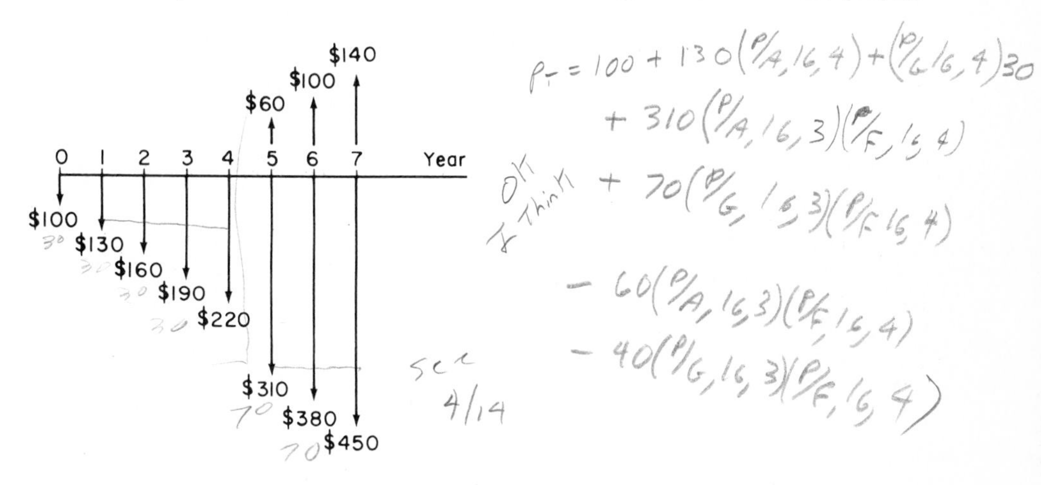

P4.50 Find the present worth (time 0) of the cash flow shown in the figure below. Assume i = 12% per year.

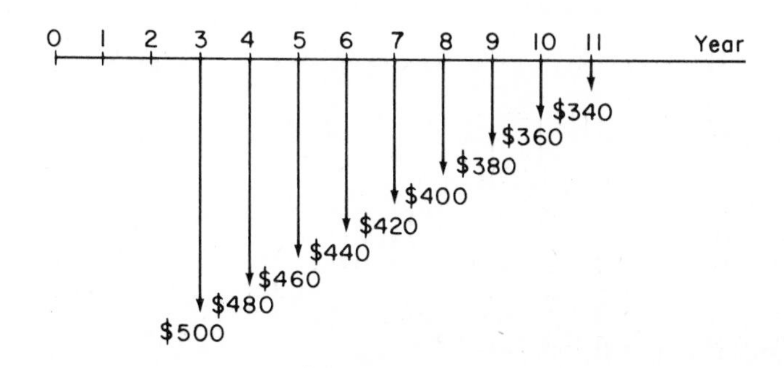

P4.51 If you start a bank account now by depositing $2000, how long will it take for you to deplete the account if you start withdrawing money $1\frac{1}{2}$ years from now by withdrawing $500 the first month, $450 the second month, $400 the next month, and amounts decreasing by $50 per month until the account is depleted? Assume that the account earns interest at a rate of a nominal 12% per year compounded monthly.

P4.52 Compute the present worth of the following cash flows at i = 12%.

Year	0	1–4	5	6	7	8	9	10
Amount, $	5000	1000	900	800	700	600	500	400

P4.53 Compute the present worth and equivalent annual cost series at $i = 10\%$ per year for the cash flows below.

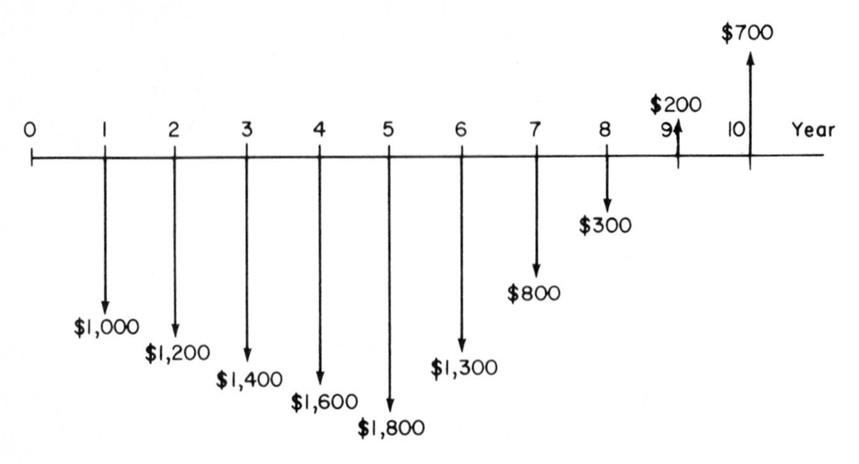

P4.54 Find the present worth of the cash flows shown below at 20% per year interest.

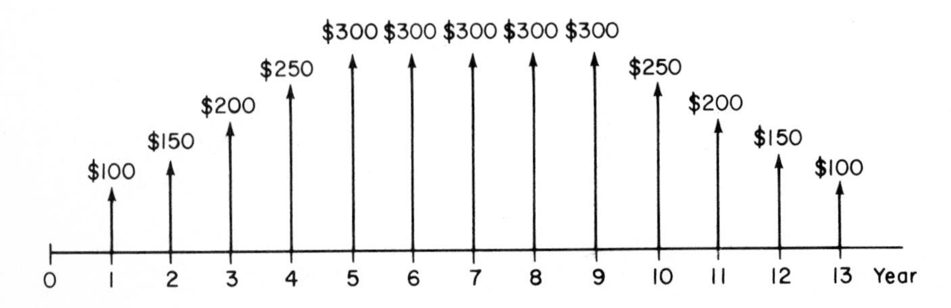

P4.55 The Chisel-Em Company plans to purchase a new piece of construction equipment now. Realizable income is $15,000 the first year, $12,000 in year 2, $9000 in year 3, and so on. If the company plans to sell the equipment after 7 years and interest is 15% per year, compute the present worth and equivalent annual series of the incomes.

P4.56 Compute the present worth and future worth of the cash flows shown below if $i = 16\%$ per year.

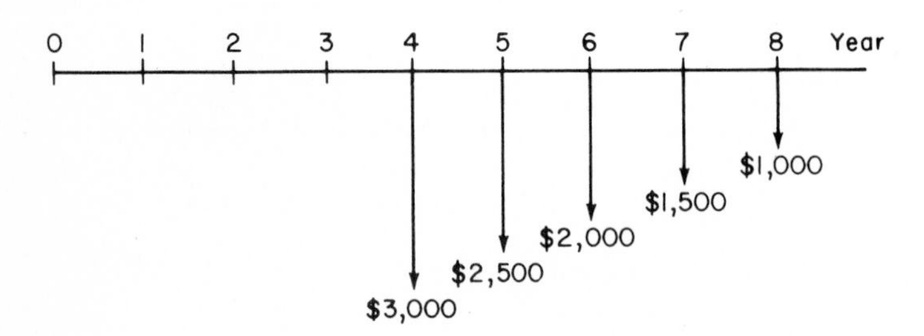

LEVEL TWO

At this point you have most of the computational skills necessary to perform an engineering-economy analysis. The chapters in this level give you a chance to learn and use the three most basic methods of alternative evaluation. All methods must give identical decisions about which alternative is best when the methods are used to compare the same alternatives.

Chapter	Subject
5	Present-worth and capitalized-cost evaluation
6	Equivalent uniform annual cost evaluation
7	Rate of return for one project
8	Rate-of-return evaluation

CHAPTER

FIVE

PRESENT-WORTH AND CAPITALIZED-COST EVALUATION

The objective of this chapter is to teach you how to compare alternatives on the basis of present-worth or capitalized-cost determinations. As you will see in later chapters, the present-worth method is probably the most versatile of all alternative evaluation procedures.

SECTION OBJECTIVES

In order to complete this chapter, you must be able to do the following:

5.1. Select the better of two alternatives using present-worth calculations for alternatives that have equal lives, given the costs and their respective dates, the lives and salvage values of the alternatives, and the interest rate.
5.2. Select as in objective 5.1, except for alternatives that have different lives.
5.3. Define *capitalized cost* and calculate the capitalized cost of a series of disbursements, given the disbursements, their respective dates, and the interest rate.
5.4. Select the better of two alternatives on the basis of capitalized cost, given the disbursements, dates, lives, and salvage values for each alternative, and the interest rate.

STUDY GUIDE

5.1 Present-Worth Comparison of Equal-Lived Alternatives

The present-worth (PW) method of alternative evaluation is very popular because future expenditures or receipts are transformed into *equivalent dollars now*. In this form, it is very easy, even for a person unfamiliar with economic analysis, to see the economic advantage of one alternative over one or more other alternatives.

Since the present worth is always less than the future worth (in terms of dollars, not equivalently) of a disbursement or receipt when the interest rate is greater than zero, the present-worth amount is known as the *discounted cash flow*. Similarly, the interest rate used in present-worth calculations is sometimes referred to as the *discount rate*, particularly in financial institutions. These terms will not be used in this text except when they more appropriately reflect everyday usage.

The comparison of alternatives having equal lives by the present-worth method is straightforward. If both alternatives are used in identical capacities, they are termed *equal-service* alternatives and the annual incomes have the same numerical value. Therefore, the cash flow involves disbursements only, in which case it is generally convenient to omit the minus sign from the disbursements. Then the alternative with the *lowest* present-worth value should be selected. On the other hand, when disbursements *and* incomes must be considered, if the above sign convention is used, the alternative selected will be the one with the *highest* present-worth value *provided that incomes exceed disbursements*, and vice versa. While it does not matter which sign is used for disbursements, it is important to be consistent in assigning the proper sign to each cash-flow element. For single-project evaluation, $PW < 0$ indicates a net loss at a certain rate of return and $PW > 0$ implies a net gain greater than the stated rate of return.

It should be pointed out that a present-worth analysis can be conducted when multiple alternatives are under consideration, using the same procedures presented here for two alternatives, which is one of the advantages of this method over the rate-of-return method, as you will see in Chaps. 7 and 8.

Example 5.1 Make a present-worth comparison of the equal-service machines for which the costs are shown below, if $i = 10\%$.

	Type *A*	Type *B*
First cost, *P*	\$2500	\$3500
Annual operating cost, AOC	900	700
Salvage value, SV	200	350
Life, years	5	5

SOLUTION The cash-flow diagram is left to the reader. The present worth of each machine is calculated as follows:

$$P_A = 2500 + 900(P/A, 10\%, 5) - 200(P/F, 10\%, 5) = \$5788$$

$$P_B = 3500 + 700(P/A, 10\%, 5) - 350(P/F, 10\%, 5) = \$5936$$

Type *A* should be selected, since $P_A < P_B$.

COMMENT Note the minus sign on the salvage value, since it is a negative cost. Also, when alternatives are evaluated by the present-worth method it is common to use PW rather than P. In this case, then, PW_A = \$5788 and PW_B = \$5936.

Solved Problem 5.5
Probs. P5.1 to P5.8

5.2 Present-Worth Comparison of Different-Lived Alternatives

When the present-worth method is used for comparing alternatives that have different lives, the procedure of the previous section is followed with this exception: *The alternatives must be compared over the same number of years.* That is, the cash flow for one "cycle" of an alternative must be duplicated for the least common multiple of years, so that service is compared over the same total life for each alternative. For example, if it is desired to compare alternatives which have lives of 3 years and 2 years, respectively, the alternatives must be compared over a period of 6 years, with reinvestment assumed at the end of each life cycle. It is important to remember that when an alternative has a terminal salvage value, this must also be included and shown as an income on the cash-flow diagram at the time reinvestment is made. Example 5.2 is a present-worth comparison of alternatives having different lives.

Example 5.2 A plant superintendent is trying to decide between the machines detailed below.

	Machine A	Machine B
First cost	\$11,000	\$18,000
Annual operating cost	3,500	3,100
Salvage value	1,000	2,000
Life, years	6	9

Determine which one should be selected on the basis of a present-worth comparison using an interest rate of 15%.

SOLUTION Since the machines have different lives, they must be compared over their least common multiple of years, which is 18 years in this case. The cash-flow diagram is shown in Fig. 5.1. Thus,

$$\begin{aligned} PW_A &= 11{,}000 + 11{,}000(P/F, 15\%, 6) - 1000(P/F, 15\%, 6) \\ &\quad + 11{,}000(P/F, 15\%, 12) - 1000(P/F, 15\%, 12) \\ &\quad - 1000(P/F, 15\%, 18) + 3500(P/A, 15\%, 18) \\ &= \$38{,}559 \end{aligned}$$

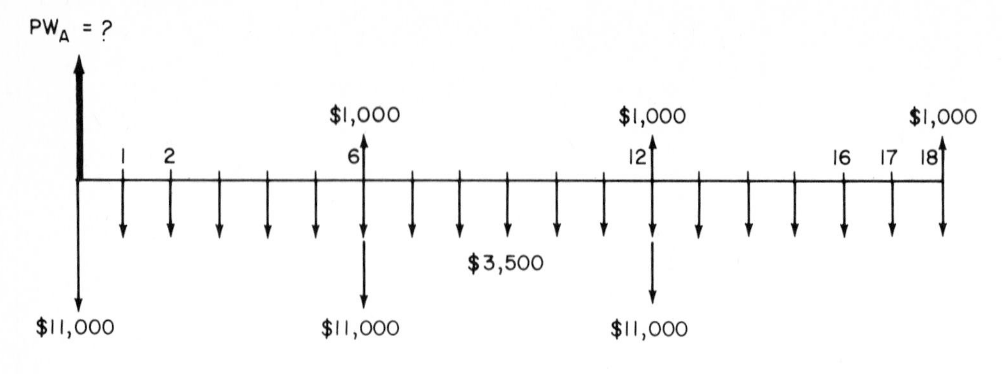

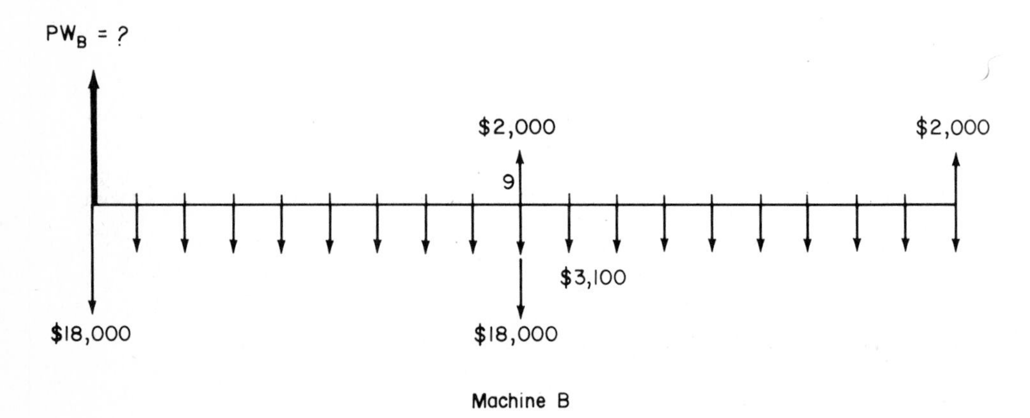

Figure 5.1 Cash-flow diagram for unequal-life assets, Example 5.2.

$$\begin{aligned}\text{PW}_B &= 18{,}000 + 18{,}000(P/F, 15\%, 9) - 2000(P/F, 15\%, 9) \\ &\quad - 2000(P/F, 15\%, 18) + 3100(P/A, 15\%, 18) \\ &= \$41{,}384\end{aligned}$$

Machine A should be selected, since $\text{PW}_A < \text{PW}_B$.

COMMENT Note that the salvage value of each machine must be recovered *after each life cycle* of the asset. In this example, the salvage value of machine A was recovered in years 6, 12, and 18, and that for machine B was recovered in years 9 and 18.

Solved Problems 5.6 and 5.7
Probs. P5.9 to P5.21

5.3 Capitalized-Cost Calculations

Capitalized cost refers to the present-worth value of a project that is assumed to last forever. Certain public works projects such as dams and irrigation systems fall into this category. In addition, permanent university or charitable organization endowments must be handled by capitalized-cost methods.

In general, the procedure that should be followed in calculating the capitalized cost or initial cost of a permanent endowment is as follows:

1. Draw a cash-flow diagram showing all nonrecurring (one-time) expenditures or receipts and at least two cycles of all recurring (periodic) expenditures or receipts.
2. Find the present worth of all nonrecurring expenditures (receipts).
3. Find the equivalent uniform annual cost (EUAC) through one cycle of all recurring expenditures and uniform annual cost series.
4. Divide the EUAC obtained in step 3 by the interest rate to get the capitalized cost of the EUAC.
5. Add the value obtained in step 2 to the value obtained in step 4.

The purpose for beginning the solution by drawing a cash-flow diagram should be evident from previous chapters. However, the cash-flow diagram is probably more important in this calculation than it is anywhere else, because it facilitates the differentiation between nonrecurring and periodic expenditures. In step 2, the present worth of all nonrecurring expenditures (receipts) should be determined. Since the capitalized cost is the *present worth* of a perpetual project, the reason for this step should be obvious. In step 3 the EUAC (which has been called A thus far) of all recurring and uniform annual expenditures should be calculated. This is done to compute the present worth of a perpetual annual cost (capitalized cost) using

$$\text{Capitalized cost} = \frac{\text{EUAC}}{i} \tag{5.1}$$

This can be illustrated by considering the time value of money. If \$100 is deposited into a savings account at 6% interest compounded annually, the maximum amount of money that can be withdrawn at the end of every year for *eternity* is \$6, or the amount equal to the interest that accumulated in that year. This leaves the original \$100 deposit to earn interest so that another \$6 will be accumulated in the next year. Mathematically, the amount of money that can be accumulated and withdrawn each year is

$$A = Pi \tag{5.2}$$

Thus, for the example,

$$A = 100(0.06) = \$6 \text{ per year}$$

The capitalized-cost calculation proposed in step 4 is the reverse of the one just made; that is, Eq. (5.2) is solved for P:

$$P = \frac{A}{i} \tag{5.3}$$

For the example just cited, if it were desired to withdraw \$6 every year for eternity at an interest rate of 6%, from Eq. (5.3),

$$P = \frac{6}{0.06} = \$100$$

After the present worths of all cash flows have been obtained, the total capitalized cost is simply the sum of these present worths. Capitalized-cost calculations are illustrated in Example 5.3.

Example 5.3 Calculate the capitalized cost of a project that has an initial cost of \$150,000 and an additional investment cost of \$50,000 after 10 years. The annual operating cost will be \$5000 for the first 4 years and \$8000 thereafter. In addition, there is expected to be a recurring major rework cost of \$15,000 every 13 years. Assume that $i = 5\%$.

SOLUTION The format outlined above will be used:

1. Draw cash flows for two cycles (Fig. 5.2).
2. Find the present worth (P_1) of the nonrecurring costs of \$150,000 now and \$50,000 in year 10:

$$P_1 = 150{,}000 + 50{,}000(P/F, 5\%, 10) = \$180{,}695$$

3. Convert the recurring cost of \$15,000 every 13 years to an EUAC (A_1) for the first 13 years:

$$A_1 = 15{,}000(A/F, 5\%, 13) = \$847$$

4. The capitalized cost for the annual-cost series can be computed in two ways, which are (*a*) consider a series of \$500 from now to infinity and find the present worth of \$8000 − \$5000 = \$3000 from year 5 on or (*b*) find the present worth of \$5000 for 4 years and the present worth of \$8000 from year

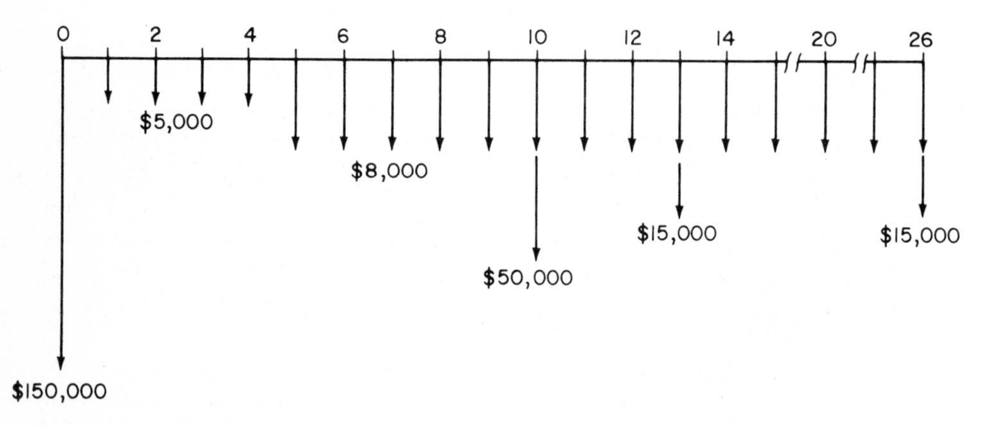

Figure 5.2 Diagram used to compute capitalized cost, Example 5.3.

5 to infinity. Using the first method, we find that the annual cost (A_2) is $5000 and the present worth (P_2) of $3000 from year 5 to infinity, using Eq. (5.3) and the P/F factor, is

$$P_2 = \frac{3000}{0.05}(P/F, 5\%, 4) = \$49{,}362$$

The two annual costs are converted to a capitalized cost (P_3):

$$P_3 = \frac{A_1 + A_2}{i} = \frac{847 + 5000}{0.05} = \$116{,}940$$

5. The total capitalized cost (P_T) can now be obtained by addition:

$$P_T = P_1 + P_2 + P_3 = \$346{,}997$$

COMMENT In calculating P_2, $n = 4$ was used in the P/F factor because the present worth of the annual $3000 cost is computed in year 4, since P is always one year ahead of the first A. You should rework the problem using the second method suggested for calculating P_2.

Probs. P5.22 to P5.27

5.4 Capitalized-Cost Comparison of Two Alternatives

When two or more alternatives are compared on the basis of their capitalized cost, the procedure of Sec. 5.3 is followed. Since the capitalized cost represents the present total cost of financing and maintaining a given alternative forever, the alternatives will automatically be compared for the same number of years (i.e., infinity). The alternative with the smaller capitalized cost will represent the most economical one. As in present-worth and all other alternative evaluation methods, it is only the differences in cash flow between the alternatives which must be considered. Therefore, whenever possible, the calculations should be simplified by eliminating the elements of cash flow which are common to both alternatives. Example 5.4 shows the procedure for comparing two alternatives on the basis of their capitalized cost.

Example 5.4 Two sites are currently under consideration for a bridge to cross the Ohio River. The north site would connect a major state highway with an interstate loop around the city and would alleviate much of the local through traffic. The disadvantages of this site are that the bridge would do little to ease local traffic congestion during rush hours, and the bridge would have to stretch from one hill to another to span the widest part of the river, railroad tracks, and local highways below. This bridge would therefore be a suspension bridge. The south site would require a much shorter span allowing for construction of a truss bridge, but would require new road construction.

The suspension bridge would have a first cost of $30 million with annual inspection and maintenance costs of $15,000. In addition, the concrete deck would have to be resurfaced every 10 years at a cost of $50,000. The truss bridge and approach roads are expected to cost $12 million and would have annual maintenance costs of $8000. The bridge would have to be painted every 3 years at a cost of $10,000. In addition, the bridge would have to be sandblasted and painted every 10 years at a cost of $45,000. The cost of purchasing right-of-way is expected to be $800,000 for the suspension bridge and $10.3 million for the truss bridge. Compare the alternatives on the basis of their capitalized cost if the interest rate is 6%.

SOLUTION Construct the cash-flow diagrams before you attempt to solve the problem. You should do this *now*.

Capitalized cost of suspension bridge

$$P_1 = \text{present worth of initial cost} = 30.0 + 0.8 = \$30.8 \text{ million}$$

The recurring operating cost is $A_1 = \$15{,}000$, while the annual equivalent of the resurface cost is

$$A_2 = 50{,}000(A/F, 6\%, 10) = \$3794$$

$$P_2 = \text{capitalized cost of recurring costs} = \frac{A_1 + A_2}{i}$$

$$= \frac{15{,}000 + 3794}{0.06}$$

$$= \$313{,}233$$

Finally, the total capitalized cost (P_S) is

$$P_S = P_1 + P_2 = \$31{,}113{,}233 \qquad (\$31.1 \text{ million})$$

Capitalized cost of truss bridge

$$P_1 = 12.0 + 10.3 = \$22.3 \text{ million}$$

$$A_1 = \$8000$$

$$A_2 = \text{annual cost of painting} = 10{,}000(A/F, 6\%, 3)$$

$$= \$3141$$

$$A_3 = \text{annual cost of sandblasting} = 45{,}000(A/F, 6\%, 10)$$

$$= \$3414$$

$$P_2 = \frac{A_1 + A_2 + A_3}{i} = \$242{,}583$$

The total capitalized cost (P_T) is

$$P_T = P_1 + P_2 = \$22{,}542{,}583 \qquad (\$22.5 \text{ million})$$

Since $P_T < P_S$, the truss bridge should be constructed.

Solved Problem 5.8
Probs. P5.28 to P5.31

SOLVED PROBLEMS

Example 5.5 A traveling saleswoman expects to purchase a used car this year. She has collected or estimated the following data: first cost is \$4800; trade-in value will be \$500 after 4 years; annual maintenance and insurance costs are \$350; and annual income due to ability to travel is \$1500. Will the woman be able to make a rate of return of 20% on her investment?

SOLUTION Compute the PW value of the investment at $i = 20\%$. (A cash-flow diagram will aid you.)

$$PW = -4800 - 350(P/A, 20\%, 4) + 1500(P/A, 20\%, 4) + 500(P/F, 20\%, 4)$$

$$= \$-1582$$

Indeed, she would not make 20%, since the PW is much less than zero.

COMMENT If the PW value had been greater than zero, an excess of 20% would be returned. In Chap. 7, calculations similar to those above will be made to determine the rate of return on project investments.

Sec. 5.1

Example 5.6 A cement plant plans to open a new rock pit. Two plans have been devised for movement of raw material from the quarry to the plant. Plan *A* requires the purchase of two earth movers and construction of an unloading pad at the plant. Plan *B* calls for construction of a conveyor system from the quarry to the plant. The costs for each plan are itemized in Table 5.1. Which plan should be selected if money is presently worth 15%?

SOLUTION Evaluation will take place over 24 years, since we plan to use present-worth (PW) analysis. Reinvestment in the two movers will occur in years 8 and 16 and the unloading pad must be repurchased in year 12. No reinvestment is necessary for plan *B*. You are advised to construct your own cash-flow diagram for each plan to follow the PW analysis.

Table 5.1 Details of plans to move rock from quarry to cement plant

	Plan *A*		Plan *B*
	Mover	Pad	Conveyor
P	\$45,000	\$28,000	\$175,000
AOC	6,000	300	2,500
SV	5,000	2,000	10,000
n, years	8	12	24

To simplify computations, we can use the fact that plan A will have an extra AOC in the amount of \$6300 - \$2500 = \$3800 per year.

PW of plan A

$$PW_A = PW_{movers} + PW_{pad} + PW_{AOC}$$

$$PW_{movers} = 2(45{,}000)[1 + (P/F, 15\%, 8) + (P/F, 15\%, 16)]$$

$$- 2(5000)[(P/F, 15\%, 8) + (P/F, 15\%, 16) + (P/F, 15\%, 24)]$$

$$= \$124{,}355$$

$$PW_{pad} = 28{,}000[1 + (P/F, 15\%, 12)] - 2000[(P/F, 15\%, 12) + (P/F, 15\%, 24)]$$

$$= \$32{,}790$$

$$PW_{AOC} = 3800(P/A, 15\%, 24) = \$24{,}448$$

$$PW_A = \$181{,}593$$

PW of plan B

$$PW_B = PW_{conveyor} = 175{,}000 - 10{,}000(P/F, 15\%, 24)$$

$$= \$174{,}651$$

Since $PW_B < PW_A$, the conveyor should be constructed.

Sec. 5.2

Example 5.7 A restaurant owner is trying to decide between two different garbage disposals. A regular steel (RS) disposal has an initial cost of \$65 and a life of 4 years. The alternative is a corrosion-resistant disposal constructed primarily of stainless steel (SS). The initial cost of the SS disposal is \$110, but it is expected to last 10 years. Because the SS disposal has a slightly larger motor, it is expected to cost about \$5 per year more to operate than the RS disposal. If the interest rate is 6%, which disposal should be selected, assuming both have a negligible salvage value?

SOLUTION The cash-flow diagram (Fig. 5.3) uses a comparison period of 20 years with reinvestment in year 10 for the SS disposal and in years 4, 8, 12, and 16 for the RS disposal. Inflation is not considered in the reinvestments. The present-worth calculations are as follows:

$$PW_{RS} = 65 + 65(P/F, 6\%, 4) + 65(P/F, 6\%, 8) + 65(P/F, 6\%, 12) + 65(P/F, 6\%, 16) = \$215$$

$$PW_{SS} = 110 + 110(P/F, 6\%, 10) + 5(P/A, 6\%, 20) = \$229$$

The regular steel disposal should be purchased, since $PW_{RS} < PW_{SS}$.

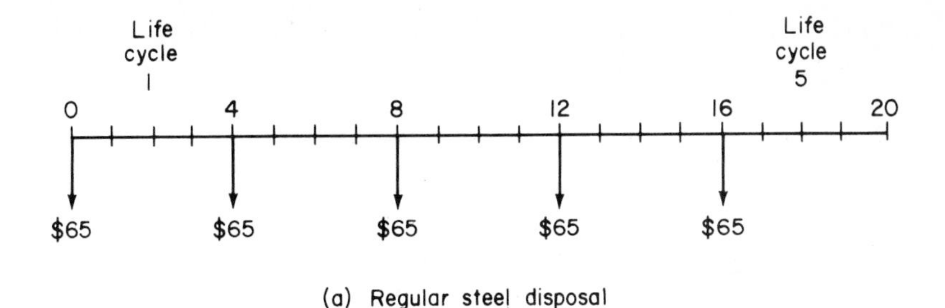

(a) Regular steel disposal

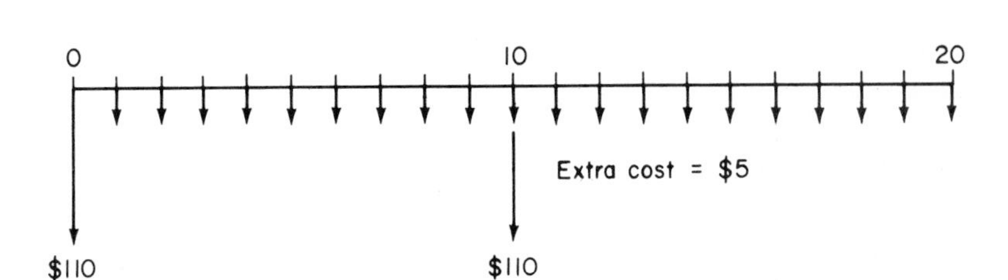

(b) Stainless steel disposal

Figure 5.3 Present-worth comparison of two unequal-life assets, Example 5.7.

COMMENT In the solution presented, the extra operating cost of \$5 per year was regarded as an expense for the SS disposal. However, the same decision would have been reached if the \$5 per year had been shown as an income for the RS disposal, but the present worths of both would have been lower by $\$5(P/A, 6\%, 20)$. This illustrates that unless the absolute money values are sought, it is only important to consider *differences* in cash flow for alternative evaluation.

Sec. 5.2

Example 5.8 A city engineer is considering two alternatives for the local water supply. The first alternative would involve construction of an earthen dam on a nearby river, which has a highly variable flow. The dam would serve as a reservoir so that the city would have a dependable source of water indefinitely. The initial cost of the dam is expected to be \$8 million and will require annual upkeep costs of \$25,000. The dam is expected to last indefinitely.

Alternatively, the city can drill wells as needed and construct pipelines for transporting the water to the city. The engineer estimates that an average of 10 wells will be required initially at a cost of \$45,000 per well, including the pipeline. The average life of a well is expected to be 5 years with an annual operating cost of \$5000 per well. If the city uses an interest rate of 5%, determine which alternative should be selected on the basis of their capitalized costs.

SOLUTION The cash-flow diagram is left to the reader. The capitalized cost of the dam is calculated as follows:

$$PW_{dam} = 8{,}000{,}000 + \frac{25{,}000}{0.05} = \$8{,}500{,}000$$

The capitalized cost of the wells can be calculated by first converting the recurring costs and annual operating costs to an EUAC and then dividing by the interest rate. Thus,

$$\begin{aligned} EUAC_{wells} &= \text{EUAC of investment} + \text{annual operating costs} \\ &= 45{,}000(10)(A/P, 5\%, 5) + 5000(10) \\ &= \$153{,}941 \end{aligned}$$

The capitalized cost of the wells, using Eq. (5.3), is

$$PW_{wells} = \frac{153{,}941}{0.05} = \$3{,}078{,}820$$

The wells should be constructed instead of the dam.

COMMENT The capitalized cost of the wells could also have been obtained by using the A/F factor for calculating the EUAC. The value obtained should then be divided by i and added to the initial investment cost. Thus,

$$\begin{aligned} EUAC_{wells} &= \frac{450{,}000(A/F, 5\%, 5) + 50{,}000}{0.05} \\ &= \$2{,}628{,}820 \end{aligned}$$

$$PW_{wells} = 2{,}628{,}820 + 450{,}000 = \$3{,}078{,}820$$

Sec. 5.4

PROBLEMS

P5.1 Two machines are under consideration by a metal fabricating company. Machine A will have a first cost of \$15,000, an annual maintenance and operation cost of \$3000, and a \$3000 salvage value. Machine B will have a first cost of \$22,000, an annual cost of \$1500, and a \$5000 salvage value. If both machines are expected to last for 10 years, determine which machine should be selected on the basis of present-worth values using an interest rate of 12% per year.

P5.2 A public utility is trying to decide between two different sizes of pipe for a new water main. A 250-millimeter line will have an initial cost of \$35,000, whereas a 300-millimeter line will cost \$55,000. Since there is less head loss through the 300-millimeter pipe, the pumping cost for the larger line is expected to be \$3000 per year less than for the 250-millimeter line. If the pipes are expected to last for 20 years, which size should be selected if the interest rate is 15% per year? Use a present-worth analysis.

P5.3 A couple are trying to decide between purchasing a house and renting one. They can purchase a new house with a down payment of $15,000 and a monthly payment of $750, beginning one month from now. Taxes and insurance are expected to amount to $100 per month. In addition, they expect to paint the house every 4 years at a cost of $600. Alternatively, they can rent a house for $700 per month payable in advance with a $600 deposit, which will be returned when they vacate the house. The utilities are expected to average $135 per month whether they purchase or rent. If they expect to be able to sell the house for $10,000 more than they paid down for it in 6 years, should they buy a house or rent one, if the interest rate is a nominal 12% per year compounded monthly? Use a present-worth analysis.

P5.4 A consulting engineer is trying to determine which of two methods should be specified for screening sewage. A manually cleaned bar screen will have an initial installed cost of $400. The labor cost for cleaning is expected to be $800 the first year, $850 the second year, and $900 the third year and to increase by $50 each year. An automatically cleaned bar screen will have an initial cost of $2500 with an annual power cost of $150. In addition, the motor will have to be replaced every 2 years at a cost of $40 per motor. General maintenance is expected to cost $100 the first year and increase by $10 per year. If the screens are expected to last for 10 years, which method should be selected if the interest rate is 10% per year? Use the present-worth method.

P5.5 A consulting engineering firm is trying to decide between purchasing and leasing cars. It estimates that medium-sized cars will cost $8300 and will have a probable trade-in value in 4 years of $2800. The annual cost of such items as fuel and repairs is expected to be $950 the first year and to increase by $50 per year. Alternatively, the company can lease the same cars for $3500 per year payable at the beginning of each year. Since some maintenance is included in the rental price, the annual maintenance and operation expenses are expected to be $100 lower if the cars are leased. If the company's minimum rate of return is 20%, which alternative should be selected?

P5.6 A building contractor is trying to determine if it would be economically feasible to install rainwater drains in a large shopping center currently under construction. Since the project is being built in the arid southwest, the total annual amount of rainfall is slight, but the rain that does occur is in the form of brief but heavy thundershowers. The thundershowers tend to cause erosion of soil in the project site, which was formed by filling in a large arroyo. In the 3 years required for construction, 12 heavy thundershowers are expected. If no drains are installed, the cost of refilling the washed-out area is expected to be $1000 per thunderstorm. Alternatively, a corrugated steel drainpipe could be installed which will prevent the soil erosion. The installation cost of the pipe would be $6.50 per meter, with a total length of 2000 meters required. After the 3-year construction period, some of the pipe could be recovered with an estimated value of $3000. Assuming that the thunderstorms occur at 3-month intervals, determine which alternative should be selected, if the interest rate is a nominal 20% per year compounded quarterly.

P5.7 A southwestern university is considering installing electric valves with automatic timers on some of their sprinkler systems. They estimate that they need 45 valves and timers at a cost of $85 per set. The initial installation cost is expected to be $2000. At the present time, there are four employees who are in charge of maintaining these lawns. These employees, each of whom earns $12,000 per year, spend 25% of their time in watering. The present cost of water for these lawns is $2200 per year. If the automatic system is installed, the manpower cost for watering could be reduced by 80% and the water bill by 35%. However, extra maintenance on the automatic system is expected to cost $450 per year. If the timers and valves are expected to last for 8 years, which system should be used if the interest rate is 16% per year? Use a present-worth analysis.

P5.8 A manufacturing company is in need of 1000 square meters of storage space for 3 years. The company is considering the purchase of land for $8000 and is erecting a temporary metal structure at a cost of $70 per square meter. At the end of the 3-year use period, the company expects to be able to sell the land for $9000 and the building for $12,000. Alternatively, the company can lease storage space for $1.50 per square meter per month payable at the beginning of each year. If the company's minimum attractive rate of return is 20% per year, which type of storage space should be used? Use the present-worth method of analysis.

P5.9 Machines that have the following costs are under consideration for a continuous production process.

	Machine *G*	Machine *H*
First cost	$62,000	$77,000
Annual operating cost	15,000	21,000
Salvage value	8,000	10,000
Life, years	4	6

Using an interest rate of 15% per year, determine which alternative should be selected on the basis of a present-worth analysis.

P5.10 Rework Prob. P5.9 assuming that machine *G* requires an extensive overhaul at the end of 2 years that costs $10,000.

P5.11 Which screen should be selected in Prob. P5.4 if the manually cleaned screen will last 20 years and the automatically cleaned screen will last only 10 years? Assume the automatic screen maintenance cost will increase by $10 per year through year 20.

P5.12 Compare the machines below on the basis of their present worths, using an interest rate of 18% per year.

	Machine *P*	Machine *Q*
First cost	$29,000	$37,000
Salvage value	4,000	5,000
Life, years	3	5
Annual maintenance cost	3,000	3,500
Overhaul after 2 years	3,700	2,000

P5.13 A small strip-mining coal company is trying to decide whether it should purchase or lease a new clamshell. If purchased, the shell will cost $150,000 and is expected to have a $65,000 salvage value in 8 years. Alternatively, the company can lease the clamshell for $30,000 per year, but the lease payment will have to be made at the *beginning* of each year. If the clamshell is purchased, it will be leased to other strip-mining companies whenever possible, an activity that is expected to yield revenues of $10,000 per year. If the company's minimum attractive rate of return is 22%, should the clamshell be purchased or leased? Make calculations on the basis of a present-worth analysis.

P5.14 A production plant manager has been presented with two proposals for automating an assembly process. Proposal *A* involves an initial cost of $15,000 and an annual operating cost of $2000 per year for the next four years. Thereafter, the operating cost is expected to be $2700 per year. This equipment is expected to have a 20-year life with no salvage value. Proposal *B* requires an initial investment of $28,000 and an annual operating cost of $1200 per year for the first 3 years. Thereafter, the operating cost is expected to increase by $120 per year. This equipment is expected to last for 20 years and have a $2000 salvage value. If the company's minimum attractive rate of return is 10%, which proposal should be accepted on the basis of a present-worth analysis?

P5.15 An environmental engineer is trying to decide between two operating pressures for a wastewater irrigation system. If a high-pressure system is used, fewer sprinklers and less pipe will be required, but the pumping cost will be higher. The alternative is to use lower pressure with more sprinklers. The pumping cost is estimated to be $3 per 1000 cubic meters of wastewater pumped at the high pressure. Twenty-five sprinklers will be required at a cost of $30 per unit. In addition, 1000 meters of aluminum pipe will be required at a cost of $9 per meter. If the lower pressure system is used, the pumping cost will be $2 per 1000 cubic meters of wastewater. Also required will be 85 sprinklers and 4000 meters of pipe. The aluminum pipe is expected to last 10 years and the sprinklers 5 years. If the volume of wastewater is expected to be 500,000 cubic meters per year, which pressure should be selected if the company's minimum attractive rate of return is 20% per year? The aluminum pipe will have a 10% salvage value.

P5.16 The owner of the Good Flick Drive-In Theatre is considering two proposals for upgrading the parking ramps. The first proposal involves asphalt paving of the entire parking area. The initial cost of this proposal would be $35,000, and it would require annual maintenance of $250 beginning 3 years after installation. The owner expects to have to resurface the theater in 15 years. Resurfacing will cost only $8000, since grading and surface preparation are not necessary, but the $250 annual maintenance cost will continue. Alternatively, gravel can be purchased and spread in the drive areas and grass planted in the parking areas. The owner estimates that 29 metric tons of gravel will be needed per year starting one year from now at a cost of $90 per metric ton. In addition, a riding lawn mower, which will cost $800 and have a life of 10 years, will be needed. The cost of labor for spreading gravel, cutting grass, etc., is expected to be $900 the first year and $950 the second and will increase by $50 per year thereafter. The owner figures that a gravel surface would not be used for more than 30 years. If the interest rate is 12% per year, which alternative should be selected? Use a present-worth analysis and a 30-year study period.

P5.17 An automobile owner is trying to decide between purchasing four new radial tires or having the worn-out tires recapped. Radial tires for the car will cost $85 each and will last 60,000 kilometers. The old tires can be recapped for $25 each, but they will last for only 20,000 kilometers. Since this is a second car, it probably will register only 10,000 kilometers per year. If the radial tires are purchased, the gasoline mileage will increase by 10%. If the cost of gasoline is assumed to be $0.42 per liter and the car gets 10 kilometers per liter, what type tires should be purchased if the interest rate is 12% per year? Use the present-worth method and assume that the salvage value of the tires is zero.

P5.18 A state highway department is trying to decide between "hot patching" an existing road and resurfacing it. If the hot-patch method is used, approximately 300 cubic meters of material will be required at a cost of $25 per cubic meter (in place). Additionally, the shoulders will have to be improved at the same time at a cost of $3000. The annual cost of routine maintenance on the patched-up road would be $4000. These improvements will last 2 years, at which time they will have to be redone. Alternatively, the state could resurface the road at a cost of $65,000. This surface will last 10 years if the road is maintained at a cost of $1500 per year beginning 4 years from now. No matter which alternative is selected now, the road will be completely rebuilt in 10 years. If the interest rate is 13% per year, which alternative should the state select on the basis of a present-worth comparison?

P5.19 The Bee-Low Mining Company is considering purchasing a machine which costs $30,000 and is expected to last 12 years, with a $3000 salvage value. The annual operating expenses are expected to be $9000 for the first 4 years; but owing to decreased use, the operating costs will decrease by $400 per year for the next 8 years. Alternatively, the company can purchase a highly automated machine at a cost of $58,000. This machine will last only 6 years because of its high technology and delicate design, and its salvage value will be $15,000. Because it is so automated, its operating cost will be only $4000 per year. If the company's minimum attractive rate of return is 20% per year, which machine should be selected on the basis of a present-worth analysis?

P5.20 Two metal fabricating machines are presently under consideration by the Heat 'N' Beat Metal fabricating company. The manual model will cost $25,000 to buy with an 8-year life and a $5000 salvage value. Its annual operating cost will be $15,000 for labor and $1000 for maintenance. A computer-controlled model will cost $95,000 to buy and it will have a 12-year life if upgraded at the end of year 6 for $15,000. Its terminal salvage value will be $23,000. The annual costs for the computer-controlled model will be $7500 for labor and $2500 for maintenance. If the company's minimum attractive rate of return is 25%, which machine would be preferred on the basis of the equivalent present cost of each?

P5.21 The Board-Stiff lumber company is considering whether it should provide disposable or reusable plates and utensils for its employee cafeteria. Disposable utensils will cost $4700 for a

2-year supply. Because of the trash created by the throwaway items, the refuse disposal costs will be \$48 per month higher. Alternatively, the company could purchase reusable utensils which will cost initially \$10,000. Their life will be 8 years, but due to breakage, another \$2000 will have to be spent in 5 years for replacements. After 8 years, the usable items remaining can be sold for \$1500. The cost of hiring a part-time dishwasher, buying detergents, obtaining hot water, etc., is expected to be \$350 per month. If the interest rate is a nominal 18% per year compounded monthly, should the company purchase the disposable or the reusable items? Use the present-worth method of analysis.

P5.22 A local planning commission has estimated the first cost of a new city-owned amusement park to be \$35,000. They expect to improve the park by adding new rides every year for the next 5 years at a cost of \$6000 per year. Annual operating costs are expected to be \$12,000 the first year; these will increase by \$2000 per year until year 5. After that time, the operating expenses will remain at \$20,000 per year. The city expects to receive \$11,000 in profits the first year, \$14,000 the second, and amounts increasing by \$3000 per year until year 8, after which the net profit will remain the same. Calculate the capitalized cost of the park if the interest rate is 6% per year.

P5.23 How much additional uniform annual cost can the city incur for the amusement park in Prob. P5.22 to break even?

P5.24 What is the capitalized cost of \$75,000 now, \$60,000 five years from now, and a uniform annual amount of \$700 per year for year 10 and every year thereafter, if the interest rate is 8% per year?

P5.25 What is the capitalized cost of \$200,000 now, \$300,000 four years from now, \$50,000 every 5 years, and a uniform annual amount of \$8000 beginning 15 years from now, if the interest rate is 16% per year?

P5.26 A wealthy alumnus of a small university wants to establish a permanent fund for tuition scholarships. He wants to support three students for the first 5 years after the fund is established and five students thereafter. If tuition alone is expected to cost \$1000 per year, how much money must the alumnus donate now if the university can earn 10% per year on the fund?

P5.27 If the tuition in Prob. P5.26 increases by \$20 per year for the first 20 years, how much money must the alumnus donate?

P5.28 Machines with costs shown below are presently under consideration by the Go-For-It Company. Using an interest rate of 15% per year, compare the alternatives on the basis of their capitalized costs.

	Machine *Y*	Machine NOT
First cost	\$31,000	\$43,000
Annual operating cost	18,000	19,000
Salvage value	5,000	7,000
Life, years	4	6

P5.29 Compare the machines shown below on the basis of their capitalized cost using an interest rate of 20% per year.

	Machine *X*	Machine *Z*
First cost	\$50,000	\$200,000
Annual operating cost	62,000	24,000
Salvage value	10,000	0
Overhaul after 6 years	...	4,000
Life, years	7	∞

P5.30 Compare the machines in Prob. P5.29 on the basis of their capitalized costs if the life of machine Z is 10 years instead of infinite. Use an interest rate of a nominal 14% per year compounded semiannually.

P5.31 A city planning commission is considering two proposals for a new civic center. Proposal F requires an initial investment of $10 million now and an expansion cost of $4 million 10 years from now. The annual operating cost is expected to be $250,000 per year. Income from conventions, shows, etc., is expected to be $190,000 the first year and to increase by $20,000 per year for four more years and then remain constant until year 10. In year 11 and thereafter income is expected to be $350,000 per year. Proposal G requires an initial investment of $18 million now and an annual operating cost of $300,000 per year. However, income is expected to be $260,000 the first year and increase by $30,000 per year to year 7. Thereafter income will remain at $400,000 per year. Determine which proposal should be selected on the basis of capitalized cost if the interest rate is 6% per year.

440,000

CHAPTER

SIX

EQUIVALENT-UNIFORM-ANNUAL-COST EVALUATION

The objective of this chapter is to teach you the primary methods of calculating the equivalent uniform annual cost (EUAC) of an asset and how to select the better of two alternatives on the basis of an annual-cost comparison. The alternative selected using EUAC must be the same as that chosen using present-worth or any other evaluation method; that is, all methods should result in identical decisions, just in a different manner.

SECTION OBJECTIVES

In order to complete this chapter you must be able to do the following:

6.1. State why the EUAC must be calculated for only one cycle of each alternative, when the alternatives have different lives.

6.2. Calculate the EUAC of an asset having a salvage value, using the *salvage sinking-fund* method, given the asset initial cost, salvage value, life, and interest rate.

*6.3. Calculate the EUAC as in 6.2, except using the *salvage present-worth* method.

*6.4. Calculate the EUAC as in 6.2, except using the *capital-recovery-plus-interest* method.

6.5. Select the better of two alternatives on the basis of their EUAC, given their initial costs, salvage values, lives, amount and time of the operating costs, and interest rate.

6.6. Calculate the EUAC of a perpetual investment, given the initial cost of the asset, amount and timing of disbursements, and interest rate.

STUDY GUIDE

6.1 Study Period for Alternatives Having Different Lives

The EUAC (equivalent uniform annual cost) is another method that is commonly used for comparing alternatives. As illustrated in Chap. 4, the EUAC means that all disbursements (irregular and uniform) must be converted to an equivalent uniform annual cost, that is, a year-end amount which is the *same each year*. The major advantage of this method over all the other methods is that it does not require making the comparison over the same number of years when the alternatives have different lives. When the EUAC method is used, the equivalent uniform annual cost of the alternative must be calculated for *one life cycle only*. Why? Because, as its name implies, the EUAC is an equivalent annual cost over the life of the project. If the project is continued for more than one cycle, the equivalent annual cost for the next cycle and all succeeding cycles would be exactly the same as for the first, assuming all cash flows were the same for each cycle. The EUAC for one cycle of an alternative therefore represents the equivalent uniform annual cost of that alternative *forever*.

Problem P6.1

6.2 Salvage Sinking-Fund Method

When an asset of a given alternative has a terminal salvage value (SV), there are several ways by which the EUAC can be calculated. This section presents the salvage sinking-fund method, probably the simplest of the three discussed in this chapter. This is the method that is used hereinafter. In the salvage sinking-fund method, the initial cost (P) is first converted to an equivalent uniform annual cost using the A/P (capital-recovery) factor. The salvage value, after conversion to an equivalent uniform cost via the A/F (sinking-fund) factor, is *subtracted* from the annual-cost equivalent of the first cost. The calculations can be represented by a general equation:

$$\text{EUAC} = P(A/P, i\%, n) - \text{SV}(A/F, i\%, n) \tag{6.1}$$

Naturally, EUAC is nothing more than an A value, but it is referred to as EUAC here. The calculations are illustrated in Example 6.1.

Example 6.1 Calculate the EUAC of a machine that has an initial cost of \$8000 and a salvage value of \$500 after 8 years. Annual operating costs (AOC) for the machine are estimated to be \$900, and the interest rate of 6% is applicable.

SOLUTION The cash-flow diagram (Fig. 6.1) requires us to compute

$$\text{EUAC} = A_1 + A_2$$

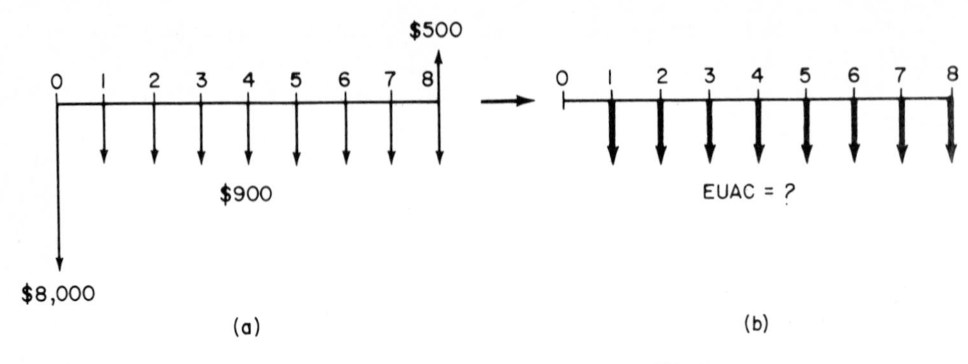

Figure 6.1 (*a*) Diagram for machine costs and (*b*) conversion to an EUAC.

where A_1 = annual cost of initial investment less salvage value, Eq. (6.1)
A_2 = annual maintenance cost = \$900

$$A_1 = 8000(A/P, 6\%, 8) - 500(A/F, 6\%, 8) = \$1238$$

$$\text{EUAC} = 1238 + 900 = \$2138$$

COMMENT Since the maintenance cost was already expressed as an annual cost over the life of the asset, no conversions were necessary. The simplicity of the salvage sinking-fund method should be obvious from the straightforward calculations shown in this example. The steps in this method are the following:

1. Annualize the initial investment cost over the life of the asset using the A/P factor.
2. Annualize the salvage value using the A/F factor.
3. Subtract the annualized salvage value from the annualized investment cost.
4. Add the uniform annual costs to the value from step 3.

Solved Problem 6.6
Probs. P6.2 and P6.3

*6.3 Salvage Present-Worth Method

The salvage present-worth method is the second method by which investment costs having salvage values can be converted into an EUAC. The present worth of the salvage value is subtracted from the initial investment cost, and the resulting difference is annualized for the life of the asset. The general equation is

$$\text{EUAC} = [P - \text{SV}(P/F, i\%, n)]\,(A/P, i\%, n) \tag{6.2}$$

The steps that must be followed in this method are the following:

1. Calculate the present worth of the salvage value via the P/F factor.
2. Subtract the value obtained in step 1 from the initial cost P.

3. Annualize the resulting difference over the life of the asset using the A/P factor.
4. Add the uniform annual costs to the result of step 3.

Example 6.2 Compute the EUAC for the machine detailed in Example 6.1 using the salvage present-worth method.

SOLUTION Using the steps outlined above and Eq. (6.2),

$$\text{EUAC} = [8000 - 500(P/F, 6\%, 8)](A/P, 6\%, 8) + 900 = \$2138$$

Prob. P6.4

*6.4 Capital-Recovery-Plus-Interest Method

The final procedure that will be presented here for calculating the EUAC of an asset having a salvage value is the capital-recovery-plus-interest method. The general equation for this method is

$$\text{EUAC} = (P - \text{SV})(A/P, i\%, n) + \text{SV}(i) \tag{6.3}$$

In subtracting the salvage value from the investment cost *before* multiplying by the A/P factor, it is recognized that the salvage value will be recovered. However, the fact that the salvage value will not be recovered for n years must be taken into account by adding the interest ($\text{SV}i$) lost during the asset's life. Failure to include this term would assume that the salvage value was obtained in year 0 instead of year n. The steps to be followed for this method are as follows:

1. Subtract the salvage value from the initial cost.
2. Annualize the resulting difference with the A/P factor.
3. Multiply the salvage value by the interest rate.
4. Add the values obtained in steps 2 and 3.
5. Add the uniform annual costs to the result of step 4.

Example 6.3 Use the values of Example 6.1 to compute the EUAC using the capital-recovery-plus-interest method.

SOLUTION From Eq. (6.3) and the steps above,

$$\text{EUAC} = (8000 - 500)(A/P, 6\%, 8) + 500(0.06) + 900 = \$2138$$

While it makes no difference which method is used to compute the EUAC, it would be good procedure hereafter to use only the one method you prefer in order to avoid unnecessary errors caused by mixing the methods. We will use the salvage sinking-fund method (Sec. 6.2).

Prob. P6.5

6.5 Comparing Alternatives by EUAC

The equivalent-uniform-annual-cost method of comparing alternatives is probably the simplest of the alternative evaluation techniques presented in this book. Selection is made on the basis of EUAC with the alternative having the lowest cost being the most favorable. Obviously, nonquantifiable data must also be considered in arriving at the final decision, but in general, the alternative having the lowest EUAC should be selected.

Perhaps the most important rule to remember when making EUAC comparisons is that *only one cycle* of the alternative must be considered. This assumes, of course, that the costs in all succeeding periods will be the same. While it is true that the cost of an asset today will probably be much lower than the cost of the same asset 10 years from today, because of inflation, it must be remembered that, in general, the costs of the other alternatives would increase as well. Inasmuch as the analytical methods presented here are mainly for the purpose of making comparisons and not for determining actual costs, the same conclusions would be reached at any future date as long as all costs increased proportionately. Obviously, when information is available which would indicate that the costs of certain assets will increase or decrease considerably, because of technical improvements or increased competition, these factors must be taken into consideration in arriving at a final decision.

Example 6.4 The following costs are proposed for two equal-service tomato-peeling machines in a food canning plant:

	Machine A	Machine B
First cost	\$26,000	\$36,000
Annual maintenance cost	800	300
Annual labor cost	11,000	7,000
Extra income taxes	. . .	2,600
Salvage value	2,000	3,000
Life, years	6	10

If the minimum required rate of return is 15%, which machine should be selected?

SOLUTION The cash-flow diagram for each alternative is shown in Fig. 6.2. The EUAC of each machine using the salvage sinking-fund method, Eq. (6.1), is calculated as follows:

$$\text{EUAC}_A = 26{,}000(A/P, 15\%, 6) - 2000(A/F, 15\%, 6) + 11{,}800 = \$18{,}442$$

$$\text{EUAC}_B = 36{,}000(A/P, 15\%, 10) - 3000(A/F, 15\%, 10) + 9900 = \$16{,}925$$

Select machine B, since $\text{EUAC}_B < \text{EUAC}_A$.

It is common to use the annual cash flows CF_t and expected salvage value in comparing alternatives by the *payback period*, which is a determination of the number of

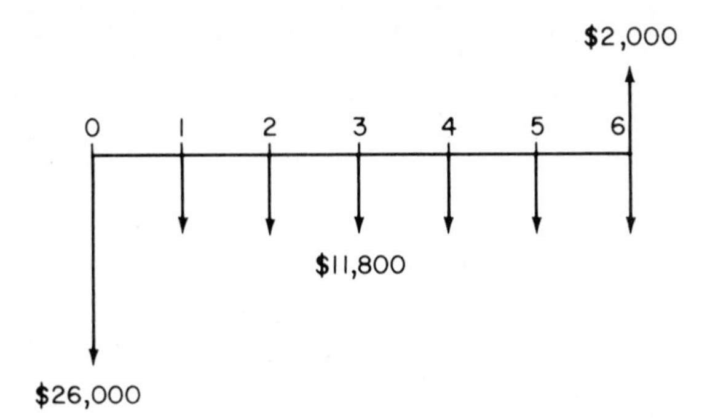

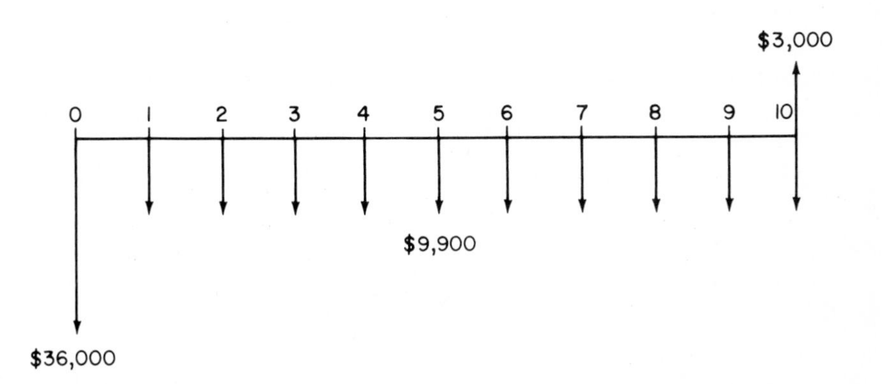

Figure 6.2 Cash flows for two alternative tomato-peeling machines, Example 6.4.

years n before the first cost and a stated return is realized. That is, for each alternative determine the n value such that

$$0 = -P + \sum_{t=1}^{n} \text{CF}_t(\text{P/F}, i\%, t)$$

The alternative with the smaller n value is selected. If $i = 0\%$, the no-return payback period is computed. Comparison using payback-period analysis may give a different result than does present-worth or EUAC analysis because it neglects all cash flows after the time n and it also overlooks the time value of money if $i = 0\%$. Payback-period analysis and its use are discussed in detail in Chap. 16.

Solved Problem 6.7
Probs. P6.6 to P6.23

6.6 EUAC of a Perpetual Investment

It is sometimes necessary to compare alternatives that can be expected to have a perpetual life, such as flood-control dams or irrigation projects. For this type of analysis, it is important to recognize that the annual cost of a perpetual *initial* investment is simply the annual interest on the initial lump sum. That is, if the federal government were to invest \$10,000 in a certain public works project, the EUAC of the investment would be \$10,000(0.04) = \$400, if the interest rate were 4%. This calculation is easy to understand when it is realized that the government could spend either \$10,000 now or \$400 per year forever. By spending \$10,000 now, the government is losing the \$400-per-year interest it would make on the \$10,000 if it were kept in a bank at 4%. On the other hand, the federal government could let an investor pay the \$10,000 for the public works project and then pay the investor the \$400 per year lost interest. From another point of view, a person receiving \$10,000 now would have an amount of money equivalent to that of another person who was to receive \$400 per year forever if the interest rate were 4%, since both persons would receive only \$400 per year perpetually.

Costs recurring at regular or irregular intervals are handled exactly as in conventional EUAC problems. That is, all other costs must be converted into equivalent uniform annual costs for *one cycle*. They are thus automatically annual costs forever, as discussed in Sec. 6.1. Example 6.5 illustrates EUAC calculations for a perpetual project.

Example 6.5 The U.S. Bureau of Reclamation is considering two proposals for increasing the capacity of the main canal in their Lower Valley irrigation system. Proposal A would involve dredging the canal in order to remove sediment and weeds which have accumulated during previous years' operation. Since the capacity of the canal will have to be maintained near its design peak flow because of increased water demand, the bureau is planning to purchase the dredging equipment and accessories for \$65,000. The equipment is expected to have a 10-year life with a \$7000 salvage value. The annual labor and operating costs for the dredging operation is estimated to be \$22,000. In order to control weeds in the canal itself and along the banks, herbicides will be sprayed during the irrigation season. The yearly cost of the weed-control program, including labor, is expected to be \$12,000.

Proposal B would involve lining the canal with concrete at an initial cost of \$650,000. The lining is assumed to be permanent, but minor maintenance will be required every year at a cost of \$1000. In addition, lining repairs will have to be made every 5 years at a cost of \$10,000. Compare the two alternatives on the basis of equivalent uniform annual cost using an interest rate of 5%.

SOLUTION The cash-flow diagram is left to the reader. The EUAC of each proposal is determined as follows:

Proposal *A*	
EUAC of dredging equipment:	
65,000(*A*/*P*, 5%, 10) − 7000(*A*/*F*, 5%, 10)	$ 7,861
Annual cost of dredging	22,000
Annual cost of weed control	12,000
	$41,861

Proposal *B*	
EUAC of initial investment: $650,000(0.05)	$32,500
Annual maintenance cost	1,000
Lining repair cost: $10,000(*A*/*F*, 5%, 5)	1,810
	$35,310

Proposal *B* should be selected.

COMMENT For proposal *A*, it was necessary to consider only one cycle. No calculations were necessary for the dredging and weed-control costs since they were already expressed as annual costs. For proposal *B*, the EUAC of the initial investment was obtained by multiplying by the interest rate, which is nothing more than Eq. (5.3), that is, $P = A/i$, solved for *A* and renamed EUAC.

If nonrecurring single or series costs are involved, they must be converted to a present worth and then multiplied by the interest rate. Note the use of the *A*/*F* (sinking-fund) factor for the lining repair cost. The *A*/*F* factor is used instead of the capital-recovery (*A*/*P*) factor because the lining repair cost began in year 5 instead of year 0 and continued indefinitely at 5-year intervals.

Solved Problems 6.8 and 6.9
Probs. P6.24 to P6.33

SOLVED PROBLEMS

Example 6.6 A drugstore chain has just purchased a fleet of five pickup trucks to be used for delivery in a particular city. Initial cost was $4600 per truck and the expected life and salvage value is 5 years and $300, respectively. The combined insurance, maintenance, gas, and lubrication costs are expected to be $650 the first year and to increase by $50 per year thereafter, while delivery service will bring an extra $1200 per year for the company. If a return of 10% is required, use the EUAC method to determine if the purchase should have been made.

SOLUTION The cash-flow diagram is shown in Fig. 6.3. If we compute the EUAC by the salvage sinking-fund method, we can first compute

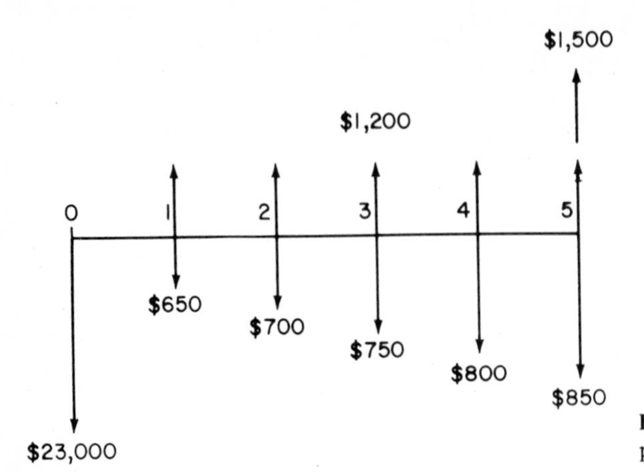

Figure 6.3 Diagram used to compute EUAC, Example 6.6.

$$
\begin{aligned}
A_1 &= \text{annual cost of fleet purchase, Eq. (6.1)} \\
&= -5(4600)(A/P, 10\%, 5) + 5(300)(A/F, 10\%, 5) \\
&= \$-5822
\end{aligned}
$$

The minus signs are used for costs since incomes and disbursements are involved. The annual disbursement and income can be combined into an annual-income equivalent (A_2) so that the net income conveniently follows a decreasing gradient.

$$A_2 = 550 - 50(A/G, 10\%, 5) = \$460$$

Now the EUAC is equal to the ***algebraic*** sum of the annual disbursement and annual income.

$$\text{EUAC} = -5822 + 460 = \$-5362$$

Since EUAC < 0, a return less than 10% will be made and the purchase is therefore not justified.

COMMENT Try one of the EUAC methods of the Secs. 6.3 and 6.4 to solve the problem. They all work.

Secs. 6.2 to 6.4

Example 6.7 Compare the two plans proposed in Example 5.6 using the EUAC method.

SOLUTION Even though the two component parts of plan A, movers and pad, have different lives, the EUAC analysis must be conducted for only one life cycle. For the salvage sinking-fund method, Eq. (6.1),

$$\text{EUAC}_A = \text{EUAC}_{\text{movers}} + \text{EUAC}_{\text{pad}} + \text{EUAC}_{\text{AOC}}$$

where $\text{EUAC}_{\text{movers}} = 90{,}000(A/P, 15\%, 8) - 10{,}000(A/F, 15\%, 8) = \$19{,}328$
$\text{EUAC}_{\text{pad}} = 28{,}000(A/P, 15\%, 12) - 2000(A/F, 15\%, 12) = \$5096$
$\text{EUAC}_{\text{AOC}} = \$3800$

Then,

$$\begin{aligned}\text{EUAC}_A &= 19{,}328 + 5096 + 3800\\ &= \$28{,}224\end{aligned}$$

$$\begin{aligned}\text{EUAC}_B &= \text{EUAC}_{\text{conveyor}}\\ &= 175{,}000(A/P, 15\%, 24) - 10{,}000(A/F, 15\%, 24)\\ &= \$27{,}146\end{aligned}$$

As was also shown in the present-worth analysis of Example 5.6, select plan B.

COMMENT You should recognize a fundamental relation between the PW and EUAC values for the two examples discussed here. If you have the PW of a given plan, you can get EUAC by EUAC = PW(A/P, $i\%$, n) or with an EUAC, PW = EUAC(P/A, $i\%$, n). The question is: What value does n assume? What would you use? We vote for the least-common-multiple value used in the present-worth method, since this method of evaluation must take place over an equal time period for each alternative. Therefore, the present-worth values are

$$\text{PW}_A = \text{EUAC}_A(P/A, 15\%, 24) = \$181{,}588$$

$$\text{PW}_B = \text{EUAC}_B(P/A, 15\%, 24) = \$174{,}652$$

as found in Example 5.6.

Sec. 6.5

Example 6.8 If an investor deposits \$1000 now, \$3000 three years from now, and \$600 per year for 5 years starting 4 years from now, how much money can be withdrawn every year forever beginning 12 years from now, if the rate of return on the investment is 8%?

SOLUTION The cash-flow diagram is shown in Fig. 6.4. The uniform amount of money that can be withdrawn every year forever is equal to the amount of interest that accumulates each year on the principal amount. To solve the problem, therefore, it is necessary to determine the total amount that would be accumulated in year 11 (*not* year 12) and then multiply by the interest rate (i) to obtain A. The future amount in year 11 would be

$$\begin{aligned}F_{11} &= 1000(F/P, 8\%, 11) + 3000(F/P, 8\%, 8) + 600(F/A, 8\%, 5)(F/P, 8\%, 3)\\ &= \$12{,}319\end{aligned}$$

The perpetual withdrawal can now be found by multiplying F_{11} (which is now a P value with respect to the perpetual withdrawal) by the interest rate.

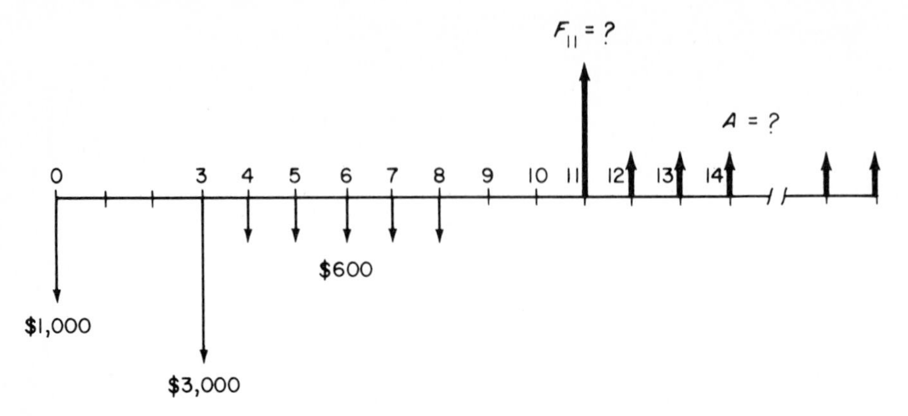

Figure 6.4 Diagram to determine perpetual annual withdrawal, Example 6.8.

$$A = Pi = 12{,}319(0.08) = \$986$$

Sec. 6.6

Example 6.9 If an investor deposits $10,000 now at an interest rate of 7% per year, how many years must the money accumulate before the investor can withdraw $1400 per year forever?

SOLUTION The cash-flow diagram is shown in Fig. 6.5. The first step is to find the total amount of money that must be accumulated in year n (P_n), which is one year prior to the first withdrawal, to permit the perpetual $1400-per-year withdrawal using

$$P_n = \frac{A}{i} = \frac{1400}{0.07} = \$20{,}000$$

When $20,000 is accumulated, the investor can withdraw $1400 per year forever. The next step is to determine when the initial $10,000 deposit will accumulate to

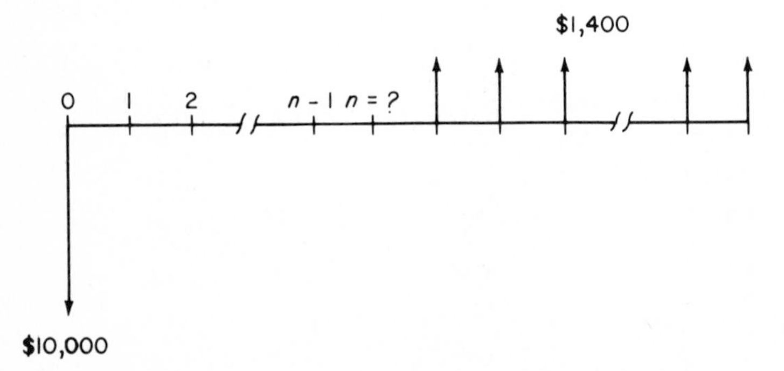

Figure 6.5 Diagram to determine n for a perpetual withdrawal, Example 6.9.

\$20,000. This can be done with the F/P factor:

$$20{,}000 = 10{,}000(F/P, 7\%, n)$$

By interpolation, $(F/P, 7\%, n) = 2.0000$ when $n = 10.24$ years.

Sec. 6.6

PROBLEMS

P6.1 Why must the least common multiple of years be used for assets that have different lives when a present-worth comparison is made, whereas an EUAC analysis requires annualization of costs over only one life cycle of each asset?

P6.2 If a woman purchased a new car for \$6000 and sold it 3 years later for \$2000, what was the equivalent uniform annual cost if she spent \$750 per year for upkeep and operation? Use an interest rate of 15% per year and the salvage sinking-fund method.

P6.3 An entrepreneur purchased a dump truck for the purpose of offering a short-haul earth-moving service. He paid \$14,000 for the truck and sold it 5 years later for \$3000. His operation and maintenance expense while he owned the truck was \$3500 per year. In addition, he had the truck engine overhauled for \$220 at the end of the third year. Calculate his equivalent uniform annual cost using the salvage sinking-fund method, if the interest rate was 14% per year.

***P6.4** Work the following problems by the salvage present-worth method: (*a*) P6.2 and (*b*) P6.3.

***P6.5** Work the following problems by the capital-recovery-plus-interest method: (*a*) P6.2 and (*b*) P6.3.

P6.6 The manager in a canned-food processing plant is trying to decide between two labeling machines. Their respective costs are as follows:

	Machine *A*	Machine *B*
First cost	\$15,000	\$25,000
Annual operating cost	1,600	400
Salvage value	3,000	6,000
Life, years	7	10

(*a*) Determine which machine should be selected using a minimum attractive rate of return of 12% per year and an EUAC analysis.

(*b*) If a present-worth analysis were used, over how many years would you make the comparison?

P6.7 Compare the following machines on the basis of their equivalent uniform annual cost. Use $i = 18\%$ per year.

	New machine	Used machine
First cost	\$44,000	\$23,000
Annual operating cost	7,000	9,000
Annual repair cost	210	350
Overhaul every 2 years	...	1,900
Overhaul every 5 years	2,500	
Salvage value	4,000	3,000
Life, years	15	8

P6.8 Compare the two processes below on the basis of their equivalent uniform annual costs at an interest rate of 18% per year.

	Process *M*	Process *R*
First cost	$80,000	$120,000
Salvage value	10,000	18,000
Life, years	10	15
Annual operating cost	15,000	13,000
Annual revenues	39,000	55,000 for year 1, decreasing by $2000 per year

P6.9 A moving and storage company is considering two possibilities for warehouse operations. Proposal 1 requires the purchase of a fork lift for $5000 and 500 pallets that cost $5 each. The average life of a pallet is assumed to be 2 years. If the fork lift is purchased, the company must hire an operator for $9000 annually and spend $600 per year in maintenance and operation. The life of the fork lift is expected to be 12 years, with a $700 salvage value. Alternatively, proposal 2 requires that the company hire two people to operate power-driven hand trucks at a cost of $7500 per person. One hand truck will be required at a cost of $900. The hand truck will have a life of 6 years with no salvage value. If the company's minimum attractive rate of return is 12% per year, which alternative should be selected?

P6.10 The supervisor of a country club swimming pool is trying to decide between two methods used for adding chlorine. If gaseous chlorine is added, a chlorinator, which has an initial cost of $800 and a useful life of 5 years, will be required. The chlorine will cost $200 per year and the labor cost will be $400 per year. Alternatively, dry chlorine can be added manually at a cost of $500 per year for chlorine and $800 per year for labor. If the interest rate is 6% per year, which method should be used?

P6.11 A carpenter is trying to determine how much insulation should be put into a ceiling. The higher the R rating of the insulation, the better the insulation. The choices are limited to either R-11 or R-19 insulation. The R-11 insulation costs $2.50 per square meter and the R-19 costs $3.50 per square meter. The annual saving in heating and cooling costs is estimated to be $25 per year greater with R-19 than with R-11. If the house has 250 square meters and the owner expects to keep the house for 25 years, which insulation should be installed at an interest rate of 10% per year?

P6.12 In Prob. P6.11, how much would the saving have to be per year in order for the R-19 insulation to be just as economical as the R-11?

P6.13 A meat-packing plant manager is trying to decide between two different methods for cooling cooked hams. The spray method involves spraying water over the hams until the ham temperature is reduced to 30 degrees Celsius. With this method, approximately 80 liters of water are required for each ham. Alternatively, an immersion method can be used in which only 16 liters of water are required per ham. However, this method will require an initial extra investment of $2000 and extra overhaul expenses of $100 per year, with the equipment expected to last 10 years. The company cooks 10 million hams per year and pays $0.12 per 1000 liters for water. The company must also pay $0.04 per 1000 liters for wastewater discharged. If the company's minimum attractive rate of return is 15%, which method of cooling should be used?

P6.14 Two environmental chambers (*A* and *B*) are being considered for a government project which is to last for 6 years. Pertinent data are listed below.

	Chamber *A*	Chamber *B*
First cost	$4000	$2500
Annual operating cost	400	300
Salvage value	1000	−100
Estimated life, years	3	2

(*a*) What chamber should be selected if money is worth 12% per year?

(*b*) What must the difference in annual operating cost be to make the equivalent annual cost of both chambers equal?

P6.15 Compare the two plans below at $i = 15\%$ per year.

	Plan *A*	Plan *B*	
		Machine 1	Machine 2
First cost	$10,000	$30,000	$5,000
Annual operating cost	500	100	200
Salvage value	1,000	5,000	−200
Life, years	40	40	20

P6.16 The Mighty Mouse Company is considering the purchase of a trap system to rid the plant of stray cats. Compare the two systems below at 10%-per-year interest.

	Scram-um	Catch-um
First cost	$25,000	$50,000
Annual operating cost	500	200
Salvage value	1,000	500
Life, years	20	40

P6.17 The Toe-Main Food Processing Company is evaluating various methods for disposing of the sludge from the wastewater treatment plant. Currently under consideration is land disposal of the sludge by spraying or incorporating into the soil. If the spraying alternative is selected, an underground distribution system will be constructed at a cost of $600,000. The salvage value after 20 years is expected to be $20,000. Operation and maintenance of the system is expected to cost $26,000 per year. Alternatively, the company can use Big Foot trucks to transport and dispose of the sludge by incorporation below the soil surface. Three trucks will be required at a cost of $220,000 per truck. The operating cost of the trucks, including driver, routine maintenance, overhauls, etc., is expected to be $42,000 per year. The used trucks can be sold after 10 years for $30,000 each. If the trucks are used, field corn can be planted and sold for $20,000 per year. For spraying, grass must be planted and harvested; and because of the presence of the "contaminated" sludge on the cuttings, the grass will have to be landfilled at a cost of $14,000 per year. If the company's minimum attractive rate of return is 20% per year, which method should be selected based on an equivalent-uniform-annual-cost analysis?

P6.18 Two methods can be used for producing a certain machine part. Method 1 costs $20,000 initially and will have a $5000 salvage value after 3 years. The operating cost with this method is $8500 per year. Method 4 has an initial cost of $15,000, but it will last only 2 years. Its salvage value is $3000. The operating cost for method 4 is $7000 per year. If the minimum attractive rate of return is 16% per year, which method should be used on the basis of an equivalent-uniform-annual-cost analysis?

P6.19 A transport company based in El Paso is trying to decide between purchasing diesel- or gasoline-powered truck. A diesel-powered truck will cost $1000 more to purchase, but the kilometers per liter will be a respectable 6. Diesel fuel can be purchased in Juarez for 4 cents a liter, but two times per year the fuel system will have to be cleaned of paraffins at a cost of $190 per cleaning. The gasoline-powered truck will average 4 kilometers per liter and gasoline can be purchased for 14 cents a liter (also in Juarez). The diesel-powered truck can be used for 270,000 kilometers if the engine is overhauled for $5500 after 150,000 kilometers. The gasoline-powered truck can be used for 180,000 kilometers if overhauled at a cost of $2200 after 120,000 kilometers. If the company's trucks average 30,000 kilometers per year, which type should be purchased at a minimum attractive rate of return of 20% per year? Assume zero salvage values for both trucks.

P6.20 Two types of materials can be used for roofing a commercial building which has 1500 square meters of roof. Asphalt shingles will cost $14 per square meter installed and are guaranteed for 15 years. Fiber-glass shingles will cost $17 per square meter installed, but they are guaranteed for 20 years. If the fiber-glass shingles are selected, the owner will be able to sell the building for $1500 more than if the asphalt shingles are used. If the owner plans to sell the building in 8 years, which shingles should be used if the minimum attractive rate of return is 17% per year and the equivalent-uniform-annual-cost method of analysis is to be used?

P6.21 The R-Sun Specialty Company is considering two types of siding for its proposed new building (the previous building was destroyed by fire). Anodized metal siding will require very little maintenance and minor repairs will cost only $500 every 3 years. The initial cost of the siding will be $250,000. If a concrete facing is used, the building will have to be painted now at a cost of $80,000 and every 5 years at a cost of $8000 more than the previous time. The building is expected to have a useful life of 23 years, and the "salvage value" will be $25,000 greater if the metal siding is used. Compare the equivalent uniform annual costs of the two methods at an interest rate of 15% per year.

P6.22 The warehouse for a large furniture manufacturing company currently requires too much energy for heating and cooling because of poor insulation. The company is trying to decide between urethane foam and fiber-glass insulation. The initial cost of the foam insulation will be $35,000, with no salvage value. The foam will have to be painted every 3 years at a cost of $2500. The energy saving is expected to be $6000 per year. Alternatively, fiber-glass batts can be installed for $12,000. The fiber-glass batts would not be salvageable either, but there would be no maintenance costs. If the fiber-glass batts would save $2500 per year in energy costs, which method of insulation should be company use at an interest rate of 15% per year? Use a 24-year study period and an equivalent uniform annual-cost analysis.

P6.23 Compare the alternatives below on the basis of an equivalent-uniform-annual-cost analysis, using an interest rate of 20% per year.

	Plan *A*	Plan *B*
First cost	28,000	36,000
Installation cost	3,000	4,000
Annual maintenance cost	1,000	2,000
Annual operating cost	2,200 + 75*k**	800 + 50*k*
Life, years	10	10

**k* = years, 1 through 10.

P6.24 Calculate the perpetual equivalent uniform annual cost of $14,000 now, $55,000 six years from now, and $5000 per year thereafter if the interest rate is 8% per year.

P6.25 Rework Prob. P6.24 using an interest rate of a nominal 18% per year compounded semi-annually.

P6.26 The first cost of a small dam is expected to be $3 million. The annual maintenance cost is expected to be $10,000 per year; a $35,000 outlay will be required every 5 years. If the dam is expected to last forever, what will be its equivalent uniform annual cost at an interest rate of 12% per year?

P6.27 A city that is attempting to attract a professional football team is planning to build a new football stadium costing $12 million. Annual upkeep is expected to amount to $25,000 per year. In addition, the artificial turf will have to be replaced every 10 years at a cost of $150,000. Painting every 5 years will cost $65,000. If the city expects to maintain the facility indefinitely, what will be its equivalent uniform annual cost? Assume that i = 12% per year.

$EUAC_p = 12\times10^6(.12) + 25K + 150{,}000(A/F, 12, 10) + 65K(A/F, 12, 5)$

How much does ALL This COST/yr.

P6.28 An alumnus of Watsa Matta University desires to establish a permanent university scholarship in his name. He plans to donate $20,000 per year for 10 years starting one year from now and leave $100,000 when he dies. If the university expects the alumnus to die 15 years from now, how much money can be given to each of five students beginning one year from now and continuing forever if the interest rate is 8% per year?

P6.29 As a grateful alumnus of Ima-Wanta University, Mr. B. G. Spender would like to make an endowment which will provide scholarships to needy skydivers who want to study to become engineers. Mr. Spender would like the scholarships to be in the amount of $13,000 per year, with the first scholarship to be given 15 years from now. Mr. Spender wants to deposit enough money into the fund in 14 years so that scholarships can be given in his name forever thereafter (i.e., from year 15 on). If Mr. Spender plans to make his first deposit one year from now, how much should he deposit each year if the fund will earn interest at a rate of 12% per year.

P6.30 Another grateful alumnus of Watsa Matta University, Mr. I. S. Rich, would also like to establish a perpetual scholarship fund for would-be engineers. Mr. Rich would like to provide one scholarship per year in the amount of $20,000 for an infinite time, with the first scholarship to be given 10 years from now. Mr. Rich plans to make his first deposit one year from now and then increase each succeeding one by $1000 through year 9, at which time no more money will be donated. If the fund earns interest at a rate of 14% per year, how much must Mr. Rich deposit in each of the first 2 years?

P6.31 For the cash-flow sequence shown below, determine the amount of money that can be withdrawn annually for an infinite time if the first withdrawal is to be made in year 13 and the interest rate is 15% per year.

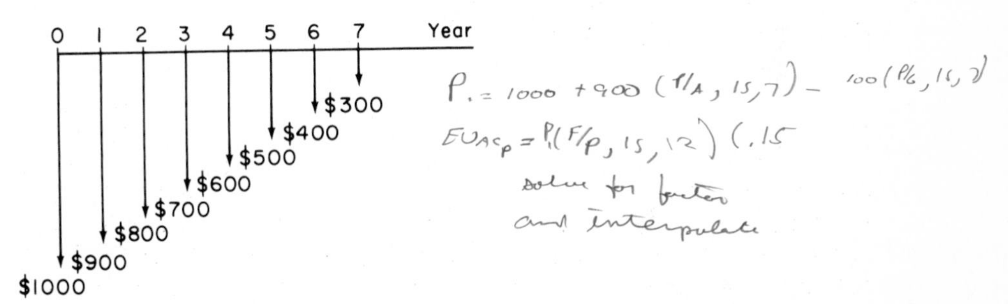

P6.32 For the cash-flow sequence shown below, how much time must elapse between the last payment (in year 9) and the first withdrawal of $4000 per year for an infinite time, if the interest rate is 13% per year?

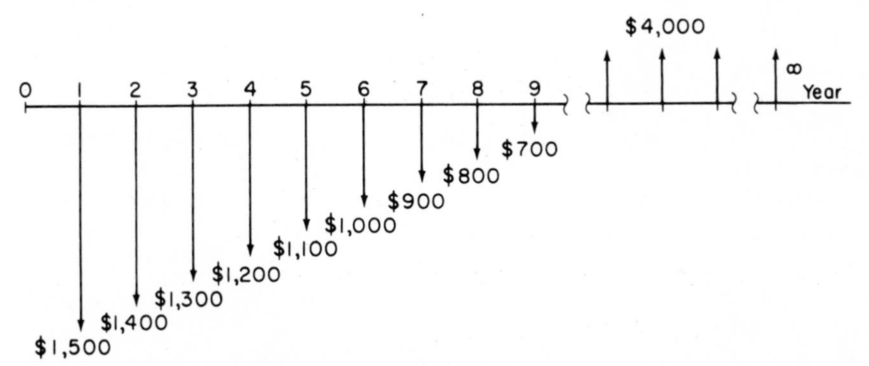

P6.33 For the deposits shown in Prob. P6.32, how much money could be withdrawn for an infinite time beginning in year 12, if the interest rate is 13% per year?

CHAPTER

SEVEN

RATE-OF-RETURN COMPUTATIONS FOR A SINGLE PROJECT

This chapter will teach you how to correctly compute the rate of return for one project using the present-worth and equivalent-annual-cost methods. Sections on the presence of multiple rates of return and the concept of rates of return external to the project are included. Computerized rate-of-return analysis is discussed in Appendix E. Rate-of-return analysis for alternative evaluation in the next chapter is an application of the principles discussed here.

SECTION OBJECTIVES

In order to complete this chapter you must be able to do the following:

7.1. Write the definition of the *rate of return* of a project and state the general equations used to compute the rate of return, given the receipts and disbursements.

7.2. Calculate the rate of return for a single project using the present-worth method and the steps to initially estimate the rate of return, given the amounts and times of project disbursements and receipts.

7.3. Calculate the rate of return as in 7.2, except using the equivalent-uniform-annual-cost method.

*7.4. State whether a given cash-flow sequence is *conventional* or *nonconventional* and determine the *multiple rates of return*, given the amounts and times of nonconventional cash flows.

*7.5. State the definitions of *internal* and *composite* rate of return and compute both rates, given the amounts and times of the cash flows and a reinvestment rate for all receipts (positive cash flows) which are an excess to the project.

STUDY GUIDE

7.1 Overview of Rate-of-Return Computation

If money is borrowed, the interest rate is applied to the *unpaid balance* so that the total loan amount and interest are paid exactly with the last payment. If money is lent or invested in a project, there is an *unrecovered balance* at each time period. The interest rate is the return on this unrecovered balance so that the total loan and interest are recovered exactly with the last receipt. Rate of return defines both these situations.

> *Rate of return* is the rate of interest paid on the unpaid balance of borrowed money or the rate of interest earned on the unrecovered balance of an investment (loan) so that the final payment or receipt brings the balance to zero with interest considered.

The rate of return is expressed as a percent—for example, $i = 10\%$. The rate is always positive, $i > 0$; that is, the fact that interest paid on a loan is actually a negative rate of return is not considered. Note that the definition above does not state that the rate of return is on the initial amount of the investment; it is rather on the *unrecovered* balance, which varies with time. The example below illustrates the difference between these two concepts.

Example 7.1 A $1000 investment is expected to produce a net cash flow of $315.47 for each of 4 years. This represents a 10% rate of return on the unrecovered balance. Compute the amount of the unrecovered investment for the 4 years using (*a*) the rate of return on the unrecovered balance and (*b*) the rate of return on the initial $1000 investment. (*c*) Explain why not all the investment is recovered in part *b*.

SOLUTION

(*a*) Table 7.1 presents the unrecovered balance figures for each year using the 10% rate on the unrecovered balance at the beginning of the year. After 4 years the total $1000 investment is recovered and the balance in column 6 is zero.

(*b*) Table 7.2 shows the unrecovered balance figures if the 10% return is always figured on the initial investment of $1000. Column 6 in year 4 shows a remaining unrecovered amount of $138.12, because only $861.88 is recovered in the 4 years.

(*c*) A total of $400 in interest must be earned if the 10% return each year is figured on the initial investment. However, only $261.88 in interest must be earned if a 10% return on the unrecovered balance is used. There is more of

Table 7.1 Unrecovered balances using a rate of return of 10%

(1) Year	(2) Beginning unrecovered balance	(3) = 0.10(2) Interest on unrecovered balance	(4) Cash flow	(5) = (4) - (3) Removal of unrecovered balance	(6) = (2) - (5) Ending unrecovered balance
0	...	...	$-1000.00	...	$-1000.00
1	$-1000.00	$100.00	+315.47	$ 215.47	-784.53
2	-784.53	78.45	+315.47	237.02	-547.51
3	-547.51	54.75	+315.47	260.72	-286.79
4	-286.79	28.68	+315.47	286.79	0
		$261.88		$1000.00	

Table 7.2 Unrecovered balances using a 10% return on the initial investment

(1) Year	(2) Beginning unrecovered balance	(3) = 0.10($1000) Interest on initial investment	(4) Cash flow	(5) = (4) - (3) Removal of unrecovered balance	(6) = (2) - (5) Ending unrecovered balance
0	...	...	$-1000.00	...	$-1000.00
1	$-1000.00	$100	315.47	215.47	-784.53
2	-784.53	100	315.47	215.47	-596.06
3	-569.06	100	315.47	215.47	-353.25
4	-353.25	100	315.47	215.47	-138.12
		$400		$861.88	

the annual cash flow available to reduce the remaining investment when the rate is applied to the unrecovered balance.

COMMENT As defined, rate of return is the interest rate on the unrecovered balance; therefore, the computations in Table 7.1 for part ***a*** present a correct interpretation of a 10% rate of return.

To determine the rate of return value i^* of a project, the present worth of disbursements D is equated to the present worth of receipts R. That is,

$$P_D = P_R$$

Equivalently,

$$\begin{aligned} 0 &= P_R - P_D \\ &= -P_D + P_R \end{aligned} \tag{7.1}$$

In this analysis, investments are disbursements and incomes are receipts. The equivalent-uniform-annual-cost method can also be used.

$$\text{EUAC}_D = \text{EUAC}_R$$

or

$$0 = \text{EUAC}_R - \text{EUAC}_D$$

$$= -\text{EUAC}_D + \text{EUAC}_R \tag{7.2}$$

In either case the i^* value which makes the relations correct may be referred to by several titles—rate of return, internal rate of return, break-even rate of return, profitability index, or return on investment (ROI).

Probs. P7.1 and P7.2

7.2 Rate-of-Return Calculations by the Present-Worth Method

In Sec. 2.7 the method for calculating the rate of return on an investment was illustrated when only one factor was involved. In this section the present-worth method for calculating the rate of return on an investment when several factors are involved is demonstrated. To understand rate-of-return calculations more clearly, remember that the basis for engineering-economy calculations is *equivalence*, or time value of money. In previous chapters, we have shown that a present sum of money is equivalent to a larger sum of money at some future date when the interest rate is greater than zero. In rate-of-return calculations, the objective is to find the interest rate at which the present sum and future sum are equivalent; in other words, the calculations that will be made here are simply the reverse of calculations made in previous chapters, where the interest rate was known.

The backbone of the rate-of-return method is a rate-of-return equation, which is simply an expression equating a present sum of money to the present worth of future sums. For example, if you invest \$1000 now and are promised receipts of \$500 three years from now and \$1500 five years from now, the rate-of-return equation is

$$1000 = 500(P/F, i^*\%, 3) + 1500(P/F, i^*\%, 5) \tag{7.3}$$

where the value of i^* to make the equality correct is to be computed (see Fig. 7.1). If the \$1000 is moved to the right side of Eq. (7.3), we have

$$0 = -1000 + 500(P/F, i^*\%, 3) + 1500(P/F, i^*\%, 5) \tag{7.4}$$

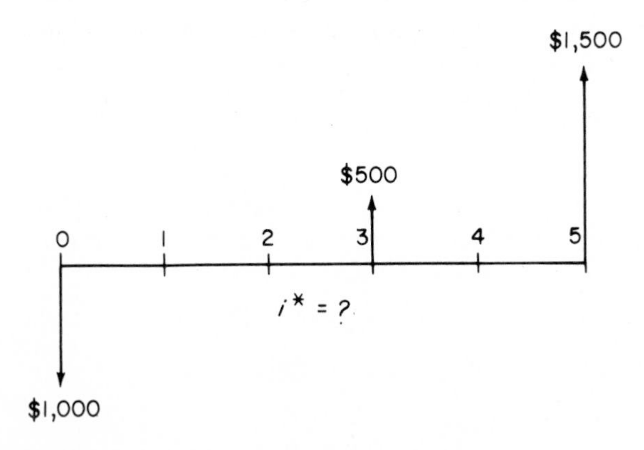

Figure 7.1 Cash flow for which a value of i^* is to be determined.

Equation (7.4) is in the general form of Eq. (7.1), $0 = -P_D + P_R$, which will be used in setting up all rate-of-return calculations for the present-worth method. The equation must then be solved for i^* by trial and error to obtain $i^* = 16.95\%$. Since there are always some receipts and some disbursements involved in any project, some value of i^* can be found; however, the rate of return will be greater than zero only if the total amount of receipts is greater than the total amount of disbursements.

It should be evident that rate-of-return calculations are merely the reverse of present-worth calculations. That is, if the above interest rate were given (16.95%) and it were desired to find the present worth of $500 three years from now and $1500 five years from now, the equation would be

$$P = 500(P/F, 16.95\%, 3) + 1500(P/F, 16.95\%, 5) = \$1000$$

which is easily rearranged to the form of Eq. (7.4). This illustrates that rate-of-return and present-worth equations are set up in exactly the same fashion. The only difference is in what is given and what is sought.

The general procedure used to make a rate-of-return calculation by the present-worth method is the following:

1. Draw a cash-flow diagram.
2. Set up the rate-of-return equation in the form of Eq. (7.1).
3. Select values of i by trial and error until the equation is balanced. It will probably be necessary to find i^* using linear interpolation.

When using the trial-and-error method to determine i^*, it is advantageous to get fairly close to the correct answer on the first trial. If the cash flows are combined in such a manner that the income and disbursements are used to compute a *single factor* such as P/F, P/A, and so forth, as in Chap. 2, it is possible to look up the interest rate corresponding to the value of that factor for n years. The problem is then to combine the cash flows into the format of only one of the standard factors. This may be done through the following procedure:

1. Convert all *disbursements* to either single amounts (P or F) or uniform amounts (A) by neglecting the time value of money. For example, if it is desired to convert an A into an F value, simply multiply the A by the number of years n. The movement of cash flows should minimize the error caused by neglecting the time value of money.
2. Convert all *receipts* to either single or uniform values, as in step 1.
3. Having reduced all disbursements and receipts to either a P/F, P/A, or A/F format, use the interest tables to find the approximate interest rate at which the P/F, P/A, or A/F value, respectively, is satisfied for the proper n value. The rate obtained is a good ball-park figure to use in the first trial.

It is important to recognize that the rate of return obtained in this manner is only an *estimate* of the actual rate of return, because the time value of money is neglected. This procedure is illustrated in Example 7.2.

Example 7.2 If \$5000 is invested now in common stock that is expected to yield \$100 per year for 10 years and \$7000 at the end of 10 years, what is the rate of return?

SOLUTION The rate-of-return procedure above is used to compute i^*.

1. Figure 7.2 shows the cash flows.
2. From Eq. (7.1),

$$0 = -5000 + 100(P/A, i^*\%, 10) + 7000(P/F, i^*\%, 10)$$

3. Use the estimation procedure above to determine the interest rate for the first trial. All income will be regarded as a single F in year 10 so that the P/F factor can be used. The P/F factor is selected because most of the cash flow already fits this factor and errors created by neglecting the time value of money will be minimized. Thus

$$P = \$5000$$

$$F = 10(100) + 7000 = 8000$$

$$n = 10$$

Now we can state that

$$5000 = 8000(P/F, i\%, 10)$$

$$(P/F, i\%, 10) = 0.625$$

The interest rate is between 4 and 5%. Therefore, use $i = 5\%$ in the equation of step 2.

$$0 = -5000 + 100(P/A, 5\%, 10) + 7000(P/F, 5\%, 10)$$

$$0 \neq \$69.46$$

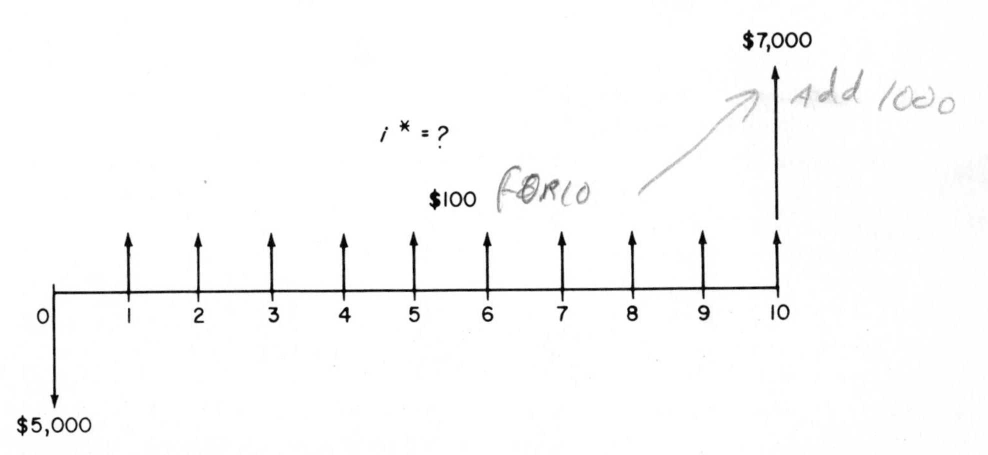

Figure 7.2 Cash flow for a stock investment, Example 7.2.

We are too large on the positive side, indicating that the receipts return more than 5%. Therefore, try $i = 6\%$.

$$0 = -5000 + 100(P/A, 6\%, 10) + 7000(P/F, 6\%, 10)$$

$$0 \neq -\$355.19$$

Since the interest rate of 6% is too high, interpolate (Sec. 2.5) using Eq. (2.10):

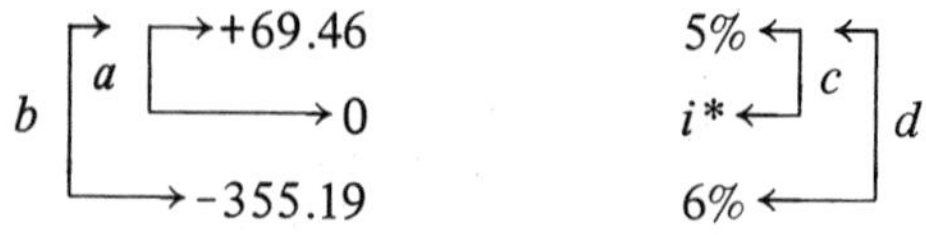

a, b	+69.46	5%	c, d
	0	i^*	
	-355.19	6%	

$$c = \frac{a}{b}(d) = \frac{(69.49 - 0)}{69.49 - (-355.19)}(1.0) = 0.16$$

$$i^* = 5.00 + 0.16 = 5.16\%$$

COMMENT Note that 5% rather than 4% was used for the first trial. The higher value was used because, by assuming that the ten \$100 amounts were equivalent to a single \$1000 in year 10, the approximate rate from the P/F factor was *lower* than the true value. This is owing to the neglect of the time value of money. Therefore, the first-trial i value used was above that indicated by the P/F factor in order to improve the accuracy.

Solved Problem 7.6
Probs. P7.3 to P7.13

7.3 Rate-of-Return Calculations by the Equivalent-Uniform-Annual-Cost Method

Just as i^* can be found by the present-worth method, it may also be determined using the EUAC relation in Eq. (7.2). The procedure is as follows:

1. Draw a cash-flow diagram.
2. Set up the relations for the EUAC of disbursements and the EUAC of receipts. This is equivalent to determining an A value for disbursements and receipts.
3. Set up the rate-of-return equation in the form of Eq. (7.2), that is,

$$0 = -\text{EUAC}_D + \text{EUAC}_R$$

4. Select values of i by trial and error until the equation is balanced. If necessary, interpolate to determine i^*.

The estimation procedure in Sec. 7.2 for the first i value is used here also. Example 7.3 illustrates the EUAC method.

Example 7.3 Use EUAC computations to find the rate of return for the investment situation in Example 7.2.

SOLUTION

1. Figure 7.2 gives the cash flow.
2. The EUAC relations for disbursements and receipts are

$$EUAC_D = -5000(A/P, i\%, 10)$$

$$EUAC_R = 100 + 7000(A/F, i\%, 10)$$

3. The EUAC formulation using Eq. (7.2) is

$$0 = -5000(A/P, i^*\%, 10) + 100 + 7000(A/F, i^*\%, 10)$$

Results are as follows: $i = 5\%$, $0 \neq \$+9.02$; and $i = 6\%$, $0 \neq \$-48.26$. Interpolation gives equation balance at $i^* = 5.16\%$, as before.

Thus, for rate-of-return calculations you can choose either the present-worth or equivalent-uniform-annual-cost method. It is generally better to get accustomed to using only one of the two methods in order to avoid unnecessary errors.

The computer program ROIDS (Return On Investment Determination System) is introduced in Appendix E. It computes the i^* value for any sequence of cash flows.

Probs. P7.3 to P7.13

*7.4 Multiple Rate-of-Return Values

In the two previous sections a unique i^* value was determined for the given cash-flow sequences. Investigation shows that the signs on the cash flows changed only once, usually from minus in year 0 to plus for the rest of the investment's life. This is called a *conventional* cash flow. If there is more than one sign change, the series is called *nonconventional*. As shown in the examples in Table 7.3, the runs of cash-flow signs may be one or more in length. A thorough discussion of cash-flow classifications is given in Ref. 1.

When there is more than one sign change, that is, when the cash flow is nonconventional, it is possible to determine multiple i^* values which will balance the rate-

Table 7.3 Examples of conventional and nonconventional cash-flow sequences for six year-lived projects

	Sign of cash flow							Number of sign changes
Type	0	1	2	3	4	5	6	
Conventional	–	+	+	+	+	+	+	1
Conventional	–	–	–	+	+	+	+	1
Conventional	+	+	+	+	+	–	–	1
Nonconventional	–	+	+	+	–	–	–	2
Nonconventional	+	+	–	–	–	+	+	2
Nonconventional	–	+	–	–	+	+	+	3

of-return equation [Eqs. (7.1) or (7.2)]. The total number of real i^* values is less than or equal to the number of sign changes in the sequence. (It is possible to determine that imaginery values or infinity will also satisfy the equation, but these are of little value to the analyst.) Example 7.4 presents the determination and graphical interpretation of multiple rate-of-return values.

Example 7.4 A new synthetic lubricant has been marketed for 3 years, with the following net cash flows in thousands of dollars.

Year	0	1	2	3
Cash flow (÷ $1000)	$+2000	-500	-8100	+6800

(*a*) Plot the present worth versus the rate of return for i values of 5, 10, 20, 30, 40, and 45%.

(*b*) Determine whether the cash-flow series is conventional or nonconventional and estimate the rate of return from the plot in *a*.

SOLUTION

(*a*) The present-worth values are found in Table 7.4 using the P/F factor for each i value. Figure 7.3 shows that the present worth has a parabolic shape and crosses the i axis two times, because there are two sign changes in the given cash-flow sequence.

(*b*) The sequence is nonconventional and the two i^* values may be graphically determined from Fig. 7.3 to be approximately

$$i_1^* = 8\% \qquad \text{and} \qquad i_2^* = 41\%$$

COMMENT If the i^* values are solved for mathematically, the more-exact values are found to be $i_1^* = 7.47\%$ and $i_2^* = 41.35\%$. If there had been three sign reversals

Table 7.4 Computation of present worth for several rate-of-return values, Example 7.4

		Year				Present worth for different i values (÷ $1000)
		0	1	2	3	
	Cash flow (÷ $1000)	$+2000	$-500	$-8100	$+6800	
i	5%	$+2000	$-476.20	$-7346.70	$+5873.84	$ +51.44
	10%	+2000	-454.55	-6693.84	+5108.84	-39.55
	20%	+2000	-416.65	-5624.64	+3935.16	-106.13
	30%	+2000	-384.60	-4792.77	+3095.36	-82.01
	40%	+2000	-357.15	-4132.62	+2477.92	-11.85
	45%	+2000	-344.85	-3852.36	+2230.40	+33.19

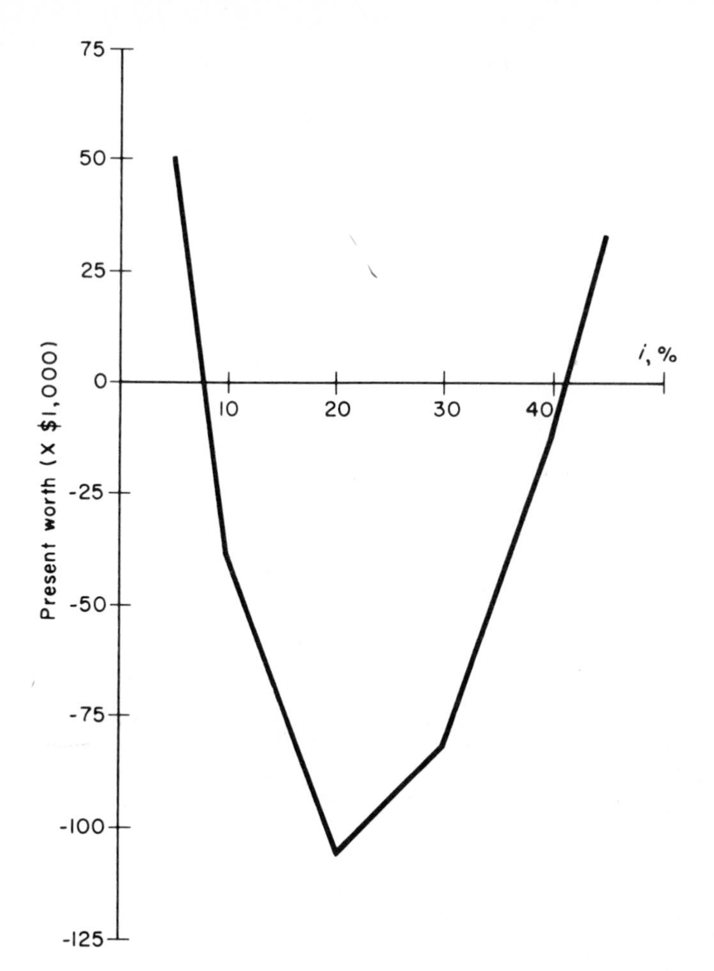

Figure 7.3 Present worth of cash flows at several different interest rates, Example 7.4.

in the cash-flow sequence, it would have been possible for there to be three i^* values, and the i_1^* and i_2^* values above would have changed.

In many cases some of the multiple i^* values will seem ridiculous because they are too large or too small (negative). For example, i^* values of 10, 150, and 750% for a sequence with three sign changes is difficult to explain. It is common to neglect the large values or to simply never compute them. However, there is an interpretation for each i^* value, as you will learn in the next section. Whenever possible you should use the present-worth or EUAC method for alternative analysis to ensure that unrealistic rates don't enter and confuse the analysis.

For the interested reader, further discussion of multiple rate-of-return computations is presented in the texts listed in Further Information at the end of the chapter. Appendix E includes a discussion on ROIDS, a computer program written in BASIC which may be used to determine all real roots for a specified cash-flow sequence.

ROIDS (Return On Investment Determination System) allows the user to interactively input the cash-flow sequence and determine all i^* values within any specified range to any desired accuracy. The procedure for determining a correct i value is presented in the next section.

Solved Problem 7.7
Probs. P7.14 to P7.16

*7.5 Internal and Composite Rates of Return

The rate-of-return values we have computed thus far have assumed that any net positive cash flows (receipts) are reinvested immediately at the rate of return that balances the rate-of-return equation. Therefore, if the balancing rate is, say, 40%, any receipt prior to the end of the project is assumed to earn 40% for the remaining years. Of course, this assumption may be unrealistic when the balancing rate is much greater or less than the minimum attractive rate of return (MARR). The rate, which is computed by Eqs. (7.1) and (7.2), is called the *internal rate of return* (IRR) because it does not consider any of the economic factors external to the project. By definition,

> The *internal rate of return* i^* is an interest rate of a project which assumes that all positive cash flows are reinvested at a rate of return that balances the rate-of-return equation.

It is the reinvestment assumption of the internal rate of return, coupled with the reversals in cash-flow signs, that allows there to be multiple rates of return for nonconventional cash-flow sequences.[1] However, if the reinvestment rate is explicitly used to compute the future worth of all positive cash flows that can be invested external to the project, return to a conventional cash-flow sequence is accomplished and the problem of muliple rates of return is eliminated. (The cumulative cash-flow sequence, which is obtained by successively adding cash-flow values, must also be conventional to ensure a single rate of return.) The reinvestment rate, symbolized by c, is often set equal to the MARR. The interest rate determined in this fashion to satisfy the rate-of-return equation will be called the composite rate of return and will be symbolized by i'. By definition

> The *composite rate of return* i' is the interest rate of a project which assumes that net positive cash flows which represent funds not immediately needed in the project are reinvested at the rate c, which is explicitly stated and has been determined by considering factors external to the project cash flow.

The term *composite* is used for i' because it is determined conditional upon the reinvestment rate c. (If c happens to equal any one of the i^* values, then i' will equal that i^* value.)

The way in which the reinvestment rate is applied to net positive cash flows must be correctly done to obtain the correct i' value for the project. If positive net cash flows are considered overrecovery of the project investments (the negative cash flows)

[1] It is actually possible to perform tests of the cash-flow signs, the accumulated cash-flow signs, the algebraic sum of all cash flows, and the project unrecovered investment balance to determine the number and values of rates of return for any nonconventional cash-flow sequence. A good summary of these tests is given in Ref. 2.

and the i' value must cause the net overall investment for the project itself to be exactly zero at the end of the project, the *project-net-investment* technique may be utilized. This procedure (under different titles), that is explained well by Bussey [1], pp. 232-236, and others [6, 7], is summarized here. For each year find the future worth F of the project net investment one year in the future; that is, F_{t+1}, using F_t and the cash flow in year t, C_t. The interest rate in the F/P factor is c if the net investment F_t is positive and it is i' if F_t is negative. Mathematically, for each year set up the relation

$$
\begin{aligned}
F_0 &= C_0 \\
F_{t+1} &= F_t(F/P, i\%, 1) + C_{t+1} = F_t(1+i) + C_{t+1} \qquad t = 1, 2, \ldots, n-1
\end{aligned} \tag{7.5}
$$

where n = total years in the project

$$
i = \begin{cases} c & \text{if } F_t > 0 \quad \text{(net positive investment)} \\ i' & \text{if } F_t < 0 \quad \text{(net negative investment)} \end{cases}
$$

Form the relation $F_n = 0$ and use trial and error to find the unique i' value. The rate c is used in Eq. (7.5) when the project investment has been recovered and excess cash flow returns the reinvestment rate; while i' (to be determined) is the return on the project to be realized for the cash-flow sequence. The program ROIDS (Appendix E) uses this procedure to find i'.

The development of F_0 through F_3 for the cash-flow sequence below (Fig. 7.4*a*) using a 15% reinvestment rate is as follows:

	Year			
	0	1	2	3
Cash flow, \$	50	−200	50	100
Cumulative cash flow, \$	50	−150	−100	0

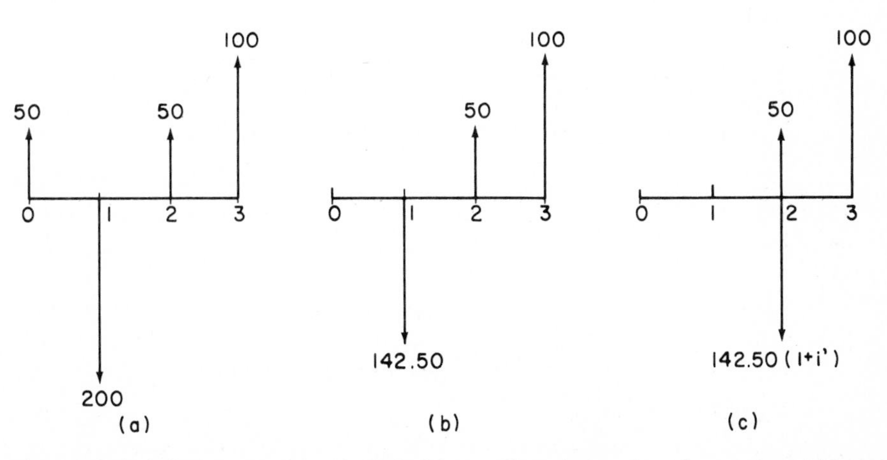

Figure 7.4 Cash-flow sequence in (*a*) original form, (*b*) equivalent form in year 1, and (*c*) equivalent form in year 2, used to compute the composite rate of return i'.

The net investment for year $t = 0$ is $F_0 = C_0 = 50$. By Eq. (7.5) in year $t = 1$, the \$50 returns $c = 15\%$, since $F_0 > 0$ and for $t = 1$ (Fig. 7.4*b*),

$$F_1 = 50(1 + 0.15) - 200 = -\$142.50$$

Since the project value is negative in $t = 1$, the value F_1 earns at the rate i' for the next year.

$$F_2 = -142.50(1 + i') + 50$$

Since this result must be negative for all $i' > 0$, use i' to find F_3 (Fig. 7.4*c*).

$$F_3 = F_2(1 + i') + C_3 = (-142.50(1 + i') + 50)(1 + i') + 100 \qquad (7.6)$$

Setting Eq. (7.6) equal to zero and solving for i' will result in the unique composite rate of return for the cash-flow sequence. If there had been more F_t expressions, i' would be used in all subsequent equations (see Example 7.8).

This project-net-investment procedure to find i' may be summarized as follows:

1. Draw a cash-flow diagram of the original cash-flow sequence.
2. Develop the series of project net investments using Eq. (7.5) and the stated c value. The result is the F_n expression in terms of i'.
3. Set the F_n expression equal to 0 and find the i' value to balance the equation. If necessary interpolate to determine i'.

Several comments are in order before we show an example. If the reinvestment rate c equals i', the value i' will be the same as i^*, the internal rate of return, because the i^* value assumes reinvestment at the internal rate (see the definition of IRR). The closer the c value is to i', the smaller the difference between i' and i^*; but since i' is unknown when c must be determined–prior to solution for i'–it may be difficult to minimize the impact on i'. It is common to use c = MARR, realizing that all receipts can realistically be reinvested at the minimum acceptable rate of return.

Example 7.5 Compute the composite rate of return for the synthetic lubricant investment in Example 7.4 if the reinvestment rate is (*a*) 7.47% and (*b*) 20%.

SOLUTION

(*a*) Use the procedure above to determine i' for $c = 7.47\%$.

1. Figure 7.5 shows the original cash flow.
2. The first-project-net-investment expression is $F_0 = \$2000$. Since $F_0 > 0$, use $c = 7.47\%$ to get F_1 by Eq. (7.5).

$$F_1 = 2000(1.0747) - 500 = \$1649.40$$

Again $F_1 > 0$, so for $c = 7.47\%$,

$$F_2 = 1649.40(1.0747) - 8100 = -\$6327.39$$

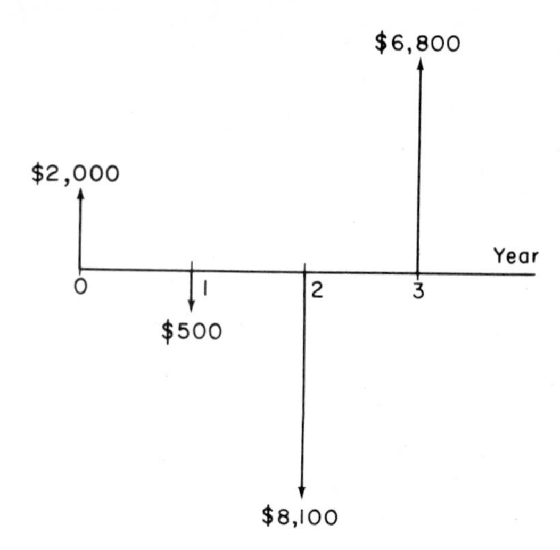

Figure 7.5 Original cash flow (÷ $1000) for Example 7.5.

Figure 7.6 shows the equivalent cash flow at this time. Since $F_2 < 0$, use i' to express F_3.

$$F_3 = -6327.29(1 + i') + 6800$$

3. Let $F_3 = 0$ and solve for i'

$$-6327.39(1 + i') + 6800 = 0$$

$$1 + i' = \frac{6800}{6327.39} = 1.0747$$

$$i = 0.0747 \qquad (7.47\%)$$

The composite rate of return is 7.47%, which is the same as c, the reinvestment rate, and the i_1^* value discussed in Example 7.4. Note that $i = 41.35\%$; the other i^* value no longer balances the rate-of-return equation. The equiva-

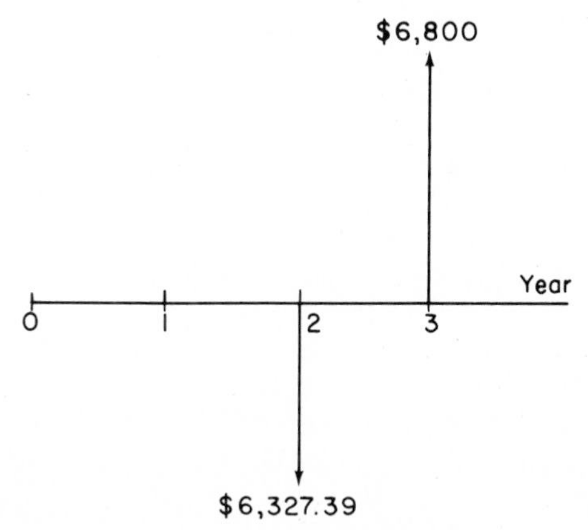

Figure 7.6 Equivalent cash flow (÷ $1000) of Fig. 7.5 at a reinvestment rate of 7.4%.

lent future worth result for $i' = 41.35\%$ and the cash flow in Fig. 7.6 is

$$6327.39(F/P, 41.35\%, 1) = 8943.77 \neq 6800$$

which indicates a return much lower than 41.35%.

(*b*) In summary for $c = 20\%$, the net investment series is

$$F_0 = 2000 \qquad (F_0 > 0, \text{use } c)$$

$$F_1 = 2000(1.20) - 500 = \$1900 \qquad (F_1 > 0, \text{use } c)$$

$$F_2 = 1900(1.20) - 8100 = -\$5820 \qquad (F_2 < 0, \text{use } i')$$

$$F_3 = -5820(1 + i') + 6800$$

Set $F_3 = 0$ and solve for i' directly

$$(1 + i') = \frac{6800}{5820} = 1.1684$$

$$i' = 0.1684 \qquad (16.84\%)$$

The composite rate of return is $i' = 16.84\%$ at a reinvestment rate of 20%, which is a marked increase from $i' = 7.47\%$ at $c = 7.47\%$.

COMMENT If the value $c = 41.35\%$ is used the equation will be balanced with $i' = i_2^* = 41.35\%$ where i_2^* is the second internal rate of return in Example 7.4. This is possible because the assumption that receipts are reinvested at the i^* value is correct if $c = i_2^* = 41.35\%$.

It is possible to summarize the relations between c, i', and any i^* as follows:

Relation between reinvestment c and IRR i^*	Relation between i' and IRR i^*
$c = i^*$	$i' = i^*$
$c < i^*$	$i' < i^*$
$c > i^*$	$i' > i^*$

These relations can be found to be correct in Example 7.5.

Solved Problem 7.8
Probs. P7.17 to P7.21

SOLVED PROBLEMS

Example 7.6 Assume that a couple invest \$10,000 now and \$500 three years from now and will receive \$500 one year from now, \$600 two years from now, and

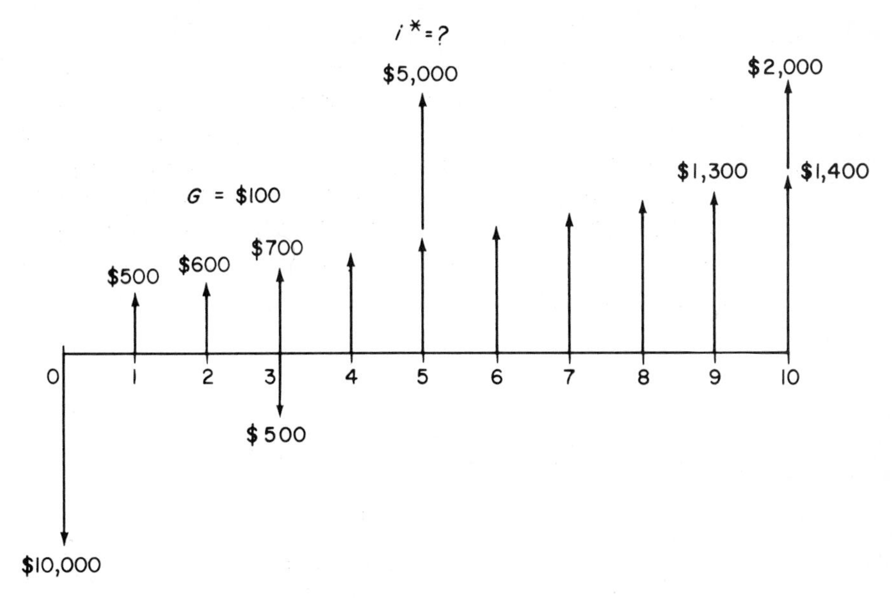

Figure 7.7 Cash flow, Example 7.6.

amounts increasing by \$100 per year for a total of 10 years. They will also receive lump-sum payments of \$5000 in 5 years and \$2000 in 10 years. Calculate the rate of return on their investment.

SOLUTION The cash-flow diagram in Fig. 7.7 is used to set up a rate-of-return relation using the present-worth method.

$$0 = -10{,}000 - 500(P/F, i^*\%, 3) + 500(P/A, i^*\%, 10) + 100(P/G, i^*\%, 10)$$
$$+ 5000(P/F, i^*\%, 5) + 2000(P/F, i^*\%, 10)$$

Solving by trial and error and interpolating between $i = 7\%$ and $i = 8\%$, we find $i^* = 7.8\%$.

COMMENT Note that the single values at years 3, 5, and 10 were handled separately so that the P/A and P/G factors could be used. Use the procedure of Sec. 7.2 to estimate the interest rate by the P/A factor for $n = 10$ years. Assume that the gradient term is an A with an average value of \$500. Your estimate should show that i is in the neighborhood of 7%.

Sec. 7.2

Example 7.7 Assume a series of cash flows as shown for an ongoing project. (Data are adapted from an article by McLean [3] and some results of Barish and Kaplan [4].) The negative net cash flow in year 4 is the result of a major alteration to the project. Compute the rate of return for the project.

Year	Net cash flow	Year	Net cash flow
1	$ 200	6	$500
2	100	7	400
3	50	8	300
4	-1800	9	200
5	600	10	100

SOLUTION The present-worth method, Eq. (7.1), can be used to set up the rate-of-return equation.

$$0 = 200(P/F, i\%, 1) + 100(P/F, i\%, 2) + \cdots + 100(P/F, i\%, 10) \qquad (7.7)$$

Tabulation of the results of the right side of Eq. (7.7) at several values of i yield the following values, which are plotted in Fig. 7.8.

i, %	10	20	30	40	50
Results of Eq. (7.7), $	+198	+42	-2	-8	+1

As you can see, the two values which satisfy Eq. (7.7) are at approximately $i_1^* = 29\%$ and $i_2^* = 49\%$. There are two i^* values because the cash flow is nonconventional and two sign changes are present.

Sec. 7.4

Example 7.8 Determine the composite rate of return in Example 7.7 if $c = 15\%$ is the stated reinvestment rate.

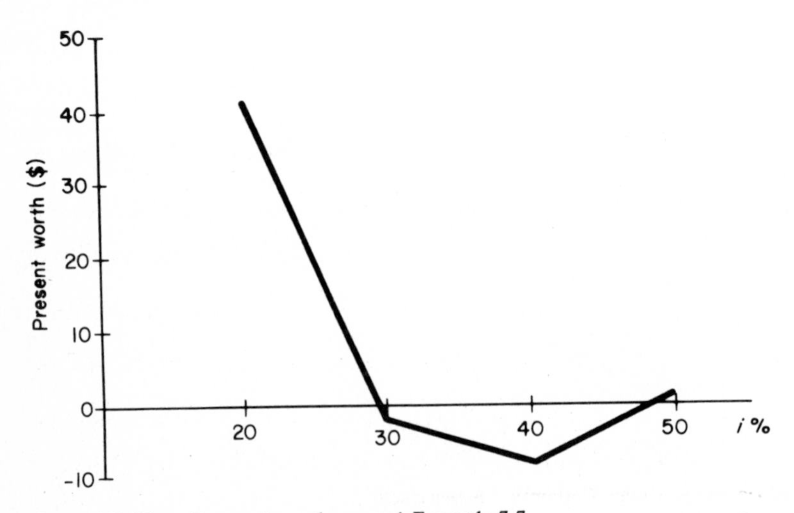

Figure 7.8 Plot of present worth versus i, Example 7.7.

SOLUTION Use the steps in Sec. 7.5 to write the net investment series for $t = 1$ to $t = 10$.

$$F_1 = C_0 = \$200 \qquad (F_0 > 0, \text{use } c)$$

$$F_2 = 200(1.15) + 100 = \$330 \qquad (F_1 > 0, \text{use } c)$$

$$F_3 = 330(1.15) + 50 = \$429.50 \qquad (F_2 > 0, \text{use } c)$$

$$F_4 = 429.50(1.15) - 1800 = -\$1306.08 \qquad (F_3 < 0, \text{use } i')$$

$$F_5 = -1306.08(1 + i') + 600$$

Since F_5 may be greater than zero or less than zero, all remaining expressions use i'.

$$F_6 = F_5(1 + i') + 500 = [-1306.08(1 + i') + 600](1 + i') + 500$$

$$F_7 = F_6(1 + i') + 400$$

$$F_8 = F_7(1 + i') + 300$$

$$F_9 = F_8(1 + i') + 200$$

$$F_{10} = F_9(1 + i') + 100$$

To find i', the rather cumbersome expression $F_{10} = 0$ must be solved. Computerized solution using ROIDS or its equivalent is desirable, but manual solution is possible to find $i' = 21.25\%$.

COMMENT At this time you might want to rework this problem with a reinvestment rate of 29 or 49%, as found in Example 7.7, to see that the i' value will be the same as these reinvestment rates; that is, if $c = 29\%$, then $i' = i^* = 29\%$. In the example above, $c = 15\%$, which is less than $i^* = 29\%$ so $i' = 21.25\% < i^*$, as discussed at the end of Sec. 7.5.

Sec. 7.5

REFERENCES

1. L. E. Bussey, *The Economic Analysis of Industrial Projects*, Prentice-Hall, Englewood Cliffs, NJ, 1978, pp. 220–238.
2. D. G. Newnan, *Engineering Economic Analysis*, rev. ed., Engineering Press, San Jose, CA, 1980, pp. 373–389.
3. J. G. McLean, "How to Evaluate New Capital Investments," *Harvard Business Review*, November–December 1958, pp. 59–69.
4. N. N. Barish and S. Kaplan, *Economic Analysis for Engineering and Managerial Decision Making*, 2d ed., McGraw-Hill, New York, 1978, p. 204.
5. J. Lorie and L. J. Savage, "Three Problems in Capital Rationing," *Journal of Business*, vol. 28, no. 4, October 1955, pp. 229–239.
6. D. Teichroew, A. A. Robichek, and M. Montalbano, "Mathematical Analysis of Rates of Return Under Certainty," *Management Science*, vol. 11, no. 3, January 1965, pp. 395–403.
7. D. Teichroew, A. A. Robichek, and M. Montalbano, "An Analysis of Criteria for Investment and Financial Decisions Under Certainty," *Management Science*, vol. 12, no. 3, November, 1965, pp. 151–179.

FURTHER INFORMATION

Multiple rate of return considerations (Secs. 7.4 and 7.5): Barish and Kaplan [2], pp. 203-206; Bussey [3], pp. 212-238; Grant, Ireson, and Leavenworth [7], pp. 591-610; Newnan [11], pp. 128-140; Stevens [15], pp. 74-79; Thuesen, Fabrycky, and Thuesen [17], pp. 151-154.

PROBLEMS

P7.1 State a definition for rate of return.

P7.2 Derive a formula to compute the beginning unrecovered balance B_{t+1} for year $t + 1$ computed in Table 7.1 in terms of B_t, the interest rate i, and the cash flow C_t for year t. Demonstrate that your formula works.

P7.3 A real estate investor purchased a piece of property for \$6000 and sold it 17 years later for \$21,000. The property taxes were \$80 the first year and \$90 the second year and increased by \$10 per year until the property was sold. What was the rate of return on the investment?

P7.4 If the property taxes in Prob. P7.3 increased by \$10 per year for the first 6 years and then increased by \$20 per year thereafter, what was the rate of return on the investment?

P7.5 A family purchased a rundown house for \$25,000 with the idea of making major improvements and then selling for a profit. In the first year that they owned the house, they spent \$5000 on improvements. They spent \$1000 the second year and \$800 the third year. In addition, they paid property taxes of \$500 per year for 3 years and then sold the house for \$35,000. What rate of return did they make on their investment?

P7.6 The Karry-Mor Trucking Company purchased a new dump truck for \$54,000. The total operating expenses were \$36,000 the first year, \$39,000 the second, and amounts increasing by \$3000 per year thereafter. The income for the first year was \$66,000 and decreased by \$500 per year thereafter. If the company kept the truck for 10 years and then sold it for \$15,000, what was the rate of return on the investment?

P7.7 If a company spends \$6000 now and \$900 per year for 17 years with the first disbursement 5 years from now, what rate of return would the company receive if the income during the 21 years was \$5000 at the end of year 3 and \$1200 per year thereafter?

P7.8 A real estate investor purchased a building for \$130,000 and received \$2000 per month in rent. The taxes were \$2500 per year and maintenance costs were \$700 every 3 years. If the building was sold for \$180,000 twelve years after it was bought, what nominal annual and effective montly rate of return was made on the investment?

P7.9 Rework Prob. P7.3 assuming that the property taxes increased by \$50 per year.

P7.10 An investor purchased three types of stocks (identified here as A, B, and C). The investor purchased 200 shares of A at \$13 per share, 400 shares of B at \$4 a share, and 100 shares of C at \$18 per share. The dividends were \$0.50 a share from stock A for 3 years and then the stock was sold for \$15 a share. There were no dividends from stock B, but the stock was sold for \$5.50 a share 2 years after it was purchased. Stock C resulted in dividends of \$2.10 a share for 10 years; but due to a depressed market at the time it was sold, the stock sold for only \$12 a share. Calculate the rate of return on each stock as well as the overall rate of return on the stock investments.

P7.11 Firewood can be purchased during July for \$55 a cord. If the purchaser waits until November, the cost of the same kind of wood is \$70 a cord. What rate of return would the purchaser receive if the firewood was purchased in July instead of November?

P7.12 A careful shopper is trying to decide between buying an artificial Christmas tree and continuing to buy cut trees. The artificial tree costs \$34 and can be used for 8 years, at which time it will be thrown away for no salvage value. Alternatively, the shopper can continue to buy cut trees at a cost of $8 now, $9 next year, $10 two years from now, and so on for 8 years. If the artificial tree is purchased, what rate of return is made on the investment?

P7.13 A homeowner who plans to build a large brick patio is expecting the price of bricks to increase considerably in the next 2 years. The cost is expected to increase from the present price of $160 per 1000 bricks to $175 per 1000 next year and $195 the following year. If 3000 bricks are purchased now instead of 1000 now and 1000 each year for the following 2 years, what will be the rate of return on the investment?

Note: For Probs. P7.14 to P7.21 a computer program may be developed to solve the problem more rapidly. (The ROIDS program introduced in Appendix E may be used if desired or the reader may develop his or her own program.)

***P7.14** A currently funded project is expected to have the following cash flow for the next 5 years.

Year	0	1	2–3	4	5
Cash flow, $	-10,000	10,000	0	30,000	-30,000

(*a*) Compute and plot the present worth at the following interest rates: 0, 10, 20, 30, and 40%.
(*b*) Determine the approximate rate-of-return values from the graph in part *a*.
(*c*) Use the rate-of-return equation to find the rate of return for the cash-flow sequence. Which value is correct?

***P7.15** Find all rate-of-return values between 0 and 100% for the following cash flows.

Year	0	1	2	3–6
Cash flow, $	500	-1000	50	200

If hand solution rather than computerized solution is used, factor values for $i > 50\%$ must be computed from the formulas since tables do not extend above 50%.

***P7.16** Write a computer program to find multiple rates of return and use it to find the i^* values for the cash flows in Example 7.4.

***P7.17** A commonly referenced multiple rate-of-return problem is the "pump problem" [5]. The cash flow generated by installing a new pump is observed to be

Year	0	1	2
Cash flow, $	1600	10,000	-10,000

(*a*) Determine the rate-of-return equation solutions and (*b*) interpret their meaning in terms of internal and composite rates of return.

***P7.18** Compute (*a*) the internal rates of return and (*b*) the composite rate of return if $c = 15\%$ for the following cash flow (same data as used in the discussion of Sec. 7.5).

Year	0	1	2	3
Cash flow, $	50	-200	50	100

***P7.19** (*a*) A company invested $5000 in auxiliary equipment and procedure changes for ore removal one year ago. Increased income was observed to be $25,000 this year. What is the calculated rate of return on the investment?

(*b*) If the cash flow is −$23,000 next year, the cash-flow sequence changes signs twice. Compute the rate-of-return values for the three cash-flow values and interpret their meaning.

(*c*) If a reinvestment rate of 30% is used, find the composite rate of return and discuss its relation to the answers in *b*.

(*d*) Find the composite rate of return if a reinvestment rate of 15% is used.

***P7.20** Explain why the composite rate of return (i') will be equal to a multiple rate-of-return value (i^*) if the assumption is made that the reinvestment rate (c) is equal to i^*.

***P7.21** (*a*) Use a computer program to find i' for different values of the reinvestment rate c between 15 and 50% for the three cash-flow values in P7.19*a* and *b*.

(*b*) Construct a plot of i' versus c values using the results of part *a*.

CHAPTER

EIGHT

RATE-OF-RETURN EVALUATION FOR MULTIPLE ALTERNATIVES

This chapter presents the methods by which two alternatives can be evaluated using rate-of-return comparison. This type of evaluation will result in the same selection as the present-worth and EUAC analysis. Selection of one alternative from several mutally exclusive alternatives is also discussed in this chapter.

SECTION OBJECTIVES

To complete this chapter you must be able to:

8.1. Prepare a tabulation of net cash flow for two alternatives having equal or different lives, given the details of each alternative.
8.2. State the rationale used to evaluate two alternatives using the incremental rate-of-return method.
8.3. Select the better of two alternatives on the basis of the incremental rate of return computed by the present-worth method, given the initial cost, life, and salvage value of each alternative, times and amounts of cash flows, and minimum attractive rate of return.
8.4. Select the alternative as in 8.3, except using the EUAC method.
8.5. State the criteria used to select one alternative from several *mutually exclusive alternatives*, and use the given procedure to compute the incremental rates of return and select the one alternative, given the initial cost, salvage values, life, and cash flow for each alternative, and the minimum attractive rate of return.

STUDY GUIDE

8.1 Tabulation of Net Cash Flow

The preparation of a cash-flow tabulation was discussed briefly in Chap. 7 with respect to net cash flow for a single alternative. In this chapter, it will be necessary to prepare a cash-flow tabulation for each of two alternatives as well as the net cash flow that results when the annual cash flows of the two alternatives are compared. The column headings for a cash-flow tabulation involving two alternatives are shown in Table 8.1. If the alternatives have equal lives, the years column will go from 0 to n, the life of the alternatives. If the alternatives have unequal lives, the years column will go from 0 to the least common multiple of the two lives. The use of the least-common-multiple rule is necessary because rate-of-return analysis on the net cash-flow values must always be done over the same number of years for each alternative (as is the case with present-worth comparisons). If the least common multiple of lives is tabulated, reinvestment in each alternative is shown at appropriate times (as was done in Chap. 5 for cash flow in present-worth analysis).

You will see in this chapter that a cash-flow tabulation is an integral part of the procedure for selecting one of two alternatives on the basis of incremental rate of return. Therefore, a standardized format for the tabulation will simplify interpretation of the final results. In this chapter, the alternative with the *higher initial investment* will always be regarded as *alternative B*. That is,

$$\text{Net cash flow} = \text{cash flow}_B - \text{cash flow}_A$$

The next two examples demonstrate cash-flow tabulation for equal-life and unequal-lived alternatives.

Table 8.1 Format for cash-flow tabulation

	(1)	(2)	(3) = (2) − (1)
	Cash flow		
Year	Alternative *A*	Alternative *B*	Net cash flow
0			
1			
2			
. . .			

Example 8.1 A tool and die company is considering the purchase of an additional milling machine. The company has the opportunity to buy a slightly used machine for \$15,000 or a new one for \$21,000. Because the new machine is a more sophisticated model with some automatic features, its operating cost is expected to be \$7000 per year, while the old machine is expected to cost \$8200 per year. The machines are expected to have a 25-year life with 5% salvage values. Tabulate the net cash flow of the two alternatives.

Table 8.2 Cash-flow tabulation for Example 8.1

Year	Cash flow		Net cash flow (new-old)
	Old mill	New mill	
0	$ -15,000	$ -21,000	$ -6,000
1-25	-8,200	-7,000	+1,200
25	+750	+1,050	+300
Total	$-219,250	$-194,950	$+24,300

savings comparing

SOLUTION Net cash flow is tabulated in Table 8.2. The salvage values in year 25 are separated from ordinary cash flow for clarity. Note that a sign must be included to indicate a disbursement (minus) or an income (plus).

COMMENT Note that when the cash-flow columns are subtracted, the difference between the totals of the two alternatives should equal the total of the net cash-flow column. This will provide a check of your addition and subtraction in preparing the tabulation.

When disbursements are the same for a number of consecutive years, it saves time to make a single cash-flow listing, as is done for years 1 to 25 of the example. However, remember that several years were combined when adding to get the column totals.

Example 8.2 The Fresh-Pak Tomato Cannery has under consideration two different types of conveyors. Type *A* has an initial cost of $7000 and a life of 8 years. The initial cost of type *B* is $9500 and has a life expectancy of 12 years. The operating cost for type *A* is expected to be $900, while the cost for type *B* is expected to be $700. If the salvage values are $500 and $1000 for type *A* and type *B* conveyors, respectively, (*a*) tabulate the cash flows of each alternative and (*b*) tabulate the net cash flow using the least common multiple of lives for present-worth analysis.

SOLUTION

(*a*) The tabulation of each asset (Table 8.3) shows the cash flows for the respective lives–8 years for *A* and 12 for *B*.

Table 8.3 Cash flow for respective asset life for Example 8.2*a*

Year	Cash flow	
	Type *A*	Type *B*
0	$-7000	$-9500
1-7	-900	-700
8	-900 + 500	-700
9-11	...	-700
12	...	-700 + 1000

(*b*) The least common multiple of years between 8 and 12 is 24 years. The net cash-flow tabulation for 24 years is given in Table 8.4. Note that the reinvestment and salvage values are shown in years 8 and 16 for type *A* and in year 12 for type *B*.

Table 8.4 Cash flow for 24 years for unequal-lived assets, Example 8.2*b*

Year	Cash flow Type *A*	Cash flow Type *B*	Net cash flow (*B* – *A*)
0	$ -7,000	$ -9,500	$-2,500
1-7	-900	-700	+200
8	-7,000 -900 +500	-700	+6,700
9-11	-900	-700	+200
12	-900	-9,500 -700 +1,000	-8,300
13-15	-900	-700	+200
16	-7,000 -900 +500	-700	+6,700
17-23	-900	-700	+200
24	-900 +500	-700 +1,000	+700
	$-41,100	$-33,800	$+7,300

Probs. P8.1 to P8.4

8.2 Interpretation of Rate of Return on Extra Investment

The first step in calculating the rate of return on the extra investment between two alternatives is the preparation of a cash-flow tabulation similar to that of Table 8.2 or 8.4. When the evaluation is conducted by the present-worth method, the least common multiple of lives must be used for the study period. The net cash-flow column then reflects the *extra investment* that would be required if the alternative with the larger first cost were selected. Thus, in Example 8.1 the new milling machine would require an extra investment of $6000, as shown in the last column of Table 8.2. Additionally, if the new machine were purchased, there would be a "savings" of $1200 per year for 25 years, plus $300 in year 25 as a result of the difference in salvage values. The decision about whether to buy the old or the new milling machine can be made on the basis of the profitability of investing the extra $6000 in the new machine. If the present worth of the savings is greater than the present worth of the extra investment using the company's minimum acceptable rate of return (MARR), then the extra investment should be made (i.e., the higher first-cost proposal should be accepted).

On the other hand, if the present worth of the savings is less than the present worth of the extra investment, then the lower first-cost proposal should be accepted.

Note that if the new mill in Table 8.2 is selected, there will be a net savings of \$24,300. Keep in mind that this figure does not take into account the time value of money, since this total was obtained by adding the values for the various years without using the interest factors and cannot therefore be used as a basis for the decision. The totals at the bottom of the table serve only as a check against the additions and subtractions for the individual years. In fact, the \$24,300 is the present worth of net cash flow at $i = 0\%$.

The rationale for making the decision is the same as if only *one alternative* were under consideration–that alternative being the one represented by the difference (net cash-flow) column in the cash-flow tabulation. When viewed in this manner, it is obvious that unless this investment yields a rate of return greater than the MARR, the investment should not be made (meaning that the lower-priced alternative should be selected *to avoid this extra investment*). However, if the rate of return on the difference investment is greater than MARR, the investment should be made (meaning that the higher-priced alternative should be selected).

Prob. P8.5

8.3 Incremental-Rate-of-Return Evaluation Using the Present-Worth Method

The information in Chap. 7 and the previous sections of this chapter are used to evaluate two alternatives by the incremental-rate-or-return method. The basic procedure given here assumes that all cash flows are negative (except salvage value) and that one of the two alternatives will be selected. Therefore, only the incremental investment is analyzed. The method involving alternatives with positive cash flows is detailed in Sec. 8.5.

The procedure is as follows:

1. Order the alternatives and call the one with the smaller initial investment alternative A.
2. Prepare the cash-flow and net cash-flow tabulation using the least common multiple of years.
3. Draw a net cash-flow diagram.
4. Set up and find the incremental return i^*_{B-A} using the present-worth method, Eq. (7.1), or a computer program such as the one discussed in Appendix E. [Check for sign changes in the net cash-flow sequence which indicate the possible presence of multiple rates of return (Secs. 7.4 and 7.5).]
5. If $i^*_{B-A} <$ MARR, select alternative A. If $i^*_{B-A} \geqslant$ MARR, select alternative B.

This method may be faster if the i^*_{B-A} value is estimated manually rather than achieved using precise linear interpretation, provided that an exact rate-of-return value is not required. For example, if MARR is 15% and you have established that i^*_{B-A} is in the 15-to-20% range, an exact value is not necessary to accept B, since $i^*_{B-A} \geqslant$ MARR. The procedure for the rate-of-return analysis is illustrated in Examples 8.3 and 8.4.

Example 8.3 A manufacturer of boys' pants is considering purchasing a new sewing machine, which can be either semiautomatic or fully automatic. The estimates for each are as follows:

	Semiautomatic	Fully automatic
First cost	\$8,000	\$13,000
Annual disbursements	3,500	1,600
Salvage value	0	2,000
Life, years	10	5

Determine which machine should be selected if the MARR is 15%.

SOLUTION Use the procedure above.

1. Alternative A is the semiautomatic (s) and alternative B is the fully automatic (f) machine.
2. The cash flows for 10 years are tabulated in Table 8.5.
3. The net cash-flow diagram is given in Fig. 8.1.
4. The incremental-rate-of-return equation for net cash flow is

$$0 = -5000 + 1900(P/A, i\%, 10) - 11{,}000(P/F, i\%, 5) + 2000(P/F, i\%, 10)$$

   Solution shows that i^*_{f-s} is between 12 and 15%. By interpolation $i^*_{f-s} = 12.65\%$.
5. Since the rate of return on the extra investment is less than the 15% minimum attractive rate, the lower-cost or semiautomatic machine should be purchased. If $i^*_{f-s} \geqslant 15\%$ had been determined, the fully automatic machine would have been selected.

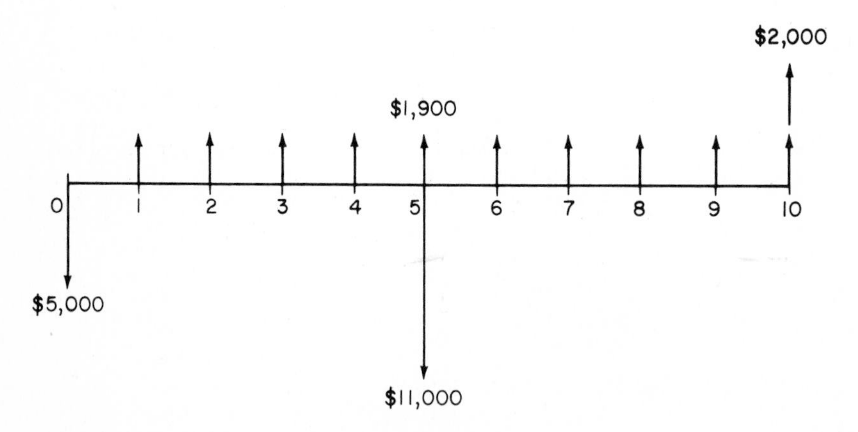

Figure 8.1 Diagram of net cash flow, Example 8.3.

Table 8.5 Cash-flow tabulation for Example 8.3

	(1)	(2)	(3) = (2) − (1)
	Cash flow		
Year	Semiautomatic	Fully automatic	Difference
0	$ −8,000	$−13,000	$ −5,000
1–5	−3,500	−1,600	+1,900
5	. . .	{ +2,000 −13,000 }	−11,000
6–10	−3,500	−1,600	+1,900
10	. . .	+2,000	+2,000
	$−43,000	$−38,000	$ +5,000

COMMENT In step 4 a check of the net cash-flow sign sequence in Table 8.5 indicates that there may be multiple (up to 3) incremental-rate-of-return values. The analysis above, according to the discussion in optional Sec. 7.5, assumes that the positive net cash flows of $1900 in years 1 to 5 are reinvested at an external rate of c = 12.65%. If this is not a reasonable assumption, the procedure in Sec. 7.5 should be performed using an appropriate reinvestment rate to determine a different i^*_{f-s} value which is compared with MARR = 15%.

The incremental rate of return obtained above can actually be interpreted as a *break-even* value of i, that is, the rate of return at which either alternative might be selected. If the i^* found by the rate-of-return equation is greater than the MARR, the larger-investment alternative is selected. As an illustration, the break-even rate for the incremental investment in Example 8.3 is 12.65%. Figure 8.2 is a general plot of the present-worth values of net cash flow for different rates of return. At values of $i < 12.65\%$, the present worth for the fully automatic is less than that of the semiautomatic machine. For $i > 12.65\%$, the fully automatic present worth is larger. Thus if MARR = 10%, select the fully automatic machine; whereas if MARR = 15%, as in Example 8.3, select the semiautomatic, because the break-even i value is less than 15%.

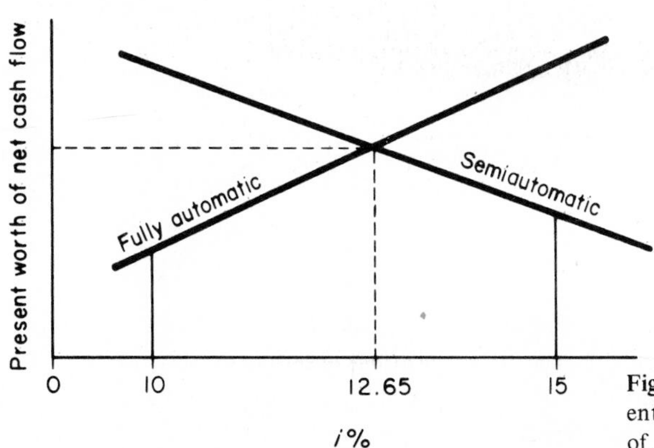

Figure 8.2 Break-even chart of present worth of net cash flow versus rate of return, Example 8.3

Example 8.4 Determine which milling machine should be purchased in Example 8.1 using MARR = 15%.

SOLUTION Since the alternatives are equal-lived, the net cash flows of Table 8.2 are used with the procedure above and Eq. (7.1) to set up the rate-of-return equation.

$$0 = -6000 + 1200(P/A, i\%, 25) + 300(P/F, i\%, 25)$$

The interpolated break-even rate of return is 19.79%. (Try solving for the value yourself *now*, using the procedure of Sec. 7.2 to estimate the first-trial interest rate.) Since 19.79% > MARR = 15%, the purchase of the new milling machine is justified.

COMMENT You should realize that you *can* compare two alternatives, A and B, using the actual (rather than the incremental) cash flows with the help of the general relation

$$0 = \text{PW}_B - \text{PW}_A$$

where alternative B has the larger first cost. In the milling-machine trade-off analysis, this approach results in the following:

$$\text{PW}_{\text{new}} = -21{,}000 - 7000(P/A, i\%, 25) + 1050(P/F, i\%, 25)$$

$$\text{PW}_{\text{old}} = -15{,}000 - 8200(P/A, i\%, 25) + 750(P/F, i\%, 25)$$

Then

$$\begin{aligned} 0 &= \text{PW}_{\text{new}} - \text{PW}_{\text{old}} \\ &= (-21{,}000 + 15{,}000) + (-7000 + 8200)(P/A, i\%, 25) \\ &\quad + (1050 - 750)(P/F, i\%, 25) \\ &= -6000 + 1200(P/A, i\%, 25) + 300(P/F, i\%, 25) \end{aligned}$$

Note that the reduced form is identical to that used in the solution to this example. This method is not advised for present-worth analysis, since assets with different lives require reinvestment for comparison purposes. Try, for example, the data of Example 8.2. You will find it gets messy and can result in possible mistakes when the rate of return is computed.

Probs. P8.6 to P8.10

8.4 Incremental-Rate-of-Return Evaluation Using the EUAC Method

Even though the use of present-worth computations to obtain i^* for alternative evaluation is recommended, the conclusions must be identical whether you use the present-worth or EUAC method. On some problems, however, you might find EUAC computationally simpler. Remember that for the present-worth method the least common multiple of years must be used in the analysis. For the EUAC this is not always necessary, depending on whether the lives are equal or unequal. Only if lives are *equal* can the EUAC of the net cash flow be used, in which case the salvage sinking-fund method (Sec. 6.2) is applied to solve for i^*_{B-A}. The rate-of-return equation takes the general form

$$0 = \pm \Delta P(A/P, i\%, n) \pm \Delta \text{SV}(A/F, i\%, n) \pm \Delta A \qquad (8.1)$$

where the Δ (delta) symbol identifies P, SV, and A as differences between the alternatives in the net cash-flow tabulation. Manual interpolation in the tables or a computer program such as ROIDS (Appendix E) may be used to determine i^*_{B-A}.

If lives are *unequal* the EUAC for one cycle of each alternative's cash flows is determined and i^*_{B-A} is found from

$$0 = \text{EUAC}_B - \text{EUAC}_A \qquad (8.2)$$

Note that the net cash flow is not used (and need not be determined) in this analysis, but the rate of return is on the incremental *equivalent uniform* annual cost between the alternatives.

The procedure is the same as in Sec. 8.3 except that in step 4 either Eq. (8.1) or (8.2) is used to solve for the rate of return. The next two examples demonstrate the EUAC method.

Example 8.5 Compare the two milling machines in Example 8.1 using the EUAC method to compute the incremental rate of return. Assume that MARR = 15%.

SOLUTION Since the lives are equal (25 years), Eq. (8.1) can be used to solve for the incremental rate of return between new (n) and old (o) mills using the net cash flows of Table 8.2.

$$0 = -6000(A/P, i\%, 25) + 300(A/F, i\%, 25) + 1200$$

Balance occurs at $i^*_{n-o} = 19.79\%$, so the extra investment is justified for the new milling machine (as it was with the present-worth method).

COMMENT The value $i = 19.79\%$ is the break-even rate of return. A graph similar to Fig. 8.2 may be constructed in which EUAC replaces present worth. For values of MARR > 19.79%, the old milling machines should be purchased; for values of MARR < 19.79%, as is the case here, select the new milling machine.

Example 8.6 Compare the sewing machines of Example 8.3 using the EUAC method with a MARR of 15%.

SOLUTION Present-worth analysis on the net cash flow in Example 8.3 shows that the semiautomatic machine should be purchased. For the EUAC we can use the salvage sinking-fund method, Eq. (6.1), and the general EUAC form in Eq. (8.2), since lives are unequal. For the respective lives of 5 years for fully automated (f) and 10 years for semiautomatic (s),

$$\text{EUAC}_f = -13{,}000(A/P, i\%, 5) + 2000(A/F, i\%, 5) - 1600$$

$$\text{EUAC}_s = -8000(A/P, i\%, 10) - 3500$$

Using the rate-of-return relation $0 = \text{EUAC}_f - \text{EUAC}_s$,

$$0 = -13{,}000(A/P, i\%, 5) + 2000(A/F, i\%, 5) + 8000(A/P, i\%, 10) + 1900$$

At $i = 12\%$, $0 \neq \$+24.33$; for $i = 15\%$, $0 \neq \$-87.52$. Interpolation yields i^*_{f-s} = 12.65% (as for the present-worth method); and the semiautomatic should be purchased, since 12.65% is less than the MARR of 15%.

COMMENT It is important to remember that if an EUAC analysis is to be made on the *net cash flow*, the cash-flow tabulation must be extended for the least common multiple of lives (10 years in this example), as in the present-worth method.

Probs. P8.11 to P8.15

8.5 Selection from Mutually Exclusive Alternatives Using Rate-of-Return Analysis

The analysis in this section, referred to as *multiple-alternative evaluation*, involves more than two alternatives. In the cases discussed, one alternative precludes the acceptance of other alternatives, so the alternatives are called *mutually exclusive*. Thus, if a particular part is to be purchased and four vendors are available, only one vendor can be selected to supply the part.

As in any selection problem in engineering economics, there are several correct solution techniques. The present-worth and EUAC methods are the simplest and most straightforward techniques. Using a specified MARR, we compute the total present-worth or EUAC value for each mutually exclusive alternative. The alternative that has the most favorable present-worth or EUAC value is the one selected. The EUAC method is illustrated in Example 8.9. Another method frequently used is rate of return. When the rate-of-return method is applied, the firm requires that the entire investment return at least the minimum attractive rate of return (MARR). When the returns on several alternatives equal or exceed the MARR, at least the alternative requiring the lowest investment is justified. However, since more capital can be invested in other acceptable alternatives, the incremental investment required must also be justified. If the return on the extra investment exceeds the MARR, the entire investment should be made in order to maximize the return. Thus, for rate-of-return analysis, the following criteria are used to select one mutually exclusive project: select the one alternative that (1) requires the *largest investment and* (2) indicates that the *incremental investment over another acceptable alternative is justified* in that the return is at least the MARR. Therefore, the most important rule to remember when evaluating alternatives by the incremental-rate-of-return method is that *an alternative can never be compared with one for which the incremental investment has not been justified*. The analysis procedure is:

1. Order the alternatives in terms of increasing *initial* investment.

2. Considering the "do-nothing" alternative as a defender, compute the overall rate of return i^* for the alternative with the lowest initial investment.
3. If $i^* < \text{MARR}$, remove the lowest investment alternative from further consideration and compute the overall rate of return for the next higher-investment alternative. Repeat this step until $i^* \geqslant \text{MARR}$ for one of the alternatives. When $i^* \geqslant \text{MARR}$, the lowest-investment alternative becomes the defender and the next higher-investment alternative is the challenger.
4. Determine the net (incremental) cash flow between the challenger and the defender.
5. Calculate the rate of return on the incremental investment required by the challenger using the net cash flow.
6. If the rate of return calculated in step 5 is greater than the MARR, the challenger becomes the defender and the previous defender is removed from further consideration. Conversely, if the rate of return in step 5 is less than the MARR, the challenger is removed from further consideration and the defender remains as the defender against a new challenger.
7. Repeat steps 4 to 6 until only one alternative remains.

Note that in the incremental analysis (steps 4 to 6), only *two* alternatives are compared at any one time. It is very important, therefore, that the correct alternatives be compared. Unless the procedure is followed as presented above, the wrong alternative can be selected from the incremental analysis. The procedure detailed above is illustrated in Examples 8.7 and 8.8.

Example 8.7 Four different building locations have been suggested, of which only one will be selected. Data for each site are detailed in Table 8.6. Annual cash flow varies because of different tax structures, labor costs, and transportation charges, resulting in different annual receipts and disbursements. If the MARR is 10%, use incremental-rate-of-return analysis to select a building location.

SOLUTION All alternatives have a 30-year life. The procedure outlined above results in the following analysis:

1. Order the alternatives by initial investment (Table 8.7, first line)
2. Find the overall i_C^* for location C. Table 8.7 shows a rate of return of 9.63%, compared with the do-nothing alternative.

Table 8.6 Four alternative building locations

	Location			
	A	*B*	*C*	*D*
Building cost	$-200,000	$-275,000	$-190,000	$-350,000
Cash flow	+22,000	+35,000	+19,500	+42,000
Life, years	30	30	30	30

Table 8.7 Computation of incremental rate of return for mutually exclusive equal-lived projects

	C	*A*	*B*	*D*
Building cost	\$-190,000	\$-200,000	\$-275,000	\$-350,000
Cash flow	19,500	22,000	35,000	42,000
Projects compared	*C* to none	*A* to none	*B* to *A*	*D* to *B*
Incremental cost	-190,000	-200,000	-75,000	-75,000
Incremental cash flow	19,500	22,000	13,000	7,000
$(P/A, i\%, 30)$	9.7436	9.0909	5.7692	10.7143
Incremental i	9.63%	10.49%	17.28%	8.55%
Increment justified?	No	Yes	Yes	No
Project selected	None	*A*	*B*	*B*

3. Since 9.63% < 10%, location *C* is eliminated and the value $i_A^* = 10.49\%$ for *A* indicates the defender is *A* and the challenger is now *B*.
4. The incremental cash flow (Table 8.7) of *B* to *A* for $n = 30$ years is computed.
5. Compute the incremental i_{B-A}^* from

$$0 = -\text{incremental cost} + \text{incremental cash flow } (P/A, i\%, 30) \qquad (8.3)$$

   Note that $(P/A, 10\%, 30) = 9.4269$; thus any P/A value resulting from Eq. (8.3) greater than 9.4269 indicates that the return is less than 10% and is therefore unacceptable. Comparing *B* incrementally to *A* using Eq. (8.3) results in the equation $0 = -75{,}000 + 13{,}000(P/A, i\%, 30)$.
6. A rate of return of 17.28% on the extra investment justifies *B*, thereby eliminating *A*.
7. Comparing *D* to *B* (steps 4 to 6) results in $0 = -75{,}000 + 7000(P/A, i\%, 30)$ with an incremental $i_{D-B}^* = 8.55\%$, which is less than 10%, thereby eliminating location *D*. Only *A* and *B* are justified and *B* is selected, since it requires the larger initial investment.

COMMENT We should mention here again as a word of warning that an alternative should *always* be compared with an acceptable alternative, noting that the do-nothing alternative may be the acceptable one. Since *C* was not justified, location *A* was *not* compared with *C*. Thus, *if* the *B*-to-*A* comparison had not indicated that *B* was incrementally justified, then the *D*-to-*A* comparison instead of *D*-to-*B* would have been made.

It is important to understand the use of incremental-rate-of-return selection because if it is not properly applied in mutually exclusive alternative evaluation, the wrong alternative may be selected. If the overall rate of return of each alternative in this example is computed, the results are as follows:

Location	*C*	*A*	*B*	*D*
Overall i^*, %	9.63	10.49	12.40	11.59

If we were now to apply *only* the first criterion stated earlier, that is, make the largest investment that has a MARR of 10% or more, we would choose location *D*. But, as shown above, this is the wrong selection, because the extra investment of $75,000 between locations *B* and *D* will not earn the MARR. In fact, it will earn only 8.55% (Table 8.7). Remember, therefore, that incremental analysis is necessary for selection of one alternative from several when the rate-of-return evaluation method is used.

When the alternatives under consideration consist of disbursements only, the "income" is the difference between costs for two alternatives. In this case, there is no need to compare any of the alternatives against the do-nothing alternative. The lowest-investment-cost alternative is the defender against the next-lowest-investment-cost alternative (challenger). This procedure is illustrated in Example 8.8.

Example 8.8 Four machines can be used for a certain stamping operation. The costs for each machine are shown in Table 8.8. Determine which machine should be selected if the company's MARR is 13.5%.

SOLUTION The machines are already ordered according to increasing investment cost; and since no incomes are involved, incremental comparisons can be made because all machines are assumed to be acceptable. Comparing machine 2 (challenger) with machine 1 (defender) on an incremental basis, we find that

$$0 = -1500 + 300(P/A, i\%, 8) + 400(P/F, i\%, 8)$$

Solution of the equation yields $i = 14.6\%$. Therefore, eliminate machine 1 from further consideration. (If you had trouble obtaining the rate-of-return equation above, prepare a tabulation of cash flow for machines 1 and 2, as in Sec. 8.1.) The remaining calculations are summarized in Table 8.9. When machine 3 is compared with machine 2, the rate of return on the increment is less than 0%; therefore, machine 3 is eliminated. The comparison of machine 4 with machine 2 shows that the rate of return on the increment is greater than the MARR, favoring machine 4. Since no additional alternatives are available, machine 4 represents the optimum selection.

Table 8.8 Four mutually exclusive alternatives

	Machine			
	1	2	3	4
First cost	$-5,000	$-6,500	$-10,000	$-15,000
Annual operating cost	-3,500	-3,200	-3,000	-1,400
Salvage value	+500	+900	+700	+1,000
Life, years	8	8	8	8

Table 8.9 Comparison using incremental rate of return, Example 8.8

	Machine			
	1	2	3	4
Initial investment	$-5,000	$-6,500	$-10,000	$-15,000
Annual operating cost	-3,500	-3,200	-3,000	-1,400
Salvage value	+500	+900	+700	+1,000
Plans compared	...	2 to 1	3 to 2	4 to 2
Incremental investment	...	-1,500	-3,500	-8,500
Incremental annual savings	...	+300	+200	+1,800
Incremental salvage	...	+400	-200	+100
Incremental i^*	...	14.6%	<0%	13.7%
Increment justified?	...	Yes	No	Yes
Alternative selected	...	2	2	4

COMMENT You should remember that when no annual incomes are present in the analysis, it is implied that one of the machines *must* be selected. This situation could arise when the alternatives under consideration are part of a larger project which has been shown to be economical regardless of which alternative is selected. You should now calculate the present worth and equivalent uniform annual cost of each machine to satisfy yourself that machine 4 would be selected by all the evaluation methods.

When the alternatives under consideration have different lives, it is necessary to make the comparison over the least common multiple of years between *all* alternatives when using the present-worth method; however, if the incremental rate of return is found using the present-worth method, the analysis is done using the least common multiple between the *two* alternatives compared. Since this is not necessary in a straight EUAC analysis at MARR, this method is generally preferred when the alternatives have different lives. Example 8.9 illustrates the calculations for alternatives having different lives. Of course, you can use incremental present-worth or incremental EUAC analysis at the MARR to solve the problem but it is still necessary to make the comparison over the least common multiple of lives.

Often the lives of the alternatives are so long that they can be considered to be infinite, in which case the capitalized-cost method is used. See Example 8.10 for this special case.

Example 8.9 The three mutually inclusive alternatives in Table 8.10 are available. If MARR = 15% and the alternatives have different lives, as shown, select the one best alternative using (*a*) incremental-rate-of-return analysis and (*b*) straight EUAC analysis at the MARR.

SOLUTION

(*a*) Use the procedure at the beginning of this section to compute the incremental rates of return.

Table 8.10 Three mutually exclusive different-lived alternatives

	A	*B*	*C*
Initial cost	\$-6000	\$-7000	\$-9000
Salvage value	0	+200	+300
Cash flow	+2000	+3000	+3000
Life, years	3	4	6

1. Ordering is already done.
2. Compare *A* to the do-nothing alternative and compute $i_A^* = 0\%$ using

$$0 = -6000 + 2000(P/A, i\%, 3)$$

3. Since $i_A^* <$ MARR = 15%, delete *A* and compute $i_B^* = 26.4\%$ from

$$0 = -7000 + 3000(P/A, i\%, 4) + 200(P/F, i\%, 4)$$

   Now *B* is the defender and *C* is the challenger.
4. The net cash flow between *C* and *B* is shown in Table 8.11 for 12 years, the least common multiple.
5. Calculate $i_{C-B}^* = 19.4\%$ using the net cash flow in Table 8.11.
6. Since 19.4% > 15%, select alternative *C* over *B*.

(*b*) For the EUAC analysis use MARR = 15% and the respective lives.

$$\text{EUAC}_A = -6000(A/P, 15\%, 3) + 2000 = \$-628$$

Table 8.11 Net cash-flow tabulation for Example 8.9

Year	Cash flow *B*	Cash flow *C*	Net cash flow (*C* – *B*)
0	\$ -7,000	\$ -9,000	\$-2,000
1	+3,000	+3,000	0
2	+3,000	+3,000	0
3	+3,000	+3,000	0
4	-3,800	+3,000	+6,800
5	+3,000	+3,000	0
6	+3,000	-5,700	-8,700
7	+3,000	+3,000	0
8	-3,800	+3,000	+6,800
9	+3,000	+3,000	0
10	+3,000	+3,000	0
11	+3,000	+3,000	0
12	+3,200	+3,300	+100
	\$+15,600	\$+18,000	\$+3,000

$$\text{EUAC}_B = -7000(A/P, 15\%, 4) + 3000 + 200(A/F, 15\%, 4) = \$588$$

$$\text{EUAC}_C = -9000(A/P, 15\%, 6) + 3000 + 300(A/F, 15\%, 6) = \$656$$

Alternative C is again selected because it offers the largest positive EUAC, indicating a return in excess of 15%.

COMMENT When comparing different-lived alternatives by incremental analysis, you must use the least common multiple of years between only the *two* alternatives being compared, and *not* the common multiple of all alternative lives.

Example 8.10 The U.S. Army Corps of Engineers wants to construct a dam on the Sacochsi River. Six different sites have been suggested. The construction and average annual dollar benefits (income) to the area are tabulated below. If a MARR of 6% is required and dam life is long enough to be considered infinite for analysis purposes, select the best location from the economic point of view.

Site	Construction cost P, millions	Annual income, A
A	\$ 6	\$350,000
B	8	420,000
C	3	125,000
D	10	400,000
E	5	350,000
F	11	700,000

SOLUTION After ordering the projects by first cost, we can use the capitalized-cost equation, Eq. (5.3), $P = A/i$, in the form of Eq. (8.4) to determine if the incremental investment is justified.

$$0 = \frac{\Delta A}{i} - \Delta P \tag{8.4}$$

Here ΔP is the incremental investment and ΔA is the incremental (net) income or cash flow. If the right side of Eq. (8.4) is greater than zero, the incremental invest-

Table 8.12 Capitalized cost comparison of mutually exclusive dam sites

	C	E	A	B	D	F
P(million)	\$ 3	\$ 5	\$ 6	\$ 8	\$ 10	\$ 11
A(\$1000)	125	350	350	420	400	700
Comparison	C to none	E to none	A to E	B to E	D to E	F to E
ΔP(million)	3	5	1	3	5	6
ΔA(\$1000)	125	350	0	70	50	350
$\Delta A/i - P$(million)	−0.92	0.83	−1.0	−1.83	−4.17	−0.17
Site selected	None	E	E	E	E	E

ment is justified. Table 8.12 indicates that only site E is justified, so this is the most economical dam site.

Probs. P8.16 to P8.26

FURTHER INFORMATION

Mutually exclusive alternatives (Sec. 8.5): DeGarmo, Canada, and Sullivan [5], pp. 187-191; Grant, Ireson, and Leavenworth [7], pp. 287-297; Kasner [8], pp. 83-88; Stevens [15], pp. 157-164; Theusen, Fabrycky, and Theusen [17], pp. 168-187.

PROBLEMS

P8.1 Prepare a tabulation of net cash flow for the following alternatives:

	Alternative *R*	Alternative *S*
First cost	$8,000	$12,000
Annual operating cost	900	1,400
Salvage value	1,000	2,000
Life, years	5	10

P8.2 Prepare a tabulation of net cash flow for the following alternatives:

	Alternative *A*	Alternative *B*
First cost	$15,000	$11,000
Annual operating cost	2,300	2,600
Annual income	4,000	3,100
Salvage value	2,000	1,500
Life, years	3	2

P8.3 Two different machines are being considered for a certain process. Machine X has a first cost of $12,000 and annual operating expenses of $3000 per year. It is expected to have a life of 12 years with no salvage value. Machine Y can be purchased for $21,000, and it will have an annual operating cost of $1200 per year. However, a factory overhaul will be required every 4 years at a cost of $2500. It will have a useful life of 12 years with a $1500 salvage value. Perpare a tabulation of net cash flow for the two alternatives.

P8.4 The owners of a new home are trying to decide between two types of landscape. They can purchase grass seed for $2.45 per kilogram and plant it themselves at no cost. They estimate that they will need 1 kilogram for 25 square meters. The front yard is 30 × 10 meters and the back yard 30 × 7 meters. They will also have to purchase 3 metric tons of compost when the lawn is seeded and 1 metric ton every 3 years thereafter at a cost of $20 per metric ton. If they elect to plant grass, they will also install a sprinker system with a $350 installation cost, a $15-per-year maintenance cost, and a life of 18 years. The cost of such items as water, fertilizer, and chemicals is expected to be $120 per year. Alternatively, they can have desert landscaping installed in the front and back yards at a cost of $2800. If they elect this alternative, the annual cost of water, etc., is expected to be only $25 per year. However, they will have to replace the grass barrier under the gravel every 6 years at a cost of $350. Prepare a tabulation of cash flows and net cash flow for the two alternatives if they are to be compared for an 18-year period.

P8.5 After preparing a tabulation of cash flow, what would you know immediately if the total of the difference (net cash-flow) column was a negative amount?

P8.6 Determine which alternative should be selected in Prob. P8.1 if the company's MARR is 15%. Use the incremental-investment method.

P8.7 Determine which alternative should be selected in Prob. P8.2 on the basis of incremental rate of return if the company's MARR is 20%.

P8.8 Determine which alternative should be selected in Prob. P8.3 if the company's MARR is 12%. Use the incremental-investment method.

P8.9 Determine which alternative should be selected in Prob. P8.4 if the purchaser's MARR is 6%. Use the incremental-investment analysis.

P8.10 A food-processing company is considering plant expansion. Under the current setup, the company can increase profits $25,000 per year by extending the workday 2 hours through overtime. No investment will be required for this alternative. On the other hand, if additional cookers and freezers are added at a cost of $175,000, the company's profits will increase by $50,000 per year. If the company expects to use the current process for 10 years, would the plant expansion be justified if the company's MARR is 25%? Solve this problem two ways.

P8.11 Which alternative should be selected in Prob. P8.3 if EUAC rate-of-return analysis and a MARR of 12% is used? Compare your answer and values with those of Prob. P8.8.

P8.12 Select the more economic alternative in Prob. P8.4 by EUAC rate-of-return analysis if the homeowner's MARR is 6% and the initial cost of desert landscaping will be $1700 rather than $2800.

P8.13 The engineer at the Smoke Ring Cigar Company wants to do a rate-of-return analysis using annual cost of two wrapping machines. The details below are available; however, the engineer does not know what value to use for a MARR figure since some projects are evaluated at 8% and some at 10%. Determine whether this difference in MARR would change the decision of which machine to buy. Use rate-of-return on the incremental-investment method.

	Machine *A*	Machine *M*
First cost	$10,000	$9,000
Annual labor cost	5,000	5,000
Annual maintenance cost	500	300
Salvage value	1,000	1,000
Life, years	6	4

P8.14 Would the answer to Prob. P8.13 change if the lives of both machines were 6 years?

P8.15 A family need a new roof on their home. They have estimates from roofer *A* and roofer *B*. Roofer *A* wants $2400 for shingle roofing, material, and labor. If the tin on the eves and valleys and old boards are replaced, an extra $300 cost is incurred. The roof itself will reasonably last 15 years if $20 per year for the first 5 years and $5 per year more each year after year 5 (gradient of $5) is spent on preventive maintenance. The replacement of the tin and board will increase the life to 18 years. Roofer *B* will put a gravel roof on the house for $2200. It is estimated that the same annual maintenance expenditure as with shingles will give the roof an expected life of 12 years. Use rate-of-return analysis to compare (*a*) shingles without tin and board replacement with gravel and (*b*) shingles with the replacements and gravel. Assume a MARR of 16%.

P8.16 What is the basic difference between an engineering-economy problem that requires selection from several mutually exclusive alternatives and one that requires selection from independent projects?

P8.17 What criteria are used to select a mutually exclusive alternative by the rate-of-return method?

P8.18 Five different methods can be used for recovering by-product heavy metals from a stream. The investment costs and incomes associated with each method are shown below. Assuming all methods have a 10-year life with zero salvage value and the company's MARR is 15%, determine which one should be selected by (*a*) the EUAC method and (*b*) the incremental rate-of-return method.

	Method				
	1	2	3	4	5
First cost	$15,000	$18,000	$25,000	$35,000	$52,000
Salvage value	+1,000	+2,000	-500	-700	+4,000
Annual income	5,000	6,000	7,000	9,000	12,000

P8.19 If method 2 in Prob. P8.18 has a life of 5 years and method 3 has a life of 15 years, which alternative should be selected?

P8.20 Select the best alternative using the incremental rate-of-return method from the proposals shown below if the MARR is 14% and the projects will have a useful life of 15 years. Assume that the cost of the land will be recoverd when the project is terminated.

	Proposal						
	1	2	3	4	5	6	7
Land cost	$ 50,000	$ 40,000	$ 70,000	$ 80,000	$ 90,000	$ 65,000	$ 75,000
Construction cost	200,000	150,000	170,000	185,000	165,000	175,000	190,000
Annual maintenance	15,000	16,000	14,000	17,000	18,000	13,000	12,000
Annual income	52,000	49,000	68,000	50,000	81,000	77,000	45,000

P8.21 Any one of five machines can be used in a certain phase of a canning operation. The costs of the machines are shown below and all are expected to have a 10-year life. If the company's minimum attractive rate of return is 18%, determine which machine should be selected using (*a*) the incremental-rate-of-return method, and (*b*) the present-worth method.

	Machine				
	1	2	3	4	5
First cost	$28,000	$33,000	$22,000	$51,000	$46,000
Annual operating cost	20,000	18,000	25,000	12,000	14,000

P8.22 An oil and gas company is considering five sizes of pipe for a new pipeline. The costs for each size are shown below. Assuming all pipes will last 15 years and the company's MARR is 8%, determine which size pipe should be used according to (*a*) the present-worth method and (*b*) the incremental-rate-of-return method.

	Pipe size, millimeters				
	140	160	200	240	300
Initial investment	$9,180	$10,510	$13,180	$15,850	$30,530
Installment cost	600	800	1,400	1,500	2,000
Annual operating cost	6,000	5,800	5,200	4,900	4,800

P8.23 An independent dirt contractor is trying to determine which size dump truck to buy. The contractor knows that as the bed size increases, the net income increases, but it is uncertain whether the incremental expenditure required in the larger trucks could be justified. The cash flows associated with each size truck are shown below. If the contractor's MARR is 18% and all trucks are expected to have a useful life of 8 years, determine which size truck should be purchased using (*a*) the incremental-rate-of-return method and (*b*) the EUAC method.

	Truck size, square meters				
	8	10	15	20	40
Initial investment	$10,000	$12,000	$18,000	$24,000	$33,000
Annual operating cost	5,000	5,500	7,000	11,000	16,000
Salvage value	2,000	2,500	3,000	3,500	4,500
Annual income	9,000	10,000	10,500	12,500	14,500

P8.24 Five processes can be used for producing a certain part. If the company's MARR is 15%, determine which process should be selected by (*a*) the present-worth method and (*b*) the incremental-rate-of-return method.

	Process				
	1	2	3	4	5
First cost	$15,000	$22,000	$27,000	$31,000	$42,000
Annual operating cost	6,000	5,000	4,500	3,000	2,000
Salvage value	500	1,000	1,100	600	3,000
Life, years	3	4	6	6	6

P8.25 One phase of a meat-packing operation requires the use of separate machines for the following functions: pressing, slicing, weighing, and wrapping. All machines under consideration are expected to have a life of 6 years with no salvage value. There are two alternatives for each of the functions as follows:

	Alternative 1		Alternative 2	
	First cost	Annual cost	First cost	Annual cost
Pressing	$ 5,000	$13,000	$10,000	$11,000
Slicing	4,000	10,000	17,000	4,000
Weighing	12,000	15,000	15,000	13,000
Wrapping	3,000	9,000	11,000	7,000

(*a*) If the company's MARR is 20%, use the incremental-rate-of-return method to determine which machine should be selected for each function (identify them as pressing 1, pressing 2, slicing 1, etc.).

(*b*) For the machines selected in part *a*, determine the total investment and operating cost for the entire operation.

P8.26 A third alternative can be added in Prob. P8.25: one machine to do the pressing and slicing and another machine to do the weighing and wrapping. The machine that will do the pressing and slicing (identified as pressing-slicing 3) will cost $29,000 and will have an annual operating cost of $9,000. The machine that will do the weighing and wrapping (identified as weighting-wrapping 3) will cost $26,000 and will have an annual operating cost of $18,000.

(*a*) Which machines should be selected for the entire operation?

(*b*) Determine the total investment and operating cost for the entire operation.

(*c*) If a fourth alternative can be considered—a single machine to perform all four functions (identified as machine 4) having an initial cost of $45,000 and an annual operating cost of $32,000—which machine(s) should be selected?

LEVEL THREE

In this level you will learn how to evaluate alternatives using the benefit/cost ratio method and how to perform an economic comparison of an asset that is currently owned with one being considered as its replacement. Additionally the procedure to determine a minimum-cost life for an asset is studied from the replacement and retirement viewpoint. After a treatment of bonds—their types and economic considerations —some optional material on inflation effects in engineering-economic analysis is presented.

CHAPTER

NINE

BENEFIT/COST RATIO EVALUATION

The objective of this chapter is to teach you how to compare two alternatives on the basis of a benefit/cost ratio. This method is sometimes regarded as *supplementary*, since it is used in conjunction with a present-worth, annual-cost, or rate-of-return analysis.

SECTION OBJECTIVES

To complete this chapter, you must be able to do the following:

9.1. State the definition used to classify specified expenditures or savings as benefits, costs, or disbenefits.
9.2. Determine whether a single project should be undertaken by comparing its benefits and costs, given values for the benefits, disbenefits, and costs, and the interest rate.
9.3. Select the better of two alternatives on the basis of a benefit/cost analysis, given the initial cost, life, salvage value, and disbursements for each alternative and the required rate of return.
9.4. State the procedure for selecting the best alternative from three or more projects using a benefit/cost ratio analysis.
9.5 Select one alternative from several *mutually exclusive alternatives* using the incremental B/C ratio method, given initial cost, salvage value, life, and cash flows for each alternative, and the minimum attractive rate of return.

STUDY GUIDE

9.1 Classification of Benefits, Costs, and Disbenefits

The method for selecting alternatives that is most commonly used by federal agencies for analyzing the desirability of public works projects is the benefit/cost ratio (B/C ratio). As its name suggests, the B/C method of analysis is based on the ratio of the benefits to costs associated with a particular project. Therefore, the first step in a B/C analysis is to determine which of the elements are benefits and which are costs. In general, *benefits* are advantages, expressed in terms of dollars, which happen to the *owner*. On the other hand, when the project under consideration involves disadvantages to the owner, these are known as *disbenefits*. Finally, the *costs* are the anticipated expenditures for construction, operation, maintenance, etc. Since B/C analysis is always used in economy studies by federal agencies, it is helpful to think of the *owner* as the *public* and the one who incurs the costs as the *federal government*. The determination of whether an item is to be considered as a benefit, disbenefit, or cost, therefore, depends on *who is affected* by the consequences. Some examples of each are illustrated in Table 9.1.

While the examples presented in this chapter are straightforward with regard to identification of benefits, disbenefits, or costs, it should be pointed out that in actual situations, judgments must sometimes be made which are subject to interpretation, particularly when it is necessary to determine whether an element of cash flow is a disbenefit or a cost. In other instances, it is not possible to simply place a dollar value on all benefits, disbenefits, or costs that are involved. These nonquantifiable considerations must be included in the final decision, as they are in other methods of analysis. In general, however, dollar values are available, or obtainable, and the results of a proper B/C analysis would agree with the methods studied in preceding chapters (such as present worth, equivalent uniform annual cost, or rate of return on incremental investment).

Probs. P9.1 and P9.2

9.2 Benefits, Disbenefits, and Cost Calculations of a Single Project

Before a B/C ratio can be computed, all the benefits, disbenefits, and costs that are to be used in the calculation must be converted to common dollar units, as in present-

Table 9.1 Examples of benefits, disbenefits, and costs

Item	Classification
Expenditure of $11,000 for new interstate highway	Cost
$50,000 annual income to local residents from tourists because of new reservoir and recreation area	Benefit
$150,000 per year upkeep cost for irrigation canals	Cost
$25,000 per year loss by farmers because of highway right-of-way	Disbenefit

worth calculations, or dollars per year, as in annual-cost comparisons. It is irrelevant whether the present-worth or annual-cost method is used so long as the procedures learned in Chaps. 5 and 6 are followed. With the use of either the present-worth or EUAC values, the B/C ratio can be calculated as

$$\text{B/C} = \frac{\text{benefits} - \text{disbenefits}}{\text{costs}} \tag{9.1}$$

Note that the *disbenefits are subtracted from the benefits*, not added to the costs. It is important to recognize that the B/C ratio could change considerably if disbenefits are regarded as costs. For example, if the numbers 10, 8, and 8 are used to represent benefits, disbenefits, and costs, respectively, the correct procedure would result in a B/C ratio of $(10 - 8)/8 = 0.25$, while the incorrect procedure would yield a B/C ratio of $10/(8 + 8) = 0.625$, which is over twice as large. Clearly then, the method by which disbenefits are handled is very important. When the proper procedure is followed, a B/C ratio greater than or equal to 1.0 indicates that the project under consideration is economically advantageous.

An alternative method that can be used to evaluate the feasibility of federal projects is to subtract the costs from the benefits, that is, B - C. In this base, if B - C is greater than or equal to zero, the project is acceptable. This method has the obvious advantage of eliminating the discrepancies noted above when disbenefits are regarded as costs, since B represents the *net benefits*. Thus, for the numbers 10, 8, and 8 the same result is obtained regardless of how disbenefits are treated.

$$\text{Subtracting disbenefits: } \text{B} - \text{C} = [(10 - 8) - 8] = -6$$

$$\text{Adding disbenefits to costs: } \text{B} - \text{C} = [10 - (8 + 8)] = -6$$

Before calculating the B/C ratio, check to be sure that the proposal with the higher EUAC is the one that yields the higher benefits *after the benefits and costs have been expressed in common units*. Thus, a proposal having a higher initial cost may actually have a lower EUAC or present worth when all other costs are considered. Example 9.1 illustrates this point.

Example 9.1 Alternative routes are being considered by the state highway department for location of a new highway. Route *A*, costing $4,000,000 to build, will provide annual benefits of $125,000 to local businesses. Route *B* will cost $6,000,000 but will provide $100,000 in benefits. The annual cost of maintenance is $200,000 for *A* and $120,000 for *B*, respectively. If the life of each road is 20 years and an interest rate of 8% is used, which alternative should be selected on the basis of a benefit/cost analysis?

SOLUTION The benefits in this example are $125,000 for route *A* and $100,000 for route *B*. The EUAC of costs for each alternative is as follows:

$$\text{EUAC}_A = 4{,}000{,}000(A/P, 8\%, 20) + 200{,}000 = \$607{,}400$$

$$\text{EUAC}_B = 6{,}000{,}000(A/P, 8\%, 20) + 120{,}000 = \$731{,}100$$

Route *B* has a *higher* EUAC than route *A* by $123,700 per year, and *less benefits* than *A*. Therefore, there would be no need to calculate the benefit/cost ratio for route *B*, since this alternative is obviously inferior to route *A*. Furthermore, if the decision had been made that either route *A* or *B* *must* be accepted (which would be the case if there were no other alternatives), then no other calculations would be necessary and route *A* would be accepted.

Solved Problem 9.4
Prob. P9.3 to P9.7

9.3 Alternative Comparison by Benefit/Cost Analysis

In computing the benefit/cost ratio by Eq. (9.1) for a given alternative, it is important to recognize that the benefits and costs used in the calculation represent the *increments* or *differences* between two alternatives. This will always be the case, since sometimes doing nothing is an acceptable alternative. Thus, when it seems as though only one proposal is involved in the calculation, such as whether or not a flood-control dam should be built to reduce flood damage, it should be recognized that the construction proposal is being compared against another alternative–the do-nothing alternative. Although this is also true for the other alternative evaluation techniques previously presented, it is emphasized here because of the difficulty often present in determining the benefits and costs between two alternatives when only costs are involved. See Example 9.2 for an illustration.

Once the B/C ratio on differences is computed, a $B/C \geqslant 1.0$ means that the extra benefits of the higher-cost alternative justify this higher cost. If $B/C < 1.0$, the extra cost is not justified and the lower-cost alternative is selected. Note that this lower-cost project may be the do-nothing alternative, if the B/C analysis is for only one project.

Example 9.2 Two routes are under consideration for a new interstate highway. The northerly route *N* would be located about 5 miles from the central business district and would require longer travel distances by local commuter traffic. The southerly route *S* would pass directly through the downtown area and, although its construction cost would be higher, it would reduce the travel time and distance for local commuters. Assume that the costs for the two routes are as follows:

	Route *N*	Route *S*
Initial cost	$10,000,000	$15,000,000
Maintenance cost per year	35,000	55,000
Road-user cost per year	450,000	200,000

If the roads are assumed to last 30 years with no salvage value, which route should be accepted on the basis of a benefit/cost analysis using an interest rate of 5%?

SOLUTION Since most of the costs are already annualized, the EUAC method will be used to obtain the equivalent annual cost. The *costs* to be used in the B/C ratio are the initial cost and maintenance cost:

$$\text{EUAC}_N = 10{,}000{,}000(A/P, 5\%, 30) + 35{,}000 = \$685{,}500$$

$$\text{EUAC}_S = 15{,}000{,}000(A/P, 5\%, 30) + 55{,}000 = \$1{,}030{,}750$$

The *benefits* in this example are represented by the road-user costs, since these are costs "to the public." The benefits, however, are not the road-user costs themselves but the *difference* in road-user costs if one alternative is selected over the other. In this example, there is a \$450,000 - \$200,000 = \$250,000 per year benefit if route S is chosen instead of route N. Therefore, the benefit (B) of route S over route N is \$250,000 per year. On the other hand, the costs (C) associated with these benefits are represented by the difference between the annual costs of routes N and S. Thus,

$$\text{C} = \text{EUAC}_S - \text{EUAC}_N = \$345{,}250 \text{ per year}$$

Note that the route that costs more (route S) is the one that provides the benefits. Hence, the B/C ratio can now be computed by Eq. (9.1).

$$\text{B/C} = \frac{250{,}000}{345{,}250} = 0.724$$

The B/C ratio of less than 1.0 indicates that the extra benefits associated with route S are less than the extra costs associated with this route. Therefore, route N would be selected for construction. Note that there is no do-nothing alternative in this case, since one of the roads *must* be constructed.

COMMENT If there had been disbenefits associated with each route, the difference between the disbenefits would have to be added or subtracted from the net benefits (\$250,000) for route S, depending on whether the disbenefits for route S were less than or greater than the disbenefits for route N. That is, if the disbenefits for route S were less than those for route N, the difference between the two would have to be added to the \$250,000 benefit for route S, since the disbenefits involved would also favor route S. However, if the disbenefits for route S were greater than those for route N, their difference should be subtracted from the benefits associated with route S, since the disbenefits involved would favor route N instead of route S.

Solved Problem 9.5
Probs. P9.8 to P9.13

9.4 Benefit/Cost Analysis for Multiple Alternatives

When only one alternative must be selected from three or more mutually exclusive (stand-alone) alternatives, a multiple alternative evaluation is required. In this case, it is necessary to conduct an analysis on the *incremental* benefits and costs similar to the procedure used in Chap. 8 for incremental rates of return. The do-nothing alternative may be one of the considerations.

There are two situations which must be considered with regard to multiple alternative analysis by the benefit/cost method. In the first case, if funds are available so that *more than one* alternative can be chosen from among several, it is necessary only to compare the alternatives against the do-nothing alternative. The alternatives are referred to as *independent* in this situation. For example, if several flood-control dams could be constructed on a particular river and adequate funding is available for all dams, the B/C ratios should be those associated with a particular dam versus no dam. That is, the result of the calculations could show that three dams along the river would be economically justifiable on the basis of reduced flood damage, recreation, etc., and should therefore, be constructed.

On the other hand, when only *one* alternative can be selected from among several, it is necessary to compare the alternatives against each other as well as against the do-nothing alternative. This procedure was discussed in Chap. 8. It is important for you to understand the difference between the procedure to be followed when multiple projects are mutually exclusive and when they are not. In the case of mutually exclusive projects, it is necessary to compare them against each other; in the case of projects that are not mutually exclusive (independent projects), it is necessary to compare them only against the do-nothing alternative.

Prob. P9.14

9.5 Selection from Mutually Exclusive Alternatives Using Incremental Benefit/Cost Ratio Analysis

In order to use the B/C ratio as an evaluation technique for mutually exclusive alternatives, an incremental B/C ratio must be computed in a fashion similar to that used for the incremental rate of return (Chap. 8). The project that has the incremental $B/C \geqslant 1.0$ and requires the largest *justified* investment is selected. The procedure to be followed is similar to that used for rate-of-return analysis; however, in a B/C analysis it is generally convenient to compute an overall B/C ratio for each alternative, since the total present-worth or EUAC values must be computed in preparation for the incremental analysis. Those alternatives that have an overall $B/C < 1.0$ can be eliminated immediately and need not be considered in the incremental analysis. Example 9.3 presents a complete application of the incremental B/C ratio to mutually exclusive alternatives.

Example 9.3 Using the four alternatives of Example 8.7 (Table 8.6), apply incremental B/C ratio analysis to select the most advantageous alternative (MARR = 10%).

SOLUTION The alternatives are first ranked from smallest to largest initial investment cost. The next step is to calculate the overall B/C ratio and eliminate those

Table 9.2 Incremental B/C ratio analysis for mutually exclusive alternatives, Example 9.3

	C	*A*	*B*	*D*
Building cost	\$–190,000	\$–200,000	\$–275,000	\$–350,000
Cash flow (CF)	19,500	22,000	35,000	42,000
Present worth of CF	\$ 183,826	\$ 207,394	\$ 329,945	\$ 395,934
Overall B/C	0.97	1.03	1.20	1.13
Projects compared	...	...	*B* to *A*	*D* to *B*
Incremental benefit	...	...	\$ 122,551	\$ 65,989
Incremental cost	...	...	75,000	75,000
Incremental B/C	...	...	1.64	0.88
Project selected	...	...	*B*	*B*

alternatives that have B/C $<$ 1.0. As shown in Table 9.2, location *C* can be eliminated on the basis of its overall B/C ratio (0.97). All other alternatives are acceptable and must therefore be compared on an incremental basis. The incremental benefits and costs can be determined as follows:

Incremental benefits: increase in present worth between alternatives
Incremental cost: increase in building cost between alternatives

A summary of the incremental B/C analysis is presented in Table 9.2. Using the acceptable alternative that has the lowest investment cost as the defender (*A*) and the next lowest acceptable alternative (*B*) as the challenger, the incremental B/C ratio is 1.64, indicating that location *B* should be selected over location *A* (therefore eliminating *A* from further consideration). Using *B* as the defender and *D* as the challenger, the incremental analysis yields incremental B/C = 0.88, favoring location *B*. Since location *B* has an incremental ratio greater than 1.0 *and* is the largest justified investment, it is selected; this, of course, is the same conclusion reached with the incremental rate-of-return method in Table 8.7.

COMMENT Note that alternative selection should not be made on the basis of the overall B/C ratio, even though location *B* would still be selected, *coincidentally in this case*. The incremental investment must also be justified in order to select the best alternative.

Although present-worth values were used in this example, EUAC values can also be used to compare the investments; in fact, this method is generally simpler if lives are unequal.

Solved Problem 9.6
Probs. P9.15 to P9.20

SOLVED PROBLEMS

Example 9.4 The Wartol Foundation, a nonprofit educational research organization, is contemplating an investment of $1.5 million in grants to develop new ways to teach people the rudiments of a profession. The grants would extend over a 10-year period and would create an estimated savings of $500,000 per year in professors' salaries, student tuition, and other expenses. The foundation uses a rate of return of 6% on all grant investments. In this case the program would be an addition to ongoing and planned activities. An estimated $200,000 a year would thus have to be released from other programs to support the educational research. Use (*a*) B/C ratio and (*b*) B − C analysis to determine the advisability of the program over a 10-year period.

SOLUTION The following definitions are in order:

Benefit: $500,000 per year
Cost: 1,500,000(*A*/*P*, 6%, 10) = $203,805 per year
Disbenefit: $200,000 per year

(*a*) Using B/C ratio analysis, Eq. (9.1), yields

$$\text{B/C} = \frac{500{,}000 - 200{,}000}{203{,}805} = 1.47$$

The project is justified, since B/C > 1.0.

(*b*) If B is the net benefit, then

$$\text{B} - \text{C} = (500{,}000 - 200{,}000) - 203{,}805 = \$96{,}195$$

Again, since (B − C) > 0, investment is wise.

COMMENT In *a*, if you happened to add the disbenefits to costs incorrectly, you would have

$$\text{B/C} = \frac{500{,}000}{203{,}805 + 200{,}000} = 1.238$$

which still justifies the investment. But the $200,000 is not a direct cost to this program and should be subtracted from B, not added to C.

Sec. 9.2

Example 9.5 Assume the same situation as in Example 9.2 for the routing of a new interstate highway. The B/C analysis showed that the northerly route *N* was to be constructed. However, this route will go through an agricultural region and the local farmers have complained about the great loss in revenue they and the economy will suffer. Likewise, the downtown merchants have complained about route *S* because of the loss in revenue due to reduced merchandising ability, parking problems, etc. To consider these eventualities, the state highway department has undertaken a study and predicted that the loss to state agriculture for route

N will be about \$500,000 per year and that route S will cause an estimated reduction in retail sales and rents of \$400,000 per year. What effect does this new information have on the B/C analysis?

SOLUTION These new "costs" should be considered as disbenefits. Since the disbenefits of route S are \$100,000 less than those of route N, this difference is *added* to the benefits of route S to give a total net benefit of \$250,000 + \$100,000 = \$350,000 to the downtown alternative. Now we have

$$\text{B/C} = \frac{350{,}000}{345{,}250} = 1.01$$

and route S is to be slightly favored. In this case the inclusion of disbenefits has reversed the earlier decision.

Sec. 9.3

Example 9.6 The U.S. Army Corps of Engineers still wants to construct a dam on the Sacochsi River as in Example 8.10. The construction and average annual dollar benefits (income) are repeated below. If a MARR of 6% is required and dam life is infinite for analysis purposes, select the best location using the B/C ratio method.

Site	Construction cost, millions	Annual income
A	\$ 6	\$350,000
B	8	420,000
C	3	125,000
D	10	400,000
E	5	350,000
F	11	700,000

SOLUTION We make use of the capitalized-cost equation, Eq. (5.2), $A = Pi$, to obtain EUAC values for capital recovery (cost), as shown in the first row of Table 9.3. Since site E is justified and has the largest investment, it is selected.

Table 9.3 Use of incremental B/C ratio analysis for Example 9.6

	C	*E*	*A*	*B*	*D*	*F*
Capital recovery (\$1000)	\$180	\$300	\$360	\$480	\$600	\$660
Annual benefits (\$1000)	125	350	350	420	400	700
Comparison	*C* to none	*E* to none	*A* to *E*	*B* to *E*	*D* to *E*	*F* to *E*
Δ Capital recovery	\$180	\$300	\$ 60	\$180	\$300	\$360
Δ Annual benefits	125	350	0	70	50	350
Δ B/C ratio	0.70	1.17	0	0.39	0.17	0.97
Site selected	None	*E*	*E*	*E*	*E*	*E*

COMMENT Suppose that site G is added with a constuction cost of \$10 million and an annual benefit of \$700,000. What site should G be compared with? What is the ΔB/C ratio? If you determine that a comparison of G to E is to be made and $\Delta B/C = 1.17$ in favor of G, you are correct! Now site F must be incrementably evaluated with G, but since the annual benefits are the same (\$700,000), the ΔB/C ratio is zero and the added investment is not justified. Therefore, site G is chosen.

Sec. 9.5

PROBLEMS

P9.1 Why should disbenefits be subtracted from benefits rather than added to costs?

P9.2 Classify the following cash flows as either benefits, costs, or disbenefits:
- (*a*) Drive in theater had to be destroyed because of highway right-of-way
- (*b*) \$10 million expenditure for new highway
- (*c*) Less disbursement by motorists because of new highway
- (*d*) Archeological sites inundated by new reservoir
- (*e*) \$2 million paid for right-of-way for new highway

P9.3 The U.S. Army Corps of Engineers is considering the feasiblity of constructing a small flood-control dam in an existing arroyo. The initial cost of the project will be \$2.2 million, with inspection and upkeep costs of \$10,000 per year. In addition, minor reconstruction will be required every 15 years at a cost of \$65,000. If flood damage will be reduced from the present cost of \$90,000 per year to \$10,000 annually, use the benefit/cost method to determine if the dam should be constructed. Assume that the dam will be permanent and the interest rate is 12% per year.

P9.4 A state highway department is considering the construction of a new highway through a scenic rural area. The road is expected to cost \$6 million, with annual upkeep estimated at \$20,000 per year. The improved accessibility is expected to result in additional income from tourists of \$350,000 per year. If the road is expected to have a useful life of 25 years, use the (*a*) B – C method and (*b*) B/C method at an interest rate of 6% per year to determine if the road should be constructed.

P9.5 If the highway in Prob. P9.4 would result in agricultural-income losses of \$15,000 the first year, \$16,000 the second, and amounts increasing by \$1000 per year, by how much would the tourist income have to increase each year (starting in year 2) in order for the highway to become economically feasible?

P9.6 The U.S. Bureau of Reclamation is considering a project to extend irrigation canals into a desert area. The initial cost of the project is expected to be \$1.5 million, with annual maintenance costs of \$25,000 per year. If agricultural revenue is expected to be \$175,000 per year, make a B/C analysis to determine whether the project should be undertaken, using a 20-year study period and an interest rate of 6% per year.

P9.7 Calculate the B/C ratio for Prob. P9.6 if the canal must be dredged every 3 years at a cost of \$60,000 and there is a \$15,000-per-year disbenefit associated with the project.

P9.8 Two routes are under consideration for a new interstate highway. The long route would be 22 miles in length and would have an initial cost of \$21 million. The transmountain route would be 10 miles long and would have an initial cost of \$45 million. Upkeep costs are estimated at \$40,000 per year for the long route and \$65,000 per year for the transmountain route. Regardless of which route is selected, the volume of traffic is expected to be 400,000 vehicles per year. If the vehicle operating expense is assumed to be \$0.12 per mile, determine which route should be selected by (*a*) B/C analysis and (*b*) B – C analysis. Assume a 20-year life for each road and an interest rate of 6% per year.

P9.9 The U.S. Army Corps of Engineers is considering three sites for flood-control dams (designated as sites *A*, *B*, and *C*). The construction costs are $10 million, $12 million, and $20 million, and maintenance costs are expected to be $15,000, $20,000, and $23,000, respectively, for sites *A*, *B*, and *C*. In addition, a $75,000 expenditure will be required every 10 years at each site. The present cost of flood damage is $2 million per year. If only the dam at site *A* is constructed, the flood damage will be reduced to $1.6 million per year. If only the dam at site *B* is constructed, the flood damage will be $1.2 million per year. Similarly, if the site *C* dam is built, the damage will be reduced to $0.77 million per year. Since the dams would be built on different branches of a large river, either one or all of the dams could be constructed and the decrease in flood damages would be additive. If the interest rate is 5% per year, determine which ones, if any, should be built on the basis of their B/C ratios. Assume that the dams will be permanent.

P9.10 Highway department officials are considering the economics of either resurfacing an existing highway or constructing a new one. The existing highway is 12 miles long and would cost $2 million to resurface. Annual upkeep cost is expected to be $5000 the first year, $10,000 the second, and amounts increasing by $5000 per year until year 10, at which time the road would have to be resurfaced again. If a new road is constructed, the initial cost would be $15 million for a road 10 miles long. The maintenance is expected to cost $5000 the first year, $7000 the second, and amounts increasing by $2000 per year until year 10, after which the cost will be $23,000 per year. If the new road is constructed, the cost of auto accidents is expected to decrease by $500,000 per year. If vehicle operating cost is assumed to be $0.10 per mile and 600,000 vehicles per year travel the road, use the benefit/cost method to determine which road should be constructed at an interest rate of 6% per year.

P9.11 The U.S. Forest Service is considering two locations for a new federal park. Location *E* would require an investment cost of $3 million and $50,000 per year in maintenance. Location *W* would cost $7 million to construct, but the forest service would receive an additional $25,000 per year in park-use fees. The operating cost of location *W* will be $65,000 per year. The revenue to park concessionaires will be $500,000 per year at location *E* and $700,000 per year at *W*. The disbenefits associated with each location are $30,000 per year for location *E* and $40,000 per year for location *W*. Use (*a*) the B/C method and (*b*) the B – C method to determine which location if either should be selected, using an interest rate of 12% per year. Assume that the park will be maintained indefinitely.

P9.12 The Bureau of Reclamation is considering the lining of the main canals of its irrigation ditches. The initial cost of lining is expected to be $4 million, with $25,000 per year required for maintenance. If the canals are not lined, a weed-control and dredging operation will have to be instituted, which will have an initial cost of $700,000 and a cost of $50,000 the first year, $52,000 the second, and amounts increasing by $2000 per year for 25 years. If the canals are lined, less water will be lost through infiltration so that additional land can be cultivated for agricultural use. The agricultural revenue associated with the extra land is expected to be $120,000 per year. Use (*a*) the B/C and (*b*) the B – C method to determine if the canals should be lined. Assume that the project life is 25 years and the interest rate is 6% per year.

P9.13 A city trying to attract professional athletic teams is considering constructing a domed playing arena or a conventional stadium. The domed arena would cost $300 million to construct and would have a useful life of 50 years. The maintenance and operation would be $300,000 the first year, with costs increasing by $10,000 per year. Every 10 years, an expenditure of $800,000 would be required for "remodeling" the interior. The conventional stadium would cost only $50 million to construct and would also have a useful life of 50 years. The cost of maintenance would be $75,000 the first year, increasing by $8000 per year. Periodic costs for repainting, resurfacing, etc., would be $100,000 every 4 years. Revenue from the domed arena is expected to be greater than that from the conventional stadium by $500,000 the first year, with amounts increasing by $200,000 per year through year 15. Thereafter, the extra revenue from the dome would remain the same at $3.3 million per year. Assuming that both structures would have a salvage value of

$5 million, use an interest rate of 8% per year and a B/C analysis to determine which structure should be built.

P9.14 Why must an incremental B/C analysis be conducted when only one proposal can be selected from three or more proposals?

P9.15 Which dam in Prob. P9.9 should be built if the dams are mutually exclusive (i.e., only one could be constructed)?

P9.16 Five methods could be used to recover grease from a rendering plant wastewater stream. The investment costs and incomes associated with each one are shown below. Assuming that all methods have a 10-year life with zero salvage value, determine which one should be selected using a minimum attractive rate of return of 15% per year and the B/C analysis method. Consider operating costs as disbenefits.

	Method				
	1	2	3	4	5
First cost, $	15,000	19,000	25,000	33,000	48,000
Annual operating cost, $	10,000	12,000	9,000	11,000	13,000
Annual income, $	15,000	20,000	19,000	22,000	27,000

P9.17 Which alternatives in Prob. P9.16 would be selected if they were not mutually exclusive? Use a B/C analysis.

P9.18 Select the best mutually exclusive alternative using the B/C ratio method from the proposals shown below if the MARR is 10% per year and the projects will have a useful life of 15 years. Assume that the cost of the land will be recovered when the project is terminated. Treat maintenance costs as disbenefits.

	Proposal						
	1	2	3	4	5	6	7
Land cost, $	50,000	40,000	70,000	80,000	90,000	65,000	75,000
Construction cost, $	200,000	150,000	170,000	185,000	165,000	175,000	190,000
Annual maintenance, $	15,000	16,000	14,000	17,000	18,000	13,000	12,000
Annual income, $	52,000	49,000	68,000	50,000	81,000	77,000	45,000

P9.19 An oil and gas company is considering five sizes of pipe for a new pipeline. The costs for each size are shown below. Assuming that all pipes will last 15 years and the company's MARR is 8%, which size of pipe should be used according to the B/C method?

	Pipe size, millimeters				
	140	160	200	240	300
Initial investment, $	9,180	10,510	13,180	15,850	30,530
Installation cost, $	600	800	1,400	1,500	2,000
Annual operating cost, $	6,000	5,800	5,200	4,900	4,800

P9.20 Which pipe would be selected in Prob. P9.19 if the lives of the 240- and 300-millimeter pipes could be extended to 30 years by installing corrosion protection equipment costing $1200 at purchase and $300 per year to maintain? Use the B/C method of analysis.

CHAPTER

TEN

REPLACEMENT ANALYSIS

The objective of this chapter is to aid you in the economic comparison of two assets: one which you presently own and the other which can be considered a replacement of the owned asset.

The replacement of an asset is usually contemplated prior to the time of its anticipated sale because of physical deterioration or obsolescence (planned or unplanned). Planned obsolescence is usually the result of "bigger and better" equipment, a fact that has become accepted in today's business world. Unplanned obsolescence usually results from an unanticipated decrease in product demand or the possibility of leasing, rather than owning, the equipment.

SECTION OBJECTIVES

When you are able to accomplish the following, you will have completed this chapter:

10.1. Describe the consultant's viewpoint for replacement analysis; determine the value of a sunk cost; and state the values of first cost, life, annual cost, and salvage value for a defender and challenger, given the appropriate data.

10.2. Select the better of a defender and challenger plan, given their initial costs, lives, operating costs, interest rate, market value of the defender, and the planning horizon.

*10.3. State the difference between the *conventional* and *cash-flow approaches* to replacement analysis and use both to perform an analysis, given the appropriate data for each plan.

10.4. Use the *one-additional-year replacement analysis* procedure to decide to retain a defender for one more year or obtain the challenger next year, given the defender value and annual costs for next year, MARR, and appropriate data for each plan.

*10.5. Compute the replacement value of a defender given the selected challenger plan, defender's remaining life, salvage value, operating costs, and the MARR.

10.6. Determine the minimum cost life of an asset using the EUAC method, given the first cost or market value now, future salvage values, annual operating costs, and the required rate of return.

STUDY GUIDE

10.1 The Defender and Challenger Concepts in Replacement Analysis

Here, as in previous chapters, we are comparing two alternatives; however, we now own one of the assets, referred to as the *defender*, and are considering the purchase of its replacement or *challenger*. This use of the defender-challenger terminology is borrowed from the publications by George Terbough of the Machinery and Allied Products Institute (MAPI) [1, 2].

In the actual comparison we must take the *consultant's viewpoint*, as in previous chapters. Thus, we *pretend* that we own neither asset. In order to "purchase" the defender, therefore, we must pay the going market value for this used asset. We then use the present fair-market value as the first cost (*P*) of the defender. Likewise, there will be an associated salvage value (SV), economic life (*n*), and annual operating cost (AOC) for the defender. Even though the values may all differ from the original data, it makes no difference to us, because we are using the consultant's viewpoint, thus making all previous data irrelevant to the present economic evaluation.

In replacement analysis it is important that the role of a *sunk cost* be understood and treated correctly. The sunk cost is defined as

$$\text{Sunk cost} = \text{present book value} - \text{present realizable value} \qquad (10.1)$$

The present book value is the remaining investment after the total amount of depreciation has been charged to date; that is, the book value is the current worth of the asset in the accounting records (see Chap. 13 for a complete discussion).

If incorrect estimates have been made about the utility or market value of an asset (as is usual, since no one can be perfect in their estimation of the future), there is a positive sunk cost, *which cannot be recovered*. However, some analysts try to "recover" the sunk cost of the defender by adding it to the first cost of the challenger. The sunk cost, rather, should be charged to an account entitled unrecovered capital, or the like, which will ultimately be reflected in the company's income statement for the year in which the sunk cost was incurred. From the tax viewpoint the value of this sunk cost will be important in that a capital gain or loss is involved (Chaps. 14 and 15). Therefore, *for replacement analysis the sunk cost should not be included in the economic comparison*. The following example illustrates the correct data to use in a replacement analysis in which a sunk cost is involved.

Example 10.1 A dump truck was purchased 3 years ago for $12,000 with an estimated life of 8 years, salvage of $1600, and annual operating cost of $3000. The current book value is $8100.

A challenger is now offered for $11,000 and a trade-in value of $7500 on the old truck. The company estimates challenger life at 10 years, salvage at $2000, and annual operating costs at $1800 per year. New estimates for the old truck are made as follows: realizable salvage, $2000; remaining life, 3 years; same operating costs.

What values should be used for P, n, SV, and AOC for each asset?

SOLUTION When using the consultant's viewpoint only the most current information is applicable.

Defender	Challenger
P = $7,500	P = $11,000
AOC = 3,000	AOC = 1,800
SV = 2,000	SV = 2,000
n = 3 years	n = 10 years

The original cost of $12,000, estimated salvage of $1600, and remaining 5 years of life are not used for the defender; only the current data applies.

COMMENT A sunk cost is incurred for the defender if it is replaced. By Eq. (10.1),

$$\text{Sunk cost} = 8100 - 7500 = \$600$$

The $600 is not added to the first cost of the challenger, since this action would (1) try to "cover up" past mistakes of estimation and (2) penalize the challenger because the capital to be recovered each year would be higher due to the increased first cost.

Prob. P10.1 to P10.5

10.2 Replacement Analysis Using a Specified Planning Horizon

The planning horizon (also called a study period) is the number of years in the future which is to be used in comparing the defender and the challenger. Typically, one of two situations is present: (1) the anticipated remaining life of the defender equals the life of the challenger or (2) the life of the challenger is greater than that of the defender. We will discuss both possibilities in order.

If the defender and challenger have equal lives, any of the evaluation methods can be used with the *most current data*. The example below compares owning with leasing to clearly depict this situation.

Example 10.2 The Pak-Mor Transport Company owns two vans but has always leased more vans on a yearly basis as needed. The two vans were purchased 2 years ago for $60,000 each. The company plans to keep the vans for 10 more years. Fair market value for a 2-year-old van is $42,000 and for a 12-year-old van, $8000. Annual fuel, maintenance, tax, etc., costs are $12,000 per year. Lease cost is $9000 per year with annual operating charges of $14,000. Should the company lease all its vans if a 12% rate of return is required?

SOLUTION Consider a 10-year life of the owned van (defender) and the leased van (challenger).

Defender	Challenger
P = $42,000	Lease cost = $ 9,000 per year
AOC = 12,000	AOC = 14,000
SV = 8,000	
n = 10 years	n = 10 years

The EUAC_D for the defender by Eq. (6.1) is

$$\begin{aligned}\text{EUAC}_D &= P(A/P, i\%, n) - \text{SV}(A/F, i\%, n) + \text{AOC}\\ &= 42{,}000(A/P, 12\%, 10) - 8000(A/F, 12\%, 10) + 12{,}000\\ &= \$18{,}977\end{aligned}$$

EUAC_C for the challenger is

$$\text{EUAC}_C = 9000 + 14{,}000 = \$23{,}000$$

Clearly, the firm should retain ownership of the two vans.

In many instances, an asset is to be replaced by another having an estimated life different from that of the defender's remaining life. For analysis, the length of the planning horizon must first be selected, usually coinciding with the life of the longer-lived asset. Selection of the planning horizon makes the assumption that the EUAC value of the shorter-lived asset is the same throughout the planning horizon. This implies that the service performed by the shorter-lived asset can be acquired at the same EUAC as presently computed for its expected service life.

Example 10.3 A company has owned a particular machine for 3 years. Based on current market value the asset has an EUAC of $5210 per year and an anticipated remaining life of 5 years due to rapid technological growth. The possible replacement for the asset has a first cost of $25,000, salvage value of $3800, life of 12 years, and an annual operating cost of $720 per year. If the company uses a minimum rate of return of 10% on asset investments and plans to retain the new machine for its full anticipated life, should the old asset be replaced?

SOLUTION A planning horizon of 12 years to correspond with the challenger's life is appropriate.

$$\text{EUAC}_D = \$5210$$

$$\text{EUAC}_C = 25{,}000(A/P, 10\%, 12) - 3800(A/F, 10\%, 12) + 720$$
$$= \$4211$$

Thus, purchase of the new asset is less costly than is retention of the presently owned machine.

COMMENT When using the planning horizon for different-lived assets and the present-worth value is desired, you must realize that a horizon of 12 years assumes that you plan to purchase a new, similar asset if the shorter-lived asset is retained. In other words, in this problem a defender-similar asset would be purchased at the end of 5 years. Thus, present-worth computation would be as follows:

$$\text{PW}_D = 5210(P/A, 10\%, 12) = \$35{,}500$$

$$\text{PW}_C = 4211(P/A, 10\%, 12) = \$28{,}693$$

Of course, the decision to purchase the new machine is still made.

Often management is skeptical about the future. Reflection of this skepticism is a management desire to use abbreviated periods of time for the planning horizon. In this situation it is assumed that only the time in the planning horizon is allowed for the recovery of invested capital and a specified return; that is, n values in the computations will reflect the shortened horizon. The following example indicates what occurs when a shortened horizon is specified.

Example 10.4 Consider the data of Example 10.3, except use a 5-year planning horizon. Management specifies 5 years because it is leery of the technological progress being made in this area, progress that has already called into question retention of presently owned, operational equipment. Assume that the challenger's salvage value will remain at $3800.

SOLUTION The approach is as in the preceding example, except a capital-recovery period of only 5 years is used for the challenger.

$$\text{EUAC}_D = \$5210$$

$$\text{EUAC}_C = 25{,}000(A/P, 10\%, 5) - 3800(A/F, 10\%, 5) + 720$$
$$= \$6693$$

Now, retention of the defender is less costly, thus reversing the decision made with a 12-year horizon.

COMMENT By not allowing the full anticipated life of the challenger to be used, management has ruled out its use. However, the decision to not consider the use of this new asset past 5 years is one of management responsibility. The reason why the decision is reversed in this example is quite simple, actually. The challenger is given only 5 years to recover the same investment and a 10% return, whereas in the previous example 12 years is allowed. Reasonably, $EUAC_C$ must increase drastically. It would be possible to recognize unused value in the challenger by increasing the salvage value from $3800 to the estimated fair market value after 5 years of service, if such a value can be predicted.

Selection of the planning horizon is a difficult decision, one which must be based on sound judgment and data. The use of a short horizon may often bias the economic decision in that the capital-recovery period for the challenger may be abbreviated to much less than the anticipated life. This is the case in Example 10.4, where only 5 years were allowed for recovery of invested capital plus a 10% return. However, use of a large horizon is also often detrimental due to the uncertainty of the future and its estimate. In this case, the direction of bias is less certain than in the case of a too-short horizon. A common practice is augmentation; that is, the defender is augmented with a newly purchased asset to make it comparable in ability (speed, volume, etc.) with the challenger. Since the analysis is similar to that covered here, a sample solution is included in Solved Problems.

Solved Problem 10.9
Prob. P10.6 to P10.14

*10.3 Conventional and Cash-Flow Approach to Replacement Analysis

There are two equally correct and equivalent ways to handle the first cost of alternatives in a replacement analysis. The *conventional approach* (used in Example 10.1) uses the defender current trade-in value as the first cost of the defender and uses the initial cost of the replacement as the challenger first cost. This approach is cumbersome when there is more than one challenger with each one offering a different trade-in value for the defender, thus causing a different P value for the defender when compared with each challenger.

Another approach is to realize that if a challenger is selected, the respective defender trade-in is realized as a cash inflow to the challenger alternative; and if the defender is selected, there is no actual outlay of cash. This is the *cash-flow approach* to replacement analysis. In this approach, set the defender first cost to zero and *subtract* the trade-in value from the challenger first cost. A planning horizon must be selected so that the comparison period will be the same for all alternatives.

Either method gives the same answer in the analysis, as shown in the next example.

Example 10.5 A 7-year-old asset may be replaced with either of two new assets. Current data for each alternative are given below. Use the cash-flow approach and MARR = 18% to determine the most economic decision.

	Current asset, defender	Possible replacements	
		Challenger 1	Challenger 2
First cost	...	\$10,000	\$18,000
Defender trade-in	...	3,500	2,500
Annual cost	\$3,000	1,500	1,200
Salvage value	500	1,000	500
Life estimate, years	5	5	5

SOLUTION The first-cost value used for the defender (D) in the replacement analysis is different for challenger 1 (C_1) and challenger 2 (C_2). Using the cash-flow approach subtract the trade-in values from the respective challenger first cost and compute the EUAC over the respective life of each alternative. With this approach, if D is selected, the first cost is zero because the asset is already owned.

$$\text{Defender: } \quad \text{EUAC}_D = 3000 - 500(A/F, 18\%, 5) = \$2930.11$$

$$\text{Challenger 1: EUAC}_{C_1} = (10{,}000 - 3500)(A/P, 18\%, 5) + 1500 - 1000(A/F, 18\%, 5) = \$3438.79$$

$$\text{Challenger 2: EUAC}_{C_2} = (18{,}000 - 2500)(A/P, 18\%, 5) + 1200 - 500(A/F, 18\%, 5) = \$6086.70$$

Since the defender has the smallest EUAC, it should be retained.

COMMENT Had the conventional approach been used, two analyses would be performed: D versus C_1, with the defender first cost of \$3500; and D versus C_2, with the defender first cost of \$2500. Results (that you should verify) are as follows:

D versus C_1	D versus C_2
$\text{EUAC}_D = \$4049.34$	$\text{EUAC}_D = \$3729.56$
$\text{EUAC}_{C_1} = \$4558.02$	$\text{EUAC}_{C_2} = \$6886.15$

As expected the decision is to retain the defender because it offers the smallest equivalent annual cost.

Prob. P10.15 to P10.18

10.4 Replacement Analysis for One-Additional-Year Retention

When a currently owned asset is close to the end of its useful life or has demonstrated deteriorating usefulness to a company, a common problem is to determine if it should

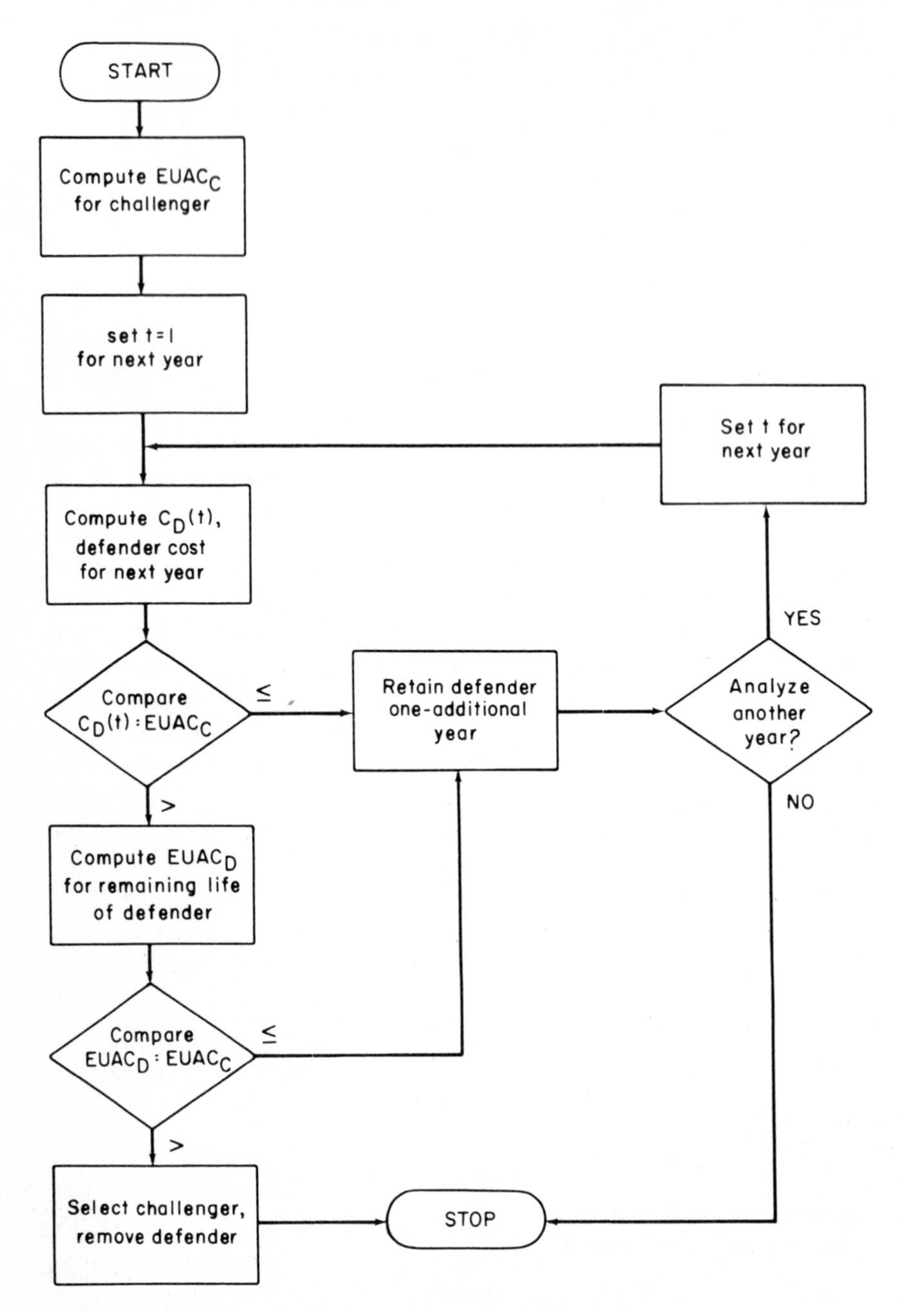

Figure 10.1 Procedure used for one-additional-year replacement analysis.

be replaced with a challenger or retained in service for only one more year. In such a case it is not correct to compare the defender cost and challenger cost over their remaining anticipated lives. Rather, the equivalent-annual-cost procedure presented in Fig. 10.1 may be used to first calculate EUAC_C and $C_D(t)$ for $t = 1$, where

$$\text{EUAC}_C = \text{challenger EUAC value}$$

$$C_D(t) = \text{defender cost for the next year } t = 1$$

If $C_D(1) \leqslant \text{EUAC}_C$, retain the defender one more year because its cost is less. If $C_D(1) > \text{EUAC}_C$, the challenger cost must also be less than the defender EUAC_D for its remaining life. The Fig. 10.1 procedure states that if $\text{EUAC}_D \leqslant \text{EUAC}_C$, the defender is still retained the next year. In either case it is possible to continue the analysis for future years $t = 2, 3, \ldots,$ one year at a time until the challenger is selected or the defender life is reached. Note that only when the challenger cost is less that the next-year defender cost *and* the defender EUAC_D is the challenger selected. Example 10.6 demonstrates the use of one-additional-year replacement analysis.

Example 10.6 The U1-Likit Sales Company commonly retains a salesperson's car for 5 years. Due to the purchase of autos exactly 2 years ago that have deteriorated much more rapidly than expected, management has asked you if it is more economic to retain and maintain the fleet for one more year and then replace the fleet, retain it for 2 more years then replace, or keep it for 3 more years. Or is it cheaper to replace the fleet this year with new, more reliable autos? Perform the replacement analysis at $i = 20\%$ for a typical defender car and challenger car having the estimated costs detailed here.

Currently owned (defender)	Value at beginning of year	Annual operating cost	Possible replacement (challenger)	
Next year (3)	\$3800	\$4500	First cost	\$8700
Next year (4)	2800	5000	Operating cost	3900 per year
Last year (5)	500	5500	Life, years	5
Remaining life, years	3		Salvage value	1800
Salvage after 3 more years	500			

SOLUTION Following the procedure in Fig. 10.1, compute the EUAC for the challenger over 5 years and the defender's cost for next year only ($t = 1$).

$$\text{EUAC}_C = 8700(A/P, 20\%, 5) - 1800(A/F, 20\%, 5) + 3900 = \$6567.22$$

$$C_D(1) = 3800(A/P, 20\%, 1) - 2800(A/F, 20\%, 1) + 4500 = \$6260.$$

Since $C_D(1) < \text{EUAC}_C$, retain the defending auto for the next year. Note that in $C_D(1)$ the salvage value of \$2800 for next year is the expected value (initial cost) for the year after, that is, for $C_D(2)$.

To determine if the auto should be kept yet another year, follow Fig. 10.1 and set $t = 2$ for

$$C_D(2) = 2800(A/P, 20\%, 1) - 500(A/F, 20\%, 1) + 5000 = \$7860$$

Now $C_D(2) > \text{EUAC}_C = \$6567.22$, so we compute the EUAC_D value for the remaining 2 years of the defender's life.

$$\text{EUAC}_D = 2800(A/P, 20\%, 2) - 500(A/F, 20\%, 2) + 5000 + 500(A/G, 20\%, 2) = \$6832.97$$

Since the challenger is also cheaper for the remaining two years, select it and replace the defender after one more year of service. Had $\text{EUAC}_C > \text{EUAC}_D$ for 2 years the defender would be retained and a similar analysis done for the last year using $C_D(3)$ and EUAC_D for one year.

COMMENT If rather than comparing $C_D(1)$ and EUAC_C only the EUAC_D value for 3 years is used in the replacement analysis, the wrong decision is made because the 3-year EUAC_D exceeds the 5-year EUAC_C calculated above.

$$\text{EUAC}_D = 3800(A/P, 20\%, 3) - 500(A/F, 20\%, 3) + 4500 + 500(A/G, 20\%, 3) = \$6606.11$$

$$\text{EUAC}_C = \$6567.22$$

Here the challenger is purchased immediately, whereas the one-additional-year analysis has shown that it is more economic to keep the defender one year and then replace it.

Probs. P10.19 to P10.22

*10.5 Computation of Replacement Value for a Defender

Once a challenger is defined, the market value of the defender is of critical importance to the replacement analysis. Rather than initially obtaining market-value quotes, it may be advisable to compute a break-even market value of the defender in terms of the contemplated challenger. The replacement value is computed most easily using the general form

$$0 = \text{EUAC}_D - \text{EUAC}_C \tag{10.2}$$

where EUAC_D is the EUAC for the defender and EUAC_C is for the challenger. After calculating the replacement value, the company can then go into the field to see if the defender market value is greater or less than the replacement value. If greater, the asset should be replaced, since EUAC_D will be greater than EUAC_C.

Example 10.7 A 3-year-old asset has annual operating costs of \$9500, salvage of \$3500, and a useful life of 7 more years. The selected challenger will cost \$28,000, have a life of 14 years, salvage of \$2000, and annual operating costs of

\$5500. What is the minimum trade-in deal that the owner can accept and still purchase the new asset, if this class of asset has a 15% rate or return?

SOLUTION To compute the replacement value of the defender we need the EUAC values. If we let RV represent the replacement value of the defender, we have

$$\begin{aligned}\text{EUAC}_D &= \text{RV}(A/P, 15\%, 7) - 3500(A/F, 15\%, 7) + 9500\\ &= 0.24036(\text{RV}) + 9184\\ \text{EUAC}_C &= 28{,}000(A/P, 15\%, 14) - 2000(A/F, 15\%, 14) + 5500\\ &= \$10{,}342\end{aligned}$$

Equation (10.2) results in a replacement value of RV = \$4818. If a trade-in value greater than \$4818 can be obtained, the challenger should be purchased.

COMMENT To verify the results, assume that the trade-in offer is \$5000. Then

$$\begin{aligned}\text{EUAC}_D &= 5000(A/P, 15\%, 7) - 3500(A/F, 15\%, 7) + 9500\\ &= \$10{,}386\end{aligned}$$

Thus, because EUAC_D is greater than EUAC_C, replacement is economically justified.

Probs. P10.23 to P10.27

10.6 Determination of Minimum Cost Life

There are several situations in which an analyst desires to know how long an asset should be used before it is removed from service. Determination of this time (an n value) is given several names. If the asset is a defender to be replaced by a potential challenger, the n value is the *remaining life of the defender*. If the function performed by the asset will be discontinued or assumed by some other facility, the n value is referred to as the *retirement* life. Finally, a similar analysis may be performed for an anticipated asset purchase, in which case the n value may be called *expected life*. Regardless of what it is called, this value should be found by determining the number of years that will yield a minimum present-worth or EUAC value. This approach is often called a *minimum-cost-life analysis*. In this chapter we consistently use EUAC analysis.

To find the minimum cost life, increase the life value, which we call k, from 1 to the maximum expected value for the asset, that is, $k = 1, 2, \ldots, n$; and determine EUAC_k for each k value using the relation

$$\text{EUAC}_k = P(A/P, i\%, k) - \text{SV}_k(A/F, i\%, k) + \left[\sum_{j=1}^{k} \text{AOC}_j(P/F, i\%, j)\right](A/P, i\%, k) \tag{10.3}$$

where SV_k = salvage value if the asset is retained k years

AOC_j = annual operating cost for year j ($j = 1, 2, \ldots, k$)

The minimum cost life is determined by selecting

$$\min_k \text{EUAC}_k$$

and the corresponding life value ($k = n$) and EUAC_k should be used in the replacement and other economic analyses. This approach is illustrated in Example 10.8.

Example 10.8 An asset purchased 3 years ago is now challenged by a new piece of equipment. Market value of the defender is \$13,000. Anticipated salvage values and annual operating costs for the next 5 years are given in columns 2 and 3, respectively, of Table 10.1. What is the minimum cost life to be used when comparing this defender with a challenger if capital is worth 10%?

SOLUTION Equation (10.3) is used to determine EUAC_k for $k = 1, 2, \ldots, 5$. Column 4 in Table 10.1 gives the capital necessary and return required using the first two terms of Eq. (10.3) and column 5 gives the equivalent operating costs for k years using the last term in the EUAC_k equation. The sum is EUAC_k shown in column 6. As an example the computations for $k = 3$ may be determined using Eq. (10.3) as follows.

$$\begin{aligned}\text{EUAC}_3 &= 13{,}000(A/P, 10\%, 3) - 6000(A/F, 10\%, 3) + [2500(P/F, 10\%, 1) \\ &\quad + 2700(P/F, 10\%, 2) + 3000(P/F, 10\%, 3)]\ (A/P, 10\%, 3) \\ &= \$6132\end{aligned}$$

The minimum cost in Table 10.1 is \$6132 per year for $k = 3$, which indicates that 3 years should be the anticipated remaining life of this asset when compared with a challenger.

COMMENT It should be realized that the approach presented in this example is general. It can be utilized to find the minimum cost life of any asset, whether it is owned presently and is to be replaced or retired, or whether it is a con-

Table 10.1 Computation of minimum cost life for a presently owned asset

(1) Life, k years	(2) SV_k	(3) AOC_j ($j = 1, 2, \ldots, k$)	(4) Capital recovery and return	(5) Equivalent operating costs	(6) = (4) + (5) EUAC_k
1	\$9000	\$2500	\$5300	\$2500	\$7800
2	8000	2700	3681	2595	6276
3	6000	3000	3415	2717	6132
4	2000	3500	3670	2886	6556
5	0	4500	3429	3150	6579

templated purchase. A further example of minimum cost life is given in the Solved Problems.

Appendix E includes a discussion on the computer program MINCL which is specifically designed to compute MINimum Cost Life for any given values of P, SV_k and $AOC_j (j = 1, 2, \ldots, k)$.

Solved Problem 10.10
Probs. P10.28 to P10.32

SOLVED PROBLEMS

Example 10.9 Three years ago the city of Water purchased a new fire truck. Due to expanded growth in a certain portion of the city, new fire-fighting capacity is needed. An additional identical truck can be purchased now or a double-capacity truck can replace the presently owned asset. Data for each asset are presented in Table 10.2. Compare the assets at $i = 12\%$ using (*a*) a 12-year study period and (*b*) a 9-year period, which the city management believes to be more realistic due to population growth.

SOLUTION Plan A is the retention of the previously owned truck and *augmentation* with the new identical-capacity vehicle; plan B is purchase of the double-capacity truck. Details of each plan are below.

Plan A		Plan B
Presently owned	Augmentation	Double capacity
P = $18,000	P = $58,000	P = $72,000
AOC = 1,500	AOC = 1,500	AOC = 2,500
SV = 5,100	SV = 6,960	SV = 7,200
n = 9 years	n = 12 years	n = 12 years

(*a*) For a full-life 12-year horizon,

$$\begin{aligned} EUAC_A &= (\text{EUAC of presently owned}) + (\text{EUAC of augmentation}) \\ &= [18{,}000(A/P, 12/, 9) - 5100(A/F, 12\%, 9) + 1500] \\ &\quad + [58{,}000(A/P, 12\%, 12) - 6960(A/F, 12\%, 12) + 1500] \\ &= 4533 + 10{,}575 \\ &= \$15{,}108 \end{aligned}$$

$$\begin{aligned} EUAC_B &= 72{,}000(A/P, 12\%, 12) - 7200(A/F, 12\%, 12) + 2500 \\ &= \$13{,}825 \end{aligned}$$

Table 10.2 Data for fire-truck replacement analysis

	Presently owned	New purchase	Double capacity
P	\$51,000	\$58,000	\$72,000
AOC	1,500	1,500	2,500
Trade-in	18,000	...	...
SV	10% of P	12% of P	10% of P
n	12	12	12

Purchase the double-capacity truck (plan B) with an advantage of \$1283 per year.

(*b*) The analysis for a truncated 9-year horizon is identical, except that $n = 9$ in each factor; that is, 3 fewer years are given to the augmentation and double-capacity truck to recover investment plus a 12% return. Here

$$\text{EUAC}_A = \$16{,}447 \qquad \text{EUAC}_B = \$15{,}526$$

and plan B is again selected but now only by a margin of \$921.

If the planning horizon were trunacted more severely, at some point the decision would be reversed.

Sec. 10.2

Example 10.10 Assume that an asset can be purchased for \$5000 and will have a negligible salvage value. The annual operating costs are expected to follow a gradient of \$200 per year with a base amount of \$300 in year 1. Find the number of years that the asset should be kept if interest is not considered important.

SOLUTION The main effects of $i = 0\%$ will be to decrease the total annual cost values and make computations simpler; however, cost patterns will be similar to those for $i > 0\%$, except that the minimum cost life may change. Table 10.3 presents the entire solution to the problem. Column 3 gives cumulative AOC according to the gradient, while the average AOC is given in column 4 [= (3)/n]. A sample computation is given below the table, which indicates that $n^* = 7$ is the minimum cost life with a total annual cost of \$1614. The number of years is indicated by n here rather than k, as in Sec. 10.6. The minimum cost life is n^*.

COMMENT Due to the tremendous regularity of this type of problem and the fact that $i = 0\%$, a quick formula can be derived to find the minimum cost life. We can write

Total annual cost (TAC) = average operating cost + average first cost

For year n this may be expressed as

$$\text{TAC}_n = \frac{\sum_{j=1}^{n} \text{AOC}_j}{n} + \frac{P}{n}$$

Table 10.3 Computation of minimum cost life for $i = 0\%$

(1)	(2)	(3) Operating cost	(4)	(5) Average first cost	(6) Total annual cost
Year	Annual	Cumulative	Average		
1	$ 300	$ 300	$ 300	$5000	$5300
2	500	800	400	2500	2900
3	700	1500	500	1667	2167
4	900	2400	600	1250	1850
5	1100	3500	700	1000	1700
6	1300	4800	800	833	1633
7	1500	6300	900	714	1614*
8	1700	8000	1000	625	1625
9	1900	9900	1100	555	1655

*Total annual cost is computed as (6) = (4) + (5) = $(3)/n + 5000/n$. For $n = 7$, $6300/7 + 5000/7 = \$1614$.

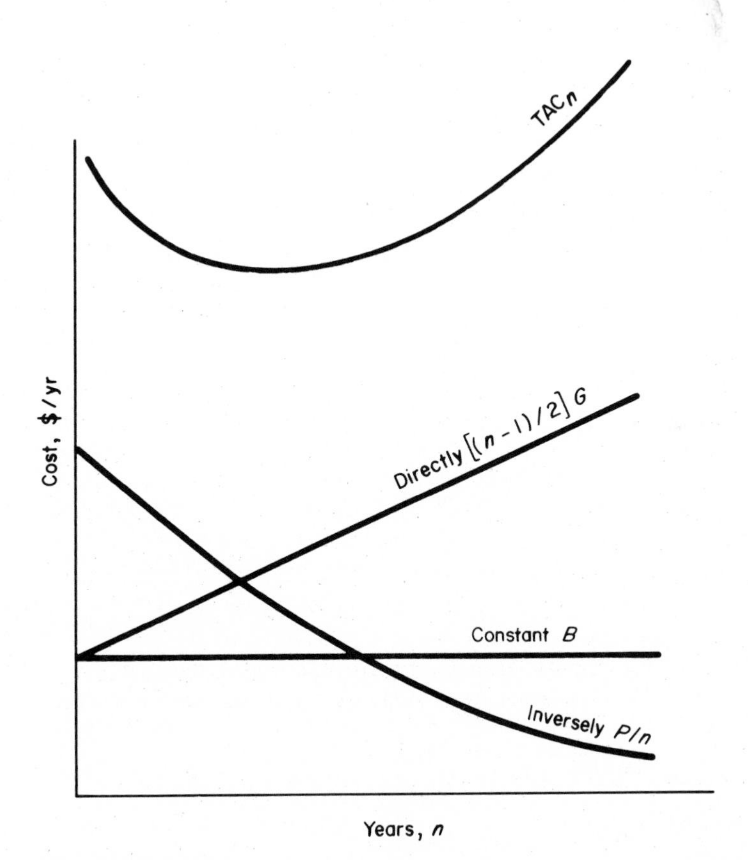

Figure 10.2 General plot of TAC_n and component terms used to compute minimum cost life, Example 10.10.

where TAC_n = total annual cost for n years of ownership
AOC_j = annual operating cost through year j $(j = 1, 2, \ldots, n)$

However, we can make the substitution

$$\frac{\sum_{j=1}^{n} AOC_j}{n} = B + \left(\frac{n-1}{2}\right) G$$

where B = base amount of the gradient
G = amount of gradient

Then

$$TAC_n = B + \left(\frac{n-1}{2}\right) G + \frac{P}{n} \tag{10.4}$$

The general shape of the terms and TAC_n itself is shown in Fig. 10.2. If we take the derivative of Eq. (10.4) and solve for an optimum life value (n^*), we have

$$\frac{dTAC_n}{dn} = \frac{G}{2} - \frac{P}{n^2} = 0$$

$$n^* = \left(\frac{2P}{G}\right)^{1/2}$$

Substitution of P = \$5000 and G = \$200 for this example yields

$$n^* = \left(\frac{10{,}000}{200}\right)^{1/2} = 7.07 \text{ years}$$

which is, for all practical purposes, the same as $n^* = 7$ obtained in Table 10.3.

Sec. 10.6

REFERENCES

1. G. D. Terbough, *Dynamic Equipment Policy*, McGraw-Hill, New York, 1949.
2. G. D. Terbough, *Business Investment Management*, Machinery and Allied Products Institute, Washington, D.C., 1967.

FURTHER INFORMATION

Replacement analysis; Canada and White [4], pp. 142–171; DeGarmo, Canada, and Sullivan [5], pp. 327–365; Grant, Ireson, and Leavenworth [7], pp. 408–443; Newman [11], pp. 254–285; Riggs [13], pp. 170–189; Taylor [16], pp. 205–207, 213–241; Thuesen, Fabrycky, and Thuesen [17], pp. 210–249; White, Agee, and Case [18], pp. 169–179.

PROBLEMS

P10.1 If the difference between present book value and realizable salvage value is negative, how would you treat this sum in a replacement analysis?

P10.2 Why is the original asset cost irrelevant in a replacement analysis? What does this fact have to do with the consultant's viewpoint in replacement analysis?

P10.3 A new meat display counter was purchased by a supermarket 4 years ago at a cost of $28,000. Book value is presently $10,000 with 5 years remaining before a salvage value of $1000 is reached. Due to lagging sales, the owners wish to trade for a new, smaller counter, which costs $13,000 and has an installation charge of $500. As estimated by the owners, the old counter will last another 8 years and has a trade-in value of $18,000 now. A review of the accounts shows annual repair costs averaging $150 for the old counter. (*a*) Determine the values of P, n, SV, and AOC for the existing counter to be used in a replacement analysis. (*b*) Is there a sunk cost involved here? If so, what is its amount?

P10.4 The owners of a downtown shoe shop are considering the possibility of moving to a rented shop in a suburban shopping center. They purchased their shop 15 years ago for $8000 cash. They estimate their annual investment in property improvement at $500 and believe the shop should have a current book value that includes the purchase price and these improvements at 8% interest. The annual insurance, utility, etc., costs have averaged $1080 per year. If they stay downtown, they hope to retire in 10 years and will give the shop to their son-in-law. They will ask $25,000 for the shop if they sell at this time. If the owners move to the shopping center, they must sign a 10-year lease for $6600 per year with no additional yearly charges. They must pay a $750 deposit when they sign the lease, but this amount is returned at the time of lease expiration. Determine (*a*) the values of P, n, SV, and AOC for the two alternatives and (*b*) the amount of the sunk cost if one exists.

P10.5 An asset purchased 2 years ago can now be traded for an "improved" version for 40% of the first cost. The asset was purchased for $18,000 and is being depreciated over 5 years for tax purposes with a current book value of $10,125 and 12 years for company income purposes with a current book value of $12,500. Compute the sunk cost (*a*) to be reported in the tax reports and (*b*) to be used by company economists.

P10.6 The Moon City Bus Lines has 20 buses purchased 5 years ago for $22,000 each. The company president plans to have a major overhaul done on these buses next year at a cost of $1800 each. However, the vice president wants to trade these 20 buses in on 25 of a new, smaller model. The trade-in value is $4000 and the new models cost $22,500 each. The president estimates a remaining life of 7 years for the old buses once the overhaul is completed and further states that annual operational costs per bus are $3000, and a $800 salvage is reasonable when sold to an individual as a "vacation van." The vice president interjects the comment that the smaller buses can maneuver in traffic more easily, will cost $1000 per year less to operate, will last for 8 years, and will have a salvage value of $500 when sold to day camps. With this wealth of knowledge, you, the "general everything" for the company, are requested to determine which officer's desire is economically correct at the firm's MARR of 10%.

P10.7 Rework Prob. P10.6 if the new buses have a life of 12 years, which happens to correspond with the planning horizon selected by the vice president.

P10.8 Rework Prob. P10.6 using a planning horizon of 5 years.

P10.9 Perform a replacement analysis for Prob. P10.3 if the new counter has an expected life of 10 years, annual cost of $30, and a salvage value of $1500. Use an interest rate of 15%.

P10.10 Determine which alternative is better in Prob. P10.4 at a 10% interest rate.

P10.11 A new earth mover was purchased by the Never Dri Cement Company 3 years ago and has been used to transport raw material from the quarry to the crushers. When purchased the mover possessed the following characteristics: P = $55,000, n = 10 years, SV = $5000, and capacity = 180,000 metric tons per year. With increased construction in industrial parks around the city, an additional mover with a capacity of 240,000 metric tons per year is needed. Such a vehicle can be purchased. If bought, this new asset will have P = $70,000, n = 10 years, SV = $8,000.

However, the company could have constructed a conveyor system to move the material from the quarry. This system will cost $115,000, have a life of 15 years, no salvage value, and carry 400,000 metric tons per year. The company will need to have some way to move material to the conveyor in the quarry. If the presently owned mover is used, it will more than suffice. However, a a new smaller-capacity mover can be purchased. A $15,000 trade-in on the old mover will be given on any new mover. This one will have P = $40,000, n = 12 years, SV = $3,500, capacity = 400,000 metric tons per year over this short distance. Monthly operating, maintenance, and insurance costs average $0.01 per ton-kilometer for the movers; similar costs for the conveyor are expected to be $0.0075 per metric ton. The company wants to make 12% on this investment. Records show that the mover must travel an average of 2.4 kilometers from the quarry to the crusher pad. The conveyor will be placed to reduce this distance to 0.32 kilometer. Should the old mover be augmented by a new mover or should the conveyor be considered as a replacement; and if so, which method of moving the material in the quarry should be used?

P10.12 Solve Prob. P10.11 under this condition: only a 4-year planning period is possible because management feels that this spurt in business is very short-lived.

P10.13 Machine A, purchased 2 years ago, is wearing out more rapidly than expected. It has a remaining life of 2 years, annual operating costs of $3000, and no salvage value. To continue the function of this asset, machine B can be purchased and a trade-in value of $9000 will be allowed for machine A. Machine B has P = $25,000, n = 12 years, AOC = $4000, and SV = $1000. As an alternative, machine C can be bought to replace A. No trade-in will be allowed for A, but it can be sold for $7000. This new asset will have P = $38,000, n = 20 years, AOC = $2500, and SV = $1000. If plan I is the retention of A, plan II is the purchase of B, and plan III is the selling of A and the purchase of C, use a 20-year period and a MARR = 8% to determine which plan is more economic.

P10.14 A family presently own a 10-year-old built-in range, which is not working properly. Repairs would cost $150, but the service shop estimates a remaining life of only 5 years with no salvage. The utility company advertises that a range costs an average of $75 per year to operate. If replaced, the range will be sold at a garage sale for $35 in its present condition. The family would like to have the range fixed and buy a microwave oven for $550. This oven will cost $150 to build in and has a life of 10 years and a salvage of $75. Its annual operating cost is $16.

The electric company promotion department personnel suggest a combination electric-microwave range for all cooking. This range costs $950, will last 10 years, has a salvage value of $125, and costs $30 per year to operate. If money is currently worth 15% in the bank and the family doesn't think that they will stay in their present house for more than 5 years, should they buy the combination range or should they repair the old range and buy the microwave oven?

***P10.15** Work Prob. P10.6 using the cash-flow approach to replacement analysis.

***P10.16** Work Prob. P10.11 using the cash-flow approach to replacement analysis, and a planning horizon of 10 years. Assume that all the values in Prob. P10.11 are correct for the 10-year lives.

***P10.17** Rework Prob. P10.13 using the cash-flow approach to replacement analysis. Assume that machine A can be made to last for a total of 12 more years with a $20,000 rework in 2 years. Also, let C have n = 12 years and SV = $1000. Use a 12-year planning horizon.

***P10.18** Dynamic Computers owns an asset (#101) used in disk-drive construction. This asset has had high annual maintenance costs and may be replaced with one of two new improved versions. Model A-1 can be installed for a total cost of $155,000 with expected characteristics of n = 5 years, AOC = $10,000, and SV = $17,500. Model B-2 has a first cost of $100,000 with n = 5 years, AOC = $13,000, and SV = $7000. If the presently owned asset is traded it will bring $31,000 from the A-1 manufacturer and $28,000 from the B-2 producer. Retention of asset #101 has been estimated to be possible for 5 more years at an AOC of $34,000 and a negative salvage value of $2000 after the 5 years. Use the cash-flow approach to determine which is the most economic decision at a required return of 16%.

P10.19 In Example 10.3 a currently owned machine has $EUAC_D$ = $5210 for 5 more years of service. The challenger, which has $EUAC_C$ = $4211 for a 12-year life, is selected because $EUAC_C < EUAC_D$. Management wants to keep the defender one more year before replacement. Make a suggestion to management if additional study indicates that the value of the defender is $3000 now with an anticipated value of $1800 one year from now. The projected operating cost for next year is $3000 and the minimum return is still 10%.

P10.20 The Harvey Paint Company owns an air compressor that should possibly be replaced. A new model which sells for $1500 will last 7 years with annual costs estimated to be $100 the first year and $50 higher each year and a zero salvage value. Mr. Harvey can sell the old compressor to his brother at the following prices: $400 this year (now), $300 next year, or $50 the third year. Harvey will keep the compressor for a maximum of 2 more years since operating costs are expected to increase to $175 next year and $350 the following year. Should Harvey trade now, next year, or two years from now if the challenging compressor will have the same costs in the future as estimated now? Let $i = 12\%$.

P10.21 A replacement study is to be performed on pressing equipment in an industrial laundry. The challenging asset has a computed $EUAC_C = \$42,000$ for its anticipated 10-year life. Thorough data collection on the defender has resulted in the following projected annual operating costs (AOC) and trade-in values for the next 5 years, after which the currently owned equipment would have to be replaced.

Additional years retained	AOC, dollars per year	Trade-in value, dollars
1	34,000	28,000
2	30,000	22,000
3	30,000	15,000
4	30,000	5,000
5	30,000	0

If the current equipment is kept for 5 more years, it will cost a net estimated $2000 to remove it from the plant. Perform one-additional-year replacement analysis at a 16% return to determine how many years to keep this asset before replacing it with the challenger.

P10.22 You and your spouse have to make the decision to keep your present car or purchase a new one. A new car will cost $10,000, last you 7 years, have annual maintenance costs of $200 the first year increasing by $100 per year thereafter, and sell for $3000 in 7 years. If you retain the currently owned car, the expected trade-in value and annual maintenance are as follows:

Additional years retained	Maintenance, dollars per year	Trade-in value, dollars
1	1800	2500
2	1500	2000
3	1500	1500

You won't consider keeping the car for more than 3 additional years, at which time you anticipate a $1000 sales price. If all other costs are considered equal for the two cars, use $i = 15\%$ to determine when to purchase a new car. (Neglect financing complications on the new car by assuming that you have just won a contest which gives you a sum of $10,000 after taxes.)

***P10.23** What is the replacement value of the old display counter described in Prob. P10.3 if, as in Prob. P10.9, the new counter has $n = 10$, AOC = $30, SV = $1500, and $i = 15\%$?

***P10.24** A construction company bought a 180,000-metric-ton-per-year-capacity earth mover 3 years ago at a cost of $55,000; the expected life at time of purchase was 10 years with a $5000 salvage value and an annual operating cost of $2700. A 480,000-metric-ton-per-year replacement mover is under consideration. This mover will cost $40,000, have a life of 12 years, a salvage of $3500, and an annual operating cost of $7200. Compute the trade-in value of the presently owned mover if the replacement mover is bought and $i = 12\%$.

***P10.25** (*a*) Solve Prob. P10.24 using a planning horizon of 4 years.

(*b*) How does this truncation of the horizon affect the replacement value of the presently owned mover?

***P10.26** Assume that for the situation of Prob. P10.14 a neighbor offers to buy the old range. What price must be asked to alter the decision made in the previous case?

***P10.27** An asset presently owned can last for 6 more years with costs of $24,000 this year and increasing by 10% per year. A desirable challenger would cost $70,000, last for 6 years, and have an annual cost of $12,000 and a salvage value of $4000. What is a trade-in value of the old asset that will make replacement economic if a 5% return is desired?

P10.28 Ms. Adams just bought a used car for $5800; her uncle financed the purchase at 5% compounded per year for 3 years and she put $400 down. The resale values for the next 6 years are $2200 after the first year, decreasing by $400 per year to year 5, after which the resale value remains at $600. Annual costs of repairs, insurance, gas, etc., are expected to be $1000 the first year and increase by 10% each year. If money is worth 7%, how many years should the car be retained? Assume that the owner will pay off the entire loan with interest if she sells the car before she has owned it 3 years.

P10.29 Rework Prob. P10.28 at $i = 0\%$ rate of return and find the difference between the two answers.

P10.30 Machine *H* was purchased 5 years ago for $40,000 and had an expected life of 10 years. The past and estimated future maintenance and operating costs and salvage values are given below. At a value of $i = 10\%$, determine the number of years the asset should be kept in service before replacement.

Year	Operating cost	Maintenance cost	Salvage value
1	$1,500	$2,000	$25,000
2	1,600	2,000	25,000
3	1,700	2,000	22,000
4	1,800	2,000	22,000
5	1,900	2,000	15,000
6	2,000	2,100	5,000
7	2,100	2,700	5,000
8	2,200	3,300	0
9	2,300	3,900	0
10	2,400	4,500	0

P10.31 One year ago the Bullwinkle Pool Company purchased a machine to blow concrete onto the walls of a new swimming pool, thereby greatly reducing the time necessary to construct a pool. The machine cost $8000 and is expected to last 14 more years. The owner has already seen newer, improved versions. To compare these challengers to the defender, knowing the most economic life of the old version would be of benefit. If costs were $500 for the first year and are expected to increase by $100 per year, compute the most economic life for $i = 0\%$. Assume that salvage value is zero for all years.

P10.32 Rework Prob. P10.31 at $i = 5\%$ and compare the answers.

CHAPTER

ELEVEN

BONDS

The objective of this chapter is to teach you how to make an economic analysis of transactions involving bonds.

SECTION OBJECTIVES

To complete this chapter, you must be able to do the following:

11.1. Define *mortgage bond, collateral bond, equipment trust bond, debenture bond, convertible debenture, municipal bond, general obligation bond, revenue bond,* and *bond rating.*
11.2. Calculate the interest payable (receivable) per period from the sale (purchase) of a bond, given the face value of the bond, the bond interest rate, and the interest payment period.
11.3. Calculate the present worth of a bond, given the face value, bond interest rate, interest payment period, date the bond matures, and the desired rate of return.
11.4. Calculate the nominal and effective rates of return that would be received from the purchase of a bond, given the face value, purchase price, bond interest rate, compounding period, and date the bond matures.

STUDY GUIDE

11.1 Bond Classifications

A *bond* is a long-term note issued by a corporation or governmental entity for the purpose of financing major projects. In essence, the borrower receives money now in return for a promise to pay later, with interest paid between the time the money was borrowed and the time it was repaid. In general, bonds may be classified as mortgage bonds, debenture bonds, and municipal bonds. These types of bonds can be further subdivided.

A *mortgage bond* is one which is backed by a mortgage on specified assets of the company issuing the bonds. If the company is unable to repay the bondholders at the time the bonds mature, the bondholders have the option of foreclosing on the mortgaged property. Mortgage bonds can be subdivided into first-mortgage and second-mortgage bonds. As their names imply, in the event of foreclosure by the bondholders, the first-mortgage bonds take precedence during liquidation. The first-mortgage bonds, therefore, generally provide the lowest rate of return. Second-mortgage bonds, when backed by collateral of a subsidiary corporation, are referred to as *collateral bonds*. An *equipment trust bond* is one in which the equipment purchased through the bond serves as collateral. These types of bonds are generally issued by railroads for purchasing new locomotives and cars.

Debenture bonds are not backed by any form of collateral. The reputation of the company is important for attracting investors to this type of bond. As further incentive for investors, debenture bonds are often *convertible* to common stock at a fixed rate as long as the bonds are outstanding. For example, a $1000 convertible debenture bond issued by the Get Rich Quik (GRQ) Company may have a conversion option to 50 shares of GRQ common stock. If the value of 50 shares of GRQ common stock exceeds the value of the bond at any time prior to bond maturity, the bondholder has the option of converting the bond to common stock. Debenture bonds generally provide the highest rate of interest because of the increased risk associated with them.

The third general type of bonds are *municipal bonds*. Their attractiveness to investors lies in their income-tax-free status. As such, the interest rate paid by the governmental entity is usually quite low. Municipal bonds can be either *general obligation bonds* or *revenue bonds*. General obligation bonds are issued against the taxes received by the governmental entity (i.e., city, county, or state) that issued the bonds and are backed by the full taxing power of the issuer. School bonds are an example of general obligation bonds. Revenue bonds are issued against the revenue generated by the project financed, as a water treatment plant or a bridge. Taxes cannot be levied for repayment of revenue bonds.

In order to assist prospective investors, all *bonds* are rated by various companies according to the amount of risk associated with their purchase. One such rating is Standard and Poor's, which rates bonds from AAA (highest quality) to DDD (bond in default). In general, first-mortgage bonds carry the highest rating, but it is not uncommon for debenture bonds of large corporations to carry a AAA rating, or ratings higher

Table 11.1 Classification and characteristics of bonds

Classification	Characteristics	Type
Mortgage	Bonds backed by mortgage or specified assets	First mortgage Second mortgage Equipment trust
Debenture	No lien to creditors	Convertible Nonconvertible
Municipal	Income tax free	General obligation Revenue

than first-mortgage bonds of smaller, less reputable companies. The concepts presented in this section are summarized in Table 11.1.

Prob. P11.1

11.2 Bond Terminology and Interest

As stated in the preceding section, a bond is a long-term note issued by a corporation or governmental entity for the purpose of obtaining needed capital for financing major projects. The conditions for repayment of the money obtained by the borrower are specified at the time the bonds are issued. These conditions include the bond face value, the bond interest rate, the bond interest-payment period, and the bond maturity date.

The bond *face value*, which refers to the denomination of the bond, is usually an even denomination starting at \$100, with the most common being the \$1000 bond. The face value is important for two reasons:

1. It represents the lump-sum amount that will be paid to the bondholder on the bond maturity date.
2. The amount of interest I paid per period prior to the bond maturity date is determined by multiplying the face value of the bond by the bond interest rate per period as follows:

$$I = \frac{\text{(face value) (bond interest rate)}}{\text{compounding periods per year}} = \frac{Fb}{c} \qquad (11.1)$$

Often a bond is purchased at a discount (less than face value) or a premium (greater than face value), but only face value, not purchase price, is used to compute bond interest (I). Examples 11.1 and 11.2 illustrate the computation of bond interest.

Example 11.1 A shirt-manufacturing company planning an expansion issued 4% \$1000 bonds for financing the project. The bonds will mature in 20 years with interest paid semiannually. Mr. John Doe purchased one of the bonds through his stockbroker for \$800. What payments is Mr. Doe entitled to receive?

SOLUTION In this example, the face value of the bond is \$1000. Therefore, Mr. Doe will receive \$1000 on the date the bond matures, 20 years from now. In addition, Mr. Doe will receive the semiannual interest the company promised to pay when the bonds were issued. The interest every 6 months will be computed using $F = \$1000$, $b = 0.04$, and $c = 2$ in Eq. (11.1):

$$I = \frac{1000(0.04)}{2} = \$20 \text{ every 6 months}$$

Example 11.2 Determine the amount of interest you would receive per period if you purchased a 6% \$5000 bond which matures in 10 years with interest payable quarterly.

SOLUTION Since interest is payable quarterly, you would receive the interest payment every 3 months. The amount you would receive would be

$$I = \frac{5000(0.06)}{4} = \$75$$

Therefore, you would receive \$75 interest every 3 months in addition to the \$5000 lump sum after 10 years.

Probs. P11.2 to P11.4

11.3 Bond Present-Worth Calculations

When a company or government agency offers bonds for financing major projects, investors must determine how much they are willing to pay for a bond of a given denomination. The amount they pay for the bond will determine the rate of return on the investment. Therefore, investors must determine the present worth of the bond that will yield a specified rate of return. These calculations are shown in Example 11.3.

Example 11.3 Ms. Jones wants to make 8% nominal interest compounded semiannually on a bond investment. How much should she be willing to pay now for a 6% \$10,000 bond that will mature in 15 years and pays interest semiannually?

SOLUTION Since the interest is payable semiannually, Ms. Jones will receive the following payment:

$$I = \frac{10{,}000(0.06)}{2} = \$300 \text{ every 6 months}$$

The cash-flow diagram (Fig. 11.1) for this investment allows us to write a present-worth relation to compute the value of the bond now, using an interest rate of 4% per 6-month period, the same as the interest payment period of the bond. Note in the following equation that I is simply an A value.

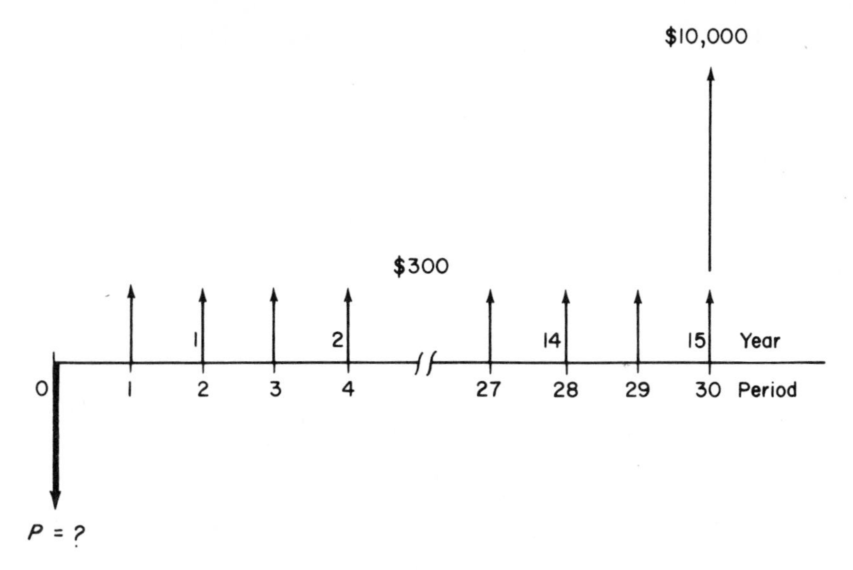

Figure 11.1 Cash flow for a bond investment, Example 11.3.

$$P = 300(P/A, 4\%, 30) + 10{,}000(P/F, 4\%, 30) = \$8270.60$$

Thus, if Ms. Jones is able to buy the bond for \$8270.60, she will receive a nominal 8% rate of return on her investment. If she were to pay more than \$8270.60 for the bond, the rate of return would be less than 8%, and vice versa.

COMMENT It is important to note that the interest rate used in the present-worth calculation is the interest rate per period that Ms. Jones *wants to receive*, not the bond interest rate. Since she wants to receive a nominal 8% per year compounded semiannually, the interest rate per 6-month period is $8\%/2 = 4\%$. The bond interest rate is used *only* to determine the amount of the interest payment. If you desire to review nominal and effective interest rates, refer to Secs. 3.1 to 3.3.

When the investor's compounding frequency is either more often or less often than the interest payment frequency of the bond, it becomes necessary to use the techniques learned in Chap 3. Example 11.4 illustrates the calculations when the investor's compounding period is less than the interest period of the bond.

Example 11.4 Calculate the present worth of a 4.5% \$5000 bond with interest paid semiannually. The bond matures in 10 years, and the investor desires to make 8% compounded quarterly on the investment.

SOLUTION The interest the investor would receive is

$$I = \frac{5000(0.045)}{2} = \$112.50 \text{ every 6 months}$$

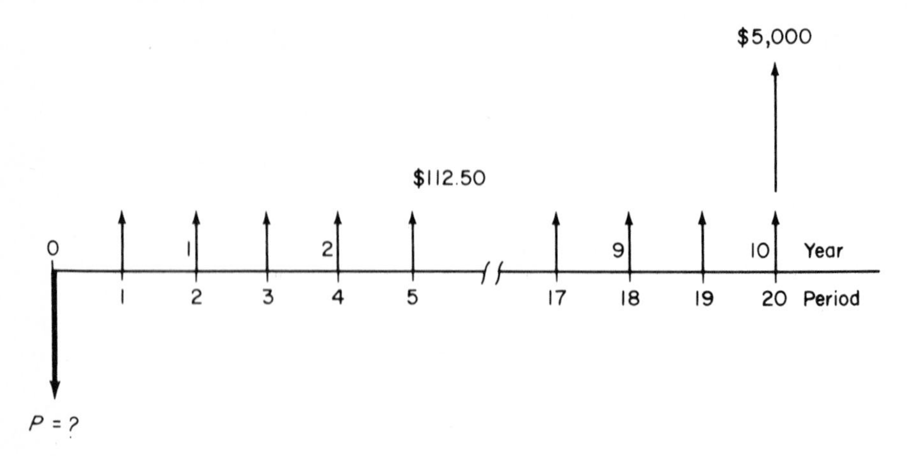

Figure 11.2 Cash flow, Example 11.4.

The present worth of the payments shown in Fig. 11.2 can be found in either of two ways:

1. Take each interest payment ($112.50) back to year 0 separately and add to the present worth of $5000. In this case, using Eq. (3.4), the interest rate would be 8%/4 = 2% per quarter and the number of periods would be double those shown in Fig. 11.2, since the interest payments are made semiannually while the desired rate of return is compounded quarterly. Thus,

$$P = 112.50(P/F, 2\%, 2) + 112.50(P/F, 2\%, 4) + 112.50(P/F, 2\%, 6)$$
$$+ \cdots + 112.50(P/F, 2\%, 40) + 5000(P/F, 2\%, 40)$$
$$= \$3788$$

2. Determine the effective interest rate compounded *semiannually* (the bond interest-payment period) that would be equivalent to the nominal 8% compounded quarterly (as stated in the problem), then use the *P/A* factor to compute the present worth of interest and add to the present worth of $5000. The semiannual rate is 8%/2 = 4%. Since there are two quarters per 6-month period, Table 3.1 indicates that the effective semiannual rate is $i = 4.04\%$. Alternatively, the effective semiannual rate can be computed from Eq. (3.3):

$$i = \left(1 + \frac{0.04}{2}\right)^2 - 1 = 0.0404$$

The present worth of the bond can now be determined with calculations similar to those in Example 11.3:

$$P = 112.50(P/A, 4.04\%, 20) + 5000(P/F, 4.04\%, 20) = \$3790$$

In summary, the steps that should be followed in calculating the present worth of a bond investment are the following:

1. Calculate the interest payment (I) per period, using the face value (F), the bond interest rate (b), and the number of interest periods (c) per year, by $I = Fb/c$.
2. Draw the cash-flow diagram of the bond receipts to include interest and face value.
3. Determine the investor's desired rate of return per period. When the bond interest period and the investor's compounding period are not the same, it is necessary to use the effective-interest-rate formula to find the proper interest rate per period (Example 11.4).

Solved Problem 11.6
Probs. P11.5 to P11.23

11.4 Rate of Return on Bond Investment

To calculate the rate of return received on a bond investment, the procedures learned in this chapter and Chap. 7 should be followed. That is, the procedures of Secs. 11.1 and 11.2 should be used to establish the timing and the magnitude of the income associated with a bond investment; the rate of return on the investment can then be determined by setting up and solving the rate-of-return equation, Eq. (7.1). The following example illustrates the general procedure for calculating the rate of return on a bond investment.

Example 11.5 In Example 11.1 it was stated that Mr. John Doe paid $800 for a 4% $1000 bond that would mature in 20 years with interest payable semiannually. What nominal and effective interest rates per year would Mr. Doe receive on his investment for semiannual compounding?

SOLUTION The income Mr. Doe will receive from the bond purchase is the bond interest every 6 months plus the face value in 20 years. The equation for calculat-

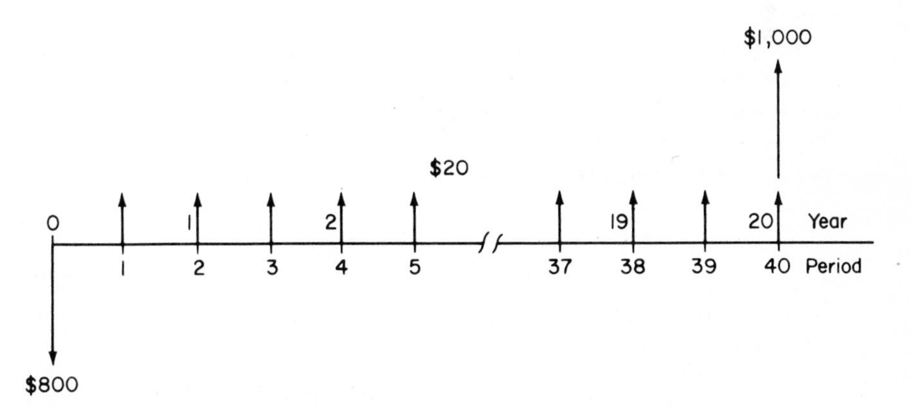

Figure 11.3 Cash flow, Example 11.5.

ing the rate of return using the cash flow of Fig. 11.3 would be

$$0 = -800 + 20(P/A, i\%, 40) + 1000(P/F, i\%, 40)$$

which can be solved (using ROIDS or a calculator) to obtain $i = 2.87\%$ compounded semiannually. The nominal interest rate is computed as interest rate per period times number of periods, which is

$$\text{Nominal } i = 2.87\%(2) = 5.74\%$$

From Table 3.1 or Eq. (3.3), the effective rate is 5.82% per year.

Solved Problem 11.7
Probs. P11.24 to P11.37

SOLVED PROBLEMS

Example 11.6 Ms. Big wants to invest in some 4% \$10,000 20-year mortgage bonds with interest paid semiannually. If she requires a rate of return of 10% per year compounded semiannually and can purchase the bonds through a broker at a discount price of \$8375, (*a*) should she make the purchase, and (*b*) if she does purchase the bond, what will be her total gain in dollars?

SOLUTION

(*a*) The interest each 6 months is

$$I = \frac{10{,}000(0.04)}{2} = \$200$$

For the nominal rate of 10%/2 = 5% per 6 months,

$$P = 200(P/A, 5\%, 40) + 10{,}000(P/F, 5\%, 40)$$
$$= \$4852$$

If Ms. Big must pay \$8375 per bond, she cannot even come close to 10% compounded semiannually, so she should not buy these bonds.

(*b*) If she buys at \$8375 per bond, we can find dollars gained by computing the future worth, assuming that Ms. Big reinvests all interest at 10% compounded semiannually:

$$F = 200(F/A, 5\%, 40) + 10{,}000 = \$34{,}159$$

Thus, she stands to gain a total of \$34,159 − \$8375 = \$25,784. However, as stated in part *a*, her rate of return would be much less than 10% per year.

Sec. 11.3

Example 11.7 In the preceding example, Ms. Big would naturally be saddened by her inability to make 10% compounded semiannually if she pays $8375 for a 4% $10,000 20-year bond. Compute the (*a*) actual nominal and effective return per year of the bond and (*b*) actual dollar gain if this rate is used for reinvestment.

SOLUTION

(*a*) The rate-of-return equation is

$$0 = -8375 + 200(P/A, i\%, 40) + 10{,}000(P/F, i\%, 40)$$

Solution or interpolation shows that $i = 5.40\%$ per year nominally (2.70% semiannually) and $i = 5.47\%$ effectively.

(*b*) Using a nominal rate of 5.40% per year compounded semiannually, we find that the future worth of the bond is

$$F = 200(F/A, 2.7\%, 40) + 10{,}000 = \$24{,}180$$

which represents a gain of $15,805.

Sec. 11.4

PROBLEMS

P11.1 What is the difference between (*a*) mortgage bonds and debenture bonds and (*b*) general obligation bonds and revenue bonds?

P11.2 What would be the interest payment and payment period on a 5% $5000 bond that is payable semiannually?

P11.3 What is the frequency and amount of the interest payments on a 6% $10,000 bond for which interest is payable quarterly?

P11.4 What are the interest payments and their frequency on a 10% $5000 bond that pays monthly interest?

P11.5 How much should you be willing to pay for a 9% $10,000 bond that is due 10 years from now if you want to make a nominal 8% per year compounded semiannually? Assume that the bond interest is payable semiannually.

P11.6 What is the present worth of a 5% $50,000 bond that has interest payable semiannually? Assume that the bond is due in 25 years and the desired rate of return is 16% per year compounded semiannually.

P11.7 A $5\frac{1}{2}$% $15,000 bond with interest payable quarterly is due 20 years from now. What is the present value of the bond if the purchaser desires to make a nominal 14% per year rate of return compounded quarterly?

P11.8 You have been offered a 6% $20,000 bond at a 4% discount. If interest is paid quarterly and the bond is due in 15 years, can you make a nominal 8% per year compounded quarterly?

P11.9 How much should you be willing to pay for a 9% $15,000 bond which has interest payable semiannually and is due in 20 years, if you desire to make an *effective* 14% per year compounded semiannually?

P11.10 How much should an investor be willing to pay for a 15-year $25,000 bond which has a bond-interest rate of 8% per year payable semiannually, if the investor desires to make a rate of return of (*a*) a nominal 12% per year compounded quarterly and (*b*) a nominal 18% per year compounded monthly?

P11.11 A company is considering issuing bonds for financing a major construction project. The company is presently trying to determine whether it should issue conventional bonds or "put" bonds (i.e., bonds which provide the holder the right to sell the bonds at a certain percentage of face value). If the company issues conventional bonds, it expects the bond purchasers to require an overall rate of return (to maturity) of 14% per year compounded quarterly. If put bonds are issued, however, investors would be satisfied with a 12.5% overall rate of return. If the bond interest rate is 12% payable quarterly and the maturity date is 20 years from now, how much more money will the company receive on each $10,000 bond if the put bonds are issued instead of the conventional bonds?

P11.12 A city wanting to repave some local streets needed to acquire $3 million through a bond issue. At the time the bond issue was approved by the voters, the bond interest rate was set at 8%. Between the time the bonds were approved and the time they were sold, however, the interest rate the market required to attract investors changed from 8 to 10%. If the bond interest rate were to remain at 8% per year payable semiannually, how much money would the city receive from the 3 million issue, provided that the bond purchasers required a nominal 10% per year compounded semiannually? Assume that the bonds would mature in 20 years.

P11.13 How much would the city receive in Prob. P11.12 if the interest rate required to attract investors were a nominal 10% per year compounded quarterly? Assume that the bond interest rate would remain at 8% per year payable semiannually for 20 years.

P11.14 If a manufacturing company needed to raise $2 million in capital to finance a small expansion, what would the face value of the bonds have to be, if the bonds were to have a bond interest rate of 12% per year, payable quarterly, and mature in 20 years. Assume that investors would require a rate of return of a nominal 16% per year compounded quarterly.

P11.15 Seven years ago, a Mr. Hughes purchased a 20-year $10,000 bond having an interest rate of 8% per year payable semiannually for $8000. Mr. Hughes would like to sell the bond now, but he will do so only if he can make a rate of return of a nominal 16% per year compounded semiannually on his investment. How much must he get for the bond in order to achieve his objective?

P11.16 A bond was purchased for $900 which has a face value of $1000 and a bond interest rate of 5% per year payable semiannually. The bond is to become due in 9 years. If the purchaser can receive a nominal 18% per year compounded semiannually on any money he invests, how much money will he have accumulated 17 years from now? Assume that all income is invested rather than spent.

P11.17 A 4% $1000 bond that has interest payable annually is to become due 6 years from now. However, the company has been having financial difficulties and has asked the bondholders to defer the due date until 10 years from now. If you could buy the bond now for $800, how much would the company have to pay 10 years from now so that you would make the same rate of return that you would receive if they paid the face value originally scheduled? The bond interest will be paid through the original bond due date, but there will be no interest payments thereafter. Assume that you would have invested the $1000 due in year 6 at the same rate of return you would have made if the company paid the $1000 when originally due.

P11.18 A small southern city was authorized by its voters to issue $8 million worth of 25-year bonds. The bonds were to have an interest rate of 10% per year payable semiannually. However, before the bonds were sold, the rate of return required by investors rose to a nominal 12% per year compounded semiannually. If the city needed to receive at least $7.5 million, by how many years would the due date have to be moved up for the bonds to yield the return desired by the investors?

P11.19 If a $9\frac{1}{2}$% $10,000 bond that has interest payable semiannually is for sale for $8500, when would the bond have to be due for the purchaser to make a nominal 14% per year compounded semiannually on the investment?

P11.20 When would the bond in Prob. P11.19 have to be due if the purchaser wanted to earn a nominal 12% per year compounded continuously on the investment?

P11.21 The Wet 'N' Wild Snowmobile Manufacturing Company must raise money for expanding its production facility. If conventional bonds are issued, the bond interest rate will have to be 16% per year compounded semiannually. The face value of the bonds will be $7 million. If convertible bonds are issued, however, the bond interest rate will have to be only 7% per year compounded semiannually. What will the face value of the convertible bonds have to be to make the present worth of both issues the same at an interest rate of 20% per year compounded semiannually? Compounded continuously? The bonds mature in 15 years.

P11.22 The Black Belch Smelter is attempting to raise $15 million for air-pollution-control equipment by issuing 30-year bonds. With the backing of the local government, industrial development bonds could be issued and, because they are income-tax-free, the bond interest rate would have to be only 13% per year payable semiannually to attract investors. If taxable bonds are issued instead of the industrial development bonds, the bond interest rate would have to be $18\frac{1}{2}\%$ per year payable semiannually. What is the present worth of the extra interest the company will have to pay if the company must issue the taxable instead of the tax-free bonds. The company's MARR is a nominal 20% per year compounded semiannually?

P11.23 The underwriters of a $10 million bond issue paid $9.5 million for bonds issued by the Stin-Key Oil Company. The bonds had an interest rate of 15% per year payable semiannually with a maturity date of 20 years. The underwriters thought the bonds could be sold at a yield to investors of a nominal 15% per year compounded semiannually, but they were unpleasantly surprised to discover that the investors required a yield of a nominal 16% per year compounded semiannually. How much money did the underwriters make or lose on their investment?

P11.24 In Prob. P11.23, what was the "effective" interest paid by the oil company to get the $9.5 million?

P11.25 A 9% $10,000 bond is offered for sale for $8000. If the bond interest is payable semiannually and the bond becomes due in 13 years, what nominal rate of return per year will the purchaser make on the investment?

P11.26 A 10% $50,000 bond is offered for sale for $43,500. If the bond interest is payable quarterly and the bond becomes due in 15 years, what nominal rate of return per year would the purchaser make on the investment?

P11.27 An investor purchased a 5% $1000 bond for $825. The interest was payable semiannually, and the bond was to become due in 20 years. The bond was kept for only 8 years and sold for $800 immediately after the sixteenth interest payment. What nominal rate of return per year was made on the investment?

P11.28 What effective rate of return per year would the purchaser of the bond in Prob. P11.27 receive?

P11.29 At what bond interest rate will a $10,000 bond yield a nominal 14% per year compounded semiannually if the purchaser pays $8000 and the bond becomes due in 15 years? Assume that the bond interest is payable semiannually.

P11.30 At what bond interest rate will a $20,000 bond that has interest payable semiannually yield an effective 8% rate of return per year if the price of the bond is $18,000 and the bond becomes due in 20 years?

P11.31 What would the bond interest rate have to be in Prob. P11.29 if the purchaser wanted to make a nominal 15% per year rate of return compounded quarterly? Compounded continuously?

P11.32 The Sli-Dog Company plans to sell 200 4% $1000 bonds. Interest will be paid quarterly and the bonds will be retired after 15 years. If the management wants to set up a fund that will be specifically used to retire the bonds, that is, pay interest and face value, what equivalent annual amount must be placed in the fund? Assume that the fund will earn a nominal 16% per year compounded quarterly. (Note that beginning-of-year deposits are required to have the quarterly interest payments available.)

P11.33 Bonds purchased for \$9000 have a face value of \$10,000 and a bond interest rate of 10% per year payable semiannually. The bonds are due in 3 years. The company that issued the bonds is contemplating a liquidity problem in 3 years and has advised all bondholders that if they will keep their bonds for another 2 years past the original due date, the bond interest for the extended 2-year period will be 16% per year payable semiannually. What nominal rate of return per year would the bondholders receive if they held the bonds for the additional 2-year period?

P11.34 What rate of return would the bondholders in Prob. P11.33 have made if the bonds were kept for only 3 years and redeemed on the original due date? Should the bondholders keep the bonds 3 years or 5, as the company has advised?

P11.35 The GRQ Corporation issued \$5 million worth of 25-year bonds, with an interest rate of 10% per year payable semiannually. The company received \$4.5 million, but semiannual expenses of \$15,000 were expected for servicing the bonds. What nominal interest rate per year did the corporation pay for getting the \$4.5 million after all expenses are taken into account?

P11.36 An investor purchased a \$1000 convertible bond for \$850 from the GRQ Company. The bond had an interest rate of 8% per year payable quarterly, and was convertible to 20 shares of GRQ common stock. If the investor kept the bond for $6\frac{1}{2}$ years and then converted it into common stock when the stock was selling for \$49 per share, what was the nominal rate of return per year on his investment?

P11.37 The Hi-Cee Steel Company issued \$5 million worth of 20-year 14%-per-year payable semiannually callable bonds (i.e., bonds which could be called in and paid off at any time). The company agreed to pay a 10% premium on the face value if the bonds were called. Seven years after the bonds were issued, the prevailing interest rate in the marketplace dropped to 11%. (*a*) What rate of return would the company make by calling the bonds and paying the \$5.5 million? (*b*) Should the company call the bonds?

CHAPTER

TWELVE

INFLATION AND ESCALATING COST CONSIDERATIONS

The purpose of this chapter is to teach you how to account for the effects of inflation and escalating costs when conducting an economic analysis of alternatives. Up to this time, inflation was philosophically ignored on the basis of an assumption that all alternatives would be equally affected. This is not necessarily true, however, since the cash flow of one alternative might occur primarily as an initial investment while the cash flow of the other alternative(s) might primarily involve deferred disbursements or receipts. This is an optional chapter.

SECTION OBJECTIVES

In order to complete this chapter, you must be able to do the following:

12.1. Define *inflated interest rate* and calculate the present worth of specified future sums, given the interest rate, inflation rate, and time period.

12.2. Define what is meant by *real interest rate* and calculate the future worth in "today's" dollars and "then-current" dollars of a specified present amount, given the interest rate, inflation rate, and time period.

12.3. Calculate the present worth or future worth of a uniform series of cash flows, given the interest rate, inflation rate, and length of the series.

12.4. Calculate the uniform annual amount of money in then-current dollars that would be equivalent to a specified present or future sum, given the interest rate, inflation rate, and time period.

12.5. Define *escalating-cost series* and derive the formula for calculating the present worth of such a series.

12.6. Calculate the present worth, equivalent uniform annual series, or future worth of an escalating series with and without consideration for inflation, given the cash-flow series, escalation rate, inflation rate, interest rate, and length of the series.

STUDY GUIDE

12.1 Present-Worth Calculations with Inflation Considered

Up to this time, whenever present-worth calculations were made for alternatives requiring future replacement costs, the replacement costs were assumed to be the same as the initial investment cost. Except in unusual circumstances, however, future replacement costs would be expected to be higher than the initial cost because of the effects of inflation. On the other hand, the inflated dollars of the future will be worth less than the dollars in use today. The higher future costs which are to be paid for with dollars that are worth less obviously have opposite effects on a present-worth analysis. Two methods which can be used to remove these effects are: (1) convert the future cash flows into *today's* dollars and then use the regular interest rate i in the interest formulas or (2) express the future cash flows in *then-current* dollars and use an interest rate which takes inflation into account (an inflated interest rate).

Method 1 is illustrated in Table 12.1, which shows the future costs of a \$5000 automobile in then-current dollars (column 3) and in today's dollars (column 4), if the inflation rate is expected to be 8% per year. Note that then-current dollars can be converted to today's dollars by dividing by $(1 + f)^n$, where f is the inflation rate per period. Column 5 shows the present-worth calculation at i = 10% per year. Observe from column 4 that when the future dollars of column 3 are converted into today's dollars, the cost is \$5000, the same as the cost at the start. This will always be true when the costs are increasing by an amount *exactly equal* to the inflation rate. The actual cost of the car 4 years from now will thus be \$6803, but in today's dollars, the cost at that time will be \$5000, which at an interest rate of 10% per year has a present worth of \$3415.

The second method of accounting for inflation in a present-worth analysis is that

Table 12.1 Present-worth calculation using today's dollars (rounded)

(1) Year, n	(2) Cost increase due to inflation	(3) Future cost in then dollars	(4) = (3)/$(1.08)^n$ Future cost in today's dollars	(5) = (4)(P/F, 10%, n) Present worth at i = 10%
0		\$5000	\$5000	\$5000
1	\$5000(0.08) = \$400	\$5400	$5400/(1.08)^1 = 5000$	\$4545
2	5400(0.08) = \$432	\$5832	$5832/(1.08)^2 = 5000$	\$4132
3	5832(0.08) = \$467	\$6299	$6299/(1.08)^3 = 5000$	\$3757
4	6299(0.08) = \$504	\$6803	$6803/(1.08)^4 = 5000$	\$3415

of adjusting the interest formulas to account for inflation. This adjusted interest rate is called the *inflated interest rate* i_f, which can be calculated from the following formula:

$$i_f = i + f + if \tag{12.1}$$

where i = interest rate
f = inflation rate
i_f = inflated interest rate

For an interest rate of 8% per year and an inflation rate of 10% per year, Eq. (12.1) gives

$$i_f = 0.08 + 0.10 + 0.08(0.10)$$
$$= 0.188 \qquad (18.8\%)$$

Equation (12.1) can be derived by considering the single-payment present-worth factor.

$$P = F\left[\frac{1}{(1+i)^n}\right] \tag{12.2}$$

F can be converted into today's dollars using division by $(1+f)^n$ to obtain

$$P = \frac{F}{(1+f)^n}\left[\frac{1}{(1+i)^n}\right]$$
$$= F\left[\frac{1}{(1+f)^n(1+i)^n}\right]$$
$$= F\left[\frac{1}{(1+i+f+if)^n}\right]$$

In order to find the inflated i_f to replace i in Eq. (12.2), substitute Eq. (12.1) into the last expression. Then, using the P/F factor definition,

$$P = F\left[\frac{1}{(1+i_f)^n}\right] = F(P/F, i_f\%, n) \tag{12.3}$$

Table 12.2 shows this second method for present-worth calculations with i_f =

Table 12.2 Present-worth calculation using an inflated interest rate

(1) Year, n	(2) Future cost in then-current dollars	(3) $(P/F, 18.8\%, n)$	(4) Present worth
0	$5000	1	$5000
1	$5400	0.8418	$4545
2	$5832	0.7085	$4132
3	$6299	0.5964	$3757
4	$6803	0.5020	$3415

18.8% for the \$5000 car considered above. As shown in column 4, the present worth of the car for each year is the same as that calculated in column 5 of Table 12.1.

To summarize, if future dollars are expressed in today's dollars (or have already been converted into today's dollars), the present worth should be calculated using the regular interest rate i in the single-payment present-worth formula. If the future dollars are expressed in then-current dollars, the inflated interest rate i_f should be used in the formula.

Example 12.1 An alumnus of Taco Tech who "made good" has decided to donate to the college's Excellence Fund and has offered the college any one of the following three plans:

Plan A: \$60,000 now
Plan B: \$16,000 per year for 12 years beginning 1 year from now
Plan C: \$50,000 three years from now and another \$80,000 five years from now

The only condition placed on the donation is that the college agree to spend the money on research related to the advancement of robotics. The college would like to select the plan which maximizes the buying power of the dollars received, so it has instructed the engineering professors evaluating the plans to account for inflation in their calculations. If the college can earn 12% per year on its ready-assets account and the inflation rate is expected to be 11% per year, which plan should the college accept?

SOLUTION The simplest method of evaluation is to calculate the present worth of each plan in today's dollars. For plans B and C, this requires the use of the inflated interest rate i_f. (Present worth discussed in Sec. 12.3). By Eq. (12.1),

$$i_f = 0.12 + 0.11 + 0.12(0.11) = 0.243 \qquad (24.3\%)$$

Compute the P value by Eq. (12.3).

$$P_A = \$60{,}000$$

$$P_B = \$16{,}000(P/A, 24.3\%, 12) = \$61{,}003.46$$

$$P_C = \$50{,}000(P/F, 24.3\%, 3) + 80{,}000(P/F, 24.3\%, 5) = \$52{,}995.84$$

Since P_B is the largest in today's dollars, accept plan B.

COMMENT The present worths of plans B and C could also have been found by first converting the cash flows into today's dollars and then using the regular i. This procedure is obviously tedious and time-consuming, but you may want to satisfy yourself that the answers would be the same. The P/A factor with inflation considered is further discussed in Sec. 12.3.

Probs. P12.1 to P12.5

12.2 Future-Worth Calculations with Inflation Considered

As in the preceding section, future-worth calculations which take into account the effects of inflation may be accomplished either by (1) converting the equivalent future dollars into dollars with today's buying power or (2) calculating the number of then-current dollars which would have the same buying power as the present amount of today's dollars.

For method 1, the number of then-current dollars which would be accumulated is simply

$$F = P(1 + i)^n = P(F/P, i\%, n)$$

The then-current dollars can be converted into dollars with today's buying power by dividing by $(1 + f)^n$. Thus,

$$F = \frac{P(1 + i)^n}{(1 + f)^n} = \frac{P(F/P, i\%, n)}{(1 + f)^n} \tag{12.4}$$

To illustrate: If $1000 is deposited into a savings account at 10%-per-year interest for 7 years and the inflation rate is 8% per year, the amount of money that will be accumulated with today's buying power will be

$$F = \frac{1000(F/P, 10\%, 7)}{(1 + 0.08)^7} = \$1137$$

If there is no inflation ($f = 0$), in 7 years the 10% rate will accumulate $1948 in dollars with today's buying power.

$$F = 1000(F/P, 10\%, 7) = \$1948$$

The future amount of money that would be accumulated with today's buying power could equivalently be determined by using a *real interest rate* i_r in the F/P factor to compensate for the decreased purchasing power of the dollar. This real interest rate can be obtained by equating the single-payment compound-amount formula (F/P factor) with the middle term in Eq. (12.4), which converts present dollars into future dollars with today's buying power.

$$P(1 + i_r)^n = P\left[\frac{(1 + i)^n}{(1 + f)^n}\right]$$

$$1 + i_r = \frac{1 + i}{1 + f}$$

$$i_r = \frac{i - f}{1 + f} \tag{12.5}$$

The real interest rate i_r represents the rate at which present dollars will expand *with their same buying power* into equivalent future dollars. The use of this interest rate is appropriate when calculating the future worth of a savings account, for example, when the effects of inflation must be taken into consideration. Thus, for the $1000 deposit mentioned previously,

$$i_r = \frac{0.10 - 0.08}{1 + 0.08} = 0.0185 \qquad (1.85\%)$$

$$F = 1000(F/P, 1.85\%, 7) = \$1137$$

Note that the interest rate of 10% per year has been reduced to only 1.85% per year because of the erosive effects of inflation. Also note that an inflation rate larger than the interest rate, $f > i$, leads to a negative real interest rate i_r in Eq. (12.5).

The second method of accounting for inflation in future-worth calculations is that of determining the then-current dollars which would be required to maintain the buying power of the present sum. This can be accomplished by using the inflated rate i_f as calculated in Sec. 12.1 to obtain the F/P factor. Using the same \$1000 deposit as above,

$$i_f = 0.10 + 0.08 + 0.10(0.08) = 0.188 \qquad (18.8\%)$$

$$F = 1000(F/P, 18.8\%, 7) = \$3340$$

This calculation shows that \$3340 of then-current dollars would be equivalent to \$1000 now when the interest rate is 10% per year and the inflation rate is 8% per year.

To summarize: The calculations made in this section reveal that \$1000 now at an interest rate of 10% per year would accumulate to \$1948 in 7 years; the \$1948 would have the purchasing power of \$1137 of today's dollars; and it would take \$3340 of then-current dollars to be equivalent to the \$1000 now when inflation is taken into account.

Example 12.2 The Shur-Kan-Do Sheet Metal Company is trying to decide whether it should pay now or pay later for upgrading its production facilities. If the company selected plan A (for Action), the necessary equipment would be purchased now for \$20,000. However, if the company selected plan I (for Inaction), the equipment purchase would be deferred for 3 years when the cost would be \$34,000. The minimum attractive rate of return is 18% per year and the inflation rate is expected to be 12% per year. (*a*) Find the real interest rate for these plans. (*b*) Determine whether the company should purchase now or purchase later when inflation is not considered, and when inflation is taken into account.

SOLUTION

(*a*) By Eq. (12.5) the real interest rate is

$$i_r = \frac{0.18 - 0.12}{1.12} = 0.0536 \qquad (5.36\%)$$

This means that an effective MARR of 5.36% per year is used when inflation is considered.

(*b*) *Inflation not considered*. The stated rate is 18% per year, so we can compute P at time 0 or F 3 years from now and select the plan with the lower cost.

Computing F as we have previously,

$$F_A = 20{,}000(F/P, 18\%, 3) = \$32{,}860.64$$

$$F_I = \$34{,}000$$

Select action plan A because it costs less.

Inflation considered. Compute the inflated rate by Eq. (12.1).

$$i_f = 0.18 + 0.12 + 0.18(0.12) = 0.322$$

Use i_f to compute the F values 3 years from now to determine the then-current dollars necessary.

$$F_A = 20{,}000(F/P, 32.2\%, 3) = \$46{,}208.77$$

$$F_I = \$34{,}000$$

Now select the inaction plan I because it will require less then-current dollars.

COMMENT This example illustrates that when inflation is taken into account, the more economic alternative may be different from the one selected if inflation is ignored.

A present-worth analysis can be used here with the same results:

No inflation considered	*Inflation considered*
	$i_f = 32.2\%$
$P_A = \$20{,}000$	$P_A = \$20{,}000$
$P_I = 34{,}000(P/F, 18\%, 3)$	$P_I = 34{,}000(P/F, 32.2\%, 3)$
$= \$20{,}693.45$	$= \$14{,}715.82$
Select plan A	Select plan B

Probs. P12.6 to P12.9

12.3 Present and Future Worth of a Uniform Series with Inflation Considered

The present worth of a uniform series can be calculated using the P/A factor with an interest rate of either i or i_f, depending on whether the cash flow is expressed in today's dollars or then-current dollars, respectively. If the series is expressed in today's dollars, then its present worth is simply the discounted cash-flow value using the regular interest rate i. If the cash flow is expressed in then-current dollars, however, the amount today that would be equivalent to the inflated then dollars would be less than the discounted cash-flow value obtained by using the regular i. The present worth could be obtained by using i_f in the uniform-series present-worth factor. Example 12.3 illustrates these calculations.

Example 12.3 Calculate the present worth of a uniform series of payments of \$1000 per year for 5 years if the interest rate is 10% per year and the inflation rate is 8% per year, assuming that the payments are in terms of (*a*) today's dollars and (*b*) then-current dollars.

SOLUTION

(*a*) Since the dollars are already expressed in today's dollars, the present worth is simply

$$P = 1000(P/A, 10\%, 5) = \$3790.80$$

(*b*) Since the dollars are expressed in then-current dollars, as shown in Fig. 12.1*a*, use the inflated interest rate in the uniform-series present-worth factor.

$$i_f = i + f + if = 18.8\%$$

$$P = 1000(P/A, 18.8\%, 5) = \$3071$$

The present worth can also be obtained by converting the future cash flows into today's dollars and then finding the present worth using the regular i. The calculations associated with this method follow. From Fig. 12.1*b*,

$$\begin{aligned} P &= 925.93(P/F, 10\%, 1) + 857.34(P/F, 10\%, 2) + 793.83(P/F, 10\%, 3) \\ &\quad + 735.03(P/F, 10\%, 4) + 680.58(P/F, 10\%, 5) \\ &= \$3071 \end{aligned}$$

This is the same result obtained in *a* using i_f.

COMMENT It is obvious from this example that use of the inflated interest rate i_f in the P/A factor is much simpler than converting the future cash flows into

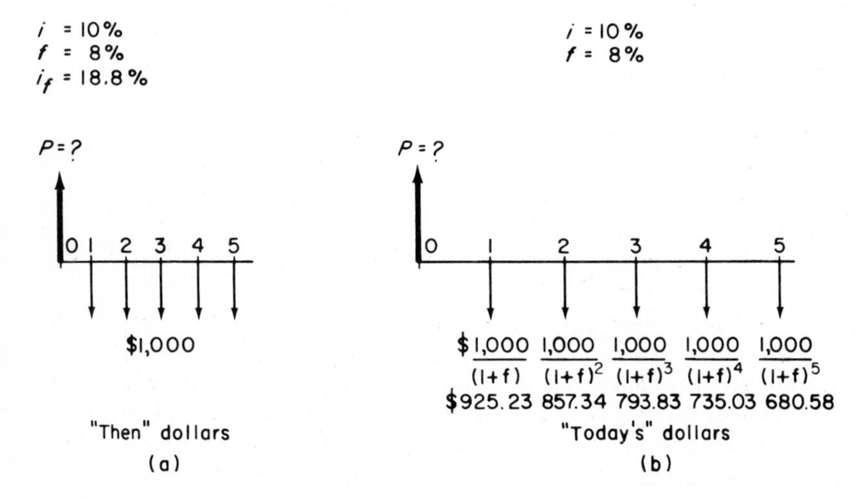

Figure 12.1 Cash-flow diagram, Example 12.3.

today's dollars and using the P/F factor for each one. Thus, only the former method will be used hereinafter.

When computing the future worth of a uniform series with inflation taken into account, the future amount can represent different equivalent amounts. From an engineering analysis viewpoint, however, the future worth most commonly of interest is the number of then-current dollars required for equivalence with the uniform series. This may always be calculated using the inflated interest rate i_f in the F/A factor. For example, with $i_f = 0.188$,

$$F = 1000(F/A, 18.8\%, 5) = \$7268$$

The same result is obtained by summing the future worth of the inflated dollars. The future worth of the \$1000 amount in year 1, for example, would be $1000(F/P, 10\%, 4)(1 + 0.08)^4 = \$1992$.

Solved Problem 12.7
Probs. P12.10 to P12.18

12.4 Capital-Recovery and Sinking-Fund Calculations with Inflation Considered

It is particularly important for capital-recovery calculations to include inflation considerations, because present dollars must be recovered with future inflated dollars. There is little significance in considering capital recovery in terms of today's dollars, so only then-current dollars will be considered in this section. Since then-current dollars have less buying power than do today's dollars, it is obvious that more dollars will be required to recover the present investment. This recovering suggests the use of the inflated interest rate in the A/P formula. For example, if \$1000 is invested today when the interest rate is 10% per year and the inflation rate is 8% per year, the annual amount of capital that must be recovered each year for 5 years in then-current dollars will be

$$A = 1000(A/P, 18.8\%, 5) = \$325.59$$

On the other hand, to accumulate a specified amount of then-current (inflated) dollars by way of a sinking fund, fewer present dollars need be invested because of the dual effect of the stated interest rate and the inflation rate. This greater worth of money through time similarly suggests using the inflated interest rate to find A given F. The annual equivalent of $F = \$1000$ five years from now in then-current dollars is thus

$$A = 1000(A/F, 18.8\%, 5) = \$137.59$$

When inflation is not considered, the equivalent annual amount to accumulate \$1000 at $i = 10\%$ is $1000(A/F, 10\%, 5) = \$163.80$. So uniform future costs should be spread out over as long a time period as possible so that inflation will have the effect of reducing the payment involved (\$137.59 versus \$163.80 here). Keep in mind that only the *buying power* of the payment is reduced, not its *amount*.

A situation which sometimes arises in economic calculations involves the determination of the amount of uniform deposit required to accumulate an amount of money with the same buying power as a specified amount *today*. Example 12.4 illustrates the calculations involved.

Example 12.4 What annual deposit will be required for 5 years to accumulate an amount of money that has the same buying power as \$680.58 today if the interest rate is 10% per year and the inflation rate is 8% per year?

SOLUTION The actual number of then-current (inflated) dollars required in 5 years is

$$F = 680.58(1.08^5) = \$1000$$

Therefore, the actual amount of the annual deposit is

$$A = 1000(A/F, 10\%, 5) = \$163.80$$

COMMENT This example shows that if \$163.80 is deposited each year for 5 years at an interest rate of 10% per year, \$1000 will be accumulated after the fifth deposit. However, from an economic analysis point of view, \$163.80 for 5 years is *not* equivalent to \$1000 in year 5 *when inflation is considered*. As shown above, \$137.59 for 5 years is equivalent to \$1000 in year 5 at i_f = 18.8%. The discrepancy between the \$163.80 and the \$137.59 is caused by the use of two different interest rates (i_f and i) in the A/F factor. The use of i_f = 18.8% accounts for the fact that the dollars spent in years 1 through 4 have greater buying power than dollars spent in year 5. We would therefore not want to spend as many of these. The use of i = 10%, on the other hand, essentially ignores the greater buying power of the dollars spent in earlier years, so the higher "equivalent" amount (\$163.80) is obtained. As implied in this example, most of the calculations associated with comparing alternatives in engineering-economic studies require the use of the inflated interest rate when inflation is considered.

Probs. P12.19 to P12.23

12.5 Derivation of Present-Worth of Escalating-Series Equation

In Sec. 2.5 uniform gradient factors were introduced which could be used for calculating the present worth or equivalent uniform annual amount of a series of payments which increase or decrease by a constant amount in consecutive payment periods. Oftentimes, cash flows change by a constant *percentage* in consecutive payment periods instead of by a constant dollar amount. This type of cash flow, called an *escalating* series, is shown in general form in Fig. 12.2, where D represents the dollar amount in year 1 and E represents the escalation rate. The derivation of the equation for calculating the present worth P_E of an escalating series is found by computing the present worth of the Fig. 12.2 cash flow using the P/F formula, $1/(1 + i)^n$.

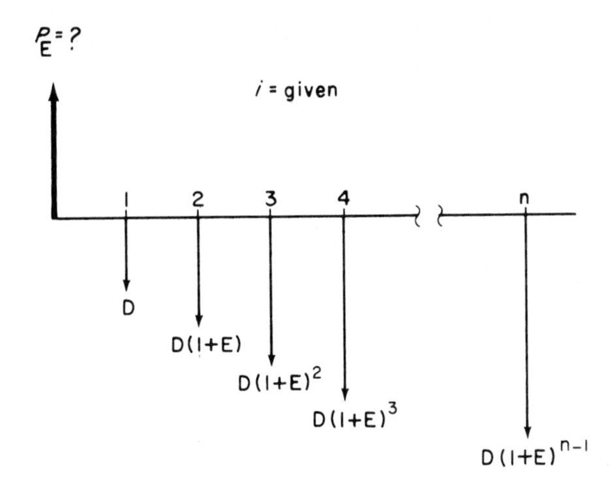

Figure 12.2 Cash-flow diagram of escalating series and its present worth P_E.

$$P_E = \frac{D}{(1+i)^1} + \frac{D(1+E)}{(1+i)^2} + \frac{D(1+E)^2}{(1+i)^3} + \cdots + \frac{D(1+E)^{n-1}}{(1+i)^n}$$

$$= D\left[\frac{1}{1+i} + \frac{1+E}{(1+i)^2} + \frac{(1+E)^2}{(1+i)^3} + \cdots + \frac{(1+E)^{n-1}}{(1+i)^n}\right] \tag{12.6}$$

Multiply both sides by $(1+E)/(1+i)$:

$$P_E\left(\frac{1+E}{1+i}\right) = D\left[\frac{1+E}{(1+i)^2} + \frac{(1+E)^2}{(1+i)^3} + \frac{(1+E)^3}{(1+i)^4} + \cdots + \frac{(1+E)^n}{(1+i)^{n+1}}\right] \tag{12.7}$$

Subtract Eq. (12.6) from Eq. (12.7) and factor out P_E to obtain

$$P_E\left(\frac{1+E}{1+i} - 1\right) = D\left[\frac{(1+E)^n}{(1+i)^{n+1}} - \frac{1}{1+i}\right]$$

Solve for P_E and simplify to obtain

$$P_E = \frac{D[(1+E)^n/(1+i)^{n+1} - 1/(1+i)]}{[(1+E)/(1+i) - 1]}$$

$$= \frac{D[(1+E)^n/(1+i)^n - 1]}{E - i} \qquad E \neq i \tag{12.8}$$

where P_E indicates the present worth of an escalating series starting in year 1 at D dollars. For the condition $E = i$, use L'Hospital's rule to modify Eq. (12.8).

$$\frac{dP_E}{dE} = \frac{D[n(1+E)^{n-1}(1+i)^{-n}]}{1} = D\left[\frac{n}{(1+E)^{1-n}(1+i)^n}\right] \tag{12.9}$$

Since $E = i, P_E$ may be written from Eq. (12.9) as

$$P_E = D\left(\frac{n}{1+E}\right) \qquad E = i \tag{12.10}$$

The equivalent value P_E occurs in the year prior to the cash flow D, as shown in Fig. 12.2.

Prob. P12.24

12.6 Calculations Involving an Escalating Series

As shown in the previous section, the present worth of an escalating series can be determined through the use of Eq. (12.8) or (12.10). The equivalent uniform annual cost or future worth of the series can be calculated by converting the present worth with the appropriate interest factor. The use of Eq. (12.8) is illustrated in Example 12.5.

Example 12.5 A new pickup truck has a first cost of \$8000 and is expected to last 6 years with a \$1300 salvage value. The operating cost of the vehicle is expected to be \$1700 the first year, increasing by 11% per year thereafter. Determine the equivalent present cost of the truck if the interest rate is 8% per year and inflation is not to be considered.

SOLUTION The cash-flow diagram is shown in Fig. 12.3. Since $E \neq i$, Eq. (12.8) is used to calculate P_E. The total P_T is

$$P_T = 8000 + P_E - 1300(P/F, 8\%, 6)$$

$$= 8000 + 1700 \frac{\{[(1 + 0.11)/(1 + 0.08)]^6 - 1\}}{0.11 - 0.08} - 1300(P/F, 8\%, 6)$$

$$= 8000 + 1700(5.9559) - 819.26 = \$17{,}305.85$$

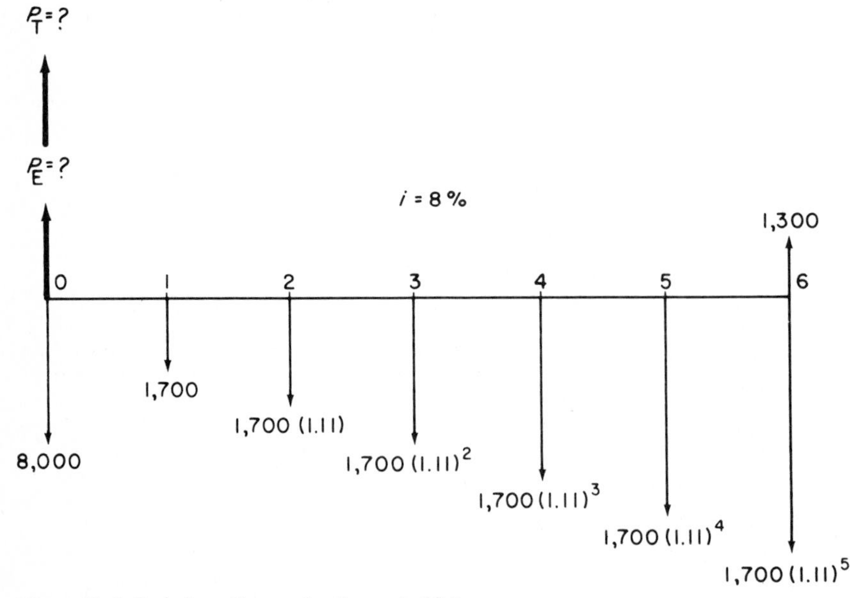

Figure 12.3 Cash-flow diagram for Example 12.5.

COMMENT The equivalent uniform annual cost of the truck could be determined by multiplying the \$17,305.85 by $(A/P, 8\%, 6)$.

Equations (12.8) and (12.10) can be used when the escalating series starts at a time other than between years 1 and 2. In this case, the P_E would be located on the diagram in a manner similar to the way P_A is located for an annual series or P_G is located for a shifted gradient. See Example 12.8 in Solved Problems.

The escalating-series equations can also be used when inflation is taken into account in the economic analysis. Although there are several methods for doing so, the simplest, as shown below, involves substitution of the inflated rate i_f for i in the escalating-series equations.

Example 12.6 Solve for the present worth of the pickup truck described in Example 12.5 if inflation of 10% per year is taken into account.

SOLUTION Table 12.3 shows the present-worth calculation for P *without* the use of the escalating series equations. The use of i_f for i in Eq. (12.8) results in the following computations.

$$i_f = i + f + if = 0.188 \qquad (18.8\%)$$

$$P = 8000 + P_E - 1300(P/F, i_f\%, 6)$$

$$= 8000 + 1700 \frac{[(1.11/1.188)^6 - 1]}{0.11 - 0.188} - 1300(P/F, 18.8\%, 6)$$

$$= 8000 + 1700(4.2906) - 462 = \$14{,}832$$

This is the same amount obtained in Table 12.3.

COMMENT The use of the escalating-series equation is obviously simpler than is converting the cash flow into today's dollars and then summing present worth values, as is done in Table 12.3.

Solved Problem 12.8
Probs. P12.25 to P12.34

Table 12.3 Present worth of an escalating series, Example 12.6

Year, n	(1) Actual cost flow	(2) = (1)/$(1+f)^n$ Cost in today's dollars	(3) $(P/F, 8\%, n)$	(4) = (3)(2) Present worth
0	\$8000	\$8000	1.0000	\$ 8,000
1	1700	1545	0.9259	1,431
2	1887	1560	0.8573	1,337
3	2095	1574	0.7938	1,249
4	2325	1588	0.7350	1,167
5	2581	1603	0.6806	1,091
6	2865 - 1300	1617 - 734	0.6302	556
				\$14,831

SOLVED PROBLEMS

Example 12.7 A \$50,000 bond which has a bond interest rate of 8% per year payable semiannually is currently for sale. The bond is due in 15 years. If the rate of return required by investors is a nominal 16% per year compounded semiannually and if the inflation rate is expected to be 4.5% per semiannual period, how much should be paid for the bond (*a*) when inflation is not taken into account and (*b*) when inflation is considered?

SOLUTION

(*a*) Without considering inflation, the dividend by Eq. (11.1) is $I = [(50{,}000)(0.08)]/2 = \$2000$ per semiannual period and the present worth at a nominal 8% per 6 months is

$$P = 2000(P/A, 8\%, 30) + 50{,}000(P/F, 8\%, 30) = \$27{,}485.60$$

(*b*) To consider inflation, use the inflated rate in the P/A factor.

$$i_f = 0.08 + 0.045 + (0.08)(0.045) = 0.1286 \text{ per semiannual period}$$

$$P = 2000(P/A, 12.86\%, 30) + 50{,}000(P/F, 12.86\%, 30) = \$16{,}466.13$$

COMMENTS The \$11,019.47 difference in present worth illustrates the tremendous negative effect of inflation on fixed-income investments. On the other hand, the entities which issue the instruments (bonds in this case) are benefactors to the same extent.

Sec. 12.3

Example 12.8 Calculate the equivalent present cost of a \$35,000 expenditure now and \$7000 per year for 5 years beginning 1 year from now with increases of 12% per year thereafter for the next 8 years. Use an interest rate of 15% per year and make the calculations (*a*) without considering inflation and (*b*) considering inflation at a rate of 11% per year.

SOLUTION

(*a*) Figure 12.4 presents the cash flows. The present worth P is found using $i = 15\%$ and Eq. (12.8) for the escalating series which has its present worth P_E in year 4.

$$P = 35{,}000 + 7000(P/A, 15\%, 4) + \left\{7000\frac{[(1.12/1.15)^9 - 1]}{0.12 - 0.15}\right\}(P/F, 15\%, 4)$$

$$= 35{,}000 + 19{,}985 + 28{,}247$$

$$= \$83{,}232$$

Note that $n = 4$ in the P/A factor because the \$7000 in year 5 is D in Eq. (12.8). The last term in the expression for P is P_E in year 4 (Fig. 12.4), which is moved to time 0 with the $(P/F, 15\%, 4)$ factor.

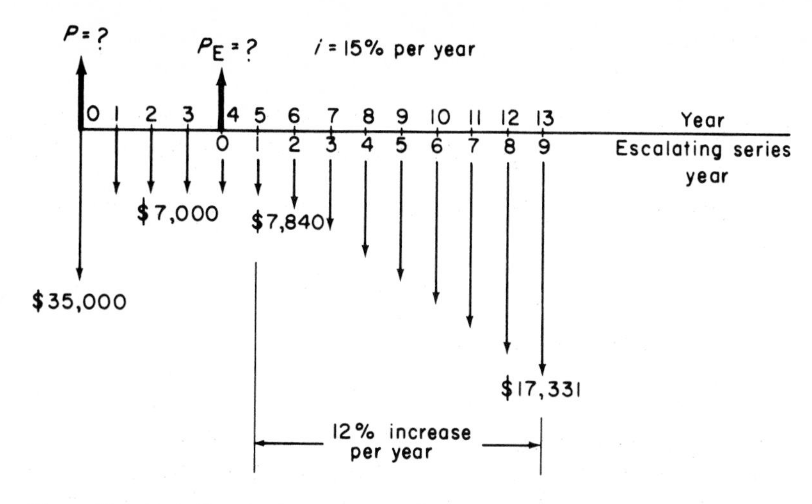

Figure 12.4 Cash-flow diagram, Example 12.8.

(*b*) With inflation taken into account, it is necessary to use the inflated interest rate from Eq. (12.1) in Eq. (12.8).

$$i_f = 0.15 + 0.11 + (0.15)(0.11) = 0.2765$$

$$P = 35{,}000 + 7000(P/A, 27.65\%, 4)$$

$$+ \left\{ \frac{7000\,[(1.12/1.2765)^9 - 1]}{0.12 - 0.2765} \right\} (P/F, 27.65\%, 4)$$

$$= 35{,}000 + 7000(2.2545) + 30{,}945(0.3766)$$

$$= \$62{,}436$$

COMMENT You can check the result of the escalating-cost factor in part *a* by multiplying each of the amounts in years 5 through 13 by the appropriate *P*/*F* factor for 15%. The same thing can be done in part *b* using the inflated interest rate.

Sec. 12.6

PROBLEMS

Note: For the problems below, all costs are in terms of then-current dollars unless otherwise specified.

P12.1 Calculate the present worth of $23,000 six years from now if the interest rate is 15% per year and the inflation rate is 10% per year.

P12.2 Calculate the present worth of $15,000 eight years from now if the interest rate is 12% per year and the inflation rate is 10% per year.

P12.3 Find the present sum of money that would be equivalent to the future sums of $5000 in year 6 and $7000 in year 8 if the interest rate is 15% per year and the inflation rate is (*a*) 10% and (*b*) 16% per year.

P12.4 How much money could the Kill-Kow Cattle Company afford to spend now for a new tractor trailer in lieu of spending $65,000 three years from now if the interest rate is 13% per year and the inflation rate is 7% per year?

P12.5 The manager of the Pick 'N' Pak Food store is trying to determine how much should be spent now to avoid spending $10,000 on freezer equipment 2 years from now. If the interest rate is $1\frac{1}{2}$% per month and the inflation rate is 1% per month, what is the maximum amount of money the manager could afford to spend?

P12.6 Calculate the number of then-current dollars that would be required in 9 years to recover a present investment of $12,000 at an interest rate of 15% per year and an inflation rate of 8% per year.

P12.7 If $23,000 is invested now at an interest rate of 20% per year, (*a*) how much money would be accumulated in 7 years if the inflation rate were 10% per year and (*b*) how many then-current dollars would be required to preserve the buying power of the original $23,000?

P12.8 What future amount of money in then-current dollars 6 years from now would be equivalent to a present sum of $80,000 at an interest rate of 18% per year and an inflation rate of 12% per year?

P12.9 Calculate the number of (*a*) today's dollars and (*b*) then-current dollars in year 10 that will be equivalent to a present investment of $33,000 at an interest rate of 15% per year and an inflation rate of 10% per year.

P12.10 If the R-Gone Sign Company invests $3000 per year for 8 years beginning 1 year from now in a new production process, how much money must be received in a lump sum in year 8 in then-current dollars in order for the company to recover its investment at an interest rate of 13% per year and an inflation rate of 10% per year?

P12.11 A woman deposited $1300 per year for 6 years in a savings account (her first deposit was made at the end of year 1). In years 7 to 12, she deposited $2000. How much money did she have in the account after the last deposit in terms of today's buying power if the interest rate she received was 12% per year and the inflation rate was 9% per year? (*Hint:* Greater buying power of earlier deposits should not be considered in this problem.)

P12.12 For the cash flow shown below, calculate the value of P by (*a*) using the inflated interest rate and (*b*) converting the annual series amounts into today's dollars and then using the appropriate P/F factors. Use an interest rate of 10% per year and an inflation rate of 8% per year.

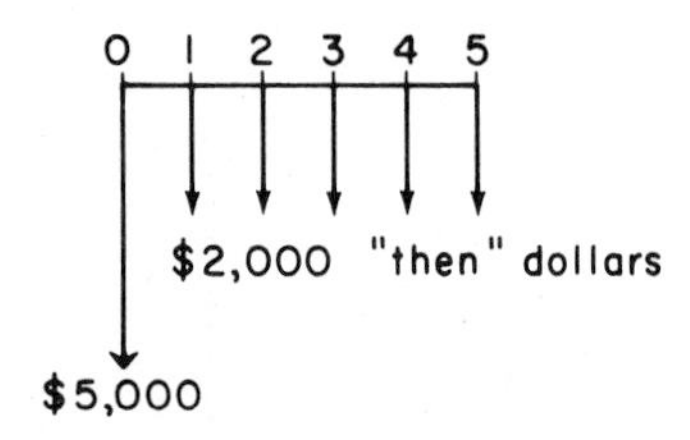

P12.13 How much money could the GRQ Company afford to spend now in order to avoid spending $5000 per year for 6 years if the interest rate is 15% per year and the inflation rate is 10% per year?

P12.14 If a program directed toward reducing product losses saves $6000 per year, what total equivalent saving would the company achieve in 5 years if the interest rate is 15% per year and the inflation rate is 8% per year?

P12.15 Two machines under consideration by May-Kit Metal Fabricating Company will have the following costs:

	Machine *A*	Machine *C*
First cost	$10,000	$20,000
Annual operating cost	8,000	3,000
Salvage value	3,000	6,000
Life, years	10	10

Which machine should be selected if $i = 15\%$ per year and inflation is 12% per year? Use PW analysis.

P12.16 If a company can buy a used computer system for $600,000 with the expectation of spending $48,000 per year for operation and maintenance, how much could it afford to spend on a new one if its annual cost would be only $38,000 per year? Assume that the salvage values will be 10% of the first cost and that both machines will have a 5-year life. Use $i = 15\%$ and $f = 10\%$ per year.

P12.17 An engineer is trying to decide which of two machines he should purchase to manufacture a certain part. He obtains estimates from two salesmen, but salesman *A* gives him the expected costs in constant value dollars (i.e., today's dollars) and salesman *B* gives him the expected costs in then-current dollars. If the company's minimum attractive rate of return is 15% per year and the company expects inflation to be 10% per year, which salesman's product should he purchase, assuming that their expected costs are correct. Use a present-worth analysis.

	Salesman *A* (today's dollars)	Salesman *B* (then dollars)
First cost	$60,000	$95,000
Annual operating cost	25,000	35,000
Life, years	5	10

P12.18 An 8% $20,000 bond with interest payable quarterly will mature in 13 years. How much should you pay for the bond if you desire to make a rate of return of a nominal 16% per year compounded quarterly and the inflation rate is 2% per quarter?

P12.19 How much money must a meat-packing company expect to save each year for 12 years through by-product recovery to justify the expenditure of $35,000 on a flotation system, if the interest rate is 20% per year and the inflation rate is 7% per year?

P12.20 How much money could the Fix-It Tool Co. afford to spend each year on maintenance if replacement costs 6 years from now will be reduced from $900,000 to $700,000 as a result of the maintenance? Assume that the interest rate is 18% per year and the inflation rate is 8% per year.

P12.21 A young couple planning for their daughter's college education 16 years from now would like to have money with an equivalent buying power of $15,000 of today's dollars. How much money will they have to deposit at the end of each year beginning 1 year from now if the interest rate they will earn is 12% per year and the inflation rate is 10% per year?

P12.22 A small building will cost $500,000 to construct and will have maintenance costs of $5000 per year for 20 years. If the maintenance costs are expressed in today's dollars, calculate the equivalent uniform annual cost of the building in then-current dollars if the interest rate is 14% per year and the inflation rate is 10% per year.

P12.23 Calculate the perpetual equivalent uniform annual cost of $50,000 now, $10,000 five years from now, and $5000 per year thereafter if the interest rate is 12% per year and the inflation rate is 9% per year.

P12.24 What is the difference between a conventional uniform gradient and an escalating series?

P12.25 Calculate the present worth of $2500 in year 1, $2750 in year 2, and amounts increasing by 10% per year thereafter until year 11, if the interest rate is 16% per year.

P12.26 Calculate the present worth of a machine which has an initial cost of $17,000, a salvage value of $2000, and an operating cost of $9000 in year 1, $9180 in year 2, and amounts increasing by 2% each year for 12 years. Use an interest rate of 17% per year.

P12.27 If you deposit $300 now into a savings account and increase your deposit by 10% each month, how much will you have after 3 years if the account earns interest at a rate of 1% per month?

P12.28 A company is planning to make deposits such that each one is 12% larger than the preceding one. How large must the first deposit be (at the end of year 1) if the company wants to accumulate $21,000 by the end of year 16? Assume that the interest rate the fund earns is 12% per year.

P12.29 What amount would the deposit in Prob. P12.28 have to be if the fund earns interest at a rate of 14% per year?

P12.30 How long will it take for a savings fund to accumulate to $15,000 if $1000 is deposited at the end of year 1 and the amount of deposit is increased by 10% each year? Assume that the interest rate is 10% per year.

P12.31 Calculate the present cost of a machine which has an initial cost of $29,000, a salvage value of $5000 after 8 years, and an annual operating cost of $13,000 for the first 3 years, increasing by 10% per year thereafter. Use an interest rate of 15% per year.

P12.32 Calculate the present cost to the A-1 Box Company of leasing a computer if the yearly cost is $15,000 for year 1 and $16,500 for year 2, with costs increasing by 10% each year thereafter. Assume that the lease payments must be made at the beginning of the year and that a 7-year study period is to be used. The company's minimum attractive rate of return is 16% per year.

P12.33 Calculate the present cost of a machine which costs $55,000 and has an 8-year life with a $10,000 salvage value. The operating cost of the machine is expected to be $10,000 in year 1 and $11,000 in year 2, with amounts increasing by 10% per year thereafter. Use an interest rate of 15% per year and account for the effects of inflation if the rate is 8% per year.

P12.34 Calculate the equivalent annual cost of a machine which costs $73,000 initially and will have a $10,000 salvage value after 9 years. The operating cost is $21,000 in year 1, $22,050 in year 2, and amounts increasing by 5% each year. The minimum attractive rate of return is 19% per year and the inflation rate is 12% per year. Make the annual cost calculation (*a*) assuming that inflation is not to be considered and (*b*) taking inflation into account.

LEVEL FOUR

The material in this level discusses important aspects of depreciation models commonly used in engineering-economic analysis including the new Accelerated Cost Recovery System (ACRS). Two chapters on corporate taxes cover all important formulas and laws current at the time this material was prepared. After-tax analysis should be used to evaluate the effects of income taxes on the outcome of the analysis.

CHAPTER

THIRTEEN

DEPRECIATION AND DEPLETION MODELS

The objective of this chapter is to acquaint you with the various methods of depreciating fixed assets. It teaches you how to calculate yearly depreciation charges and book values for the three methods and how to switch between these methods. The major changes in rates of depreciation and switching introduced in 1981 by the Accelerated Cost Recovery System (ACRS) are discussed and illustrated. In addition, you are introduced to the concept of depletion for reserves of natural resources.

SECTION OBJECTIVES

In order to complete this chapter, you must be able to do the following:

13.1. Define the terms *depreciation, book value, market value, first cost, asset life*, and *salvage value*. Identify the four common depreciation models and sketch a graph of book values versus time for each.

Given the initial cost, salvage value, and life of the asset(s) in Secs. 13.2 to 13.4, you must be able to do the following:

13.2. Calculate the depreciation charge, book value, and rate of depreciation for a specified year using the *straight-line* method.

13.3. Calculate the yearly depreciation charge, book value, and rate of depreciation for a specified year using the *sum-of-year-digits* method.

13.4. Calculate the yearly depreciation charge, book value, rate of depreciation, and implied salvage value for a specified year using the *declining-balance* and *double-declining-balance methods*.

13.5. State the rules and compute the annual depreciation charge and book value when switching from one depreciation model to another for pre-1981 purchased assets. When switching from a declining-balance method to straight line, use the specified procedure.

13.6. Calculate the annual depreciation charge and present worth of depreciation using the Accelerated Cost Recovery System (ACRS), given the year after 1980 it was placed in service and the prescribed recovery period.

*13.7. State the meaning of *Section 179 property capital expense, additional first-year depreciation,* and *investment tax credit* and determine the 179 capital-expense deduction for ACRS-depreciated assets.

*13.8. Calculate the annual depreciation charge for a class of assets using the *group* and *composite* methods.

*13.9. Define *depletion* and compute the depletion charge using factor depletion, given the initial investment and resource capacity; or compute the percentage depletion, given the depletion rate and annual gross income.

STUDY GUIDE

13.1 Depreciation Terminology

The dictionary definiton of *depreciation* is "a decrease in value of property through wear, deterioration, or obsolescence." As stated in the definition, there are several reasons why an asset can decrease in value from its original worth. Thus, even though a machine might be in perfect mechanical condition, due to obsolescence it may be worth considerably less than when it was new because of technological improvements in machine capability. Regardless of the reason for an asset's decrease in value, the depreciation should be taken into account in engineering-economy studies because of favorable income tax considerations. Taxes are paid on net income *less* depreciation for the year, thereby lowering the taxes paid. Income tax is described in detail in Chaps. 14 and 15.

The *book value* of an asset is equal to the remaining investment after the total amount of depreciation has been charged to date. That is, book value represents the current worth of an asset as shown in the books of account. Since depreciation is usually charged once a year, the book value is computed at the end of the year, which is in keeping with the end-of-year convention used previously. Book value is never taken into consideration in before-tax engineering-economy studies.

The *market value* of an asset refers to the amount of money that could be obtained for the asset if it were sold in the free market. In some cases, the market value bears very little relation to the book value. For example, commercial buildings tend to increase in market value, whereas their book value decreases due to depreciation charges. It is the market value which must be taken into consideration in engineering-economy comparisons.

Terms actually used in the computation of annual depreciation charges are defined as follows:

First cost. The initial price of the asset including purchase price, delivery charges, installation fees, and other direct costs incurred to ready the asset for use. Also called the *unadjusted basis* of the property.

Life. The anticipated useful life in years of the asset prior to disposal and/or replacement. It is also called the *recovery period*, which may be somewhat different for depreciation and tax purposes than the actual expected productive life-span. This difference occurs for several reasons: tax-law rulings, management policy, anticipated product changes, and the like.

Salvage value. The expected trade-in or net realizable value of the asset at the end of its useful life. This estimated figure is usually made at the time of purchase and may be positive, zero, or negative if dismantling or carry-away costs are expected. It is common to state the salvage value as a percentage of the first cost.

There are four commonly used depreciation methods for writing off the value of an asset. These are:

Straight-line depreciation (SL)
Sum-of-year-digits depreciation (SYD)
Declining-balance depreciation (DB)
Accelerated Cost Recovery System (ACRS)

ACRS is the method prescribed for assets put in service after 1980. This method, which is a combination of the first three, is described in Sec. 13.6. The plot of book value versus time, shown in Fig. 13.1 in general form, indicates that the SYD method gives an *accelerated* write-off compared with the SL method, and the DB method

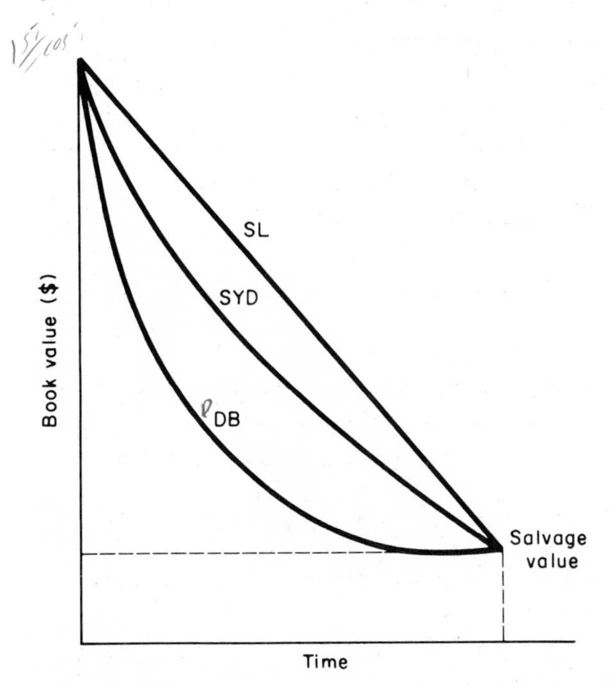

Figure 13.1 General shape of the book-value curve, for three different depreciation models.

is more accelerated than either. Accelerated used in the depreciation context means that the book value reaches the salvage value faster for some models than for others. The SL rate of depreciation is commonly used as the norm, or standard, rate.

A fifth depreciation model, the sinking-fund method, is a decelerated method compared with the SL method. Its book-value curve would be above the SL line in Fig. 13.1. The sinking-fund method is one of the older depreciation models and is no longer used because of its income tax disadvantage.

Prob. P13.1

13.2 Straight-Line (SL) Depreciation

The straight-line (SL) model has always been a popular method of depreciation. It derives its name from the fact that the book value of the asset decreases linearly with time, because the same depreciation charge is made each year. The yearly depreciation is determined by dividing the first cost or basis of the asset minus its salvage value by the life of the asset. In equation form,

$$D_t = \frac{P - \text{SV}}{n} \tag{13.1}$$

where t = year ($t = 1, 2, \ldots, n$)
D_t = annual depreciation charge
P = first cost or unadjusted basis
SV = salvage value
n = expected depreciable life or recovery period

Since the asset is depreciated by the same amount each year, the book value after t years of service BV_t will be equal to the first cost of the asset minus the annual depreciation times t. Thus,

$$\text{BV}_t = P - tD_t \tag{13.2}$$

The *rate of depreciation* d is the fraction by which the depreciable amount, $P - \text{SV}$, is decreased each year. For the SL method, this rate is the same for each year.

$$d = \frac{1}{n} \tag{13.3}$$

The relations above are illustrated in Example 13.1.

Example 13.1 If an asset has a first cost of \$50,000 with a \$10,000 salvage value after 5 years, (*a*) calculate the annual depreciation and (*b*) compute and plot the book value of the asset after each year using SL depreciation.

SOLUTION
(*a*) The depreciation charged each year can be found by Eq. (13.1).

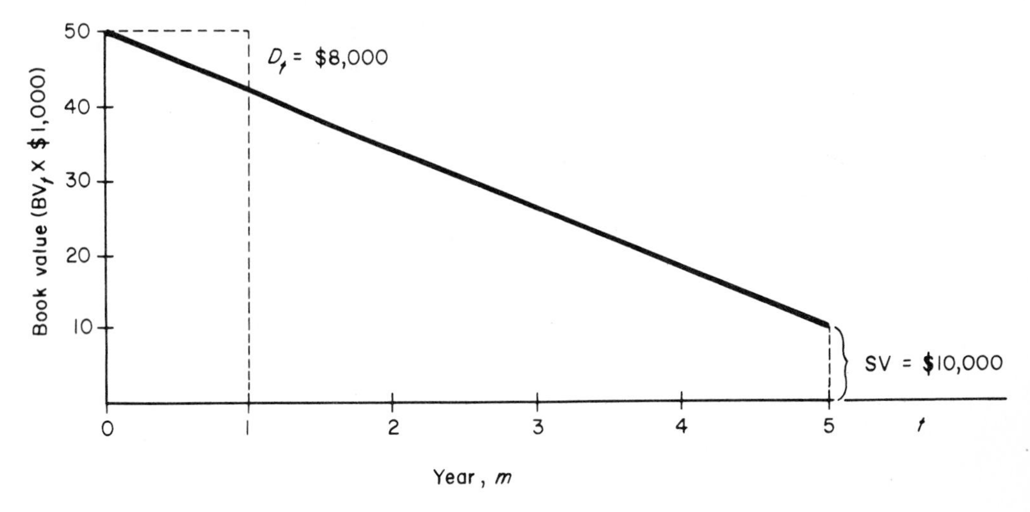

Figure 13.2 Plot of book value for SL depreciation, Example 13.1.

$$D_t = \frac{P - \text{SV}}{n} = \frac{50{,}000 - 10{,}000}{5}$$

$$= \$8000 \text{ per year for 5 years}$$

(*b*) The book value of the asset after each year can be found by Eq. (13.2).

$$\text{BV}_t = P - tD_t \qquad (t = 1, 2, 3, 4, 5)$$

$$\text{BV}_1 = 50{,}000 - 1(8000) = \$42{,}000$$

$$\text{BV}_2 = 50{,}000 - 2(8000) = \$34{,}000$$

. . . .

$$\text{BV}_5 = 50{,}000 - 5(8000) = \$10{,}000 = \text{SV}$$

A plot of BV_t versus t is given in Fig. 13.2.

Probs. P13.2 to P13.4

13.3 Sum-of-Year-Digits (SYD) Depreciation

The SYD method of depreciation is an accelerated or rapid write-off technique by which much of the value of the asset is written off in the first one-third of its life. That is, the depreciation charges are very high in the first few years but decrease rapidly in later years of the asset's life. The book value follows a convex curve, as shown in Fig. 13.1.

The mechanics of the method involve initially finding the sum of the year digits from 1 through n. The number obtained in this manner represents the sum-of-year-digits. The depreciation charge for any given year is then obtained by multiplying the first cost of the asset less its salvage value (P - SV) by the ratio of the number of years remaining in the life of the asset to the sum-of-year-digits. In equation form,

$$D_t = \frac{\text{depreciable years remaining}}{\text{sum-of-year-digits}} \text{ (first cost - salvage value)}$$

$$= \frac{n - t + 1}{S}(P - SV) \qquad (t = 1, 2, \ldots, n) \tag{13.4}$$

where D_t = depreciation charge for year t

S = sum-of-year-digits 1 to n

$$= \sum_{j=1}^{n} j = \frac{n(n+1)}{2} \tag{13.5}$$

Note that the depreciable years remaining must include the year for which the depreciation charge is desired. That is why the 1 has been included in the numerator of Eq. (13.4). For example, to determine the depreciation for the fourth year of an asset which has an 8-year life, the numerator of Eq. (13.4) must be 8 - 4 + 1 = 5. For an 8-year-life asset, using Eq. (13.5),

$$S = \sum_{j=1}^{8} j = 1 + 2 + \cdots + 8 = \frac{8(9)}{2}$$

$$= 36$$

The book value for any given year can be calculated without making the year-by-year depreciation determinations through the use of the following equation:

$$BV_t = P - \left[\frac{t(n - t/2 + 0.5)}{S}\right](P - SV) \tag{13.6}$$

The rate of depreciation d_t, which decreases each year for the SYD method, follows the multiplier in Eq. (13.4), that is,

$$d_t = \frac{n - t + 1}{S} \tag{13.7}$$

For example, if $n = 8$ the rates from Eq. (13.7) are

t	1	2	3	4	5	6	7	8
d_t	8/36	7/36	6/36	5/36	4/36	3/36	2/36	1/36

Example 13.2 illustrates the calculations for the SYD depreciation method.

Example 13.2 Calculate the depreciation charges for the first 3 years and the book value for year 3 for an asset which had a first cost of $25,000, a $4,000 salvage value, and a life of 8 years.

SOLUTION The sum-of-year-digits must be calculated first, using Eq. (13.5):

$$S = \frac{n(n+1)}{2} = \frac{8(9)}{2} = 36$$

The depreciation charges for each of the 3 years can now be calculated, using Eq. (13.4):

$$D_1 = \frac{(8-1+1)}{36}(25{,}000 - 4000) = \$4667$$

$$D_2 = \frac{7}{36}(21{,}000) = \$4083$$

$$D_3 = \frac{6}{36}(21{,}000) = \$3500$$

Note that $D_1 > D_2 > D_3$ to indicate that depreciation takes place at a decreasing rate. The book value for year 3, using Eq. (13.6), is

$$\begin{aligned} BV_3 &= 25{,}000 - \frac{3(8-3/2+0.5)}{36}(25{,}000-4000) \\ &= 25{,}000 - \frac{3(7)}{36}(21{,}000) \\ &= \$12{,}750 \end{aligned}$$

Probs. P13.5 to P13.7

13.4 Declining-Balance (DB) and Double-Declining-Balance (DDB) Depreciation

The declining-balance method of depreciation, also known as the uniform- or fixed-percentage method, is another of the rapid write-off techniques. Very simply, the depreciation charge for any year is determined by multiplying a uniform percentage by the book value for that year. For example, if the uniform-percentage depreciation rate were 10%, then the depreciation write-off for any given year would be 10% of the book value for that year. Obviously, the depreciation charge is largest in the first year and decreases each succeeding year.

The maximum percentage depreciation that is permitted is 200% (double) the straight-line rate. When this rate is used, the method is known as the double-declining-balance (DDB) method. Thus, if an asset had a useful life of 10 years, the straight-line rate would be $1/n = 1/10$. The uniform rate of 2/10 therefore could be used in the DDB depreciation method. The general formula for calculating the maximum DB depreciation rate in any year d_M is two times the straight-line rate.

$$d_M = 2d = \frac{2}{n} \tag{13.8}$$

This is the rate used for the DDB method. Other commonly used rates for the DB method are 175 and 150% of the straight-line rate, where $d = 1.75/n$ and $d = 1.50/n$, respectively. The actual depreciation rate for year t is then computed as

$$d_t = d(1 - d)^{t-1}$$

When DB or DDB depreciation is used, the estimated salvage value should not be subtracted from the first cost when calculating the depreciation charge. It is important that you remember this, since this procedure further increases the rate of write-off. Even though salvage values are not considered in the depreciation calculation, an asset may not be depreciated below the amount that would be considered a reasonable salvage value. If the salvage value is reached prior to year n, no additional depreciation may be taken thereafter. (See Sec. 13.6 for changes in this statement.) This generally occurs only for short-lived assets ($n < 5$) or assets having large salvage values ($SV > 0.2P$).

The depreciation D_t for year t is the depreciation rate d times the book value at the end of the previous year, that is,

$$D_t = (d)\, BV_{t-1} \tag{13.9}$$

If the BV_{t-1} value is not known, the depreciation charge is

$$D_t = (d)\, P(1 - d)^{t-1} \tag{13.10}$$

The book value in year t is

$$BV_t = P(1 - d)^t \tag{13.11}$$

Since the salvage value is not used directly in declining-balance methods, an implied SV after n years may be computed as

$$BV_n = SV = P(1 - d)^n \tag{13.12}$$

Finally, if not stated, the depreciation rate may be computed using the expected SV. For $SV > 0$,

$$d = 1 - \left(\frac{SV}{P}\right)^{1/n} \tag{13.13}$$

The allowed range on d is $0 \leqslant d \leqslant 2/n$. In all DB models d is stated or calculated by Eq. (13.13), and for the DDB model $d = 2/n$. Example 13.3 illustrates the DDB model and Solved Problems includes the computations for $d < 2/n$.

Example 13.3 Assume that an asset has a first cost of \$25,000 and an expected \$4000 salvage after 12 years. Calculate its depreciation and book value for (*a*) year 1, (*b*) year 4, and (*c*) the implied salvage value after 12 years using the DDB method.

SOLUTION First compute the DDB depreciation rate.

$$d = \frac{2}{n} = \frac{2}{12} = 0.1667 \text{ per year}$$

(*a*) For the first year, the depreciation and book value can be calculated using Eqs. (13.9) and (13.11), where $BV_0 = P$:

$$D_1 = (0.1667)\ 25{,}000 = \$4167.50$$

$$BV_1 = 25{,}000(1 - 0.1667)^1 = \$20{,}832.50$$

(*b*) From Eqs. (13.10) and (13.11) and $d = 0.1667$,

$$D_4 = 0.1667(25{,}000)\ (1 - 0.1667)^{4-1} = \$2411.46$$

$$BV_4 = 25{,}000(1 - 0.1667)^4 = \$12{,}054.40$$

(*c*) Using Eq. (13.12) the salvage value at $n = 12$ is

$$SV = 25{,}000(1 - 0.1667)^{12} = \$2802.57$$

Since the salvage value is anticipated to be \$4000, the lower limit on book value is \$4000.

COMMENT An important fact to remember about the DB and DDB methods is that salvage value is not subtracted from the first cost when the depreciation is calculated. However, when the book value reaches the expected salvage value (\$4000 here), no additional depreciation may be taken. In this case $BV_{10} =$ \$4036.02 and $D_{11} =$ \$672.80, making $BV_{11} =$ \$3362.22, which is less than the expected SV of \$4000. Therefore, in years 11 and 12 the depreciation is $D_{11} =$ \$36.02 and $D_{12} = 0$, respectively.

Solved Problem 13.11 presents a comparison of the three depreciation methods discussed thus far.

Solved Problems 13.10 and 13.11
Probs. P13.8 to P13.13

13.5 Switching between Depreciation Models

For assets purchased prior to 1981 and therefore not depreciated using the ACRS rates (Sec. 13.6), the procedures in this section apply. Switching is commonly done to more rapidly reduce the book value toward the salvage value, thus obtaining the tax advantage of increased depreciation deductions occurring early in the asset's life.

Switching from a DB model to the SL method is the most common procedure, because it may offer a real advantage, especially if the DB model is the DDB, i.e., twice the straight-line rate is used to compute the D_t values. It is not usually possible to demonstrate an advantage of switching from SYD to SL depreciation.

Pertinent rules of switching may be summarized as follows:

1. Switching is allowed when the depreciation for year t using the established method D_t^e is less than that for a new method, D_t^n, that is, the selected depreciation D_t^* is the maximum charge.

$$D_t^* = \max\ [D_t^e, D_t^n]$$

2. Regardless of the methods used, book value can never go below the anticipated reasonable salvage value set at purchase time.
3. The undepreciated amount or book value BV_t is used as the basis of computation to select D_t^* when switching is considered.
4. When switching from a declining-balance method, the anticipated salvage value, not the implied salvage, is used to compute the depreciation for the new method.
5. Only one switch can commonly take place during the depreciable life of the asset.

In all situations, the criteria of maximizing the present worth of the total depreciation P_D is used in switching determination. The depreciation method or methods (using switching) which results in the *maximum present worth,* is the best strategy, where

$$P_D = \sum_{t=1}^{n} D_t(P/F, i\%, t) \tag{13.14}$$

This is correct, because it minimizes tax liability in the early part of an asset's life (see Sec. 14.6 for further discussion of the effects of depreciation on taxes).

Virtually all switching occurs from a rapid write-off method to the SL method. The most promising, as mentioned earlier, is the DDB-to-SL switch. This switch is predictably advantageous if the implied salvage value computed by Eq. (13.12) is greater than the anticipated salvage value, that is, switch from DB or DDB to SL if

$$BV_n = P(1 - d)^n > \text{anticipated SV} \tag{13.15}$$

The procedure to consider the switch from DB or DDB to SL depreciation is:

1. Use Eq. (13.15) to determine if the switch to SL will be advantageous. If not, use the DB or DDB method for all n years and go to step 4. If switching is advisable, go to step 2.
2. For each year compute the two depreciation charges.

$$D_t^{DDB} = (d)\, BV_{t-1} \tag{13.16}$$

$$D_t^{SL} = \frac{BV_{t-1} - SV}{n - t + 1} \tag{13.17}$$

3. For each year select the maximum value. The depreciation for $t = 1, 2, \ldots, n$ is D_t^*.

$$D_t^* = \max\,[D_t^{DDB}, D_t^{SL}] \tag{13.18}$$

4. If required, compute the present worth of total depreciation charges using Eq. (13.14).

It is acceptable, though not usually financially advantageous, to state that a switch will take place in a particular year, for example, a mandated switch from SYD to SL in year 7 of a 10-year life. This approach is usually not taken, but the switching technique will perform correctly for any depreciation models involved with switching in any year $t \leqslant n$.

Example 13.4 A municipality has purchased a vehicle for \$10,000 with an anticipated salvage of \$500 after 8 years of service. Compute the annual depreciation schedule and calculate and compare present worth of total depreciation for (*a*) the straight-line model, (*b*) the double-declining-balance model, and (*c*) the DDB-to-SL model, if switching is advantageous. Let $i = 15\%$.

SOLUTION

(*a*) For $P = \$10,000$, SV = \$500, and $n = 8$, Eq. (13.1) gives the SL depreciation for each year.

$$D_t = \frac{10,000 - 500}{8} = \$1187.50$$

In Eq. (13.14) for present worth the P/A factor can replace P/F, since D_t is the same for all years.

$$P_D = 1187.50(P/A, 15\%, 8) = 1187.50(4.4873) = \$5328.67$$

(*b*) For the DDB model $d = 2/n = 0.25$. Equation (13.9) is used to obtain the results presented in Table 13.1. The present-worth value $P_D = \$6045.48$ is larger than the P_D for straight-line depreciation in part *a*, thus indicating that DDB is a method which writes off more of the depreciable value in the initial years of the asset's life.

(*c*) Use the DDB-to-SL switching procedure.

1. For the DDB method the implied salvage is

$$BV_8 = 10,000(1 - 0.25)^8 = \$1001.13$$

Since BV_8 is larger than \$500, the anticipated salvage value, switching to SL is recommended.

Table 13.1 Depreciation schedule and present worth using DDB ($d = 0.25$) for Example 13.4*b*

Year, t	D_t	BV_t	$(P/F, 15\%, t)$	Present worth of D_t
0	...	\$10,000.00	...	...
1	\$2,500.00	7,500.00	0.8696	\$2,174.00
2	1.875.00	5,625.00	0.7561	1,417.69
3	1,406.25	4,218.75	0.6575	924.61
4	1,054.69	3,164.06	0.5718	603.07
5	791.01	2,373.05	0.4972	393.30
6	593.26	1,779.79	0.4323	256.47
7	444.95	1,334.84	0.3759	167.25
8	333.71	1,001.13	0.3269	109.09
	\$8,998.87			\$6,045.48

Table 13.2 Depreciation schedule and present worth for DDB-to-SL switching, Example 13.4*c*

Year,	DDB model		D_t^{SL}, Eq.			Present worth
t	D_t^{DDB}	BV_t	(13.17)	D_t^*	$(P/F, 15\%, t)$	of D_t^*
0	...	$10,000.00	...	...	...	...
1	$2,500.00	7,500.00	$1,187.50	$2,500.00	0.8696	$2,174.00
2	1,875.00	5,625.00	1,000.00	1,875.00	0.7561	1,417.69
3	1,406.25	4,218.75	854.17	1,406.25	0.6575	924.61
4	1,054.69	3,164.06	743.75	1,054.69	0.5718	603.07
5	791.01	2,373.05	666.02	791.01	0.4972	393.29
6	593.26	1,779.79	624.35	624.35	0.4323	269.91
7	444.95	1,334.84	639.89	624.35	0.3759	234.69
8	333.71	1,001.13	834.83	624.35	0.3269	204.10
	$8,998.87			$9,500.00		$6,221.36

2. The DDB values of D_t are computed in Table 13.1 but repeated in Table 13.2 for comparison with the D_t^{SL} values from Eq. (13.17). Note that the SL depreciation values change each year because the remaining undepreciated value BV_{t-1} is different. Only in year $t = 1$ is $D_1^{SL} = \$1187.50$ as computed in part *a*. Two D_t^{SL} values are computed here for illustration:

Year $t = 2$:

$$BV_1 = \$7500$$

$$D_2^{SL} = \frac{7500 - 500}{8 - 2 + 1} = \$1000$$

Year $t = 6$:

$$BV_5 = \$2373.05$$

$$D_6^{SL} = \frac{2373.05 - 500}{8 - 6 + 1} = \$624.35$$

3. The D_t^* values are shown in Table 13.2. The switch occurs in year 6 and $D_6^* = D_7^* = D_8^* = \$624.35$ using the SL model. The D_t^{SL} for $t = 7$ and 8 would be used only if the switch took place in these years. Note that the total depreciation is now \$9500 rather than the lower value of \$8998.87 for DDB with no switching. This is because of the use of the \$500 anticipated salvage rather than the implied value of \$1001.13 for the DDB model. Figure 13.3 presents a graphic summary of the book value with and without switching.
4. The amount $P_D = \$6221.36$ in Table 13.2 shows that the switch to SL has again increased the present worth of total depreciation: This is the highest P_D that can be obtained using the switching rules in effect prior to the Economic Recovery Tax Act of 1981.

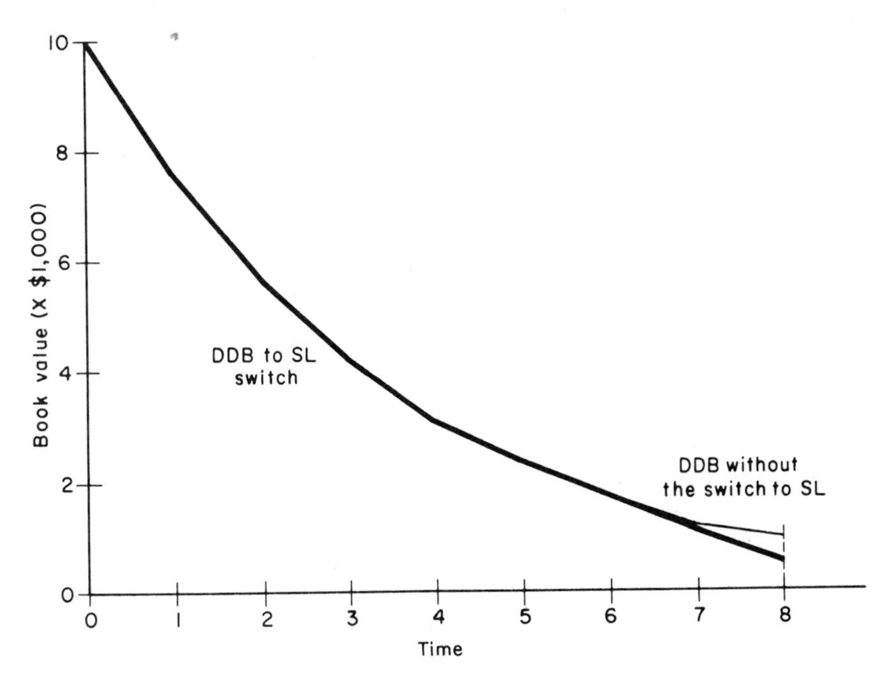

Figure 13.3 Book value using DDB depreciation with a switch to SL in year 6, Example 13.4.

Appendix E includes a discussion of a computer program named DEPSTRAT (DEPreciation techniques and STRATegies), which develops the depreciation schedule using a particular depreciation method or switching between stated methods. Selection of the optimal switching option causes the program to check all methods and use the one with the largest depreciation for each year. Depreciation strategies with income taxes considered (Sec. 14.6) are included in DEPSTRAT.

Solved Problem 13.12
Probs. P13.14 to P13.21

13.6 Accelerated Cost Recovery System (ACRS) Depreciation Method

The Economic Recovery Tax Act of 1981 made changes in the allowed methods of recovering capital invested in assets through depreciation by introducing tabulated annual rates of depreciation using the Accelerated Cost Recovery System (ACRS) [1]. Table 13.3 gives the prescribed percent of the unadjusted basis P that is to be used as the annual depreciation depending upon when the personal-property asset is placed in service. The 1981 to 1984 and 1985 rates are transitional while ACRS is being introduced. Assets placed in service prior to 1981 are depreciated using the classic methods and switching as discussed thus far. ACRS depreciation of real property is discussed later in this section.

Note that all ACRS-depreciated assets have a statuatory recovery period of either

Table 13.3 Depreciation rates (in percent) prescribed by ACRS for personal property assets placed in service after 1980

	Recovery period, years											
	3			5			10			15		
	Year placed in service (19xx)											
Depreciation year, t	1981 to 1984	1985	After 1985	1981 to 1984	1985	After 1985	1981 to 1984	1985	After 1985	1981 to 1984	1985	After 1985
1	25	29	33	15	18	20	8	9	10	5	6	7
2	38	47	45	22	33	32	14	19	18	10	12	12
3	37	24	22	21	25	24	12	16	16	9	12	12
4				21	16	16	10	14	14	8	11	11
5				21	8	8	10	12	12	7	10	10
6							10	10	10	7	9	9
7							9	8	8	6	8	8
8							9	6	6	6	7	7
9							9	4	4	6	6	6
10							9	2	2	6	5	5
11										6	4	4
12										6	4	3
13										6	3	3
14										6	2	2
15										6	1	1

3, 5, 10, or 15 years, which is determined in the first year that depreciation is charged. This recovery period, which was historically tied to the actual or expected life of the asset, is now set based on the ADR class life recommended for most major asset groups as a part of the class life asset depreciation range (CLADR) system [2]. For our purposes we will assume that the recovery period is already determined to be either 3, 5, 10, or 15 years. One of the biggest advantages of ACRS is that the recovery period is usually considerably less than the corresponding ADR life. For example, if the ADR life is 5 *or more* years, ACRS recovery occurs in 5 years, thus, most assets (especially manufacturing) fit in the category for 5-year recovery [3].

ACRS also states that no salvage value greater than zero is used; 100% of the unadjusted basis (first cost) P is instead written off. Check to see that each column in Table 13.3 adds to 100%.

It is possible to elect during the first year of depreciation to not use the standard ACRS percentages, but the only alternative is SL depreciation and the recovery period is always equal to or greater than the ACRS recovery period. For example, a 5-year recovery asset depreciated under this straight-line ACRS election must be depreciated over 5, 12, or 25 years. Thus, the possibility of accelerated depreciation by the SYD or DB methods has been eliminated. Additionally this SL election requires the use of the *half-year convention* for personal property, which includes all machinery, equipment, etc., used in a business. A half-year depreciation is claimed the first year and a half-year is claimed in the year after the selected recovery period. Thus, if n is the recovery period used for the SL election, the rate of depreciation in Eq. (13.3) is now different for some years t.

$$d_t = \begin{cases} \dfrac{1}{2n} & t = 1, n+1 \\ \dfrac{1}{n} & t = 2, 3, \ldots, n \end{cases} \tag{13.19}$$

We will neglect the half-year convention in examples and problems unless specifically stated otherwise.

For assets placed in service after 1980, switching is not allowed because ACRS sets the annual depreciation rate. The presence of built-in switching in ACRS is discussed after Example 13.5.

Example 13.5 (*a*) An asset with a 5-year life was placed in service in 1983. If P = \$10,000 and SV = \$2000, plot the book value of the asset using SL, DDB, and ACRS. (*b*) Compute the present worth of depreciation for the three methods using $i = 15\%$.

SOLUTION

(*a*) The annual depreciation rates d_t and amounts D_t are computed as follows.

Straight line: $d_t = \frac{1}{5} = 0.2 \qquad t = 1, 2, 3, 4, 5$

$$D_t = d_t(P - \text{SV})$$
$$= 0.2(8000)$$

Double declining balance:

$$d_t = \tfrac{2}{5} = 0.4 \qquad t = 1, 2, 3, 4, 5$$
$$D_t = 0.4(\text{BV}_{t-1})$$

ACRS: d_t from Table 13.3 value divided by 100

$$D_t = d_t(P)$$
$$= d_t(10{,}000) \qquad t = 1, 2, 3, 4, 5$$

Table 13.4 shows the D_t and BV_t values and Fig. 13.4 is a plot of BV_t for the 5 years. The ACRS percentages show a slight deceleration compared with SL for the first year, but the ACRS book value is reduced to zero rather than the expected salvage of \$2000 as in other methods.

(*b*) The 15% present worths of depreciation using Eq. (13.14) and the D_t values in Table 13.4 are

SL: P_D = \$5363.52

DDB: P_D = \$6331.33

ACRS: P_D = \$6593.11

Table 13.4 Computation of depreciation and book value for different methods (P = \$10,000, SV = \$2000, n = 5 years)

	SL			DDB			ACRS		
t	d_t	D_t	BV_t	d_t	D_t	BV_t	d_t	D_t	BV_t
0	...	...	\$10,000	...	...	\$10,000	...	...	\$10,000
1	0.2	\$1,600	8,400	0.4	\$4,000	6,000	0.15	\$1,500	8,500
2	0.2	1,600	6,800	0.4	2,400	3,600	0.22	2,200	6,300
3	0.2	1,600	5,200	0.4	1,440	2,160	0.21	2,100	4,200
4	0.2	1,600	3,600	0.4	160*	2,000	0.21	2,100	2,100
5	0.2	1,600	2,000	0.4	0	2,000	0.21	2,100	0

*Cannot depreciate below SV = \$2000 under old tax laws.

Since ACRS has the maximum P_D, it offers the largest tax advantage. The P_D for ACRS is largest in part because no salvage value is used to limit depreciation through the recovery period of the asset.

The annual depreciation percentages in ACRS (Table 13.3) have built-in switching between different methods. The recovery periods are 3, 5, 10, or 15 years, and the annual depreciation percentages of the unadjusted basis are determined by switching

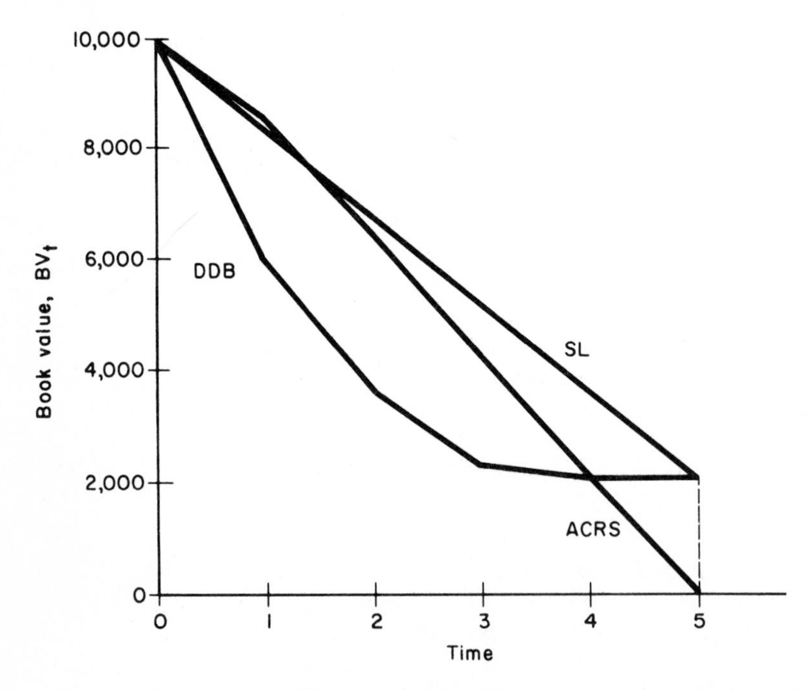

Figure 13.4 Book value for different depreciation methods including ACRS, Example 13.5.

to the SL or SYD method after starting with the DB method. The depreciation amount allowed in the first year is only one-half that computed by the DB method. This use of the half-year convention is slightly different from that in the SL election, since the standard ACRS rates allow recovery of the entire first cost in n years. The strict switching procedures discussed in Sec. 13.5 are not used, but in general the switch occurs and rounding to the nearest integer percentage value takes place using the following switching philosophy, which is built into the ACRS depreciation rates.

1981–1984 *tax years*. Start with 150% DB but allow only one-half the depreciation in year 1 and then switch to SL when the rate exceeds that of the full 150% DB rate. The n value used for SL depreciation is equal to the remainder of the recovery period. For example, for a 3-year recovery period:

Year 1. Use $n = 3$ years and one-half the 150% DB rate to obtain d_1 (as in Sec. 13.4) and D_1.

$$d_1 = \frac{1.5}{2n} = \frac{1.5}{6} = 0.25$$

$$D_1 = d_1 P = 0.25P$$

Years 2 and 3. Use $n = 2$ and the SL depreciation rate to remove the remaining $1 - d_1 = 0.75$ of P with rounding to obtain an integer percentage.

$$d_2 = \frac{1}{n}(1 - d_1) = \frac{1}{2}(1 - 0.25) = 0.375 \qquad \text{(use 38\%)}$$

$$D_2 = 0.38P$$

$$d_3 = \frac{1}{2}(0.75) = 0.375 \qquad \text{(use 37\%)}$$

$$D_3 = 0.37P$$

The switch to SL occurs in year 2 because the DB rate is smaller at a value of $d_2 = 0.5(1 - 0.5)^{2-1} = 0.25$ (Sec. 13.4).

For 10- and 15-year periods this procedure is not followed as strictly.

1985 tax year. In the first year use only one-half the 175% DB depreciation and then switch to SYD with n reduced by 1 and remove the amount $(1 - d_1)P$ over the next $n - 1$ years.

After 1985 tax years. The same procedure as the 1985 tax year is used except 200% DB (which is DDB) is used in year 1 only.

As you can see, the switching rules of Sec. 13.5 are not strictly applied in ACRS, and even though the depreciation is accelerated compared with SL over the entire recovery period, the shape of the book-value curve is generally different owing to the neglect of any salvage value in the depreciation computations. The ACRS annual depreciation amounts are larger, except in the first year, because the SL or SYD method is used for most of the recovery period and SV = 0. Therefore, it can be expected that P_D for ACRS will be greater than that for other methods, even with switching, as shown in Example 13.6.

Example 13.6 In Example 13.4*c* a $10,000 8-year asset was depreciated using the DDB-to-SL switching procedure with P_D = $6221.36 at i = 15% (Table 13.2). Use ACRS to depreciate the same asset and compute the resulting P_D values assuming it was placed in service in 1983 and 1986. The 5-year recovery period for ACRS should be used for an asset with an expected 8-year life.

SOLUTION Table 13.5 presents the depreciation and present worth of depreciation P_D using the appropriate ACRS rates from Table 13.3. The P_D values are larger than those for the DDB-to-SL switch. This makes it clear that ACRS offers a tax advantage in that P_D is larger. Note that there is no ACRS depreciation for years 6 to 8 due to the use of a 5-year recovery period.

Most real property (business real estate, buildings, and so on) has a 15-year recovery period under ACRS. The month of the tax year in which acquisition occurs is used to determine the depreciation rates for each year of the recovery period. The d_t values given in Table 13.6 are applicable to most real property. Select the appropriate column (month) and multiply the depreciation rate d_t by the unadjusted basis B for each year t ($t = 1, 2, \ldots, 16$).

The ACRS real property rates are the result of implied switching from the 175% DB method to the SL method. If the straight line ACRS election is selected for real property the recovery period is either 15, 35, or 45 years and the half-year convention is not imposed. This election must be made in the year of acquisition on a property-by-property basis.

Appendix E introduces the computer program ACRS, which develops the depreciation schedule using ACRS and any of the elections allowed under the 1981 tax act.

Probs. P13.22 to P13.25

Table 13.5 Determination of present worth of depreciation for an asset using 1983 and 1986 ACRS rates (P = $10,000)

Year,	1983 ACRS rates			1986 ACRS rates		
t	d_t	D_t	P_D at 15%	d_t	D_t	P_D at 15%
1	0.15	$ 1,500	$1,304.40	0.18	$ 1,800	$1,565.28
2	0.22	2,200	1,663.42	0.33	3,300	2,495.13
3	0.21	2,100	1,380.75	0.25	2,500	1,637.50
4	0.21	2,100	1,200.78	0.16	1,600	914.88
5	0.21	2,100	1,044.12	0.08	800	397.76
6–8	0	0	0	0	0	0
		$10,000	$6,593.47		$10,000	$7,010.55

Table 13.6 Depreciation rates for ACRS for real property placed in service after 1980 (except low-income housing)

Recovery year, t	Depreciation rates, d_t (× 100%)											
	Month property placed in service											
	1	2	3	4	5	6	7	8	9	10	11	12
1	12	11	10	9	8	7	6	5	4	3	2	1
2	10	10	11	11	11	11	11	11	11	11	11	12
3	9	9	9	9	10	10	10	10	10	10	10	10
4	8	8	8	8	8	8	9	9	9	9	9	9
5	7	7	7	7	7	7	8	8	8	8	8	8
6	6	6	6	6	7	7	7	7	7	7	7	7
7	6	6	6	6	6	6	6	6	6	6	6	6
8	6	6	6	6	6	6	5	6	6	6	6	6
9	6	6	6	6	5	6	5	5	5	6	6	6
10	5	6	5	6	5	5	5	5	5	5	6	5
11	5	5	5	5	5	5	5	5	5	5	5	5
12	5	5	5	5	5	5	5	5	5	5	5	5
13	5	5	5	5	5	5	5	5	5	5	5	5
14	5	5	5	5	5	5	5	5	5	5	5	5
15	5	5	5	5	5	5	5	5	5	5	5	5
16			1	1	2	2	3	3	4	4	4	5

*13.7 Additional First-Year Depreciation, Capital Expensing, and Investment Tax Credit

In an effort to encourage more rapid economic growth, the federal government in recent years has permitted additional depreciation or capital expensing in order to provide more funds for industrial expansion. In addition to the accelerated depreciation methods detailed in the previous sections, early write-off of depreciable assets can be increased even more through these methods.

Additional first-year depreciation, which was repealed with the Economic Recovery Tax Act of 1981, allowed a business to take, in the purchase year, an additional 20% depreciation on a maximum of $10,000 of the first cost of any qualifying assets acquired during the year provided that the lives were 6 or more years. However, it was required that the total depreciable amount be reduced by this additional first-year depreciation.

After 1981, additional first-year depreciation was replaced with the *Section 179 property capital expense deduction*, which allows as much as $10,000 of the first cost of most assets used in business (called Section 179 property) to be treated as a business expense in the year of purchase rather than a capital expenditure to be depreciated over some recovery period. Limits per tax return during transition years are:

1982, 1983	$5,000
1984, 1985	7,500
1986 and future	10,000

The capital expense amount is deducted from the first cost before computing the ACRS depreciation amount (Sec. 13.6) for the year that the asset is placed in service. Therefore, a $20,000 5-year asset purchased in 1984 could have $7500 of the first cost expensed and be depreciated in the first year by 0.15(20,000 - 7500) = $1875 using ACRS (Table 13.3). In this case the ACRS rates apply to years 1 through 5 for the $12,500 that was not expensed.

Another method of increasing capital recovery is through the investment tax credit, which applies directly to taxes and can reduce the company's income tax liability in the year that the asset is purchased. For example, if a company purchases a piece of equipment costing $8000 which is qualified for a 6% tax credit, it can deduct 8000(0.06) = $480 from its taxes. This deduction cannot be based on the original first cost if the Section 179 property capital expense is taken, but it can be used in addition to normal depreciation charges on the portion not expensed. The investment tax credit is discussed in detail in Sec. 14.4.

Probs. P13.26 and P13.27

*13.8 Group and Composite Methods of Depreciation

All previous methods of depreciation are called unit depreciation because they allow write-off on only one asset at a time. Because of the large number of assets in big companies, assets are usually grouped into classes and the depreciation charge is not computed for each individual asset but for the class.

The *group-depreciation* method requires computation of a single annual depreciation charge for a group of assets using their average life. Retirement after a number of years different from the average requires the prorating of depreciation charges to the remaining assets. Any method of depreciation, including ACRS, can be used to compute the group-depreciation charge. For example, if $40,000 is spent on the purchase of 100 serving machines with an average useful life of 5 years and no salvage value, the straight-line depreciation charge is $40,000/5 = $8000 per year. Sale prior to or after 5 years is accounted for on a machine-by-machine basis at a rate of $80 per machine per year.

Composite depreciation uses the same principle as group depreciation but accounts for varying expected lives within the same asset class. A depreciation charge for assets with the same lives is calculated, and total depreciation is then found by adding the depreciation for all assets in the class. Finally, a composite life is computed:

$$\text{Composite life} = \frac{\text{total depreciable value}}{\text{total annual depreciation}}$$

The rate determined for the composite method is continued until the composite life is reached or until the lives in the class change significantly by the addition or retirement of assets.

Table 13.7 Calculations for the composite method of depreciation

Asset	Life, n	Cost, P	Salvage, SV	Depreciable value	Annual SL depreciation
A	4	\$ 50,000	\$10,000	\$ 40,000	\$10,000
B	6	22,000	4,000	18,000	3,000
C	7	34,000	6,000	28,000	4,000
D	8	93,000	5,000	88,000	11,000
		\$199,000		\$174,000	\$28,000

Example 13.7 Four similar assets are purchased, the details of which are given in the first four columns of Table 13.6. Determine the annual depreciation charges and composite life for this class of asset using straight-line depreciation.

SOLUTION For each asset we compute the straight-line depreciation (Table 13.7). Total annual depreciation is \$28,000 for the class. The composite life is 174,000/ 28,000 = 6.21 years, which may be viewed as a weighted average life of assets in this class.

Probs. P13.28 and P13.29

*13.9 Depletion Methods

We have thus far computed depreciation for an asset which has a value that can be recovered by purchasing a replacement. Depletion is similar to depreciation; however, depletion is applicable to natural resources, which, when removed, cannot be "repurchased," as can a machine or building. Therefore, a depletion method is applicable to natural deposits removed from mines, wells, quarries, forests, and the like.

There are two methods of depletion: factor, or cost, depletion and percentage depletion. Factor depletion is based on the level of activity or usage, not time, as in the case of depreciation. The depletion factor d_t' for year t is

$$d_t' = \frac{\text{initial investment}}{\text{resource capacity}} \tag{13.20}$$

and the annual depletion charge is d_t' times the year's usage or activity volume. As is the case for depreciation, accumulated depletion by the factor method cannot exceed total cost of the resource. Example 13.8 illustrates factor depletion.

Example 13.8 The Knotty Wood Company has purchased some forest acreage for \$350,000 from which an estimated 175 million board feet of lumber are recoverable. Determine the depletion charges if 15 million and 22 million board feet are removed in the first and second years, respectively.

SOLUTION Using Eq. (13.20), we find that the depletion factor for each year $t = 1, 2, \ldots, n$ is

$$d_t' = \frac{350{,}000}{175} = \$2000 \text{ per million board feet}$$

Actual depletion charges are

First year, $t = 1$: 2000(15) = $30,000

Second year, $t = 2$: 2000(22) = $44,000

This will continue until $350,000 depletion is accumulated.

COMMENT Often the recoverable material estimate is altered once operation is begun, in which case d_t' must be changed. The Solved Problems present an illustration of this case.

The second depletion method, that of percentage depletion, is a special consideration given when natural resources are exploited. A flat percentage of the resource's gross income may be depleted each year provided it does not exceed 50% of taxable income. Using percentage depletion, total depletion charges may exceed actual costs with no limitation. Since it is possible to use the depletion figure computed either by the factor or by the percentage method, the percentage depletion method is usually chosen, because of the possibility of writing off more than the original cost of the venture. Below are listed some of the percentages for activities that can use the percentage depletion method, which is illustrated by Example 13.9. The percentages allowed for oil and gas wells were recently reviewed, resulting in programmed annual degradations from 1981 to 1984.

Activity	Percentage of gross income
Oil and gas wells (1984 and after)	15
Coal, sodium chloride	10
Gravel, sand, peat, some stones	5
Sulfur, cobalt, lead, nickel, zinc, etc.	22
Gold, silver, copper, iron ore	15

Example 13.9 A gold mine purchased for $750,000 has an anticipated gross income of $1.1 million per year for years 1 to 5 and $0.85 million per year after year 5. Compute annual depletion charges for the mine.

SOLUTION A 15% depletion applies to the gold mine. Thus, assuming that depletion charges do not exceed 50% of taxable income, depletion will be 0.15(1.1 million) = $165,000 for years 1 to 5 and 0.15(0.85 million) = $127,500 each year thereafter. At this rate, the cost of $750,000 will be recovered in approximately 4.5 years of operation.

Since there are several tax considerations to be made when depletion is used, the tax angles are detailed in Sec. 15.5.

A depreciation method similar to factor depletion, but applicable to depreciable assets, is the *unit-of-production* method. The asset can be depreciated using this method only if the rate of use or production is a measure of its rate of deterioration. The depreciation factor is computed in a fashion similar to Eq. (13.20), with the salvage value removed from the initial investment. For example, oil-producing equipment at a lease site costing \$300,000 with a salvage of \$50,000 may be depreciated on a basis of the estimated 1.0 million barrels of oil to be extracted from the lease. The depreciation factor is then (300,000 - 50,000)/1,000,000 = \$0.25 per barrel. This method is not commonly used because of the difficulty of finding a production unit which accurately measures the rate of asset deterioration.

Solved Problem 13.13
Probs. P13.30 to P13.32

SOLVED PROBLEMS

Example 13.10 The Dandy Company has just purchased an ore-crushing unit for \$80,000. The unit has an anticipated life of 10 years and a salvage of \$10,000. (*a*) Use the declining-balance method to develop a schedule of depreciation and book values for each year. (*b*) Compare computed and anticipated salvage values for this DB model and the DDB model.

SOLUTION

(*a*) The depreciation rate from Eq. (13.13) using SV = \$10,000 is

$$d = 1 - \left(\frac{10{,}000}{80{,}000}\right)^{1/10} = 0.1877$$

Note that $0.1877 < 2/n = 0.2$, so this DB model does not exceed twice the straight-line rate. Table 13.8 presents the D_t values using Eq. (13.9) and the BV_t values from $BV_t = BV_{t-1} - D_t$, rounded to the nearest dollar. For example, at $t = 2$,

$$D_2 = (d)BV_1 = 0.1877(64{,}984) = \$12{,}197$$

$$BV_2 = 64{,}984 - 12{,}197 = \$52{,}787$$

Due to the round-off to even dollars in year 10, \$2312 is calculated for depreciation; but $D_{10} = \$2318$ is deducted to make $BV_{10} = SV = \$10{,}000$ exactly.

(*b*) The computed rate of depreciation for the DB model was $d = 0.1877476$ from Eq. (13.3). Using Eq. (13.12),

$$BV_{10} = SV = 80{,}000(1 - 0.1877476)^{10} = \$10{,}000$$

Table 13.8 D_t and BV_t values using declining-balance depreciation, Example 13.10

Year, t	D_t	BV_t
0	...	\$80,000
1	\$15,016	64,984
2	12,197	52,787
3	9,908	42,879
4	8,048	34,831
5	6,538	28,293
6	5,311	22,982
7	4,314	18,668
8	3,504	15,164
9	2,846	12,318
10	2,318	10,000

This is, as it must be, the same as the expected salvage value. For the DDB model $d = 2/n = 0.2$ and the computed SV is

$$BV_{10} = 80{,}000(1 - 0.2)^{10} = \$8590$$

which is considerably less than the predicted \$10,000 value.

Sec. 13.4

Example 13.11 Graphically compare the rate of write-off for an asset with $P =$ \$80,000, SV = \$10,000, and $n = 10$ years using the following methods of depreciation: SL, SYD, and DDB.

SOLUTION Table 13.9 gives the results of the three methods, where the general expression $BV_t = BV_{t-1} - D_t$ is used for book value. A plot of BV_t versus t (Fig. 13.5) indicates that the DDB method would reduce book value below the SV value in year 10; thus, only \$738 depreciation is allowed for this year. Note that at year 5 the book value for DDB (\$26,214) is approximately only 58% of the SL method (\$45,000), attesting to the very rapid write-off of DDB. Under DDB, approximately $\frac{1}{2}P$ is removed from the books after only 3 years.

COMMENT If any accelerated method reduces book value to the SV value before year n, no additional depreciation is allowed. You can see that the SYD method is a fair approximation to the DDB method, thus accounting for some of the former method's popularity.

The data used in this problem are the same as in Example 13.10 for the DB method. A plot of the BV_t values from Table 13.8 will show that the DB curve closely follows the DDB and SYD curves in Fig. 13.5.

Secs. 13.2, 13.3, and 13.4

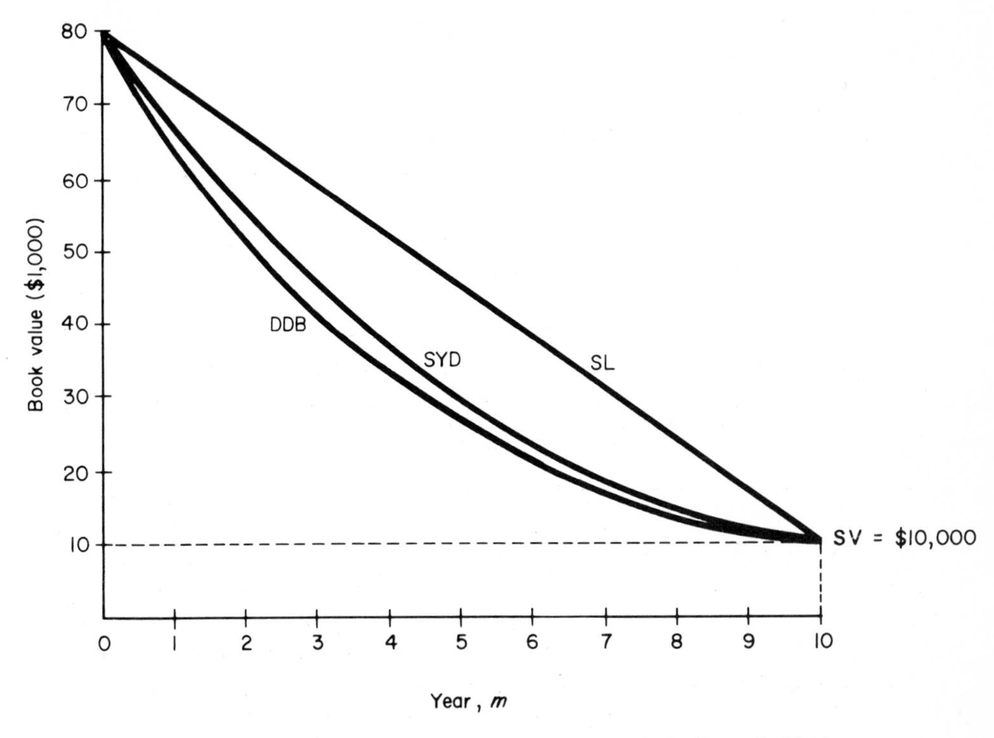

Figure 13.5 Book-value comparison for several depreciation methods, Example 13.11.

Table 13.9 Comparison of depreciation methods (P = \$80,000, SV = \$10,000, n = 10 years)

Year,	Straight-line		Sum-of-year digits		Double-declining balance	
t	D_t, (Eq. 13.1)	BV_t	D_t, (Eq. 13.4)	BV_t	D_t, (Eq. 13.9)	BV_t
0	...	\$80,000	...	\$80,000	...	\$80,000
1	\$7,000	73,000	\$12,727	67,273	\$16,000	64,000
2	7,000	66,000	11,455	55,818	12,800	51,200
3	7,000	59,000	10,182	45,636	10,240	40,960
4	7,000	52,000	8,909	36,727	8,192	32,768
5	7,000	45,000	7,636	29,091	6,553	26,214
6	7,000	38,000	6,364	22,727	5,243	20,971
7	7,000	31,000	5,091	17,636	4,194	16,777
8	7,000	24,000	3,818	13,818	3,355	13,422
9	7,000	17,000	2,545	11,273	2,684	10,738
10	7,000	10,000	1,273	10,000	738*	10,000

*Because of internal revenue tax laws, only \$738 can be claimed for depreciation.

Example 13.12 (*a*) Derive an inequality relation that will assist in predicting when the switch from DDB to SL depreciation should take place. (*b*) Use the relation to confirm the switch in year 6 found in Example 13.4*c*.

SOLUTION

(*a*) From Eqs. (13.16) through (13.18) the switch takes place in the first year t when

$$D_t^{\text{SL}} > D_t^{\text{DDB}}$$

$$\frac{\text{BV}_{t-1} - \text{SV}}{n - t + 1} > (d)\text{BV}_{t-1}$$

where d is the DDB rate of depreciation and SV is the anticipated salvage. Using Eq. (13.11) the BV_{t-1} values may be substituted.

$$\frac{P(1 - d)^{t-1} - \text{SV}}{n - t + 1} > dP(1 - d)^{t-1}$$

Division of both sides by P and some algebraic manipulations result in

$$\frac{\text{SV}}{P} < (1 - d)^{t-1}\,[1 - d(n - t + 1)]$$

or

$$\frac{\text{SV}}{P} < (1 - d)^{t-1}\,[d(t - 1) - 1] \qquad (13.21)$$

For a given SV, P, d, and n, the minimum t value to satisfy the inequality can be found easily by taking the logarithm of Eq. (13.21).

$$\log\frac{\text{SV}}{P} < (t - 1)\log(1 - d) + \log\,[d(t - 1) - 1] \qquad (13.22)$$

(*b*) From Example 13.4,

$$P = \$10{,}000 \qquad \text{SV} = \$500 \qquad n = 8 \qquad d = 0.25$$

Substitution into Eq. (13.22) for $t = 6$ results in

$$\log 0.05 < 5\log 0.75 + \log 0.25$$

$$-1.301 < -1.227$$

which is a correct statement, thus indicating that switching to the SL model at $t = 6$ is advantageous. However, if $t = 5$ is substituted into Eq. (13.22) the result is $\log 0.05 < 4\log 0.75 + \log 0$. Since $\log 0$ is undefined, the switch at $t = 5$ is not advised, since it is impossible to satisfy the inequality relation for any integer value of $t \leqslant 5$.

Sec. 13.5

Example 13.13 Consider again the forest acquisition discussed in Example 13.8. If after 2 years of operation total recoverable board feet is reestimated at 225 million, compute the new d'_t $(t = 3, 4, \ldots)$.

SOLUTION After 2 years, a total of $74,000 had been depleted; thus, the new d'_t value must be based on the remaining $350,000 - 74,000 = $276,000 of undepleted investment. Additionally, with the new estimate of 225 million, a total of 225 - 15 - 22 = 188 million board feet remain. Now, for $t = 3, 4, \ldots,$

$$d'_t = \frac{\$276{,}000}{188 \text{ million}} = \$1468 \text{ per million board feet}$$

Sec. 13.9

REFERENCES

1. *Economic Recovery Tax Act of 1981*, Commerce Clearing House, Chicago, 1981.
2. *Depreciation,* U.S. Internal Revenue Service Publication 534, published annually.
3. Blank, L., and Smith, D., "A Comparative Analysis of the Accelerated Cost Recovery System as Enacted by the 1981 Economic Recovery Tax Act," *The Engineering Economist,* 1982.

FURTHER INFORMATION

Depreciation switching (Sec. 13.5): Bussey [3], pp. 99–105; Taylor [16], pp. 337–341; White, Agee, and Case [18], pp. 206–207, 219–221.

PROBLEMS

P13.1 The ABC Company paid $52,000 for an asset in 1977 and installed it at a cost of $3000. The asset was expected to remain in service for 10 years and be sold for 10% of the original purchase price. If the asset was sold in 1982 for $8700, write the values used in depreciation analysis for the following: first cost, anticipated life; and salvage value, actual life, and market value in 1982; book value in 1982 if 75% of the first cost had been written off in depreciation.

P13.2 Smoothline Construction, Inc., has just purchased a Glop-a-de-glop machine for $275,000. A $75,000 installation charge is required to use the machine. The expected life is 30 years with a salvage value of 10% of the purchase price. Use the straight-line (SL) method of depreciation to determine (*a*) first cost, (*b*) salvage value, (*c*) annual depreciation, and (*d*) book value after 20 years.

P13.3 A machine costing $12,000 has a life of 8 years with a $2000 salvage value. Calculate the (*a*) depreciation charge and (*b*) book value of the machine for each year using the straight-line method. (*c*) What is the rate of depreciation for this method? Explain the meaning of this rate.

P13.4 Compare accelerated- and decelerated-depreciation methods to the straight-line method using the definition of rate of depreciation.

P13.5 Work Prob. P13.3 using the sum-of-year-digits (SYD) method.

P13.6 Earth-moving equipment having a first cost of $82,000 is expected to have a life of 18 years. The salvage value at that time is expected to be $15,000. Calculate the depreciation charge and book value for years 2, 7, 12, and 18 using the sum-of-year-digits method.

P13.7 If P = \$12,000, n = 6 years, and SV is 15% of P, use the SYD method to determine the (*a*) book value after 3 years, (*b*) rate of depreciation in year 4, and (*c*) depreciation amount in year 4 using the rate from *b* rather than Eq. (13.4).

P13.8 What is the basic difference between the declining-balance and the double-declining-balance methods of depreciation?

P13.9 Work Prob. P13.3 using the double-declining-balance (DDB) method and plot the book value for the SL, SYD, and DDB depreciation methods.

P13.10 A building costing \$320,000 is expected to have a 30-year life with a 25% salvage value. Calculate the depreciation charge for years 4, 9, 18, and 26 using the (*a*) straight-line method, (*b*) sum-of-year-digits method, and (*c*) double-declining-balance method.

P13.11 Calculate the book value of the building in Prob. P13.10 for year 13 using the (*a*) straight-line method, (*b*) sum-of-year-digits method, and (*c*) double-declining-balance method.

P13.12 Work Prob. P13.6 using (*a*) the double-declining-balance method and (*b*) the declining-balance method with a depreciation rate from Eq. (13.13). Compare the book values.

P13.13 Declining-balance depreciation at a rate of 1.5 times the straight-line rate is to be used for automated process-control equipment with P = \$175,000, n = 12, and expected SV = \$32,000. Compute (*a*) the depreciation and (*b*) the book value for years 1 and 12. (*c*) Compare the expected salvage value and the salvage used by the DB method.

Note: Many of the problems from P13.14 through P13.21 are efficiently solved with the program DEPSTRAT (Appendix E) or its equivalent.

P13.14 An asset has a first cost of \$45,000, a life of 5 years, and a \$3000 salvage value. Use the switching procedure from DDB to SL depreciation to maximize the present worth of depreciation. Let i = 18%.

P13.15 Rework Prob. P13.14 allowing a switch to either SL or SYD depreciation.

P13.16 Management of the Above Board Company has a new piece of machinery with P = \$110,000, n = 10 years, and SV = \$10,000. (*a*) Determine the depreciation schedule and present worth of depreciation at i = 12% using the SYD method for the first 5 years and the SL method for the last 5 years. (*b*) Was it a good idea to switch depreciation methods in year 5? When should the switch to SL have taken place? Why?

P13.17 The Zap Electric Company owns a building with a first cost of \$155,000 and an anticipated salvage of \$50,000 after 25 years. (*a*) Should the switch from DDB to SL depreciation be made? (*b*) For what rate of depreciation values of the DB method would it be advantageous to switch from DB to SL depreciation some time in the life of the building?

P13.18 A company car is purchased for \$9000 and is to be depreciated over 8 years then sold for an estimated \$750. (*a*) Start with DDB depreciation and perform an analysis to determine the maximum depreciation allowed each year if switching to SYD or SL is allowed at any time. Let i = 12%. (*b*) Repeat the analysis above, using the declining-balance method at a rate equal to 150% of the SL rate.

P13.19 If P = \$12,000, n = 8 years, and SV = \$800, (*a*) develop the depreciation schedule and present-worth value for the DDB method if i = 20%. (*b*) Allow switching to SL depreciation and develop the new schedule and present-worth value. (*c*) Plot the book values of the asset with and without switching on the same graph.

P13.20 Use a computer program to develop the depreciation schedules and present-worth value at i = 12% for the following methods: SL, SYD, DB at a rate of 1.25 times the SL rate, DDB and optimal switching from DDB to any other method. Let P = \$20,000, n = 5 years, and SV = \$2,000.

P13.21 Use Eq. (13.22) developed in Example 13.12 to determine when a switch from DDB to SL depreciation should occur in Prob. P13.19.

P13.22 Joe's Machine Shop purchased a \$30,000 asset in 1985 and must depreciate it using ACRS over 10 years. If the salvage is zero, compare (*a*) the plot of book value and (*b*) the present worth of depreciation at 20% for SL and ACRS depreciation over the 10 years.

P13.23 If P = \$45,000, SV = \$3000, and n = 5-year recovery period, use i = 18% to maximize the present worth of depreciation using the following methods: DDB-to-SYD switching (as done in Prob. P13.15) and ACRS using the after-1985 rates. Comment on the answers.

P13.24 An asset is purchased in 1986 for \$20,000. (*a*) If n = 5 years, use ACRS to compute the annual depreciation and present worth of depreciation. (*b*) Compare these results with DDB depreciation for n = 10. Use i = 15%.

P13.25 The ACRS method for 1985 with n = 5 years switches from 175% DB to SYD in year 2. (*a*) Use P = \$10,000 and SV = 0 to determine if the switch should take place at a time other than year 2 to maximize the present worth of depreciation at i = 15%. (*b*) Does the ACRS depreciation or the optimal 175% DB to SYD from part *a* give a larger value for present worth of depreciation?

***P13.26** A janitorial supply company purchases a new truck for \$26,500 in 1983. The owner wants to compute the 5-year ACRS depreciation and the maximum amount that can be capital expensed in 1983. Perform the analysis.

***P13.27** A cosmetic company acquires and uses a \$15,800 mixing machine in 1985. (*a*) What amount of capital expense deduction is allowed and for what year is it claimed? (*b*) Determine the depreciation schedule using ACRS if the asset is capital-expensed and has a 10-year recovery period.

***P13.28** Western Tool and Die owns a number of three different models of lathes. The values of P, SV, and n for each lathe are given below.

Model number	Number owned	First cost	Expected life	Salvage value
A	30	\$37,000	8	\$5,000
B	10	4,800	12	0
C	6	20,000	10	2,500

One-half the lathes of each model type will be sold after 8 years and the remainder will be replaced after 12 years. Compute the total annual straight-line depreciation charges using (*a*) the group depreciation method and (*b*) the composite depreciation method. (*c*) Compare the total annual depreciation charges for each method and compute the composite life based on the depreciation when all assets are owned by the company.

***P13.29** Rework Prob. P13.28 using the same data. However, calculate the *actually realized* annual depreciation by assuming that the company uses a life of 5 years for all models and the 1984 ACRS rates apply to all models.

***P13.30** A coal mining company has owned a mine for the past 5 years. During this time the following tonnage of ore has been removed each year: 40,000; 52,000; 58,000; 60,000; and 56,000 tons. The mine is estimated to contain a total of 2.0 million tons of coal and the mine had an initial cost of \$3.5 million. If the company had a gross income for this coal of \$15 per ton for the first 2 years and \$18 per ton for the last 3 years, (*a*) compute the depletion chages each year using the larger of the values for the two accepted depletion methods and (*b*) compute the percent of the initial cost that has been written off in these 5 years.

***P13.31** If the mine operation explained in Prob. 13.30 is reevaluated after the first 3 years of operation and estimated to contain a remaining 1.5 million tons, answer the two questions posed in Prob. P13.30.

***P13.32** Assume that the earth-moving equipment described in Prob. P13.6 is to be depreciated by the unit-of-production method. Total tons moved in a lifetime is based on an annual average of 150,000 tons per year. If the tonnage moved in the first 3 years is 200,000, 250,000, and 175,000 tons, respectively, compute the depreciation charge and book value for each year.

CHAPTER

FOURTEEN

BASICS OF TAXATION FOR CORPORATIONS

The two objectives of this chapter are first to give you a basic knowledge of tax definitions and tax rates as applied to corporations; and second to compare the depreciation methods of Chap. 13 from the tax viewpoint.

This chapter is only an introduction to the effect of taxes on corporate income and engineering-economy studies. Further analysis of tax considerations are presented in the next chapter. To perform an economy study without accounting for tax effects may be misleading, because taxes can reverse the before-tax decision. It is not necessary, however, to know all the details of corporation taxing in order to accomplish a realistic study with taxes considered. Of course, the introduction of taxes increases the complexity of the study, but they are a major consideration in any economic analysis.

We will investigate only a simplified version of corporate taxes in this text—a version useful to engineering economists. More detail is available in Refs. 1 to 3. Individual tax problems should be resolved by consulting the appropriate Internal Revenue Service (IRS) publication or personnel.

SECTION OBJECTIVES

To complete the material in this chapter, you must be able to do the following:

14.1. Define *gross income, taxable income, capital gain, recaptured depreciation, capital loss, investment tax credit, operating loss*, and *Section 1231* and *179 properties.*

14.2. Compute the income tax using the published tax rates or an effective tax rate, given taxable income and applicable tax rates.

14.3. Compute the resulting net capital gains or losses and the income tax, given the gain or loss values, federal tax laws for gains and losses, and applicable tax rates.
14.4. Compute the investment tax credit allowed on a qualified asset, given expected life, initial cost, and tax credit laws.
14.5. State the tax law as it concerns operating losses occurring in a particular year.
*14.6. Show the advantage of one depreciation method over another by computing the present worth of the taxes involved, given the depreciation methods, asset data, tax rate, and an after-tax rate of return.
*14.7. Show the advantage of a shorter expected life or recovery period by computing the present worth of the taxes involved, given the depreciation method, asset data, tax rate, and an after-tax rate of return.
*14.8. Compute an exaggerated before-tax rate of return, given the effective tax rate and after-tax rate of return.

STUDY GUIDE

14.1 Definitions Useful in Tax Computations

To help you better understand the tax rates and formulas discussed in this chapter some basic definitions are presented here.

Gross income. The total of all incomes from revenue-producing sources, including all items listed in the revenue section of an income statement. Refer to Appendix D for a review of accounting statements.

Expenses. All costs incurred while transacting business.

Taxable income. The dollar value remaining upon which taxes are to be paid, computed as follows:

$$\text{Taxable income} = \text{gross income} - \text{expenses} - \text{depreciation} \qquad (14.1)$$

Capital gain. A gain incurred when the selling price of depreciable property (assets) or real property (land) exceeds the purchase price (unadjusted basis). Thus, at sale time the computation is

$$\text{Capital gain} = \text{selling price} - \text{unadjusted basis}$$

where the capital gain > 0. If the sales date occurs within 1 year of purchase date, the capital gain is referred to as *short-term gain* (STG); if the ownership period is longer than 1 year, the gain is a *long-term gain* (LTG). An STG and an LTG are taxed differently.

Capital loss. If the selling price is less than the book value the loss is

$$\text{Capital loss} = \text{book value} - \text{selling price}$$

The terms *short-term loss* (STL) and *long-term loss* (LTL) are defined in a fashion similar to capital gains, that is, using a 1-year break point. The concept of sunk cost, briefly discussed in Sec. 10.1, results in a capital loss.

Recaptured depreciation (RD). If a depreciable property is sold for an amount greater than the current book value, the excess is depreciation recaptured by the sale and must be considered and taxed as ordinary taxable income, not as a capital gain. At sale time compute RD.

$$\text{RD} = \text{Selling price} - \text{book value}$$

where RD $\geqslant 0$. If selling price exceeds the unadjusted basis B, a capital gain is also incurred, and all previous depreciation claimed is considered recaptured. Since the Tax Act of 1981 implemented the Accelerated Cost Recovery System (ACRS) for depreciating assets to a zero salvage value in a specified recovery period, the recaptured depreciation (often called and taxed as a gain in economic analyses) or capital loss is determined relative to the current ACRS book value at sale time. If disposal occurs after the recovery period, the book value is zero, so the gain is the selling price (if it is positive). This interpretation is applied in the next chapter. It is possible that IRS rulings for gain and loss computations will have altered the situation by the time you study this material.

Investment tax credit. A tax credit given to the purchaser of new or used equipment that qualifies, that is, equipment that is tangible and integral to the production process or provides research or storage facilities for production. The tax credit is given to encourage the purchase and use of modern equipment.

Operating loss. When a corporation experiences a year of net loss rather than net profit, it has an operating loss. Special tax considerations are made in an attempt to balance the lean and fat years. Anticipation of operating losses, and thus the ability to take them into account in an economy study, is, of course, virtually impossible, but the tax treatment of past losses may be relevant in an analysis.

Section 1231 property. Corporation-owned property that is depreciated. The two important 1231 categories which encompass most assets involved in engineering economic analyses are:

1245 property. Depreciable property that is tangible and an integral part of manufacturing, production, extraction, or support services. Buildings and structural components are excluded.

1250 property. Depreciable buildings and structural components and other non-1245 property that is integral to the business of a corporation.

Section 179 property. All depreciable Section 1245 property with an ACRS recovery period of at least 3 years plus such assets as elevators, escalators, research facilities, and certain storage facilities.

Prob. P14.1

14.2 Basic Tax Formulas and Computations

Taxes are computed using the general relation

$$\text{Taxes} = (\text{gross income} - \text{expenses} - \text{depreciation})\, T \tag{14.2}$$

Table 14.1 Corporation tax rate schedule (effective for 1983 tax years)

(1) TI range	(2) Tax rate	(3) Maximum tax for this range	(4) Total TI	(5) = sum of (3) Total maximum tax charged
$1-25,000	15%	$ 3,750	$ 25,000	$ 3,750
$25,001-$50,000	18%	4,500	50,000	8,250
$50,001-$75,000	30%	7,500	75,000	15,750
$75,001-$100,000	40%	10,000	100,000	25,750
All over 100,000	46%	Unlimited	Above 100,000	Unlimited

where T is the tax rate. Since Eq. (14.2) uses the definition of taxable income (TI) from Eq. (14.1), we have

$$\text{Taxes} = (\text{TI})\,T \tag{14.3}$$

However, to give the small businesses a slight assist, corporate taxes are actually computed using the graduated tax-rate schedule in Table 14.1. You can see from this schedule that all taxable income above $100,000 is taxed at a rate of 46%, so a 46% *federal* tax rate may be used in engineering-economy studies but the tax liability will be somewhat overestimated. (These and other tax rates quoted in this text vary from year to year, but the general equations used for tax computations remain the same.)

Example 14.1 For a particular year, the Muche Company has a gross income of $2,750,000 with expenses and depreciation totaling $1,950,000. What is the amount of taxes to be paid for the year?

SOLUTION Compute the TI and taxes using the schedule in Table 14.1.

$$\text{TI} = 2{,}750{,}000 - 1{,}950{,}000 = \$800{,}000$$

$$\text{Taxes} = 25{,}750 + (800{,}000 - 100{,}000)(0.46) = \$347{,}750$$

COMMENT The graduated tax rate provides a savings of $20,250 compared with the $368,000 that would be paid in taxes if the 46% rate were applied to the entire $800,000.

It is important to realize that the total quantity which is classed as *depreciation* is deducted from gross income in computing TI. This is an advantage sought after by corporation-management tax people. More will be said about taxes and depreciation later in this chapter.

For the sake of simplicity, the tax rate used in an economy study is often a "one-figure" effective tax rate, which serves to account for federal, state, and city taxes. Commonly used effective tax rates are 50 or 52%, but the applicable rate in a particular case is easily approximated. One advantage in applying an effective rate is that state taxes are deductible from federal taxes. Thus, you can use the following relation to compute the effective tax rate as a decimal fraction

$$\text{Incremental effective rate} = \text{state rate} + (1 - \text{state rate})(\text{federal rate}) \quad (14.4)$$

where the federal rate of 46% is usually applicable, because TI is already above $100,000; that is, the incremental federal rate is 46%.

Example 14.2 Compute the income tax for the data presented in Example 14.1 using an effective tax rate, if the state rate is 8% and the company uses a 46% incremental federal rate.

SOLUTION First the effective rate is computed using Eq. (14.4):

$$\text{Effective rate} = 0.08 + (1 - 0.08)(0.46) = 0.5032$$

Now, by Eq. (14.3),

$$\begin{aligned}\text{Taxes} &= \text{TI}(\text{effective tax rate}) \\ &= 800{,}000(0.5032) \\ &= \$402{,}560\end{aligned}$$

COMMENT Do not compare this $402,560 with the results of Example 14.1, since the latter do not include a state tax.

Solved Problem 14.8
Probs. P14.2 to P14.7

14.3 Tax Laws for Capital Gains and Losses

Long- and short-term capital gains and losses as defined in Sec. 14.1 are usually treated according to the following schedule for corporations and not individual tax payers. (These rules are current with the 1982 tax year.)

Category	Effect
LTG	20%
STG	Taxed as ordinary income
LTL	Offset LTG
STL	Offset STG

The 20% LTG tax rate is called the alternative tax rate. It offers a tax advantage to corporations only if TI > $50,000, due to the graduated schedule in Table 14.1.

The exact procedure used to compute taxes in which capital gains and losses are involved is now described. A LTG is taxed at a rate of 20% of gain incurred for the year. A LTL can be used only to offset a LTG. Thus, if in a particular year LTG = $5500 and LTL = $3200, there is a net LTG = $2300 to be taxed. The treatment of STG and STL is similar, except that the net gain is taxed as *regular income*. Once the losses have been used to offset gains, the net result is obtained. This value, not the individual gain and loss values, is then used for tax computation. In any event, the net losses claimed in any one year cannot exceed the gains for the year. The net result of gain and loss offsetting is treated as follows:

Net LTG Reduce TI by LTG, compute tax on new TI, and compute tax on LTG at 20% (if original TI > \$50,000)

Net STG Tax as ordinary income

Net LTL or net STL Carry back 3 tax years and carry forward 5 tax years to offset capital gains in these years

Examples 14.3 and 14.4 will clarify the procedure.

When recaptured depreciation is involved the effective corporation tax rate is applied to this amount. In this section, the capital gain tax of 20% is assumed and no recaptured depreciation is included.

Example 14.3 A company has the following income, gain, and loss values for 1 year.

$$\begin{aligned} \text{Taxable income} &= \$50{,}000 \\ \text{STG} &= \quad 4{,}000 \\ \text{LTG} &= \quad 7{,}500 \\ \text{STL} &= \quad (500) \\ \text{LTL} &= (9{,}000) \end{aligned}$$

(Parentheses are used to indicate losses.) Compute the income taxes.

SOLUTION Using losses to offset gains, the following results are obtained:

LTG	\$ 7500	STG	\$4000
LTL	(9000)	STL	(500)
Net LTL	\$(1500)	Net STG	\$3500

Result: Net STG \$2000

Therefore, \$52,000 is actually taxed as ordinary income.

Taxable income	\$50,000
STG	2,000
Ordinary income	\$52,000

Income tax liability is determined using the Table 14.1 amount for the \$50,000 plus 30% of the excess.

$$\text{Tax} = 8250 + 0.30(52{,}000 - 50{,}000) = \$8850$$

Example 14.4 The Harvey Company has a TI of \$160,000 and the following gains and losses. Compute the income taxes for the year.

$$\text{LTG} = \$13{,}000 \qquad \text{STG} = \$2000$$

$$\text{LTL} = (3600) \qquad \text{STL} = (5400)$$

SOLUTION The net effect of the gains and losses is an LTG of \$6000 which comes from the following:

Long term	Short term
Net LTG = \$9400	Net STL = (\$3400)

The income tax is computed by reducing the TI by the net LTG of \$6000, applying the regular tax rates in Table 14.1 to the reduced TI and the 20% alternative tax to the net LTG.

$$\text{Actual taxable income} = 160{,}000 - 6000 = \$154{,}000$$

$$\text{Regular tax} = 25{,}750 + 0.46(154{,}000 - 100{,}000) = \$50{,}590$$

$$\text{LTG tax} = 0.20(6000) = \$1200$$

$$\text{Total taxes} = 50{,}590 + 1200 = \$51{,}790$$

COMMENT Since the 20% LTG tax is an alternative to the regular tax, the Harvey company can pay the smaller of the regular tax on \$160,000 of TI or the taxes above. Since the regular tax on \$160,000 is 25,750 + 0.46(60,000) = \$53,350, Harvey will use the alternative rate of 20% on the LTG. For TI $\leqslant$ \$50,000 the alternative tax does not reduce the total tax. Try this same problem using an original TI of \$50,000 to see the effect.

Special tax rules are applied to Section 1231 property after *each* asset's capital gain or loss has been determined for the year it is sold, retired, or disposed of. Group all gains and losses for such properties. If the grouped gains exceed the loss, treat each individual gain or loss as if it were derived from a long-term capital asset and therefore favorably taxed at the lower LTG rate or used to offset capital losses.

If the grouped loss exceeds the gain, treat each gain and loss as if it were *not* from a capital asset sale, so that in effect the net loss is an ordinary business expense which reduces taxable income. More information is available in Ref. 1 and other tax-law publications.

These LTG benefits and operating loss benefits are very important when corporation taxes are computed, but they are difficult to forecast and include in the engineering-economy analysis when an alternative is being evaluated.

Solved Problems 14.9 and 14.10
Probs. P14.8 to P14.11

14.4 Tax Laws for Investment Tax Credit

Corporations are encouraged to invest in modern equipment because of a tax advantage applicable in the purchase year of the asset. The advantage is termed the *investment tax credit*, first referred to in Sec. 13.7 of this book. As the name implies, it is a tax credit and in no way affects TI. The actual amount of the tax credit has varied between 7 and 10% of first cost in the past 10 years, and has been completely repealed at times, because it was considered inflationary. We will use a figure of 10% for the remainder of this chapter and text. Under the Tax Law of 1981, 100% of the tax credit is allowable only if the ACRS depreciation recovery period (Sec. 13.6) or life is more than 3 years, and 60% is allowed for 3-year assets. The tax-credit percentage is based on recovery periods used in the ACRS method and not on the expected useful life of an asset.

The investment tax credit can be applied only to a depreciable asset which is an integral member of the manufacturing process and has a recovery period and life of at least 3 years. (This type of asset is often called a Section 38 property.) To avoid confusion, we will assume that newly acquired assets are *not* qualified for the investment tax credit unless specifically stated otherwise.

Example 14.5 The following assets have been purchased within one tax year.

Asset number	Life, years	Cost, P
309	10	\$75,000
318	3	8,000

Compute the actual cash outflow if both qualify for a 10% investment tax credit. (No Section 179 capital expense deduction is to be claimed.)

SOLUTION Neglecting all other factors, we find that the first-cost cash outflow is \$83,000. The tax credits are 10% for asset 309 and 0.6(10%) = 6% for asset 318. Tax credit amounts and actual cash outflows are

Asset 309: 0.10(75,000) = \$7,500

Asset 318: 0.06(8000) = \$ 480

Cash outflow = 83,000 − (7500 + 480) = \$75,020

It is assumed that the investment tax credit is creditable at the time of asset purchase, that is, year 0. The tax credit that can be claimed in a year is limited by the income tax liability or some specified amount for each year. If the credit exceeds this amount, the unclaimed portion can be carried back for 3 and forward for 15 succeeding tax years. In addition, this tax credit does not reduce the total depreciable value of an asset, but it must be taken on the reduced first cost if a Section 179 capital expense (Sec. 13.7) is claimed. Thus in Example 14.5, the full first cost can still be written off since no Section 179 expense was claimed.

Probs. P14.12 to P14.14

14.5 Tax Laws on Operating Losses

We have discussed situations in which capital losses and unclaimable tax credits can be carried backward and forward for several tax years. Another important tax advantage is the provision that allows an operating loss to be carried backward for 3 and forward for 15 years until the loss is completely exhausted. The number of years allowed for carry-back and carry-forward may vary, but the amount of operating loss claimed in any one year cannot exceed taxable income. Since only the amount of the loss is recoverable, this and all carry-back–carry-forward laws present a question of strategy, that is, *when* to utilize the tax advantage.

Prob. P14.15

*14.6 Tax Effects of Different Depreciation Models

This section is designed to give you an idea of how ACRS, SL, and other depreciation models affect the taxes that must be paid. Here we assume that some after-tax rate of return is to be used for the analysis. Comparison of different depreciation methods shows that for a constant tax rate, annual gross income greater than or equal to annual depreciation, and write-off down to the same salvage value, the following statements are always correct:

1. The total taxes paid are equal for any depreciation model.
2. The present worth of taxes are *less* for accelerated depreciation methods.

Example 14.6 will illustrate these points.

Terminology commonly used henceforth will be as follows:

CFBT Cash flow before taxes
CFAT Cash flow after taxes

In Eq. (14.1) CFBT is gross income minus expenses, so

$$\text{TI} = \text{CFBT} - \text{depreciation} \tag{14.5}$$

Then

$$\text{CFAT} = \text{CFBT} - \text{taxes} \tag{14.6}$$

Example 14.6 Assume that an asset with the following characteristics has been purchased.

$$P = \$50{,}000 \qquad \text{CFBT} = \$20{,}000 \text{ per year} \qquad n = 5 \text{ years}$$

Table 14.2 Tax computation using the 1985 ACRS depreciation rates

		Depreciation			
Year	CFBT	Rate	Amount	TI	Taxes
0	\$−50,000				
1	+20,000	0.18	\$ 9,000	\$11,000	\$ 5,500
2	+20,000	0.33	16,500	3,500	1,750
3	+20,000	0.25	12,500	7,500	3,750
4	+20,000	0.16	8,000	12,000	6,000
5	+20,000	0.08	4,000	16,000	8,000
			\$50,000		\$25,000

If the effective tax rate is 50% and an after-tax rate of return of 8% is used, compare (*a*) 1985 ACRS, (*b*) ACRS straight-line election, and (*c*) sum-of-year-digits depreciation from the tax viewpoint.

SOLUTION

(*a*) The 1985 ACRS depreciation rates are given in Table 13.3. Table 14.2 presents the depreciation and TI values using Eq. (14.5). Taxes are computed by Eq. (14.3).

$$\text{Taxes} = T(\text{TI}) = 0.5\text{TI}$$

Figure 14.1 presents a *tax* cash-flow diagram for the total taxes of \$25,000. Present worth of taxes P_{tax} calculated below the cash-flow diagram is \$19,424.28.

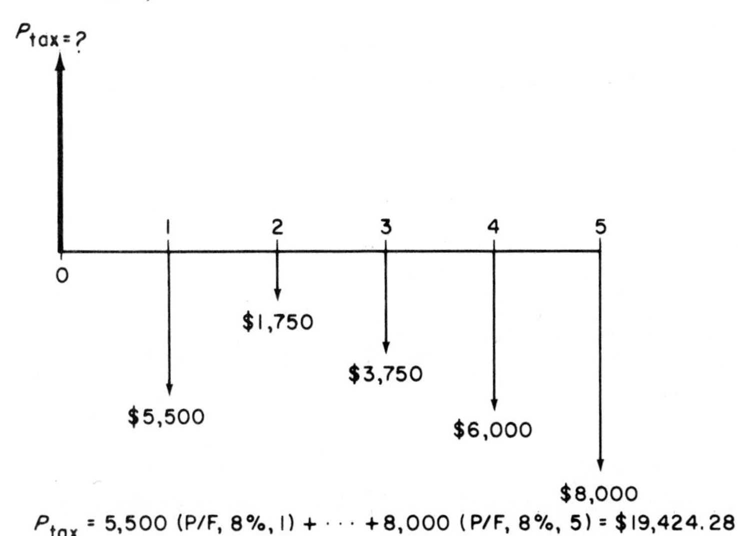

Figure 14.1 Tax cash flow using 1985 ACRS depreciation, Example 14.6.

(*b*) If the ACRS straight-line election is taken, a recovery period of 5, 12, or 25 years is allowed. Using 5 years, the depreciation is (half-year convention neglected)

$$D_t = \frac{50{,}000}{5} = \$10{,}000 \qquad (t = 1, 2, \ldots, 5)$$

Tax computations are given in Table 14.3 and Fig. 14.2 presents the tax cash flow and P_{tax} = \$19,963.50. The SL method gives a P_{tax} larger than ACRS (19,424.28), since depreciation accumulates faster because of the switch to the SYD method embedded in ACRS for years 2 through 5.

(*c*) If SYD depreciation for 5 years were used, the results in Table 14.4 and Fig. 14.3 would apply. The total taxes are again \$25,000, but the tax cash flows are smaller in the early part of the life (especially compared with ACRS in the first year) owing to the larger depreciation values. Therefore, the present worth of taxes, as shown in Fig. 14.3, is \$18,944.96, which is less than P_{tax} for straight-line depreciation and for 1985 ACRS depreciation.

COMMENT From this illustration it is clear that any depreciation that writes off faster than does straight-line depreciation gives a tax advantage, since the present-worth value will be greater than that for straight-line depreciation. This is true, since

Table 14.3 Tax computation for asset using ACRC straight-line depreciation

Year	CFBT	Depreciation	TI	Taxes
0	\$-50,000			
1-5	+20,000	\$10,000	\$10,000	\$5,000

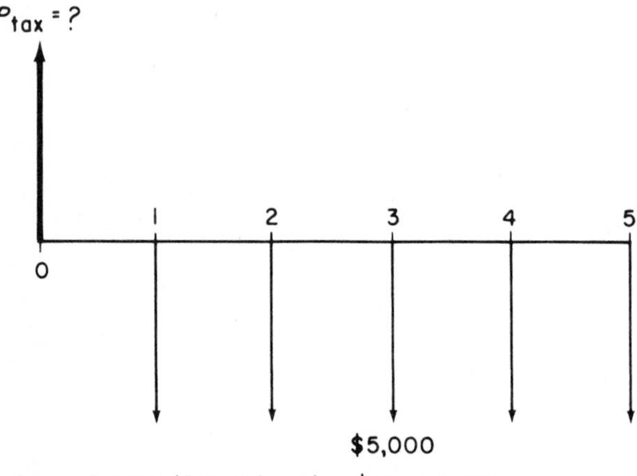

Figure 14.2 Tax cash flow using straight-line depreciation for a 5-year recovery period.

Table 14.4 Tax computation using sum-of-year-digits depreciation

Year	CFBT	Depreciation*	TI	Taxes
0	\$-50,000			
1	+20,000	\$16,667	\$ 3,333	\$ 1,667
2	+20,000	13,333	6,667	3,333
3	+20,000	10,000	10,000	5,000
4	+20,000	6,667	13,333	6,667
5	+20,000	3,333	16,667	8,333
		\$50,000		\$25,000

*From Eq. (13.4), $D_t = [(5 - t + 1)/15](50,000)$, for $t = 1, 2, \ldots, 5$.

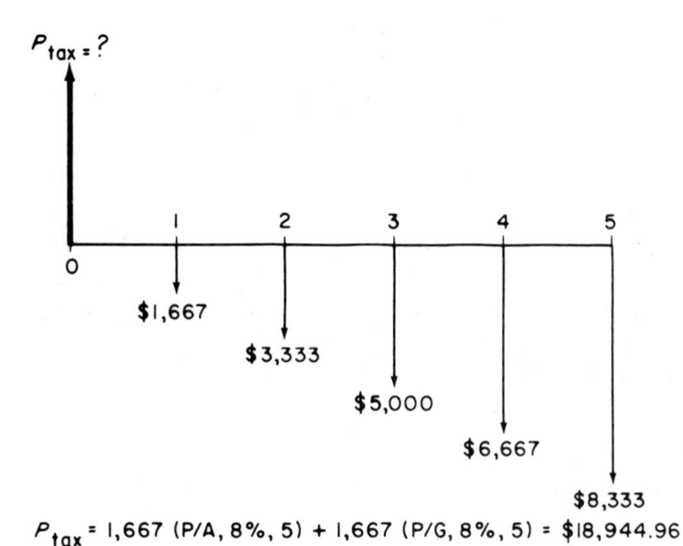

$$P_{tax} = 1{,}667\ (P/A, 8\%, 5) + 1{,}667\ (P/G, 8\%, 5) = \$18{,}944.96$$

Figure 14.3 Tax cash flow using sum-of-year-digits depreciation.

with rapid write-off, the taxes are "moved" to the later years of ownership, which decreases their present worth when compared with the constant taxes applicable for straight-line depreciation. An example with another depreciation method is presented in Solved Problems.

Solved Problem 14.11
Probs. P14.16 to P14.25

*14.7 Tax Effects of Different Recovery Periods

If in lieu of ACRS the straight-line method is elected, as discussed in Sec. 13.6, different n values may be selected as follows:

ACRS recovery period, years	Different n value allowed
3	3, 5, or 12
5	5, 12, or 25
10	10, 15, or 35
15	15, 35, or 45

For a specified after-tax rate of return, a constant tax rate and the same depreciation method, a lower n value will offer a tax advantage. Comparison of taxes for different n values will show the following:

1. The total taxes paid are equal for all n values.
2. The present worth of taxes are *less* for smaller n values.

Example 14.7 demonstrates these points for the SL method, but this can be done for each method of depreciation.

Example 14.7 The Know-It-All Manufacturing company uses the common practice of keeping two sets of books on depreciable assets, one for its own internal use and one for tax purposes. The company owns an asset with $P = \$9000$ and actual $n = 9$ years; however, a recovery period of 5 years is specified for this equipment. Show the tax advantage afforded the company by the lower n if CFBT = \$3000 per year, a tax rate of 50% applies, invested money is returning 10% after taxes, and straight-line depreciation is used. Neglect the effect of any salvage value.

SOLUTION Compute the present worth of taxes for both n values and compare to see if there is a tax advantage for $n = 5$ over $n = 9$ years.

$n = 9$ years:

$$D_t = \frac{9000}{9} = \$1000 \qquad t = 1, 2, \ldots, 9$$

$$\text{TI} = 3000 - 1000 = 2000 \text{ per year}$$

$$\text{Taxes} = 2000(0.50) = 1000 \text{ per year}$$

$$P_{\text{tax}} = 1000(P/A, 10\%, 9) = \$5759$$

$n = 5$ years:

$$D_t = \begin{cases} \dfrac{9000}{5} = \$1800 & t = 1, 2, 3, 4, 5 \\ 0 & t = 6, 7, 8, 9 \end{cases}$$

TI and taxes are shown in Table 14.5.

$$P_{\text{tax}} = 600(P/A, 10\%, 5) + 1500(P/A, 10\%, 4)(P/F, 10\%, 5)$$
$$= \$5226.77$$

Table 14.5 Tax computation for shortened life $n = 5$, Example 14.7

Year	CFBT	Depreciation	TI	Tax
0	$-9000			
1-5	+3000	$1800	$1200	$ 600
6-9	+3000	0	3000	1500
				$9000

Note that a total of $9000 in taxes is paid for both the 9- and 5-year period. However, the more rapid write-off allowed for $n = 5$ results in a present-worth tax saving of $532.23.

COMMENT As mentioned in Sec. 13.6, there is a half-year convention necessary when claiming SL depreciation in lieu of ACRS. We have neglected this computational detail for the sake of simplicity, but it allows only a half year of SL depreciation in the first year. However, this amount is recovered by taking it in the year following the end of the recovery period. The net effect is a longer recovery period for the SL method with a tax disadvantage in the first year.

Probs. P14.26 to P14.28

*14.8 Using an Exaggerated Before-Tax Rate of Return

If an engineering economist does not wish to be bothered with tax considerations, he or she may choose to increase the rate of return used in the before-tax study to include an approximation of the tax effect. A simple relation to use is

$$\text{Before-tax rate of return} = \frac{\text{after-tax rate of return}}{1 - \text{effective tax rate}} \tag{14.7}$$

Thus, if the effective tax rate is 52% and a 6% after-tax return is required, the study could be performed using a before-tax rate of return of $0.06/(1 - 0.52) = 0.125$, or 12.5%. Use of a before-tax rate of return is the reason why most of the rates employed prior to this chapter are in the 12 to 20% range.

Probs. P14.29 to P14.31

SOLVED PROBLEMS

Example 14.8 The Wa-Out Fruit Company is considering the purchase of one of two new lines of automatic sorter, cleaner, and packer machines. Details of each machine are as follows:

Wonderama		Spee-de Fix	
P =	\$145,835	P =	\$225,000
Expenses =	30,000	Expenses =	10,000
Expected SV =	15,000	SV =	0
n =	5	n =	5

If the effective tax rate is $T = 0.52$, straight-line depreciation is used with the SV neglected, and gross income is expected to be the same from either machine, compute the tax benefit for one of the assets.

SOLUTION First compute the taxes for each asset and then determine the tax difference between the two. Since the gross income (GI) is the same for both assets, the tax benefit goes to the asset with lower taxes. Annual depreciations for Wonderama (D_w) and Spee-de Fix (D_s) with salvage neglected are

$$D_w = \frac{145{,}835}{5} = \$29{,}167$$

$$D_s = \frac{225{,}000}{5} = \$45{,}000$$

We can set up a tax relation for each asset from Eq. (14.2):

$$\text{Taxes}_w = (\text{GI} - \text{expenses} - D_w)\,T = (\text{GI} - 30{,}000 - 29{,}167)\,(0.52)$$

$$\text{Taxes}_s = (\text{GI} - 10{,}000 - 45{,}000)\,(0.52)$$

If we use Wonderama as a basis for *tax difference* between the two assets,

$$\begin{aligned}\text{Taxes}_w - \text{taxes}_s &= (\text{GI} - 30{,}000 - 29{,}167 - \text{GI} + 10{,}000 + 45{,}000)\,(0.52)\\ &= (-4167)\,(0.52)\\ &= -\$2167\end{aligned}$$

The negative sign indicates that the asset which is *not* the base pays more taxes. So the Wonderama has a tax benefit of \$2167 per year.

COMMENT Logically, the asset with more deductions in computing TI will have the advantage. Since Wonderama has \$59,167 in deductions and Spee-de Fix has \$55,000, the former has the tax advantage.

Sec. 14.2

Example 14.9 The tax rates used prior to those in Table 14.1 were as follows: 22% normal rate of all TI and 26% surtax rate on TI > \$25,000. The new rates were developed to give smaller businesses a lower tax burden. Rework Example 14.3 using the old rates and compare the resulting tax figures.

SOLUTION From Example 14.3 the total TI is \$52,000 and the current tax is \$8850. The old tax computation is

$$\text{Taxes} = 52{,}000(0.22) + (52{,}000 - 25{,}000)(0.26)$$
$$= \$18{,}460$$

The decrease in tax is 52%–a significant amount. Of course, the old tax structure placed an effective 48% tax on all TI over \$25,000 and the new structure provides for an effective tax of 46% on all TI over \$100,000, so the tax break for larger businesses is much smaller.

Sec. 14.3

Example 14.10 Assume that the following losses and gain occur in one year:

$$\text{STL} = \$(3000)$$
$$\text{LTL} = \quad (500)$$
$$\text{STG} = \quad 1000$$

How would the resulting net gain or loss be treated in tax computation?

SOLUTION The result is a net capital loss of \$2500. Tax law provides that losses only up to the amount of gains can be deducted from taxable income. Therefore, of the \$2500 loss only \$1000, the amount of the gain, can be taken from TI to reduce taxes. Tax laws, however, do allow the undeducted difference of losses and gains to be carried back 3 years as a STL or forward 5 years, if not completely absorbed in carry-back. Therefore, the remaining \$1500 loss may be used to offset capital gains over a maximum 9-year period.

Sec. 14.3

Example 14.11 As another illustration of the tax advantage for a fast write-off depreciation model, compute the present worth of taxes for Example 14.6 using the double-declining-balance (DDB) method and an effective tax rate of 50%. There is no salvage for this asset.

SOLUTION The depreciation rate for DDB is $d = 2/n = 2/5 = 0.40$. Table 14.6 details the taxes for the life of the asset. The loss in year 5 indicated by a negative TI is due to the implied salvage value of \$3888 from DDB depreciation. This results in a capital loss and a tax credit shown as a negative tax in Table 14.6. This is not carried backward or forward (as is permitted by the tax laws) in this example because there are many different strategies allowed by the tax laws and the purpose of this example is to illustrate the tax advantage of rapid write-off depreciation. From Table 14.6 the present worth of the taxes is

$$P_{\text{tax}} = (4000)(P/F, 8\%, 2) + 6400(P/F, 8\%, 3) + \cdots + (-1{,}944)(P/F, 8\%, 5)$$
$$= \$18{,}872.78$$

Table 14.6 Tax computation using double-declining-balance depreciation, Example 14.11

Year	CFBT	Depreciation	Book value	TI	Taxes
0	$-50,000	...	$50,000		
1	+20,000	$20,000	30,000	$ 0	$ 0
2	+20,000	12,000	18,000	8,000	4,000
3	+20,000	7,200	10,800	12,800	6,400
4	+20,000	4,320	6,480	15,680	7,840
5	+20,000	2,592	3,888	17,408	8,704
5	0	0	3,888	-3,888	-1,944
		$46,112			$25,000

which is less that the comparable value for ACRS, straight-line, and sum-of-year-digits methods in Example 14.6. If the asset has no salvage value, DDB would give the lowest tax present worth for this asset.

COMMENT If the asset is sold for the implied DDB salvage value of $3888, no recaptured depreciation is involved and P_{tax} = $20,195.86, which is considerably higher than the 1985 ACRS value of P_{tax} = $19,424.28.

Sec. 14.6

REFERENCES

1. *U.S. Master Tax Guide*, Commerce Clearing House, Chicago, published annually.
2. *Tax Information on Corporations*, U.S. Internal Revenue Service Publication 542, published annually.
3. *Sales and Other Dispositions of Assets*, U.S. Internal Revenue Service Publication 544, published annually.

PROBLEMS

P14.1 The situations listed below were recorded for a dairy, the Pure Milk Company, in the past year. For each situation state which of the following is involved: gross income, taxable income, recaptured depreciation, capital gain, capital loss, investment tax credit, or operating loss.

(*a*) An asset with a book value of $8000 and a first cost of $15,000 was retired and sold for $8450.

(*b*) An artificial milk-making machine was purchased and had a first-year depreciation of $9600.

(*c*) The company estimates that it will report a $-75,000 net profit to IRS on the tax return.

(*d*) The asset in part *b* will have a $4200-per-year interest cost.

(*e*) An asset that had a life of 8 years has been owned for 14 years and has a book value of $800. It was sold this year for $275.

(*f*) The cost of goods sold in the past year was $468,290.

P14.2 Two small businesses have the following data on their tax returns:

	Dough Company	Broke Company
Sales	$1,500,000	$820,000
Interest revenue	31,000	25,000
Expenses	754,000	591,000
Depreciation	48,000	54,000

If both concerns do business in the state of No-Taxes, compute the federal income tax for the year, using the federal tax rates.

P14.3 Compute the taxes for the situation in Prob. P14.2 using an effective rate of 46% for the entire TI. What percentage reduction in taxes is allowed by the graduated tax rate?

P14.4 A-to-Z Car Dealers will have a $250,000 taxable income (TI) this year. If an advertising campaign is initiated, the TI is estimated to increase to $290,000 for the year. Neglecting any state and local taxes, use the federal tax rates to compute the (*a*) *effective* federal tax rate on TI = $250,000, (*b*) *effective* federal tax rate on only the additional taxable income, (*c*) *effective* federal tax rate on the entire $290,000 taxable income, and (*d*) after-tax profit on the additional taxable income.

P14.5 A company has a gross income of $3.9 million for the year. Depreciation and all expenses amount to $2.45 million. If the combined state and local tax rate amounts to 6.5% and an effective federal rate of 46% is applicable, compute the income taxes using the effective tax-rate formula.

P14.6 Wallbanger Contractors reported a taxable income of $80,000 last year. If the state tax rate is 8%, compute the (*a*) *federal* effective tax rate, (*b*) overall effective tax rate, and (*c*) total taxes to be paid by the company.

P14.7 The taxable income for a small partnership business is $150,000 this year. An effective tax rate of 46% is used by the owners. Investment in some new machinery was considered in the first quarter of the year. The equipment would have cost $35,000, have a life of 5 years, be salvaged for an estimated $5000, and be written off using the 1983 ACRS rates. The purchase would have increased taxable income by $10,000 and expenses $1000 for the year. Compute the change in income taxes for the year if the purchase had been made.

P14.8 The following capital gains and losses are present this year for a small clothing manufacturer.

LTG = $2800

LTL = 500

STL = 2000

If TI = $80,000, compute the income taxes the company must pay if state taxes are 5% and regular federal rates apply.

P14.9 Compute the recaptured depreciation, capital gains and losses for all the asset transactions below and use them to compute the annual income taxes. Total sales for the year were $80,000, while expenses and accumulated depreciation amounted to $39,400.

(*a*) A 3-year-old straight-line-depreciated asset was sold for 0.68*P*. The asset had a first cost of $50,000, no salvage value, and a life of 10 years.

(*b*) A machine that was only 5 months old was replaced because of its extreme technological obsolescence. The asset had *P* = $10,000, SV = $1000, *n* = 4 years, and was depreciated by the sum-of-year-digits method. The trade-in deal allowed the company $8000 on a new machine. (Use 50% of annual depreciation for the 5-month period.)

(*c*) Land purchased 4 months ago for $8000 was sold for a 10% profit.

(*d*) A 23-year-old asset was sold for \$500. When purchased the asset was entered on the books with P = \$18,000, SV = \$200, n = 20 years. Straight-line depreciation was used for the life of the machine.

(*e*) A capital loss of \$22,000 incurred 4 years ago is not completely exhausted. A total of \$3500 remains on the books. The company's tax specialists want to remove one-half this loss from the accounts this year.

P14.10 Referring back to Prob. P14.7, assume that the asset purchase had been made, but in December, due to lagging sales, the equipment had to be sacrificed for \$28,000. Compute the change in income taxes for the year as a result of this asset sale.

P14.11 Four similar assets were purchased 7 years ago (1976) and written off by the group method (Sec. 13.8) using straight-line depreciation. Details of the asset purchases are given below:

Asset number	First cost	Expected life, years	Salvage value
108	\$25,000	5	\$3,000
109	25,000	5	3,000
110	25,000	5	3,000
111	25,000	5	3,000

Details of asset disposal and yearly taxable income *before depreciation* are as follows:

Year	Asset sold	Actual salvage	Taxable income
1976	...	...	\$170,000
1977	...	...	190,000
1978	109	\$15,000	175,000
1979	108	16,500	210,000
1980	...	...	130,000
1981	110	1,300	90,000
1982	111	3,500	150,000

If the company uses an effective tax rate of 50%, compare the total taxes for the years 1976 through 1982 under these conditions: (*a*) assets were group-depreciated as explained; (*b*) each asset was individually straight-line-depreciated and gains or losses accounted for in the applicable year. Tax recaptured depreciation as a capital gain in this problem only.

P14.12 Rework Prob. P14.7 assuming that the machinery qualifies for an investment tax credit.

P14.13 Use the assets described in Example 14.5 to determine the first year's income taxes, if taxable income before removal of depreciation is \$250,000 and an effective federal tax rate of 46% and a state tax rate of 10% are used in economy studies. Assume 1985 ACRS depreciation is used for both assets.

P14.14 Determine the investment tax credit for the following assets:

	First cost	Recovery period, years	Salvage value
(*a*)	\$10,000	10	\$ 750
(*b*)	5,500	5	0
(*c*)	7,000	3	1,000
(*d*)	9,500	3	3,000
(*e*)	1,800	5	200

P14.15 An operating loss of $25,000 was incurred by the Bozo Health Spa in 1983. The effective tax rate is 50% and taxable income from 1980 to 1982 has been $110,000, $90,000 and $50,000, respectively. If the operating loss will be carried back only, compute the present worth in the year 1980 of the resulting taxes paid in 1980, 1981, and 1982 for the seven plans specified below (i = 10%):

(*a–c*) Recover the entire loss in 1980, 1981, or 1982.

(*d–f*) Recover one-half the loss in each of two of the years 1980, 1981, or 1982.

(*g*) Recover one-third the loss in each of the years 1980, 1981, and 1982.

***P14.16** Assume that the Wa-Out Fruit Company of Example 14.8 purchased the Wonderama sorter, cleaner, and packer machine. If annual income is expected to be $100,000 from this new asset and the effective tax rate is 48%, determine the percentage difference in taxes obtained by using (*a*) sum-of-year-digits and straight-line depreciation and (*b*) declining-balance with d = 0.15 and sum-of-year-digits methods. Take the time value of money into account by using a rate of return of 10%. Constructing the tax cash-flow diagram for each method will be helpful. Use the salvage value.

***P14.17** What is the difference in the present worth of total tax paid for the following situation? Asset A can be purchased to produce a CFBT of $65,000 or five of asset B can be purchased to produce the same CFBT. The interest rate is 12% and after-1985 ACRS or elected SL depreciation over 5 years is used, as indicated below. Neglect any capital gain or loss or recaptured depreciation at sale time and neglect the half-year depreciation convention for the SL election.

	Asset A	Five of asset B
Total first cost	$250,000	$260,000
Total salvage value	25,000	25,000
Total annual CFBT	65,000	65,000
Depreciation method	ACRS	SL
Tax rate	50%	50%
Life, years	5	5

***P14.18** Rework Prob. P14.17 if asset A is depreciated by the DDB method and asset B uses the SYD method. Use the stated salvage values.

***P14.19** An asset costing $45,000 has a life of 5 years, a salvage value of $3000, and an anticipated CFBT = $15,000 per year.

(*a*) Use the 1984 ACRS rates and compute the present worth of taxes.

(*b*) Determine the depreciation schedule for SL and DDB and for switching from DDB to SL to maximize depreciation. (This is the same as Prob. P13.14.) Use i = 18% and a tax rate of 50% to determine how the present worth of taxes decreases when ACRS or switching is used. (Assume that the asset is sold for $3000 in year 6 and that any negative TI or capital loss at sale time is a tax advantage.)

***P14.20** (*a*) Construct a cash-flow diagram of taxes for a $9000 5-year ACRS-recovery asset purchased in 1982 if the effective tax rate is T = 40% and CFBT = $10,000 per year.

(*b*) Calculate the present worth of taxes at i = 12%.

***P14.21** In 1983 the new manager of Hominy Grits Railroad had to decide whether to buy a $25,000 3-year-recovery-period asset or a $40,000 5-year asset. Both performed the same function and were to be written off by the ACRS method. Use T = 0.50 and i = 10% to determine which asset has a smaller present worth of taxes over the respective recovery period if expected CFBT = $10,000 per year for both machines.

***P14.22** If the CFBT is not considered, it is possible to compute the *tax savings* TS_t in year t due to depreciation alone using

$$TS_t = (\text{tax rate})\ (\text{depreciation}) = T(D_t)$$

The present worth of tax savings P_{TS} increases for accelerated depreciation methods and when switching between methods is allowed. Compute the present worth of tax savings for the depreciation schedules in Prob. P14.19.

***P14.23** (*a*) Compute the present worth of taxes for the depreciation schedules in Example 13.4 using $i = 15\%$, $T = 50\%$, and CFBT = $2000 per year. (Assume that any negative TI or capital loss at sale time is a tax advantage.)

(*b*) Compute the present worth of tax savings (see Prob. P14.22) due to depreciation only for the same example and compare the values.

***P14.24** The 1986 ACRS rates and the SL depreciation election are being considered for an asset with P = $10,000 and n = 5-year recovery. Use CFBT = $4000 to select the better method based on the present worth of taxes. Let $T = 0.46$ and $i = 20\%$. Neglect the half-year convention.

***P14.25** Use DEPSTRAT or the program developed for Prob. P13.20 to determine the present worth of (*a*) depreciation and (*b*) tax savings if $T = 46\%$ for the data in Prob. P13.20. (*c*) Compute the present worth of taxes if CFBT = $10,000 per year. (*d*) Select the best method of depreciation using these results.

***P14.26** Once again, review the situation discussed in Example 14.6. Compute the n value for the straight-line election that would make its present worth of taxes equal to that for the 1985 ACRS method for $n = 5$ years.

***P14.27** The Whiz Bang Computer Software Company has just bought an asset that has P = $88,000, SV = $8000, and $n = 10$ years. For tax purposes, the company is allowed to (*a*) use a life of 5 or 10 years and depreciate by straight line (salvage neglected) or (*b*) use a period of 5 years and depreciate by the 1986 ACRS method. If the interest rate is 10%, expected CFBT is $25,000 per year for 10 years, and effective tax rate is 52%, determine what n value and depreciation method should be used to minimize the time value of taxes under the following condition: all operating losses and capital losses are treated as a tax advantage and recaptured depreciation is neglected.

***P14.28** A new asset which cost $6000 is expected to last 4 years and produce a CFBT = $3000 for 4 years only. The asset can be straight-line-depreciated over 3 or 5 years. If the tax rate is 50% and $i = 15\%$, use the minimum tax present worth to select the 3- or 5-year recovery period. Assume that any depreciation in excess of cash flow is a tax advantage.

***P14.29** Compute the before-tax rate of return for Prob. P14.5 if an after-tax rate of return of 8% is required.

***P14.30** Compare the plans in Prob. P14.17 by the EUAC method with the (*a*) after-tax return and (*b*) computed inflated before-tax return. (*c*) Compare the answers using the after-tax and before-tax returns.

***P14.31** If a company uses a before-tax rate of return of 19% and an after-tax return of 8%, what percent of income is assumed to be absorbed in taxes?

CHAPTER

FIFTEEN

AFTER-TAX ECONOMIC ANALYSIS

The objective of this chapter is to study in more detail the effects of taxes on an engineering-economy study. Rather than attempt to utilize an inflated before-tax rate of return in this chapter, sufficient detail of the income tax effect is considered so that the rate of return reflects the situation after taxes.

SECTION OBJECTIVES

To complete this chapter you must be able to do the following:

15.1. Compute and tabulate cash flow after taxes, given the cash flow before taxes, effective tax rate, and depreciation method.
15.2. Select the better of two plans using present-worth or EUAC analysis, given details of the plans, the tax rate, and the after-tax rate of return.
15.3. Compute the after-tax rate of return for one asset or the breakeven rate of return for two competing assets, given details of the asset(s), the depreciation method, and the effective tax rate.
15.4. Select between a challenger and defender using after-tax replacement analysis, given the challenger and defender plans, the defender market value, depreciation methods, before- and after-tax rates of return, and the effective tax rate.
*15.5. Compute the tax effect on economy studies of the tax laws dealing with depletion, given the applicable depletion laws and details of the plans.

STUDY GUIDE

15.1 Tabulation of Cash Flow after Taxes

It is vital that you be able to correctly tabulate cash flow after taxes (CFAT) so that the present-worth, EUAC, or rate-of-return computations reflect the correct after-tax situation. The formulas to be used are no different from those used previously; however, they are reviewed here for easy reference. We will consistently use the abbreviations CFBT and CFAT for cash flow before taxes and cash flow after taxes, respectively, to conserve space. Continued use of abbreviations TI for taxable income and T for tax rate is made.

$$\text{CFBT} = \text{gross income} - \text{expenses} \quad (15.1)$$

$$\text{TI} = \text{CFBT} - \text{depreciation} \quad (15.2)$$

$$\text{Taxes} = \text{TI}(T) \quad (15.3)$$

$$\text{CFAT} = \text{CFBT} - \text{taxes} \quad (15.4)$$

If borrowed funds (debt financing) are used to finance a venture, the associated interest is tax deductible and Eq. (15.2) for TI will reflect this. The aspect of debt and equity financing is introduced in the Solved Problems and discussed thoroughly in Chap. 18.

If the numerical value of Eq. (15.3) is negative, it will be assumed that this negative tax will affect other taxes for the same year attributable to other income-producing assets owned by the company. This simplifying procedure is used in lieu of the carry-back-carry-forward tax laws (Sec. 14.5), which may be more advantageous. This assumption is illustrated in Example 15.2.

Example 15.1 A proposal has been made that a new piece of equipment be purchased this year. Characteristics of the purchase plan are:

$$P = \$50{,}000$$

$$\text{SV} = 0$$

$$n = 5 \text{ years}$$

$$\text{Expected income} = 28{,}000 - 1000k \qquad (k = 1, 2, 3, 4, 5)$$

$$\text{Expected disbursements} = 9500 + 500k$$

If the effective tax rate is 40% on this type of asset and if straight-line depreciation is used, tabulate the CFAT.

SOLUTION Table 15.1 details all tax information and CFAT for the asset using Eqs. (15.1) to (15.4).

COMMENT If some portion of the \$50,000 investment is borrowed from company-external sources (debt financing), a tax credit is offered in that the interest is tax deductible. See Solved Problems for an illustration.

Table 15.1 Tabulation of CFAT for Example 15.1

Year	(1) Income	(2) Disbursements	(3) = (1) − (2) CFBT	(4) Depreciation*	(5) = (3) − (4) TI	(6) = 0.4(5) Taxes	(7) = (3) − (6) CFAT
0	. . .	$50,000	$−50,000	. . .	. . .	. . .	$−50,000
1	$27,000	10,000	+17,000	$10,000	$7,000	$2,800	+14,200
2	26,000	10,500	+15,500	10,000	5,500	2,200	+13,300
3	25,000	11,000	+14,000	10,000	4,000	1,600	+12,400
4	24,000	11,500	+12,500	10,000	2,500	1,000	+11,500
5	23,000	12,000	+11,000	10,000	1,000	400	+10,600

*Depreciation = 50,000/5 = $10,000 per year.

Example 15.2 A production-related capital asset is purchased. When the purchase is made, the following data apply:

$$P = \$30{,}000$$

$$n = 6 \text{ years}$$

$$\text{CFBT} = \$8000 \text{ per year}$$

$$\text{Tax rate} = 52\%$$

Tax law requires this type of asset to be straight-line-depreciated over 10 years. Therefore, even though the company anticipates that the asset will have a useful life of 6 years, it must be depreciated over 10 years. The company will utilize a CFBT = $1000 for years 7 to 10, when the asset will be in a standby condition. If the company experiences an actual sales price of $5815 after 10 years, tabulate the CFAT for the 10 years. Assume that this asset will be awarded an investment tax credit of $1200 and that TI > $50,000.

SOLUTION Table 15.2 details the CFAT. Note that the computations for depreciation and the recaptured depreciation (RD) tax are detailed in the table footnotes. Recaptured depreciation, which equals the anticipated sales price because the book value is zero in year 10, is taxed at the 52% rate. (See the comment in

Table 15.2 Tabulation of CFAT for Example 15.2

Year	CFBT	Depreciation*	TI	Taxes	CFAT
0	$−30,000	. . .	. . .	$−1,200	$−28,800
1–6	+8,000	$3,000	$5,000	+2,600	+5,400
7–10	+1,000	3,000	−2,000	−1,040	+2,040
10	+5,815	. . .	. . .	+3,023†	+2,792

*Depreciation = 30,000/10 = $3000 per year.
†Income tax = RD tax = 0.52(5815) = $3023.

Sec. 14.1 about recaptured depreciation computations for ACRS-depreciated assets.)

Solved Problem 15.8
Probs. P15.1 to P15.6

15.2 After-Tax Analysis Using Present-Worth or EUAC Analysis

If the minimum after-tax rate of return is stated or known and the CFAT values are computed, present-worth or EUAC analysis can be used to select the most economic plan, in a fashion similar to that of Chaps. 5 and 6.

Example 15.3 Using a 7% after-tax return, select the more economic plan of those detailed in Examples 15.1 and 15.2 using (*a*) EUAC and (*b*) present-worth analysis. (The plans are summarized below for convenience.)

Plan *X* (Example 15.2)	Plan *Y* (Example 15.1)
P = \$30,000	P = \$50,000
Actual SV = \$5815	SV = 0
n = 10 years (depreciable)	n = 5 years
CFAT in Table 15.2	CFAT in Table 15.1

SOLUTION

(*a*) EUAC equations may be set up and solved at $i = 7\%$ as follows:

$$\begin{aligned}\text{EUAC}_X &= [-28{,}800 + 5400(P/A, 7\%, 6) + 2040(P/A, 7\%, 4)\,(P/F, 7\%, 6)\\ &\quad + 2792(P/F, 7\%, 10)]\,(A/P, 7\%, 10)\\ &= \$421.78 \qquad (15.5)\end{aligned}$$

$$\begin{aligned}\text{EUAC}_Y &= [-50{,}000 + 14{,}200(P/F, 7\%, 1) + 13{,}300(P/F, 7\%, 2)\\ &\quad + 12{,}400(P/F, 7\%, 3) + 11{,}500(P/F, 7\%, 4)\\ &\quad + 10{,}600(P/F, 7\%, 5)]\,(A/P, 7\%, 5)\\ &= \$327.01 \qquad (15.6)\end{aligned}$$

Plan X is selected, since both EUAC values are positive (profit) and EUAC_X is larger.

(*b*) Present-worth analysis must be based on a 30-year horizon to equalize the anticipated lives. Using the EUAC values above,

$$\text{PW}_X = \text{EUAC}_X\,(P/A, 7\%, 30) = \$5233.87$$

$$\text{PW}_Y = \text{EUAC}_Y\,(P/A, 7\%, 30) = \$4057.87$$

Again, plan X is selected, because PW_X is larger.

COMMENT If only disbursement values before taxes, such as annual operating costs are known, that is, if CFBT < 0, the related taxes are considered a tax advantage to be applied against other expenses of the company.

Probs. P15.7 to P15.14

15.3 After-Tax Rate-of-Return Analysis

In all cases the present worth or EUAC method is used to compute the rate of return of the CFAT values. It is common that the after-tax return is approximately one-half the before-tax return.

For a single project, as shown in Chap. 7, set the *present worth* or *EUAC* of the CFAT values equal to zero and solve for the rate of return i^*.

Present worth:

$$0 = \sum_{t=1}^{n} \text{CFAT}_t (P/F, i^*\%, t) \tag{15.7}$$

EUAC:

$$0 = \left[\sum_{t=1}^{n} \text{CFAT}_t (P/F, i^*\%, t) \right] (A/P, i^*\%, n) \tag{15.8}$$

Example 15.4 Using the asset purchase described in Example 14.6 and straight-line (ACRS-election) depreciation, compute the after-tax rate of return if no salvage value is realized. (Summary: $P = \$50{,}000$, $n = 5$, CFBT = \$20,000 per year, $T = 50\%$.)

SOLUTION Table 15.3 presents the CFAT for the asset. By Eq. (15.7) the present-worth equation for the after-tax return is

$$0 = 50{,}000 + 15{,}000(P/A, i^*\%, 5)$$

$$(P/A, i^*, 5) = 3.3333$$

Solution gives $i^* = 15.25\%$ as the after-tax rate of return.

COMMENT If as an economist you want to use an exaggerated before-tax rate to approximate the tax effect on this type of asset, you can use Eq. (14.7) to obtain $i^*/(1 - T) = 0.1525/(1 - 0.50) = 0.305$, or 30.5%. Actual before-tax return computed using the CFBT figures in Table 15.3 can be found from the equation

$$0 = \$-50{,}000 + 20{,}000(P/A, i^*\%, 5)$$

Table 15.3 Cash flow after taxes using straight-line depreciation for Example 15.4

Year	CFBT	Depreciation	TI	Taxes	CFAT
0	\$-50,000	. . .	. . .	. . .	\$-50,000
1-5	+20,000	\$10,000	\$10,000	\$5,000	+15,000

which gives a value of i^* =28.7%. Comparison shows that the tax effect is overestimated by using a 30.5% before-tax return.

The ROIDS computer program (Appendix E) may be used to compute i^* once the CFAT sequence is determined. If the CFAT sequence is nonconventional, multiple rates of return (Sec. 7.4) may be found or a reinvestment rate may be used to determine the one composite rate of return i' (Sec. 7.5) instead of the multiple i^* values.

If two (or more) alternatives are involved in the after-tax rate-of-return analysis, the procedure of either Sec. 8.3 for present worth or Sec. 8.4 for EUAC is used to determine the *incremental* after-tax return i^*_{B-A} where B is the alternative with the larger first cost. For *present-worth* analysis the rate-of-return system is set up for the least common multiple of years.

$$0 = \sum_{t=1}^{n} \Delta\text{CFAT}_t (P/F, i^*\%, t) \tag{15.9}$$

where the Δ (delta) symbol indicates the net or incremental CFAT values between the alternatives and the $B - A$ subscript has been omitted on $i^*\%$. This return may be considered the breakeven rate of return between the two alternatives, which is compared with the MARR for alternative selection. (See Sec. 8.3 for a review of this concept.) Example 15.5 compares two plans using after-tax analysis.

For *EUAC* analysis if the alternative lives are *equal*, use the incremental CFAT values in the relation

$$0 = \left[\sum_{t=1}^{n} \Delta\text{CFAT}_t (P/F, i^*\%, t)\right] (A/P, i^*\%, n) \tag{15.10}$$

If lives are *unequal*, i^* is determined on the basis of incremental EUAC values for the respective alternative lives using

$$0 = \text{EUAC}_B - \text{EUAC}_A \tag{15.11}$$

When ΔCFAT_t values have been computed, the computer program discussed in Appendix E may be used to determine the incremental i^* values. When hand solution is used, determining the i^* value will require trial-and-error computations.

Example 15.5 Let the after-tax MARR = 6% and select one of the plans below by determining the breakeven rate of return using the (*a*) EUAC and (*b*) present-worth methods. (Note that the plans are the same as those in Examples 15.1 and 15.2.)

Plan X	Plan Y
P = \$30,000	P = \$50,000
Actual SV = \$5815	SV = 0
Tax n = 10 years	n = 5 years
CFAT in Table 15.2	CFAT in Table 15.1

SOLUTION

(*a*) Since the lives are unequal, Eq. (15.11) is used to determine the i^* value. The EUAC relations in Eq. (15.5) for plan *X* and Eq. (15.6) for plan *Y* are used in $0 = \text{EUAC}_Y - \text{EUAC}_X$ to find $i^* = 6.35\%$ by interpolation. Since 6.35% is greater than MARR = 6%, the increment of plan *Y* over *X* is justified. Therefore, select plan *Y*.

Figure 15.1 is a plot of (profit) EUAC values for both plans between 5 and 8%. For any MARR select the plan with the larger equivalent annual profit value. You can see that if MARR is less than 6.35%, plan *Y* is selected because of its higher profit EUAC; and if MARR exceeds 6.35%, plan *X* has the higher profit EUAC.

(*b*) Present-worth analysis to find i^* by Eq. (15.9) requires considerably more work than the EUAC basis. For the least common multiple of 10 years the ΔCFAT_t values are determined (Table 15.4) and the interpolated breakeven $i^* = 6.35\%$ is found by using trial and error.

$$i^* = 6 + \frac{376.00}{376.00 - (-668.55)} = 6.35\%$$

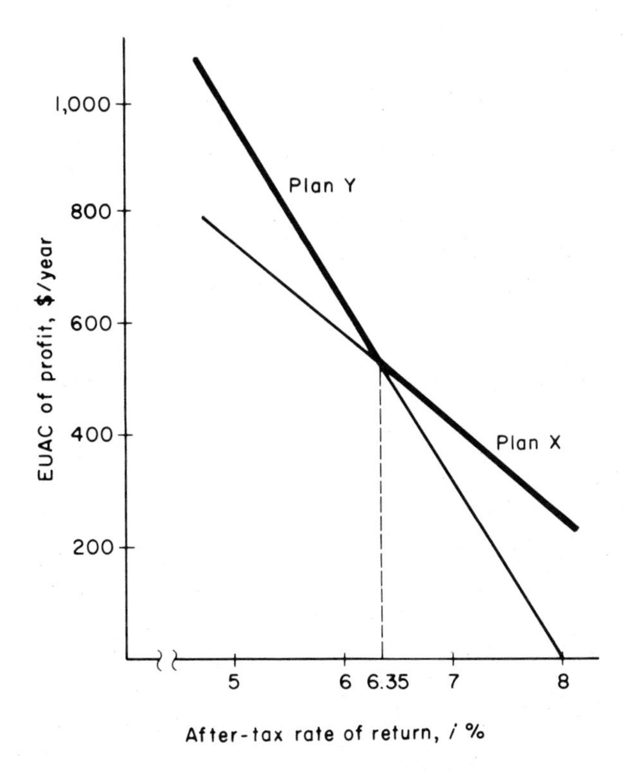

Figure 15.1 Breakeven rate-of-return chart for after-tax analysis, Example 15.5.

Table 15.4 Computation of after-tax i^* value for unequal-lived assets, Example 15.5

Year, t	$CFAT_X$ (Table 15.2)	$CFAT_Y$ (Table 15.1)	$\Delta CFAT_t$	Present worth 6%	Present worth 7%
0	$-28,800	$-50,000	$-21,200	$-21,200.00	$-21,200.00
1	5,400	14,200	8,800	8,301.92	8,224.48
2	5,400	13,300	7,900	7,031.00	6,899.86
3	5,400	12,400	7,000	5,877.20	5,714.10
4	5,400	11,500	6,100	4,831.81	4,653.69
5	5,400	-39,400	-44,800	-33,479.04	-31,942.40
6	5,400	14,200	8,800	6,204.00	5,863.44
7	2,040	13,300	11,260	7,489.03	7,011.60
8	2,040	12,400	10,360	6,499.86	6,029.52
9	2,040	11,500	9,460	5,599.37	5,145.29
10	4,832	10,600	5,768	3,220.85	2,931.87
				$ 376.00	$ -668.55

Note that there are possibly multiple rates for the $\Delta CFAT_t$ series due to three sign changes. Since 6.35% exceeds MARR = 6%, plan Y is again selected over plan X.

COMMENT As in most cases, this again illustrates that for i^* computations the EUAC method may be simpler to use than the present-worth method.

Solved Problem 15.9
Probs. P15.15 to P15.27

15.4 After-Tax Replacement Analysis

When a defending asset is challenged by a new asset, the effects of income taxes may be considerable. To account for all the tax details in *replacement analysis* is sometimes neither time- nor cost-effective; however, it is worthwhile to account for any applicable capital gain or loss which would occur if the defender were replaced. Also important is the future tax advantage stemming from deductible operating and depreciation expenses. The following example will give you an idea of the impact of taxes on replacement analysis.

Example 15.6 Thurston Mining purchased certain production-related mining equipment 3 years ago. Because of the inadequacy of the machinery, a new piece of equipment is being considered. The defender and challenger characteristics are (AOC for annual operating costs):

Defender	Challenger
$P = \$6000$	$P = \$10{,}000$
SV = 0	SV = 2000
AOC = 1000	AOC = 150
Original n = 8 years	n = 5 years

If a \$4000 trade-in is offered for the defender, perform the (*a*) before-tax and (*b*) after-tax analysis, using a 15% return before taxes and a 7% after-tax return. Tax rates are 52% on income and recaptured depreciation (RD), and straight-line depreciation with no salvage value recognized is applied to both assets.

SOLUTION

(*a*) The before-tax analysis will use $P = \$4000$ and a 5-year remaining life for the defender. The EUAC analysis appears as follows using the conventional approach to replacement analysis (Sec. 10.3).

$$\text{EUAC}_D = 4000(A/P, 15\%, 5) + 1000 = \$2193$$

$$\text{EUAC}_C = 10{,}000(A/P, 15\%, 5) - 2000(A/F, 15\%, 5) + 150 = \$2837$$

The defender is favored by the EUAC margin of \$644.

(*b*) For retention of the *defender* first compute the actual investment, since it is not sold.

$$\text{Depreciation} = \frac{6000}{8} = \$750 \text{ for next 5 years}$$

$$\text{Present book value} = 6000 - 3(750) = \$3750$$

$$\text{RD on trade-in} = 4000 - 3750 = \$250$$

$$\text{RD tax savings} = 0.52(250) = \$130$$

$$\text{Actual after-tax investment} = 4000 - 130 = \$3870$$

If trade-in is less than book value, the resulting capital loss is a foregone tax advantage when the defender is retained, so it will be added to the \$4000 trade-in value.

The after-tax adjusted annual costs for the defender are:

$$\text{Annual operating cost} = \$1000$$

$$\text{Depreciation} = \$750 \text{ for 5 years}$$

$$\text{Annual tax saving} = (1000 + 750)(0.52) = \$910$$

$$\text{Net after-tax disbursement} = 1000 - 910 = \$90 \text{ per year}$$

The after-tax EUAC for the defender is

$$\text{EUAC}_D = 3870(A/P, 7\%, 5) + 90 = \$1034$$

For the *challenger*, after-tax analysis for the next 5 years is as follows:

$$\text{Depreciation} = \frac{10{,}000}{5} = \$2000 \text{ per year}$$

$$\text{Annual tax saving} = (2000 + 150)(0.52) = \$1118$$

$$\text{Net after-tax disbursement} = 150 - 1118 = -\$968 \text{ (income)}$$

When the asset is sold for \$2000 in year 5 the book value is zero and the RD tax is \$2000(0.52) = \$1040. The EUAC is then

$$\text{EUAC}_C = 10{,}000(A/P, 7\%, 5) - (2000 - 1040)(A/F, 7\%, 5) - 968 = \$1304$$

Select the defender with an advantage of \$270 per year. Even though before- and after-tax decisions are the same, the defender advantage after taxes is less than 42% of the before-tax advantage. Since the adequacy of the defender is in question, this intangible factor may be enough to cause management to replace.

Probs. P15.28 to P15.32

*15.5 Tax Effect of Depletion Laws

As was shown in Sec. 13.9, depletion is a vehicle similar to depreciation; that is, it is used to write off the initial investment of natural-resource recovery methods. Recall that there are two methods of depletion:

1. Factor depletion method

$$\text{Annual depletion} = \frac{\text{initial investment}}{\text{resource capacity}} \text{ annual volume}$$

2. Percentage depletion, where a percentage of GI (gross income) is tax deductible, provided the result does not exceed 50% of TI (taxable income) prior to deduction of this depletion.

The question of depletion favoritism toward some industries and producers (those in the higher allowance bracket) is often present, since they are allowed to deplete the property much beyond the original investment. Either of the two methods reviewed above may be used, but naturally the larger deduction is always used in tax computations. To illustrate the impact of a 22% depletion allowance, the highest possible, Example 15.7 compares identical initial investments and corresponding write-off using depreciation and depletion.

Example 15.7 A \$240,000 investment has been made. The following income and expenses are realized for the 12-year life:

Year	Gross income	Expenses
1-10	$150,000	$50,000
11	75,000	20,000
12	25,000	5,000

The effective tax rate is 50%. Compute and compare the after-tax rate of return for the two situations below.

(*a*) The $240,000 is used to purchase an asset which is to be straight-line-depreciated over 12 years (ACRS election) and has no salvage value.
(*b*) The $240,000 is used to develop a mine subject to a 22% depletion allowance. The ore is sold for $10 a ton, and 15,000 tons are sold for each of the years 1 to 10; 7500 tons are sold in year 11 and 2500 tons in year 12. Total expected tonnage in the mine was 160,000 tons at the time of purchase.

SOLUTION
(*a*) For the asset, Table 15.5 details the after-tax situation. The after-tax return is $i^* = 22.35\%$.
(*b*) Table 15.6 presents the tax situation using the percentage depletion or factor depletion, as allowed. The depletion factor is 240,000/160,000 = $1.50 per ton. Note that the percentage depletion is allowable for all 12 years, since it never exceeds 0.5(TI). To find the after-tax return the present worth of CFAT is equated to zero; but note that a total depletion of $352,000 (Table 15.6), or 1.47 times the first cost, is claimed. The rate of return is $i^* = 25.21\%$, or approximately 3% higher than the return for a depreciable asset with the same gross income, expenses, and life. Thus, the advantage of depletion in this case is obvious—a larger after-tax return.

Solved Problem 15.10
Probs. P15.33 and P15.34

Table 15.5 Effect of *depreciation* on after-tax rate of return* for Example 15.7*a*

Year	GI	Expenses	CFBT	Depreciation†	TI	Taxes	CFAT
0	...	$240,000	$-240,000	...	...	...	$-240,000
1-10	$150,000	50,000	+100,000	20,000	$80,000	$40,000	+60,000
11	75,000	20,000	+55,000	20,000	35,000	17,500	+37,500
12	25,000	5,000	+20,000	20,000	0	0	+20,000

*Rate-of-return computation: $-240{,}000 + 60{,}000(P/A, i^*\%, 10) + 37{,}500(P/F, i^*\%, 11) + 20{,}000(P/F, i^*\%, 12) = 0 \quad i^* = 22.35\%$.

†Depreciation = 240,000/12 = $20,000 per year.

Table 15.6 Effect of *depletion* on after-tax rate of return* for Example 15.7*b*

Year	(1) Tons sold	(2) GI	(3) Expenses	(4) = (2) – (3) TI before depletion†	(5) = 0.5(4) 0.5TI
0	...	...	$240,000	$–240,000	...
1–10	15,000	$150,000	50,000	+100,000	$50,000
11	7,500	75,000	20,000	+55,000	27,500
12	2,500	25,000	5,000	+20,000	10,000

Year	(6) = 1.5(1) Factor depletion	(7) = 0.22(2) Percentage depletion	(8) TI‡	(9) = 0.5(8) Taxes	(10) = (4) – (9) CFAT
0	...	...	...	...	$–240,000
1–10	$22,500	$ 33,000§	$67,000	$33,500	+66,500
11	11,250	16,500§	38,500	19,250	+35,750
12	3,750	5,500§	14,500	7,250	+12,750
		$352,000			

*Rate-of-return computation: $-240{,}000 + 66{,}500(P/A, i^*\%, 10) + 35{,}750(P/F, i^*\%, 11) + 12{,}750(P/F, i^*\%, 12) = 0 \quad i^* = 25.21\%$.

†Also the CFBT.

‡TI = TI before depletion – depletion.

§Depletion claimed (note that the $33,000 is claimed for each of the first 10 years).

SOLVED PROBLEMS

Example 15.8 The Gutsy Cleaning Company plans to invest in a new dry cleaner. Details of the investment are:

$$P = \$15{,}000$$

$$SV = 0$$

$$\text{Income} = 7000 \text{ per year}$$

$$\text{Expenses} = 1000 \text{ per year}$$

$$\text{Effective tax rate} = 50\%$$

$$n = 5 \text{ years}$$

If straight-line depreciation is used, tabulate CFAT for the following conditions: (*a*) all $15,000 is from company funds (100% equity financing) and (*b*) one-half the investment is borrowed from a bank (50% equity – 50% debt financing) at 10% interest. Assume that the 10% is simple interest on the total amount borrowed and repayment will be in five equal payments.

SOLUTION

(*a*) For 100% equity financing, Eq. (15.1) is used to obtain CFBT = $6000 per year. Annual depreciation is $15,000/5 = $3000. Table 15.7 details CFAT.

Table 15.7 CFAT tabulation for Example 15.8*a*

Year	CFBT	Depreciation	TI	Taxes	CFAT
0	$-15,000	...	...	...	$-15,000
1-5	6,000	$3,000	$3,000	$1,500	4,500
	$ 15,000			$7,500	$ 7,500

(*b*) The 50% debt financing requires that $7500 be borrowed from outside the company or its stockholders. As in part *a*, CFBT = $6000. The loan repayment scheme will be as follows:

Principal: $$\frac{7500}{5} = \$1500 \text{ per year}$$

Interest: $$7500(0.10) = \$750 \text{ per year}$$

The $750 interest is tax deductible; however, the principal is *not deductible.* Therefore, the formulas for TI and CFAT, Eqs. (15.2) and (15.4), may be restated as

$$\text{TI} = \text{CFBT} - \text{depreciation} - \text{interest} \tag{15.12}$$

$$\text{CFAT} = \text{CFBT} - \text{taxes} - \text{interest} - \text{principal} \tag{15.13}$$

Table 15.8 presents CFAT computation for 50% debt-50% equity financing. You can see that the annual CFAT has decreased from $4500 to $2625 because of 50% debt financing. The $7500 equity cash outflow in year 0 is used because only 50% of the first cost comes from company funds while the remaining is from the lending bank.

COMMENT If only equity financing is involved, as in part *a*, we can use the relation

$$\text{CFAT} = \text{depreciation} + \text{TI}(1 - \text{tax rate}) \tag{15.14}$$

where TI(1 - tax rate) is the *profit*, that is, the portion of TI not absorbed by taxes. Thus, in Table 15.7,

$$\text{CFAT} = 3000 + 3000(1 - 0.50) = \$4500$$

Table 15.8 CFAT tabulation for Example 15.8*b*

Year	(1) CFBT	(2) Depreciation	(3) Interest	(4) Principal	(5) TI*	(6) Taxes	(7) CFAT†
0	$-7,500	...	...	...	...	...	$-7,500
1-5	6,000	$3,000	$ 750	$1,500	$2,250	$1,125	+2,625
	$22,500		$3,750	$7,500		$5,625	$ 5,625

*In column notation, (5) = (1) - (2) - (3), Eq. (15.12).
†(7) = (1) - (6) - (3) - (4), Eq. (15.13).

could be used in lieu of the method presented. But you must be careful because, if there is any debt financing whatsoever, as in part *b*, this simple approach will not work. Let's try it on Table 15.8.

$$\text{CFAT} = 3000 + 2250(1 - 0.50) = \$4125 \neq \$2625$$

Why doesn't it work? Simple. This method neglects the fact that interest is tax deductible *and* that CFAT is computed in a completely different way in the two cases. You might compare Eqs. (15.14) and (15.13) to verify this. Thus, we advise you to use the general formulas for CFAT tabulation, that is, Eqs. (15.1) to (15.4).

Sec. 15.1

Example 15.9 Compute the after-tax rate of return for the two situations of Example 15.8: (*a*) 100% equity financing and (*b*) 50% equity–50% debt financing. Compare the results.

SOLUTION

(*a*) For 100% equity financing, the CFAT values of Table 15.7 can be used to find i^* in the general form of Eq. (15.7).

$$0 = -15{,}000 + 4500(P/A, i^*\%, 5)$$

$$(P/A, i^*\%, 5) = 3.3333$$

$$i^* = 15.25\%$$

(*b*) For the 50%–50% split on financing, Table 15.8 results are used to set up the equation

$$0 = -7500 + 2625(P/A, i^*\%, 5)$$

$$(P/A, i^*\%, 5) = 2.8571$$

$$i^*\% = 22.22\%$$

Comparison of the two return values shows that debt financing increases the rate of return on company investments, simply because less of the firm's capital is tied up in investment.

COMMENT Why not use close to 100% debt financing and maximize the return? We will let you speculate about this dilemma until Sec. 18.7, where we answer this question. Please venture a guess now. What if every time you wanted to make a purchase you had to borrow? How stable would your business be?

Sec. 15.3

Example 15.10 Assume that the mine purchase discussed in Example 15.7 was 60% debt-financed by a 12-year loan with interest computed at 3% per year on the original principal. Now compute the after-tax rate of return.

Table 15.9 After-tax return* with 60% debt financing for Example 15.10

Year	Principal payment	Interest payment	TI before† depletion	0.5TI	Factor depletion
0	...	...	$-96,000	...	...
1-10	$12,000	$ 4,320	95,680	$47,840	$22,500
11	12,000	4,320	50,680	25,340	11,250
12	12,000	4,320	15,680	7,840	3,750

Year	Percentage depletion	TI	Taxes	CFAT‡
0	...	...	...	$-96,000
1-10	$33,000§	$62,680	$31,340	48,020
11	16,500§	34,180	17,090	17,270
12	5,500§	10,180	5,090	-5,730

*Rate of return computation: $-96{,}000 + 48{,}020(P/A, i^*\%, 10) + 17{,}270(P/F, i^*\%, 11) - 5730(P/F, i^*\%, 12) = 0 \quad i^* = 49.25\%$.

†TI before depletion = GI – expenses – interest (see Table 15.6).

‡CFAT = TI before depletion – taxes – principal – interest.

§Depletion claimed.

SOLUTION For the 60% debt financing, the following data apply:

$$\text{Amount of loan} = 240{,}000(0.60) = \$144{,}000$$

$$\text{Principal payment} = \frac{144{,}000}{12} = \$12{,}000 \text{ per year}$$

$$\text{Interest} = 144{,}000(0.03) = \$4320 \text{ per year}$$

Table 15.9 details the CFAT values. Note the outflow of $96,000 in year 0, not $240,000, since only 40% of the mine's first cost is due to equity funds. The factor and percentage values for depletion are taken from Table 15.6. The present-worth computation under Table 15.9 indicates that $i^* = 49.25\%$, which is 1.95 times the return for depletion with 100% equity funds. Again we see the dramatic advantage of borrowed funds.

Secs. 15.1 and 15.5

FURTHER INFORMATION

Income tax considerations (Secs. 15.1 to 15.5): Bussey [3], pp. 112-151; Canada and White [4], pp. 113-136; Grant, Ireson, and Leavenworth [7], pp. 229-266; Stevens [15], pp. 38-63; Taylor [16], pp. 319-369; White, Agee, and Case [18], pp. 215-240.

After-tax replacement (Sec. 15.4): Stevens [15], pp. 224-247.

PROBLEMS

Note: Unless specifically mentioned, the problems in this chapter do not use the ACRS depreciation rates. The problems involving SL or other depreciation should take the salvage into account and recaptured depreciation and capital losses or gains are based on current book value or expected salvage value, not ACRS or some extended-recovery-period book value as presented in the Economic Recovery Tax Act of 1981. These simplifications are made for this chapter only.

P15.1 An investment company plans to purchase an apartment complex for $350,000. Annual income before taxes of $28,000 is expected for the next 8 years, after which the property will be sold for an estimated $453,600. The applicable tax rate is 52%, the estimated annual operating cost is $3000, and the gain on property sale is taxed at 20%. Tabulate the cash flow after taxes for the years of ownership, if the property will be straight-line-depreciated over a 20-year life with a 40% salvage value. (Note that both recaptured depreciation and a capital gain are involved at sale time.)

P15.2 Tabulate the after-tax cash flows for the Wonderama machine described in Example 14.8 if it is eligible for the investment tax credit, produces a CFBT of $50,000 annually, and is actually salvaged for $29,166 after 5 years.

P15.3 Rework Prob. P15.2 assuming that 40% of the first cost of Wonderama is borrowed at 12% "simple" interest on the declining balance to be repaid in six equal installments of $13,806 per year of which $4084 is annual interest.

P15.4 An asset costing $10,000 with SV = 0 has been owned by the Renuzit Plumbers for 4 years (since 1981). ACRS depreciation with a recovery period of 5 years has been used. Management expects to retain the machine for 2 more years with the same annual CFBT ($5000) as the past 4 years.

(*a*) Tabulate the cash flow after taxes for an effective tax rate of 48% if the asset is actually salvaged for $3075.

(*b*) What net difference in total cash flow would have been realized if Renuzit had elected straight-line depreciation for 5 years?

(*c*) What is the difference in total cash flow if straight-line depreciation had been allowed for 6 years and the $3075 salvage value had been considered in computing depreciation?

P15.5 (*a*) Rework Prob. P15.2 using double-declining-balance depreciation. Neglect the expected SV of $15,000 when computing depreciation.

(*b*) Which of the two depreciation methods should be used to obtain a larger net total CFAT value? Assume that any operating loss is simply a tax savings for the company in the year of occurrence.

(*c*) Repeat this problem using 1985 ACRS depreciation and the straight-line election (neglect the half-year convention).

P15.6 Rework Prob. P15.4 *a* and *b* assuming that the purchase was 50% financed by a $5000 loan with yearly interest at 3% on the original principal and repayment in five equal annual payments of $1150 each.

P15.7 Revise Prob. P5.1 as follows: Both machines will be depreciated by the straight-line method, the tax rate is 50%, and an after-tax return of 7% is required. Select the more economic of the machines using after-tax analysis. Assume that annual maintenance and operating costs are a tax advantage.

P15.8 Rework Prob. P6.7 if the used machine is depreciated by the straight-line method over 5 years and the new machine by the 1985 ACRS method over 15 years. Assume that both machines are eligible for investment tax credit, that the effective tax rate is 50%, and an after-tax return of 10% is required. Neglect salvage values in depreciation, but assume that they actually occur after the asset's anticipated life. Use current tax laws.

P15.9 Update Prob. P6.6 as follows: Regardless of which machine is selected, a $10,000 loan will be necessary for the purchase. Repayment of this loan will be in five equal annual installments of

$2700 each ($2000 principal, $700 interest). If the food-processing company is in the 52% tax bracket, uses straight-line depreciation on all assets, and an after-tax MARR of 6% is required, determine which labeling machine is more economic.

P15.10 The Siko-so-matic Drug Company has to decide between the two pill-forming machines detailed below.

	Roundee	Scored
First cost	$24,000	$15,000
Salvage value	6,000	3,000
Annual CFBT	4,000	2,000
Life, years	12	12

The machines have an anticipated useful life of 12 years as detailed above, but the tax laws require straight-line-depreciation over 10 years with a zero salvage value in lieu of ACRS depreciation. If an effective tax of 50% applies and an after-tax return of 10% is desired, compare the two machines using (*a*) present-worth analysis and (*b*) EUAC analysis.

P15.11 If in Prob. P15.10, Siko-so-matic had used the anticipated life and salvage values, would the decision have been the same?

P15.12 Select the more economic of the two alternatives detailed below if the after-tax MARR is 8%, straight-line depreciation is used, and $T = 50\%$.

	Alternative *A*	Alternative *B*
First cost	$10,000	$15,000
Salvage value	1,000	2,000
Annual cost	1,500	600
Life, years	10	10

P15.13 Rework P15.12 if one-half the first cost of each asset will be debt-financed with 10 equal payments of $616.45 for *A* and $924.68 for *B*. For each loan, one-tenth of the principal amount is removed each year with the remainder of the payment necessary for interest.

P15.14 Compare the two plans below using an after-tax MARR of 10% and present-worth analysis.

	Plan *A*		
	Machine 1	Machine 2	Plan *B*
First cost	$5,000	$25,000	$40,000
Salvage value (after tax)	0	1,000	5,000
Annual savings			
Years 1–4	500	10,000	15,000
Years 5–8	500	10,000	20,000
Years 9–10	500	5,000	25,000
Tax rate	48%	48%	48%
Life, years	5	10	10

Assume straight-line depreciation for all assets and neglect the salvage value in depreciation charges. Further, assume that any operating loss will simply be a tax savings applicable to other taxable income for the year incurred and use the stated SV as an after-recaptured-depreciation tax salvage amount.

P15.15 Compute the after-tax rate of return for Prob. P15.1.

P15.16 Compute the after-tax rate of return for Probs. (*a*) P15.2 and (*b*) P15.3. (*c*) Explain the difference in your answers.

P15.17 Determine the difference in the after-tax rate of return made by the situations presented in Prob. P15.4.

P15.18 What is the after-tax rate of return for Prob. P15.6, where partial debt financing is involved? What difference does this financing make in the after-tax return? (See Prob. P15.17.)

P15.19 (*a*) Compute the breakeven after-tax rate of return for Prob. P15.7. (*b*) Plot the present-worth values and select the better plan if the after-tax MARR is 4%, (*c*) 6%, (*d*) 7%, and (*e*) 10%.

P15.20 Compute the after-tax return for each machine in Prob. P15.7, if gross income is $5000 per year.

P15.21 Determine the after-tax return at which the machines of Prob. P15.9 are economically indifferent.

P15.22 At what annual value of depreciation will the after-tax return be (*a*) 5% and (*b*) 10% for the Wonderama machine of Prob. P15.2?

P15.23 (*a*) Set up an after-tax relation of the form in Eq. (15.9) and (*b*) determine the breakeven after-tax return for Prob. P15.14.

P15.24 Determine the after-tax return for each alternative of Prob. P15.14.

P15.25 Consider the problem solved in Example 15.4. Assume that the asset's owner is interested in having an after-tax return of 20%. If the tax rate remains at 50%, compute the value of the (*a*) first cost, (*b*) salvage value, and (*c*) annual depreciation at which this will occur. When determining any one of the values above, assume that the remaining parameters retain the value detailed in the example.

P15.26 Compute the after-tax return for the following situation. The owners of Slightly Tipsy Underwater Diving Services (STUDS) wish to make an investment. They can purchase special diving equipment to handle a job. The equipment will cost $2500, have a life of 5 years, and have no salvage value. They will receive $1500 in year 1, but believe they can make $300 each year using the equipment in the future. The diving equipment will be straight-line depreciated and the tax rate is 45%.

P15.27 If the equipment discussed in Prob. P15.26 is not purchased, the STUDS can pay another firm–DROWN–to do the job. If the STUDS can make 5% after taxes on its money, what percent of the $1500 amount in year 1 must they have to make the same return possible as if they purchase the equipment and retain it for the full 5-year life?

P15.28 Rework Example 15.6 under the assumptions that the trade-in value of the defender is only $2000 and that new remaining life and salvage value estimates are 10 years and $750, respectively. Salvage is still not considered in depreciation computations.

P15.29 Perform an after-tax analysis for Prob. P10.13 using a tax rate of 50% and an after-tax return of 4%. Use straight-line depreciation for all assets and assume that asset *A* cost $20,000 when purchased and had an expected life of 8 years.

P15.30 If the tax rate is 50%, what is (*a*) the breakeven rate of return between plans I and II and (*b*) the replacement value of asset *A* compared to asset *B* ($i = 4\%$) for Prob. P10.13?

P15.31 (*a*) Compare the two plans detailed below using a tax rate of 48% and an after-tax return of 8%.

(*b*) Is the decision different from the before-tax result? Use $i = 15\%$.

	Defender	Challenger
First cost	$28,000	$15,000
AOC when purchased	...	1,500
Actual AOC	1,200	
Expected salvage when purchased	2,000	3,000
Trade-in value	18,000	
Depreciation	SL	SL
Life, years	10	8
Year owned	2	

P15.32 Rework Prob. P10.20 using an after-tax analysis. Assume that the currently owned compressor was purchased 4 years ago for $1000 with $n = 5$ years and no salvage value. The asset has been straight-line-depreciated and the effective tax rate for the company is 50%. When performing the analysis for the challenger assume that the *equivalent* annual costs for maintenance are used for expenses to reduce taxes.

***P15.33** In Prob. P13.30 assume that one-half the first cost of the mine is debt-financed at 4% to be repaid in 5 years with payments of $420,000 per year of which interest accounts for $70,000. Assume a tax rate of 50% and compute the after-tax cash flows for the 5 years. Is the rate of return positive for these cash flows?

***P15.34** Compute the rates of return for the investments detailed below. Assume that alternative *A* is depreciable by the straight-line method and that *B* is depletable using the larger of a 10% allowance, or factor of $1 per unit extracted.

	A	*B*
Investment	$ 50,000	$ 50,000
CFBT	15,000	15,000
Gross income	100,000	100,000
Tax rate	50%	50%
Life, years	10	10
Annual units	...	5,000

LEVEL FIVE

Some of the material in this level may be considered optional to a basic understanding of engineering-economic analysis. After a discussion of payback and breakeven analysis, a brief discussion of capital budgeting shows you how to select several projects from many in order to maximize the return without violating stated restrictions. The setting of the MARR (minimum attractive rate of return), the value of which has been assumed thus far, is investigated, as is the sensitivity of an economic decision to changes in cost, income, and life estimates.

CHAPTER

SIXTEEN

PAYBACK, BREAKEVEN, AND LIFE-CYCLE COSTING ANALYSIS

This chapter will discuss the computations and pitfalls of the payback-period technique and the general approach of breakeven analysis as commonly used in engineering-economic analysis. In addition, an overview of the life-cycle costing approach to project comparison is presented.

SELECTION OBJECTIVES

To complete this chapter, you must be able to:

16.1. Define the term and compute the value of the *payback period* for a single project, given the first cost, useful life, and expected annual cash flows. State the fallacies of payback analysis and demonstrate that payback analysis of two alternatives may result in a decision different from that of other evaluation methods.

16.2. Use the stated procedure to calculate the *breakeven point* between two alternatives and select one of them, given the necessary data for each alternative.

*16.3. Define *life-cycle costing*, state the major cost categories, and use engineering-economic analysis to determine which of several alternatives has a minimum life-cycle cost, given the data for each alternative.

STUDY GUIDE

16.1 Payback-Period Analysis

Primarily, payback analysis is used to determine the number of years an asset must be retained and used to recover its initial cost with a stated return, given the annual cash flow and salvage value. The analysis may be performed using after-tax cash-flow values (Chaps. 14 and 15), so that the results are more realistic. To find the payback period at a given rate of return the following model is used to determine the value of n' in years.

$$0 = -P + \sum_{t=1}^{n'} \text{CF}_t(P/F, i\%, t) \qquad (16.1)$$

where CF_t is the net cash flow at the end of year t ($t = 1, 2, \ldots, n'$). If the cash flows CF_t in Eq. (16.1) are the same each year the P/A factor may be used in the relation $0 = -P + \text{CF}(P/A, i\%, n')$. After n' years (not necessarily an integer), the cash flows will recover the first cost P and a return of $i\%$. Equation (16.1) may therefore be used to find the number of years necessary to recover the first cost at a stated rate of return. If the paycheck period n' is less than the time you would expect to be able to employ or retain the asset, it should be bought. If n' is greater than the expected usable life, the asset should not be bought, since there will not be enough time to recover the investment plus the stated return during the usable life.

A frequent, but incorrect, industrial practice is to determine n' at $i = 0\%$, that is, no return is required. In this case, Eq. (16.1) becomes

$$0 = -P + \sum_{t=1}^{n'} \text{CF}_t \qquad (16.2)$$

If the cash flows are the same for each year, Eq. (16.2) may be solved for n' directly.

$$n' = \frac{P}{\text{CF}} \qquad (16.3)$$

Use of the no-return payback analysis is incorrect for two reasons:

1. It neglects the time value of money, since all computations are made at $i = 0\%$.
2. All cash flows that would commonly occur after the computed payback time n' are neglected.

Example 16.1 illustrates the computation of the payback period.

Example 16.1 A semiautomatic assembly machine can be purchased for $18,000 with a salvage value of $3000 and an annual cash flow of $3000. If the company would never expect such a machine to be used for more than 10 years, should it be purchased if (*a*) a 15% return is required and (*b*) no return is required.

SOLUTION Of course, there are several ways to answer this question—present worth, EUAC, or rate-of-return analysis. But let's use the payback approach. Using Eq. (16.1),

$$0 = -18{,}000 + \sum_{t=1}^{n'} \text{CF}_t(P/F, 15\%, t)$$

We assume that the salvage value of $3000 is correct regardless of how long the asset is retained. We can therefore modify the above relation:

$$0 = -18{,}000 + \text{CF}(P/A, 15\%, n) + \text{SV}(P/F, 15\%, n)$$

where SV(P/F, 15%, n) is the present worth of the salvage value after n years and the P/A factor has been used where possible. At $n = 15$ years we have

$$P = -18{,}000 + 3000(P/A, 15\%, 15) + 3000(P/F, 15\%, 15) = \$-89.10$$

For $n = 16$, the result is $+183.30. Interpolation indicates that in $n' = 15.3$ years the first cost plus 15% will be recovered. Since a fair estimate of usability is 10 years, the machine should not be purchased.
(*b*) For a no-return payback use Eq. (16.2) with $n' = 5$.

$$0 = -18{,}000 + 5(3000) + 3000$$

Therefore, the asset must be retained at least 5 years to return the investment with *no interest.* Based on the 10-year useful life, the machine should be purchased.

COMMENT Note the tremendous difference that a required return makes. At a 15% required return, this asset would have to be retained 15.3 years, while with no return only 5 years of ownership are required. This characteristic of longer required ownership is always present with $i > 0\%$ because of the time value of money. The pattern of cash flows is the primary determinant of the difference in required retention periods.

This result indicates that the payback-period method of analysis is appropriate only when investment capital is in very short supply and management requests recovery of capital in a short period. The payback calculation will give the amount of time required to recover the invested dollars; but from the point of view of the economic and time value of money, the payback-period method of analysis is not appropriate.

If two or more alternatives are present and payback analysis is used to select one, the second fallacy of payback periods mentioned above may lead to an incorrect decision. When cash flows which would occur after n' are neglected, it is possible to favor short-lived assets even when longer-lived assets produce a higher return. In these cases present-worth or EUAC analysis should always be considered the primary decision technique and payback only an auxiliary analysis tool. Comparison of the short- and long-lived assets in Example 16.2 illustrate this use of payback analysis.

Example 16.2 Consider the two machines detailed below.

Machine 1	Machine 2
P = \$12,000	P = \$8,000
Net income per year = 3,000	Net income per year = 1,000 (years 1–5)
	= 3,000 (years 6–15)
Maximum n = 7 years	Maximum n = 15 years

The increase in cash flow for machine 2 is anticipated because of its versatility and future application in the manufacture of new products.

(*a*) Which machine should be purchased if EUAC analysis is used with i = 15%?
(*b*) Use no-return and 15%-return payback analysis to select an alternative if the machine with the shorter payback period is favored. Comment on the results.

SOLUTION
(*a*) If the two machines are compared using EUAC, the larger positive value is selected, since the associated machine generates more income above the required 15% return.

$$\text{EUAC}_1 = -12{,}000(A/P, 15\%, 7) + 3000 = \$116$$

$$\text{EUAC}_2 = -8000(A/P, 15\%, 15) + 1000 + 2000(F/A, 15\%, 10)(A/F, 15\%, 15) = \$485$$

Since the EUAC_2 is larger, considerably more than 15% is returned and machine 2 is selected.

(*b*) For the no-return payback analysis, use Eq. (16.2) or (16.3); and for the 15%-return payback analysis, use Eq. (16.1) to determine n'_1 and n'_2.

No-return payback

Machine 1:

$$n'_1 = \frac{12{,}000}{3{,}000} = 4 \text{ years}$$

Machine 2: For the time period n'_2 = 6 years

$$0 = -8000 + 5(1000) + 3000$$

Select machine 1 because of its smaller n' value.

15%-return payback

Machine 1:

$$0 = -12{,}000 + 3{,}000(P/A, 15\%, n)$$

$$(P/A, 15\%, n) = 40$$

Interpolation gives $n'_1 = 6.57$ years, which is less than the maximum allowed life of 7 years.

Machine 2: For a 9-year payback period

$$0 = -8000 + 1000(P/A, 15\%, 5) + 3000(P/A, 15\%, 4)\,(P/F, 15\%, 5)$$

$$0 \neq \$-389.28$$

For 10 years the right side is positive.

$$0 = -8000 + 1000(P/A, 15\%, 5) + 3000(P/A, 15\%, 5)\,(P/F, 15\%, 5)$$

$$0 \neq \$352.34$$

By interpolation $n'_2 = 9.52$ years. Again, machine 1 is selected for its shorter payback period.

The payback analysis results are contradictory to the EUAC analysis because the 15-year life of machine 2 increases the return on the investment of $8000 with larger net incomes for years 6 to 15. Note that machine 1 is chosen for no-return and 15%-return payback analysis.

COMMENT Remember, payback analysis may be used to determine the required retention period of single projects, but it should not be relied upon for multiple alternative evaluation. Use present worth, EUAC, or rate of return for comparing two or more alternatives.

Probs. P16.1 to P16.9

16.2 Computation of Breakeven Points between Alternatives

In some economic analyses, one or more of the cost components may vary as a function of the number of units produced or consumed in manufacture. In these cases it is convenient to express the cost relation in terms of the variable and find the value at which the alternative proposals break even. In breakeven analysis the variable is usually *common to both alternatives*, such as a variable operating cost or production cost. Figure 16.1 graphically illustrates the breakeven concept for two proposals. As shown in the figure, the fixed cost, which may be simply the initial investment cost, of proposal 2 is greater than that of proposal 1, but proposal 2 has a lower variable cost as indicated by its smaller slope. The point of intersection (B) of the two lines represents the breakeven point between the two proposals. Thus, if the variable units (such as hours of operation or level of output) are expected to be greater than the breakeven amount, proposal 2 will be selected, since the total cost of the operation will be lower with this alternative. Conversely, an anticipated level of operation below the breakeven number of variable units would favor proposal 1.

Instead of plotting the total costs of each alternative and finding the breakeven point graphically, it is generally easier to calculate the breakeven point algebraically. Although the total cost may be expressed as either a present worth or equivalent uniform annual cost, the latter is generally preferable because the variable units are oftentimes expressed on a yearly basis. Additionally, EUAC calculations are simpler when

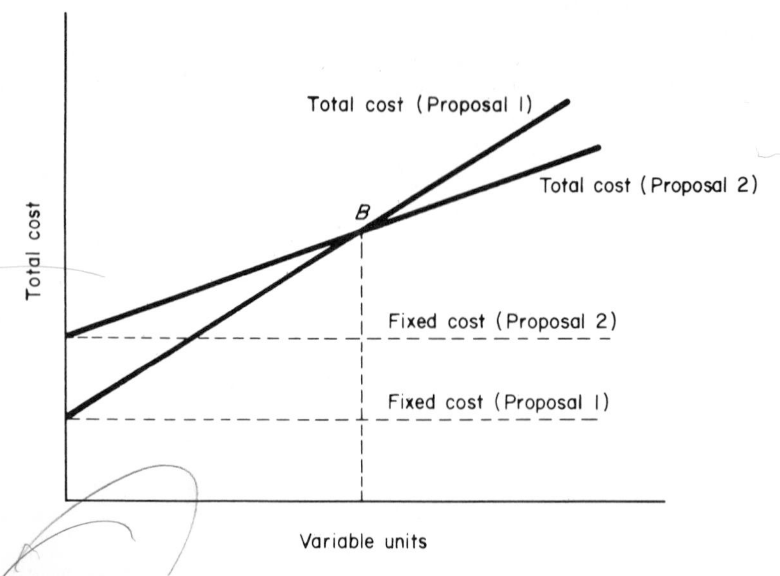

Figure 16.1 Graphical illustration of breakeven between two proposals.

the alternatives under consideration have different lives. In either case, the following steps may be used to select one of the alternatives:

1. Clearly define the variable and state its dimensional units.
2. Use EUAC or present-worth analysis to express the total cost of each alternative as a function of the defined variable.
3. Equate the two cost relations and solve for the breakeven value of the variable.
4. If the anticipated operating level is below the breakeven value, select the alternative with the higher variable cost (larger slope). If the level is above breakeven, select the alternative having the lower variable cost. Refer to Fig. 16.1.

The next two examples demonstrate typical breakeven computations.

Example 16.3 A sheet metal company is considering the purchase of an automatic feed machine for a certain phase of the finishing process. The machine has an initial cost of \$23,000, a salvage value of \$4000, and a life of 10 years. If the machine is purchased, one operator will be required at a cost of \$12 an hour. The output with this machine will be 8 tons per hour. Annual maintenance and operating cost of the machine is expected to be \$3500.

Alternatively, the company can purchase a less-sophisticated manual-feed machine for \$8000, which has no salvage value and a life of 5 years. However, with this alternative, three laborers will be required at a cost of \$8 an hour and the machine will have an annual maintenance and operation cost of \$1500. Output is expected to be 6 tons per hour for this machine. All invested capital must return 10%. (*a*) How much sheet metal must be finished per year in order to justify the purchase of the automatic-feed machine? (*b*) If management anticipates a requirement to finish 2000 tons per year, which machine should be purchased?

SOLUTION

(*a*) Use the steps above to find the breakeven point.

1. Let x be the number of tons of sheet metal per year.
2. For the automatic-feed machine the annual variable cost may be written as

$$\text{Annual variable cost} = \left(\frac{\$12}{\text{hour}}\right)\left(\frac{1 \text{ hour}}{8 \text{ tons}}\right)\left(\frac{x \text{ tons}}{\text{year}}\right) = \frac{12}{8}x$$

Note that the cost is in dollars per year, which is required for EUAC analysis. The EUAC for the automatic-feed machine is

$$\text{EUAC}_{\text{auto}} = 23{,}000(A/P, 10\%, 10) - 4000(A/F, 10\%, 10) + 3500 + \frac{12}{8}x$$

$$= \$6992 + 1.5x$$

Similarly, the annual variable cost and EUAC for the manual-feed machine is

$$\text{Annual variable cost} = \left(\frac{\$8}{\text{hour}}\right)(3 \text{ persons})\left(\frac{1 \text{ hour}}{6 \text{ tons}}\right)\left(\frac{x \text{ tons}}{\text{year}}\right) = 4x$$

$$\text{EUAC}_{\text{manual}} = 8000(A/P, 10\%, 5) + 1500 + \frac{3(8)}{6}x$$

$$= \$3610 + 4x$$

3. Equating the two costs and solving for x yields the breakeven value.

$$\text{EUAC}_{\text{auto}} = \text{EUAC}_{\text{manual}}$$

$$6992 + 1.5x = 3610 + 4x$$

$$x = 1352.8 \text{ tons per year}$$

4. If the output is expected to exceed 1352.8 tons per year, purchase the automatic-feed machine, since its variable cost slope of 1.5 is smaller than the manual-feed slope of 4.

(*b*) Select the smaller-sloped alternative, which is the automatic-feed machine in this case. Equivalently, it is possible to select an alternative by substituting the expected production level of 2000 tons per year into the EUAC relations. Then $\text{EUAC}_{\text{auto}} = \$9992$ and $\text{EUAC}_{\text{manual}} = \$11{,}610$, and purchase of the automatic-feed machine is justified.

COMMENT Work the problem on a present-worth basis to satisfy yourself that either method results in the same breakeven value. Remember how to select the correct alternative: As shown in Fig. 16.1, the alternative with the smaller slope (i.e., lower variable cost) should be selected when the variable units are above the breakeven point (and vice versa).

Example 16.4 The Jack n' Jill Toy Company currently purchases the metal parts which are required in the manufacture of certain toys, but there has been a proposal that the company make these parts themselves. Two machines will be required for the operation: Machine *A* will cost \$18,000 and have a life of 6 years and a \$2000 salvage value; machine *B* will cost \$12,000 and have a life of 4 years and a \$-500 salvage value. Machine *A* will require an overhaul after 3 years costing \$3000. The annual operating cost of machine *A* is expected to be \$6000 per year and for machine *B* \$5000 per year. A total of four laborers will be required for the two machines at a cost of \$2.50 per hour per worker. In a normal 8-day, the machines can produce parts sufficient to manufacture 1000 toys. Use a MARR of 15% per year and a purchase price of \$0.50 per toy if the parts are not manufactured.

(*a*) How many toys must be manufactured each year in order to justify the purchase of the machines?

(*b*) If the company expects to produce 75,000 toys per year, what maximum expenditure could be justified for the more expensive machine, assuming its salvage value and all other costs will be the same as stated?

SOLUTION

(*a*) Use steps 1 to 3 of the procedure above to determine the breakeven point.

1. Let x be the number of toys produced per year.
2. There are variable costs for the workers and fixed costs for the two machines. The annual variable cost is

$$\text{Variable cost/year} = (\text{cost/unit})\,(\text{units/year})$$

$$= \left(\frac{4 \text{ persons}}{1000 \text{ units}}\right)\left(\frac{\$2.50}{\text{hour}}\right)(8 \text{ hours})\,x = 0.08x$$

The fixed annual costs for the machines are

$$\text{Fixed EUAC}_A = 18{,}000(A/P, 15\%, 6) - 2000(A/F, 15\%, 6) + 6000 + 3000(P/F, 15\%, 3)\,(A/P, 15\%, 6)$$

$$\text{Fixed EUAC}_B = 12{,}000(A/P, 15\%, 4) + 500(A/F, 15\%, 4) + 5000$$

3. Equating the annual costs of the purchase option ($0.50x$) and the manufacture option yields

$$\begin{aligned} 0.50x &= \text{fixed EUAC}_A + \text{fixed EUAC}_B + \text{variable cost/year} \\ &= 18{,}000(A/P, 15\%, 6) - 2000(A/F, 15\%, 6) + 6000 \\ &\quad + 3000(P/F, 15\%, 3)(A/P, 15\%, 6) + 12{,}000(A/P, 15\%, 4) \\ &\quad + 500(A/F, 15\%, 4) + 5000 + 0.08x \qquad (16.4) \\ 0.42x &= 20{,}352.43 \\ x &= 48{,}458 \text{ units per year} \end{aligned}$$

Therefore, a minimum of 48,458 toys must be produced each year to justify the manufacture proposal.

(*b*) Substitute 75,000 for the variable x and the unknown first cost P_A for the \$18,000 in Eq. (16.4) and solve to obtain P_A = \$60,187. Thus, as much as \$60,187 is justified as the first cost for machine A if 75,000 toys are produced and other costs are as estimated.

Even though the preceding examples dealt with only two alternatives, the same type of analysis can be made for three or more alternatives. In this case, it becomes necessary to compare the alternatives with each other in order to find their respective breakeven points. The results reveal the ranges through which each alternative would be the most economical one. For example, in Fig. 16.2, if the output is expected to be

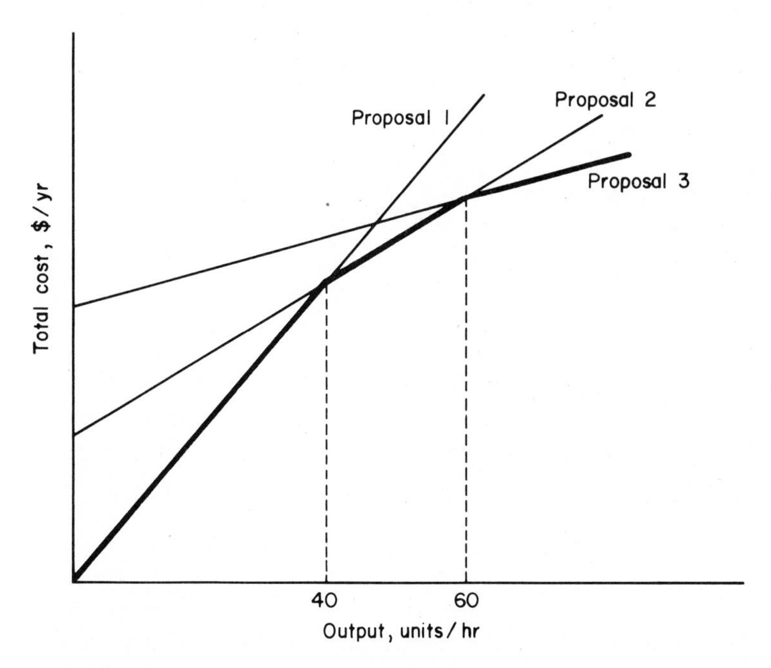

Figure 16.2 Breakeven points for three proposals.

less than 40 units per hour, proposal 1 should be selected. Between 40 and 60 units per hour, proposal 2 would be most economical; and above 60 units per hour, proposal 3 would be favored.

If the variable-cost relations are nonlinear, analysis is more complicated. If the costs increase or decrease uniformly, mathematical expressions that allow direct determination of the breakeven point can be derived. For further discussion of breakeven analysis refer to the Bibliography.

Probs. P16.10 to P16.21

*16.3 Life-Cycle Costing (LCC)

The technique of life-cycle costing analysis is primarily devoted to government work, especially for the selection of contractors to carry out defense-related projects. LCC may be defined as [1]:

> The total cost of an item including research and development, production, modification, transportation, introduction into inventory, construction, operation, support, maintenance, disposal, salvage revenue, and any other cost of ownership.

The total anticipated costs of an alternative are usually estimated using the following major categories [1, 2]:

Research and development costs. All expenditures for design, prototype fabrication, testing, manufacturing planning, engineering services, and the like of a product.

Production costs. The investment necessary to produce or acquire the product, including expenses to employ and train personnel, transport subassemblies and the final product, build new facilities, and acquire equipment.

Operating and support costs. All costs incurred to operate, inventory, maintain, and manage the product for its anticipated life.

The use of a present-worth analysis to account for the time value of money is helpful when performing a LCC. The major difference in LCC and the analyses we have performed thus far is the attempt to include *all* development costs and future maintenance costs.

The purpose of such an evaluation is to critically cost out each alternative for the entire life and select the one with the minimum LCC. Application is usually made to projects which will require research and development time to design and test a product or system intended to perform a specific task, such as radar for aircraft range detection.

Actually, an engineering-economic comparison of alternatives with every definable cost estimated for the life of each alternative is the same as the LCC analysis. However, because most applications of LCC are defense-related, the methods of cost estimation and comparison as used for government acquisition are somewhat different in presentation format. For a description of cost-estimation procedures in LCC refer to Seldon [1].

A case study using the LCC and engineering-economy approach is presented in Example 16.5.

Example 16.5 A medium-sized municipality wants to develop a computerized system which will assist in street-maintenance project selection during the next 10 years. A life-cycle cost approach has been used to categorize costs into development (that is, feasibility study, design, and procedure changes), programming and installation, and annual operating and support costs for the alternatives, each of which may be developed at either of two levels—one for information collection only using manual project selection, and a second level which provides for data collection and automatic selection of candidate street-maintenance projects.

The first alternative *A* involves the tailoring of software for the hardware currently owned by the city. This alternative would involve some procedure changes and one (level 1) or two (level 2) years to get ready. Alternative *B* is the development of a new system using the current municipal procedures. The same development times are anticipated for this alternative, but the resulting annual costs will be less.

As the final alternative *C*, the currently used manual system can be upgraded to perform the same function as the level 2 design for $100,000 per year. Assume that a $15,000-per-year cost will be incurred during the development time for each alternative to maintain the current manual system.

The costs are summarized in Table 16.1 and shown as cash-flow diagrams in Fig. 16.3. Note that the level 2 programming costs are equally spread over the 2-year development time. Use present-worth analysis and an interest rate of 10% to determine which alternative and level has the lowest LCC.

SOLUTION All possible costs throughout the 10-year life of the street-maintenance system are assumed included in the LCC. The lowest LCC with the 10% return

Table 16.1 Life-cycle cost components for selecting street-maintenance projects

		Level 1	Level 2
Alternative	Cost component	Information only	Projects selected
A (tailor software)	Development	$100,000	$200,000
	Programming	$175,000	$350,000
	Annual	$ 60,000	$ 55,000
	Time, years	1	2
B (new system)	Development	$ 50,000	$100,000
	Programming	$200,000	$500,000
	Annual	$ 45,000	$ 30,000
	Time, years	1	2
C (current)	Annual	...	$100,000

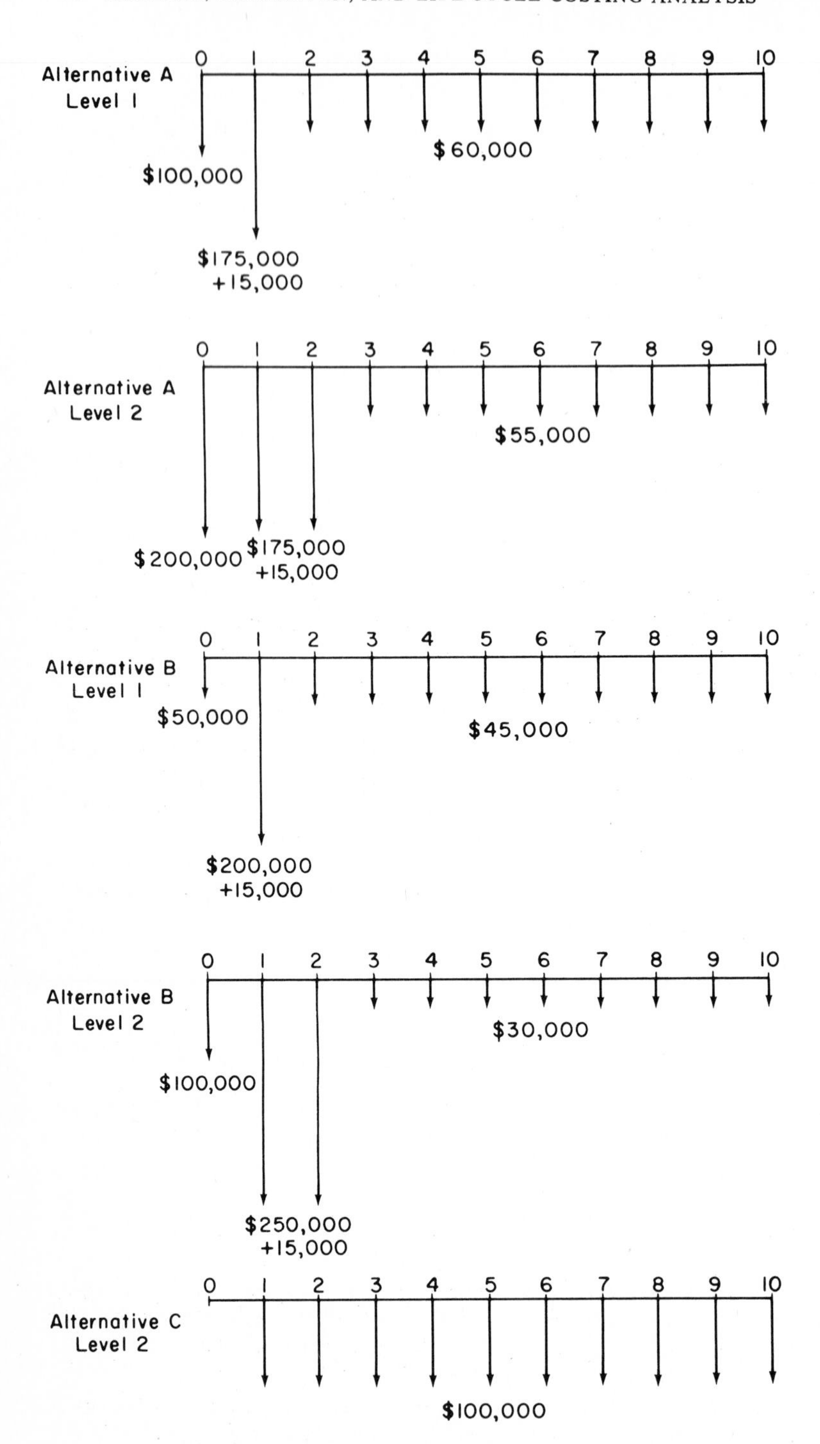

Figure 16.3 Cash-flow diagrams for Example 16.5.

considered is found by computing the PW value for each of the five competing plans. For alternative A the present worth for levels 1 and 2 are

$$PW_{A1} = 100{,}000 + 190{,}000(P/F, 10\%, 1) + 60{,}000(P/A, 10\%, 9)(P/F, 10\%, 1)$$
$$= \$586{,}859.41$$

$$PW_{A2} = 200{,}000 + 190{,}000(P/A, 10\%, 2) + 55{,}000(P/A, 10\%, 8)(P/F, 10\%, 2)$$
$$= \$772{,}226.87$$

For alternative B,

$$PW_{B1} = 50{,}000 + 215{,}000(P/F, 10\%, 1) + 45{,}000(P/A, 10\%, 9)(P/F, 10\%, 1)$$
$$= \$481{,}054.31$$

$$PW_{B2} = 100{,}000 + 265{,}000(P/A, 10\%, 2) + 30{,}000(P/A, 10\%, 8)(P/F, 10\%, 2)$$
$$= \$692{,}170.34$$

For the upgraded manual system,

$$PW_{C2} = 100{,}000(P/A, 10\%, 10)$$
$$= \$614{,}460.00$$

Since the present worth for alternative B, level 1, is smallest, the city should develop a new information gathering and recording system for its own hardware and use the manual project-selection technique. This is a sound decision economically as well as managerially because it uses the talents of computers and people in appropriate ways.

Prob. P16.22

REFERENCES

1. M. R. Seldon, *Life Cycle Costing: A Better Method of Government Procurement,* Westview, Boulder, CO, 1979.
2. D. R. Earles, "LCC-Commercial Application, Ten Years of Life Cycle Costing," Proceedings, 1975 Annual Reliability and Maintainability Symposium, January, 1975, pp. 74–85.

PROBLEMS

P16.1 Determine the number of years that an investor must keep a piece of property to make a 15% return. The purchase price was $6000 with taxes of $80 the first year increasing by $10 each year until sold. Assume a sales price of $14,000 for the first 10 years and $20,000 thereafter.

P16.2 When is it incorrect to use no-return payback-period analysis?

P16.3 (*a*) Determine the payback period for an asset that initially costs $8000, has a salvage of $500 when sold, and has a net profit of $900 per year. The required return is 8%.

(*b*) If the asset will be used for 5 years by the owner, should it be purchased?

P16.4 Determine the number of years that the two machines of plan B of Prob. P6.15 must be kept to make a 10% return if the annual income of the plan is \$4000. Assume that the estimated salvage values apply for all years.

P16.5 How many years must the owner of the artificial Christmas tree in Prob. P7.12 use the tree to make 20% on the investment?

P16.6 The R-U-There Detective Agency would like to purchase some new camera equipment. The following data are estimates

$$\text{First cost} = \$1050$$

$$\text{Annual operating cost} = 70 + 5(k) \qquad k = 1, 2, 3, \ldots$$

$$\text{Annual income} = 200 + 50(k) \qquad k = 1, 2, 3, \ldots$$

$$\text{Salvage value} = \$600$$

(*a*) Compute the payback period to make a return of 5% on the investment.
(*b*) Should the camera be purchased if the expected useful period is 7 years?
(*c*) Answer *a* and *b* above for a return of 10%.

P16.7 (*a*) Decide which alternative is more economic in Prob. P8.13 using the shorter payback-period criterion if annual revenues are expected to be \$8000 and a 10% return is required.

(*b*) Is the same alternative selected if EUAC analysis is used and the machine is kept its anticipated life?

P16.8 The lease cost of trucks is \$3300 per year payable at the beginning of each year. No refund is given for partial-year leases. As an alternative the purchase price per truck is \$1700 now with a monthly payment of \$300 for 4 years. A truck can be sold for an average of \$1200 regardless of the length of time of ownership. The prospective owner, Speede Drug Company, expects to increase net revenue by \$400 per month and requires a nominal return of 12% per year. The drug company must pay operation, maintenance, and insurance for the leased or purchased trucks, so this cost is equal and neglected.

(*a*) How many months must the lease and purchase plans be adhered to in order to make the required return?

(*b*) If a purchased truck has an expected life of 6 years, should the trucks be leased or purchased?

P16.9 (*a*) Use payback-period analysis to determine which of two machines should be selected if a 10% return is required.

(*b*) Is the answer here different for EUAC analysis if the retention period is the maximum life? Why?

Machine 1	Machine 2
$P = \$12{,}000$	$P = \$8000$
Cash flow = \$3000/year	Cash flow = \$1000 (years 1–5)
	= \$3000 (years 6–15)
Maximum n = 7 years	Maximum n = 15 years

P16.10 Two pumps can be used for pumping a corrosive liquid. A pump with a brass impeller costs \$800 and is expected to last 3 years. A pump with a stainless steel impeller will cost \$1900 and last 5 years. An overhaul costing \$300 will be required after 2000 operating hours for the brass impeller pump while an overhaul costing \$700 will be required for the stainless pump after 9000 hours. If the operating cost of each pump is \$0.50 per hour, how many hours per year must the pump be required to justify the purchase of the more expensive pump? Use an interest rate of 10% per year.

P16.11 The Sli-Dog Company is considering two proposals for improving the employees' parking area. Proposal A would involve filling, grading, and blacktopping at an initial cost of \$5000. The life of the parking lot constructed in this manner is expected to be 4 years with annual maintenance costs of $1000. Alternatively, the parking area would be paved under proposal *B*, in which case the life would be extended to 16 years. The annual maintenance cost will be negligible for the paved parking area, but the markings will have to be repainted every 2 years at a cost of $500. If the company's minimum attractive rate of return is 12% per year, how much can it afford to spend for paving the parking area so that the proposals would break even?

P16.12 The Redi-Bilt Construction Company is considering the purchase of a dirt scraper. A new scraper will have an initial cost of $75,000, a life of 15 years, a $5000 salvage value, and an operating cost of $30 a day. Annual maintenance cost is expected to be $6000 per year. Alternatively, the company can lease a scraper and driver as needed for $210 per day. If the company's minimum attractive rate of return is 12% per year, how many days per year must the scraper be required in order to justify the purchase?

P16.13 The Slo-Stitch Pant Company is considering the purchase of an automatic cutting machine. The machine will have a first cost of $22,000, a life of 10 years, and $500 salvage value. The annual maintenance cost of the machine is expected to be $2000 per year for the range of usage anticipated. The machine will require one operator at a cost of $24 a day. Approximately 1500 yards of material can be cut each hour with the machine. Alternatively, if manual labor is used, five workers each earning $18 a day can cut 1000 yards per hour. If the company's minimum attractive rate of return is 8% per year, how many yards of material must be cut each year in order to justify the purchase of the automatic machine?

P16.14 A couple have an opportunity to buy an old fire-damaged house for what they believe to be a bargain price of $28,000. They estimate that remodeling the house now will cost $12,000 and annual taxes will be approximately $800 per year. They estimate that utilities will cost $500 per year and that the house must be repainted every 3 years at a cost of $400. At the present time, resale houses are selling for $16 per square foot, but they expect this price to increase by $1.50 per square foot per year. They will lease the house for $2500 per year until it is sold. If the house has 2500 square feet and they want to make 8% on their investment, (*a*) how long must they keep the house before they can break even and (*b*) what would the selling price be at the time of the sale?

P16.15 The Rawhide Tanning Company is considering the economics of furnishing an in-house water-testing laboratory instead of sending samples to independent laboratories for analysis. If the lab were completely furnished so that all tests could be conducted in-house, the initial cost would be $25,000. A technician would be required at a cost of $13,000 per year. The cost of power, chemicals, etc., would be $5 per sample. If the lab is only partially furnished, the initial cost will be $10,000. A parttime technician will be required at an annual salary of $5000. The cost of in-house sample analysis will be $3 per sample, but since all tests cannot be conducted by the company, outside testing will be required at a cost of $20 per sample. If the company elects to continue the present condition of complete outside testing, the cost will be $55 per sample. If the laboratory equipment will have a useful life of 12 years and the company's MARR is 10% per year, how many samples must be tested each year in order to justify (*a*) the complete laboratory and (*b*) the partial laboratory? (*c*) If the company expects to test 175 samples per year, which of the three alternatives should be selected?

P16.16 The Quik-Kil slaughterhouse and packing plant currently pays the city $4000 a month in wastewater discharge fees. The company is considering constructing a treatment plant of its own. The treatment plant will require a parttime operator at $300 per month. In addition, the operating cost is expected to be $500 per month. The treatment plant will require minor repairs every 4 years costing $1500 and is expected to last for 20 years. If the company's MARR is a nominal 12% per year compounded monthly, how much could the company afford to spend on a treatment plant and not exceed its present costs?

P16.17 A city engineer is considering two methods for lining water-holding tanks. A bituminous coating can be applied at a cost of $2000. If the coating is touched up after 4 years at a cost of $600, its life can be extended 2 more years. Alternatively, a plastic lining can be installed, which will have a life of 15 years. If the city uses an interest rate of 5% per year, how much money can be spent for the plastic lining so that the two methods just break even?

P16.18 A family who are planning to build a new house are trying to decide between purchasing a lot in the city or in the suburbs. A 1000-square-meter lot in the city will cost $10,000 in the area in which they want to buy. If they purchase a lot outside the city limits, a similar parcel will cost only $2000. For the size of house they plan to build, they expect annual taxes to amount to $1200 per year if they build in the city and only $150 per year in the suburbs. If they purchase the lot outside the city limits, they will have to drill a well for $4000. With their own well, they will save $150 per year in water charges, but they expect the city to provide water to their area in 5 years, after which time they will purchase the city water. They estimate that the increased travel distance will cost $325 the first year, $335 the second year, and amounts increasing by $10 per year. Using a 25-year analysis period and an interest rate of 6% per year, how much extra could the family afford to spend on the house outside the city limits and still have the same total investment? Assume that the land can be sold for the same price as its initial cost.

P16.19 The I. M. Rite family are considering insulating their attic to prevent heat loss. They are considering R-11 and R-19 insulation. They can install R-11 for $160 and R-19 for $240. They expect to save $35 per year in heating and cooling with R-11. If the interest rate is 6%, how much money must they be able to save per year in order to justify the R-19 insulation if they want to recover their investment in 7 years?

P16.20 A waste-holding lagoon situated near the main plant receives sludge on a daily basis. When the lagoon is full, it is necessary to remove the sludge to a side located 4.95 kilometers from the main plant. At the present time, whenever the lagoon is full, the sludge is removed by pumping it into a tank truck and hauling it away. This requires a portable pump that costs $800 and has an 8-year life. The company supplies the labor to operate the pump at a cost of $25 per day, but the truck and driver must be rented at a cost of $110 per day. Alternatively, the company can install a pump and pipeline to the remote site. The pump would have an initial cost of $600 and a life of 10 years and would cost $3 per day to operate. The company's MARR is 15%. (*a*) If the pipeline would cost $3.52 per meter to construct, how many days per year must the lagoon require pumping in order to justify construction of the pipeline? (*b*) If the company expects to pump the lagoon one time per week, how much money could it afford to spend on the pipeline in order to just break even? Assume a pipeline life of 10 years.

P16.21 A building contractor is considering two alternatives for improving the exterior appearance of a commercial building that is being renovated. The building can be completely painted at a cost of $2800. The paint is expected to remain attractive for 4 years, at which time the job would have to be redone. Every time the building is painted, the cost will increase by 20% over the previous time. Alternatively, the building can be sandblasted now and every 10 years at a cost 40% greater than the previous time. The remaining life of the building is expected to be 38 years. If the company's MARR is 10%, what is the maximum amount that could be spent now on the sandblasting alternative so that the two alternatives will just break even? Use present-worth analysis to solve this problem.

***P16.22** A window manufacturing company is considering the development of a computerized-decision support system to help reduce the scarp cost of glass. The current manual system allows $150,000 in scrap each year and requires 0.5 man-year to maintain at a cost of $35,000 per man-year. Improvements in the system next year will cost $25,000 and the system will be used for 10 more years. The suggested system has the following projected costs for a life of 10 years:

$125,000 per year for 2 years to develop (years 1 and 2)
$100,000 for a share of a new minicomputer (year 2)
$17,500 per year to maintain the manual system (years 1 and 2)
$8000 per year to maintain hardware and software (years 3 to 10)
$37,000 per year personnel cost (years 3 to 10)
$30,000 per year scrap cost once installed

Sales of the system to other companies is expected to net a $15,000 revenue for years 4 through 10. If a 20% return is required, which alternative has the lower LCC for the 10-year life?

CHAPTER

SEVENTEEN

CAPITAL RATIONING UNDER BUDGET CONSTRAINTS

This chapter will give you an introduction to the selection of alternatives or capital-investment projects when the total amount of investment is limited. Both the present-worth and rate-of-return solutions to this problem, which is commonly called *capital budgeting*, are presented. The mathematical programming formulation for the problem is also discussed. This is an optional chapter.

SECTION OBJECTIVES

To complete this chapter you must be able to:

17.1. Define the *capital-budgeting* problem and state its prominent characteristics.
17.2. Use the procedure to find the present worth of project bundles to solve a capital-budgeting problem, given the MARR, budget constraint, cash-flow sequence, and life for each project.
17.3. Do the same as in 17.2 except use the rate-of-return method.
*17.4. Develop the *integer linear program* for a capital-budgeting problem given the investments, cash flows, and lives of each project, the MARR, and the budget constraint.

STUDY GUIDE

17.1 The Capital-Budgeting Problem

Most corporations have the opportunity to select from several capital-investment projects which are not mutually exclusive, that is, more than one of the projects may be selected. However, the money available for investment is limited to some amount which is generally stated by management. This is commonly known as the *capital-budgeting problem* and it has the following characteristics:

1. Several projects are available that are *independent* of each other, that is, selection of a particular project does *not* preclude selection of any other of the projects.
2. Each project is either selected or not selected, that is, partial investment in a project is not possible.
3. A budgetary constraint restricts the total investment possible in the projects. Additional budget constraints may be present for several years after the initial investment is made.
4. The objective of financial investment for the corporation is to maximize the value of the investments.

It is also likely that the candidate projects have nondeterministic cash-flow sequences and different risks associated with them. We will discuss in detail problems which are of equal risk, have only an initial-year budget constraint, and have deterministically predictable cash flows for the life of the project.

In the typical capital-budgeting problem, there are several independent alternatives, each with a first cost, life, series of yearly independent cash flows, and possibly a salvage value. If any of the alternatives are mutually exclusive, this must be specifically accounted for. Likewise, if any are dependent, that is, must be done in conjunction with some other alternative, the dependent components should be combined into *one independent* alternative. (If you need a review of mutually exclusive alternatives, read Sec. 8.5 at this time.)

17.2 Capital Budgeting Using Present-Worth Analysis

To determine the selected projects for a given budget limitation B, it is necessary first to formulate all mutually exclusive bundles of projects, each of which has a total investment that does not exceed B. Then the present worth of each bundle is found at the MARR and the bundle with the *largest* PW value is selected.

To illustrate the concept of bundling consider these four projects.

Project	Investment
A	\$10,000
B	5,000
C	8,000
D	15,000

If the budget constraint is B = \$25,000 feasible bundles are as follows:

Projects	Total investment	Projects	Total investment
A	\$10,000	AD	\$25,000
B	5,000	BC	13,000
C	8,000	BD	20,000
D	15,000	CD	23,000
AB	15,000	ABC	23,000
AC	18,000		

The number of bundles to be considered increases rapidly according to the relation $2^m - 1$, where m is the number of candidate projects.

The procedure to solve a capital-budgeting problem using PW analysis is:

1. Develop all feasible mutually exclusive bundles which have a total initial investment that does not violate the budgetary constraint B.
2. Determine the cash-flow sequence CF_{jt} for each bundle j and year t.
3. Compute the PW value P_j for each bundle j at the MARR using the relation

$$P_j = \sum_{t=1}^{n_j} \text{CF}_{jt}(P/F, i\%, t) \tag{17.1}$$

where CF_{jt} = cash flow for bundle j in year t
n_j = life of bundle j

4. Select the bundle with the largest P_j value.

Selecting the maximum P_j value means that this bundle produces a return larger than any other bundle above the MARR. Any bundle with $P_j < 0$ is discarded because it does not produce a return greater than MARR. This procedure is illustrated in the next example.

Example 17.1 The Megabuck Company has \$20,000 to invest next year on any or all of the projects detailed in Table 17.1. Select the subset of projects to accept if the MARR is 15%.

SOLUTION Use the procedure above with B = \$20,000 to select one mutually exclusive bundle of projects which maximizes present worth. In the solution a $\text{CF}_{jt} > 0$ is a net cash inflow and $\text{CF}_{jt} < 0$ is a net outflow. The initial investment for bundle j is CF_{j0}, which is negative.

1. All feasible bundles with $|\text{CF}_{j0}| \leqslant 20{,}000$ are detailed in column 2 of Table 17.2. Project E is eliminated from further consideration because the initial investment exceeds B.

Table 17.1 Independent alternatives considered for investment

Project	Initial investment	Expected annual cash flow	Project life, years
A	$10,000	$2,870	9
B	15,000	2,930	9
C	8,000	2,680	9
D	6,000	2,540	9
E	21,000	9,500	9

Table 17.2 Present-worth analysis for the capital-budgeting problem in Example 17.1

(1) Bundle j	(2) Projects	(3) Investment CF_{j0}	(4) Cash flow CF_j	(5) Present worth P_j at $i = 15\%$
1	*A*	$-10,000	$2,870	$+ 3,694.49
2	*B*	-15,000	2,930	- 1,019.21
3	*C*	- 8,000	2,680	+ 4,787.89
4	*D*	- 6,000	2,540	+ 6,119.86
5	*AC*	-18,000	5,550	+ 8,482.38
6	*AD*	-16,000	5,410	+ 9,814.36
7	*CD*	-14,000	5,220	+10,907.75

2. The cash-flow sequence, column 4, is the sum of individual project cash flows for the 9-year life of each bundle.
3. Use Eq. (17.1) to compute the present worth of each bundle. Let $CF_j = CF_{jt}$ for $t = 1, 2, \ldots, 9$ and $n = n_j = 9$ for $j = 1, 2, \ldots, 7$, since the cash flows are the same for each bundle and the lives are all 9 years. Use the relation

$$P_j = -CF_{j0} + CF_j(P/A, 15\%, 9)$$

for the present-worth values in Table 17.2. The largest present worth is $P_7 =$ $10,907.75; therefore, invest $14,000 in projects *C* and *D*.

COMMENT The return on bundle 7 exceeds 15%; in fact, solution to $0 = -14{,}000 + 5220(P/A, i^*\%, 9)$ gives $i^* = 34.8\%$ as the actual return. Note that bundle 2 does not return at least the MARR, since $P_2 < 0$. This analysis also assumes that any funds above the initial investment, for example, the $6000 that is uncommitted by selection of bundle 7, will return the MARR.

In a capital-budgeting problem it is assumed that the investment is made for the life of the longest-lived project with reinvestment of all positive cash flows at the MARR. There is no renewal of the same projects at the end of their life, so the use of a least common denominator for the n value for unequal-lived projects is not appropriate as it is in PW analysis. Rather Eq. (17.1) and the procedure above is used to select a bundle by present-worth analysis for projects of any life.

In fact, it is possible to demonstrate that PW evaluation using the factor (P/A, MARR%, n_j) for a bundle j with life n_j is the same as reinvesting from year n_j to n_L at MARR, where n_L is the longest-lived project. At the end of a bundle's life the cash flows, assumed to be the same each year, are worth $\mathrm{CF}_j(F/A, \mathrm{MARR}\%, n_j)$. Reinvestment of this sum to year n_L and computation of the present worth of all future receipts to year n_L in year 0 results in the following relation.

$$P_j = \mathrm{CF}_j(F/A, \mathrm{MARR}\%, n_j)(F/P, \mathrm{MARR}\%, n_L - n_j)(P/F, \mathrm{MARR}\%, n_L) \quad (17.2)$$

Let MARR = i and use the factor formulas (Chap. 3 or Appendix A) to simplify Eq. (17.2).

$$P_j = \mathrm{CF}_j\left[\frac{(1+i)^{n_j}-1}{i}\right][(1+i)^{n_L-n_j}]\left[\frac{1}{(1+i)^{n_L}}\right]$$

$$= \mathrm{CF}_j\left[\frac{(1+i)^{n_j}-1}{i(1+i)^{n_j}}\right] = \mathrm{CF}_j(P/A, i\%, n_j) \quad (17.3)$$

Since the bracketed expression in Eq. (17.3) is the (P/A, i%, n_j) factor, evaluation of each bundle for n_j years by present-worth analysis assumes reinvestment at MARR of all positive cash flows until the longest-lived project is completed in year n_L.

The next example demonstrates capital-budgeting problem selection for unequal-lived projects.

Example 17.2 Use MARR = 15% and B = \$20,000 to select projects from the following capital-investment possibilities.

Project	Initial investment	Expected annual cash flow	Project life, years
A	\$10,000	\$2,870	6
B	15,000	2,930	9
C	8,000	2,680	5
D	6,000	2,540	4

SOLUTION The unequal lives cause the cash flows to vary from year to year, but the present-worth solution procedure is the same as that in Example 17.1. Of the $2^4 - 1 = 15$ bundles the seven feasible ones with cash flows and P_j values by Eq. (17.1) are given in Table 17.3. For example, bundle $j = 7$ has

$$P_7 = -14{,}000 + 5220(P/A, 15\%, 4) + 2680(P/F, 15\%, 5) = \$2235.60$$

Select projects C and D since P_7 is the largest present worth.

This selection coincidentally happens to be the same as that in Example 17.1 where all lives are 9 years. It is the cash-flow values *and* the number of years over which they are received that determine the selected bundle.

Table 17.3 Present-worth analysis for an unequal-lived projects capital-budgeting problem, Example 17.2

(1)	(2)	(3)	(4)	(5)	(6)
			Cash flows		
Bundle j	Projects	Investment CF_{j0}	Year t	CF_{jt}	Present worth P_j
1	A	$-10,000	1-6	$2,870	$+ 861.52
2	B	-15,000	1-9	2,930	-1,019.21
3	C	- 8,000	1-5	2,680	+ 983.90
4	D	- 6,000	1-4	2,540	1,251.70
5	AC	-18,000	1-5	5,550	+1,845.41
			6	2,870	
6	AD	-16,000	1-4	5,410	+2,113.43
			5-6	2,870	
7	CD	-14,000	1-4	5,220	+2,235.60
			5	2,680	

COMMENT Consider the bundle $j = 7$ evaluation in Table 17.3 and its cash-flow diagram in Fig. 17.1. At 15% the future worth at year 9, life of the longest-lived bundle, is

$$F = 5220(F/A, 15\%, 4)(F/P, 15\%, 5) + 2680(F/P, 15\%, 4) = \$57,111.36$$

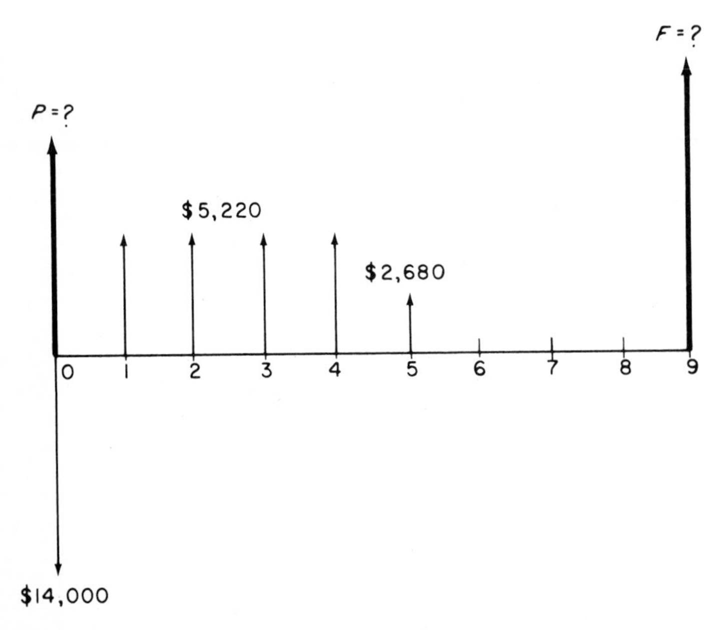

Figure 17.1 Cash-flow diagram for bundle 7, Example 17.2, for a 9-year investment period.

The present worth with CF_{70} = \$-14,000 considered is

$$P = -14{,}000 + 57{,}111.36(P/F, 15\%, 9) = \$2236.76$$

Allowing for a small round-off error from the multiple use of factors, we find that the P value is the same as P_7 in Table 17.3, thus verifying the reinvestment assumption summarized by Eq. (17.3). So we see that it is not necessary to equalize the evaluation periods of bundles in a capital-budgeting problem.

Probs. P17.1 to P17.6

17.3 Capital Budgeting Using Rate-of-Return Analysis

It is correct to solve a capital-budgeting problem by using incremental rate-of-return analysis (Sec. 8.5) on mutually exclusive bundles that have been ordered by increasing initial investment and do not require an initial investment $CF_{j0} > B$. The procedure may be summarized as follows:

1. Develop all feasible mutually exclusive bundles such that $CF_{k0} \leqslant B$ for $k = 1, 2, \ldots,$ m bundles.
2. Order the bundles by increasing initial investments and renumber the bundles $j = 1, 2, \ldots, m$.
3. Determine the cash-flow sequence CF_{jt} for each bundle j and year t.
4. Considering the do-nothing alternative as the defender, compute the overall i^* for the lowest-investment bundle ($j = 1$) using present-worth [Eq. (7.1)] or EUAC [Eq. (7.2)] analysis (or ROIDS in Appendix E).
5. If $i^* <$ MARR, remove the lowest-investment bundle and compute the return on the next bundle. Repeat this step until an $i^* \geqslant$ MARR, then the current lowest-investment bundle is the defender and the next bundle is the challenger.
6. Determine the incremental cash-flow sequence ΔCF_{jt} between the challenger and the defender.
7. Calculate i^* on the ΔCF_{jt} sequence of the challenger.
8. If the step 7 $i^* \geqslant$ MARR, the challenger becomes the defender and the old defender is discarded. Conversely, if $i^* <$ MARR in step 7 retain the defender, remove the challenger, and use the next bundle as a new challenger.
9. Repeat steps 6 to 8 until only one feasible bundle remains.

This procedure will ensure that each incremental investment is justified in that the return is at least MARR. (See Sec. 8.5 for a review.) Note that knowing that $i^* \geqslant$ MARR is sufficient in this analysis, that is, the exact i^* values must be calculated only if you desire to know the number.

As discussed by Fleischer [1], this approach and the present-worth approach will result in the correct selection of the same bundle, so it is suggested that the PW approach be used since it is easier to perform and (possibly) understand.

Example 17.3 Solve the capital-budgeting problem in Example 17.1 if the minimum required return is 15%.

Table 17.4 Ordered bundles for the capital-budgeting problem in Example 17.3

Ordered bundle j	Projects	Investment CF_{jo}	Cash flow $CF_{jt}(t = 1, 2, \ldots, 9)$
1	*D*	\$- 6,000	\$2,540
2	*C*	- 8,000	2,680
3	*A*	-10,000	2,870
4	*CD*	-14,000	5,220
5	*B*	-15,000	2,930
6	*AD*	-16,000	5,410
7	*AC*	-18,000	5,550

SOLUTION The procedure above is used to select a bundle.

1, 2, 3. All feasible bundles with an initial investment $|CF_{jo}| \leqslant B = \$20{,}000$ are developed and ordered and the cash-flow sequence $CF_{jt}(t = 1, 2, \ldots, 9)$ is shown in Table 17.4.

4. Use Eq. (7.1) in the form

$$0 = \text{incremental investment} + \text{incremental } CF_{jt}(P/A, i^*\%, 9)$$

$$= \Delta CF_{jo} + \Delta CF_{jt}(P/A, i^*\%, 9) \qquad (17.4)$$

to approximate i^* as shown in Table 17.5.

5. For bundle 1, $i^* > 40\%$, so project *C* (challenger) is to be compared with *D* (defender).
6. The incremental investment is $\Delta CF_{20} = \$-2{,}000$ and the incremental cash flow is $\Delta CF_{2t} = \$140$ for $t = 1, 2, \ldots, 9$.

Table 17.5 Incremental rate-of-return analysis of a capital-budgeting problem, Example 17.3

	Bundle j						
	1	2	3	4	5	6	7
	Projects						
	D	*C*	*A*	*CD*	*B*	*AD*	*AC*
Investment CF_{jo}	\$-6,000	\$-8,000	\$-10,000	\$-14,000	\$-15,000	\$-16,000	\$-18,000
Annual CF_{jt}	2,540	2,680	2,870	5,220	2,930	5,410	5,550
Bundles compared	*D* to none	*C* to *D*	*A* to *D*	*CD* to *D*	*B* to *CD*	*AD* to *CD*	*AC* to *CD*
ΔCF_{jo}	\$-6,000	\$-2,000	\$- 4,000	\$- 8,000	- 1,000	\$- 2,000	\$- 4,000
ΔCF_{jt}	2,540	140	330	2,680	- 2,290	190	330
$(P/A, i^*\%, 9)$	2.3622	...	...	2.9851	...	...	...
Incremental i^*	>40%	<0%	<0%	>30%	<0%	<0%	<0%
Increment justified?	Yes	No	No	Yes	No	No	No
Projects selected	*D*	*D*	*D*	*CD*	*CD*	*CD*	*CD*

7. By Eq. (17.4) we have $(P/A, i^*\%, 9) > 9.0$, which is the limit at a zero return for $n = 9$.
8. Since C is not incrementally justified, retain D as the defender.
9. Continue to compare new challengers until bundle 4 (CD) has $i^* > 30\%$ and remains the defender against all remaining bundles (see Table 17.5). Select CD as the best bundle.

This solution is, as expected, the same as in Example 17.1, but the present-worth method requires less work because the incremental analysis is not necessary.

COMMENT Had the project lives been unequal as in Example 17.2, the same procedure would be used over the respective bundle lives, to ensure that all positive (incremental) cash flows are invested at the minimum return of 15% until the longest-lived bundle has been completed. You should try to solve Example 17.2 at this time using the rate-of-return procedure to be sure you understand the method.

There are several other ways that the capital-budgeting problem has been "solved" using rate-of-return analysis, but none will consistently give correct results. The development of mutually exclusive bundles and the use of incremental i^* analysis will ensure a correct result which is always the same as the (correct) present-worth solution to the capital-budgeting problem.

Probs. P17.7 to P17.10

*17.4 Capital Budgeting Using Mathematical Programming

As soon as there are several alternative projects in a capital-budgeting problem the number of mutually exclusive bundles $2^m - 1$ gets too large to evaluate by hand or calculator. It is best then to use a computerized solution to the capital-budgeting problem using present-worth analysis written in terms of the *integer linear programming* formulation that follows.

Maximize $$Z = \sum_{k=1}^{m} P_k x_k$$

Constraints $$\sum_{k=1}^{m} |\mathrm{CF}_{ko}|\, x_k \leqslant B \qquad (17.5)$$

$$x_k = 0 \text{ or } 1 \qquad k = 1, 2, \ldots, m$$

where $P_k = \Sigma_{t=1}^{n_k} \mathrm{CF}_{kt}(P/F, i\%, t)$ is the present worth at the MARR of project k which has a life of n_k years (CF_{ko} is not included here)

B = maximum allowed investment

$$x_k = \begin{cases} 0 & \text{if project } k \text{ is not included} \\ 1 & \text{if project } k \text{ is included} \end{cases}$$

The objective function Z requires that the combination of projects included in the solution has the maximum present worth possible. Note that these are projects, not mutually exclusive bundles, as in previous sections. The first constraint ensures that the initial investments for all selected projects CF_{k0} does not exceed the budget limitation B. The last constraint, $x_k = 0$ or 1, ensures that each project is completely included ($x_k = 1$) or completely excluded ($x_k = 0$) from the solution. This is commonly called the project indivisibility constraint and it makes this a 0-1 integer programming (IP) problem.

Solution is best accomplished by a canned computerized IP solution package. Research in this area is presented in optimization texts and current journal articles.

Since solution to an IP problem is often time-consuming (even for a computer), and therefore quite expensive, it is possible to relax the project indivisibility constraint to the form

$$0 \leqslant x_k \leqslant 1 \qquad k = 1, 2, \ldots, m$$

where x_k now represents the fraction of the initial investment of project k that is committed in a particular solution. This relaxation assumes that any project can be divided for investment purposes, an assumption that may not be correct but does allow the capital-budgeting problem to now be solved using the faster technique of linear programming (LP). The answers by LP and IP will not necessarily be the same because of the removal of the indivisibility constraint to reformulate as an LP problem. Discussion of these aspects are included in advanced texts. (See Further Information references.)

Example 17.4 Develop an integer programming formulation for the capital-budgeting problem in Example 17.2.

SOLUTION First calculate the P_k values for the four projects detailed in Example 17.2 using $i = 15\%$.

Project k	Cash flow CF_{kt}	Life n_k	$(P/A, 15\%, n_k)$	Present worth P_k
A	\$2,870	6	3.7845	\$10,861.52
B	2,930	9	4.7716	13,980.79
C	2,680	5	3.3522	8,983.90
D	2,540	4	2.8550	7,251.70

Substitute the P_k values and $B = \$20,000$ into the Eq. (17.5) problem with subscripts 1, 2, 3, and 4 for projects A, B, C, and D, respectively.

Maximize

$$Z = 10{,}861.52\, x_1 + 13{,}980.79\, x_2 + 8{,}983.90\, x_3 + 7{,}251.70\, x_4$$

Constraints

$$10{,}000\,x_1 + 15{,}000\,x_2 + 8000\,x_3 + 6000\,x_4 \leqslant 20{,}000$$

$$x_1, x_2, x_3, \text{ and } x_4 = 0 \text{ or } 1$$

By inspection of the budget constraint in conjunction with the objective function It is quite easy in this small problem to see the feasible combinations of projects which parallel the mutually exclusive bundles of Example 17.2. You may refer to Table 17.3 for these combinations. The solution is, as in Example 17.2, to select projects C and D for a maximum value of Z = \$16,235.60 and an initial investment of \$14,000. This solution is written

$$x_1 = 0 \quad x_2 = 0 \quad x_3 = 1 \quad x_4 = 1$$

Of course, it becomes increasingly difficult to obtain the answer by inspection when there are more projects, so a computerized IP package should be used.

There are several other constraints that can be used in the Eq. (17.5) problem to accommodate special cases. Situations like dependent projects, complementary projects, and investment over several years are easily modeled by additional constraints. For example, if there are budget limitations for several years in the future and projects require investments for more that the first year, a series of budget constraints is used instead of the single-year restriction in Eq. (17.5). This series may be written

$$\sum_{k=1}^{m} I_{kt} x_k \leqslant B_t \qquad t = 1, 2, \ldots, n_L$$

where I_{kt} = investment in project k for year t
B_t = budget limitation in year t
n_L = life of the longest-lived project

There are as many constraints of this type as there are B_t values. Additional information on this and other constraint types is available in more advanced texts.

Probs. P17.11 to P17.13

REFERENCE

1. G. A. Fleischer, "Two Major Issues Associated with the Rate of Return Method for Capital Allocation: The 'Ranking Error' and 'Preliminary Selection,'" *Journal of Industrial Engineering*, vol. 17, no. 4, pp. 202–208.

FURTHER INFORMATION

Bussey [3], pp. 269-324; Canada and White [4], pp. 420–443; Grant, Ireson, and Leavenworth [7], pp. 557-571; Newnan [11], pp. 352-368; Stevens [15], pp. 165-172; Thuesen, Fabrycky, and Thuesen [17], pp. 187-198.

PROBLEMS

P17.1 Determine which of the following projects should be selected for investment if $30,000 is available and the required return is 10%. Use the present-worth method to make your selection.

Project	Investment	Cash flow	Life
A	$10,000	$3950	8
B	12,000	2400	8
C	18.000	5750	8
D	22,000	3530	8

P17.2 (*a*) Rework Prob. P17.1 using the following project-life values:

Project	*A*	*B*	*C*	*D*
Life, years	3	8	5	12

(*b*) How many mutually exclusive bundles may be formed if the budget restriction is ignored?

P17.3 Management of the Hoof-Power Cattle Company has decided it can invest in any three of four available projects. Each project has an initial investment of $10,000 and a specified present-worth value at i = 18%. Select the projects which offer the best investment opportunity.

Project	Life (years)	Project present worth at 18%
1	13	$ 1840
2	5	375
3	10	-1800
4	8	25

P17.4 Use the present-worth procedure to solve the following capital-budgeting problem: Select up to three of four alternatives to maximize the return if MARR after taxes is 10% and the available budget is (*a*) $16,000 and (*b*) $24,000. The last year in which a CFAT value is shown indicates the end of the alternative.

		CFAT for year				
Alternative	Investment	1	2	3	4	5
1	$ 6,000	$1,000	$1,700	$2,400	$ 3,100	$ 3,800
2	10,000	500	500	500	500	10,500
3	8,000	5,000	5,000	2,000		
4	10,000	0	0	0	15,000	

P17.5 New investment funds for the Company of the Future are restricted to $100,000 for this next year. Select any or all of the following projects using i = 15% and the present-worth values.

Project	Initial investment	Annual cash flow	Life, years	Salvage value
1	$25,000	$ 6,000	4	$ 4,000
2	30,000	9,000	4	-1,000
3	50,000	15,000	4	20,000

P17.6 Use the project bundle 34 for the data in Prob. P17.4 to demonstrate that equalization of evaluation periods is not necessary for present-worth comparison if reinvestment at the required return (10%) is assumed until the life of the longest-lived bundle is reached.

P17.7 Work Prob. P17.1 using the rate-of-return method.

P17.8 Use (*a*) the rate-of-return method and (*b*) the present-worth method to solve the following capital-budgeting problem for a budget of $5000. Let MARR = 14%.

Project	Investment	Cash flow	Life
I	$3000	$1000	5
II	4500	1800	5
III	2000	900	5

P17.9 Work Prob. P17.4*a* using the rate-of-return method.

P17.10 Use the rate-of-return procedure at $i = 15\%$ to solve Prob. P17.5.

***P17.11** Set up the integer programming formulation for Prob. P17.1.

***P17.12** (*a*) Use the data in Prob. P17.5 to develop the integer program to solve the capital-budgeting problem. (*b*) What change is necessary if management has decided to allow partial investment in the projects assuming that cash flows can also be partitioned?

***P17.13** Locate an integer programming package for your computer and solve Prob. P17.4 using it.

CHAPTER

EIGHTEEN

ESTABLISHING THE MINIMUM ATTRACTIVE RATE OF RETURN

This chapter introduces you to some of the problems and methods, both quantitative and subjective, involved in setting the minimum attractive rate of return (MARR) used in engineering-economy analysis.

The problems of an uncertain future, inflation, risk, subjectivity, and altered management policy make the establishing of a realistic MARR quite difficult. While the quantitative models illustrated here give approximate answers, the degree of error is not calculable. Nevertheless, the setting of MARR is important to a good economy study and sound financial decisions. This is an optional chapter.

SECTION OBJECTIVES

To complete this chapter, you must be able to do the following:

18.1. Define the two separate sources of capital financing available to a corporation.
18.2. Define the term *cost of capital* and state why the minimum attractive rate of return is greater than the cost of capital.
18.3. State five reasons why the minimum attractive rate of return can vary.
18.4. Compute the before- and after-tax cost of capital for *debt* financing, given the type of financing, period of financing, interest rate, repayment scheme, and tax rate.
18.5. Compute the before- and after-tax cost of capital for *equity* financing using the methods below, given market price, book value, earnings and dividend for a share of common stock, and tax rate:
    (*a*) Dividend method
    (*b*) Earnings/price ratio method
    (*c*) Gordon-Shapiro method
    (*d*) Opportunity-cost method

18.6. Compute the average cost of capital before and after taxes given the before- or after-tax cost of capital for both debt and equity financing, the proportion of each type of financing, and the tax rate.
18.7. Compute the debt-to-equity ratio given the proportion of debt and equity financing, and state what happens to the cost of capital as this ratio changes.

STUDY GUIDE

18.1 Types of Capital Financing

A corporation will accumulate capital by three different methods. These three are categorized into two sources, debt and equity financing, to correspond with the balance sheet sections of liabilities and owner's equity, respectively (Appendix D). The types of financing are defined as follows:

Debt financing. Capital borrowed from others that will be paid back at a stated interest rate by a certain specified date. The original owner (lender) takes no direct risk on the return of the funds and interest nor does he share in the profits the borrowing firm makes on the funds. Debt financing includes borrowing via bonds, mortgages, and loans and may be classed as long-term or short-term liabilities.
Equity financing. Capital owned by the corporation used to make a profit for the corporation. There are two types of equity financing: *owner's funds*, which are funds obtained from stock sales and may include funds from the company owners, if the firm is small or not a stock-issuing concern; *retained earnings*, which are sometimes referred to as plowback funds. These funds have been previously retained by the firm for investment and expansion purposes; they are owned by the stockholders, not the corporation per se.

In actual computations dealing with the setting of MARR, the cost to the corporation of each type of financing, debt or equity, is computed independently of other types. This assumption of independence will be used throughout this chapter. The proportion of debt and equity financing that should be used by a corporation is a very difficult problem to solve, a problem we will discuss only briefly in this chapter.

Prob. P18.1

18.2 The Cost of Capital

The actual interest rate paid by the corporation in developing investment capital is called the *cost of capital* (CC). Since most firms use a combination of debt and equity financing and since these two types have different interest rates, the CC is an intermediate rate. As mentioned earlier, independence between types of financing is assumed in computing the CC. If we assume a project will be financed by a $100,000 bond issue (debt financing) and the actual rate of interest paid is 8% per year, the CC is 8%. In other words, the CC is a minimum cutoff of the return required on an investment. Thus, if the $100,000 investment will return 6%, money will be lost.

For most commerical and industrial organizations, the amount of investment capital available is the limiting resource; that is, there are many investment opportunities that would yield a rate of return greater than even the MARR. Since only limited investment funds are available, however, the projects that are undertaken usually have a projected rate of return considerably higher than the CC and MARR. In addition, several projects that cannot be funded immediately because of limited capital also have projected rates of return greater than the MARR. Therefore, new projects that are under consideration are not to be undertaken unless their expected rate of return is at least as great as the rate of return *on the least attractive proposal that has not yet been funded.* Figure 18.1 is a presentation of the relation between the different rate-of-return values and the cost of capital.

Commonly, CC < MARR, the latter being the return criterion used in economy

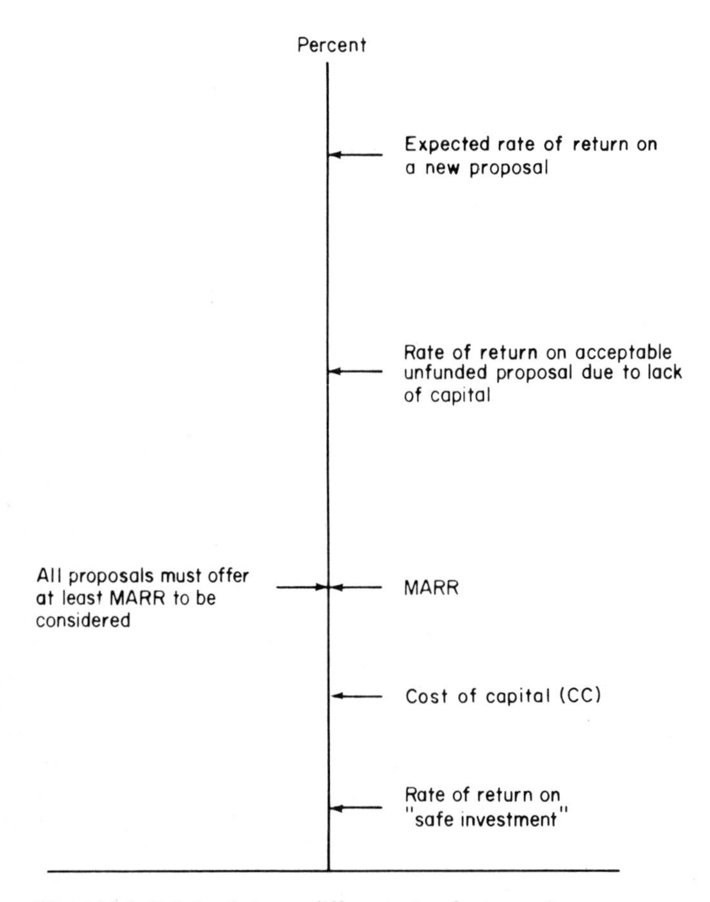

Figure 18.1 Relation between different rate-of-return values.

studies. If the CC is 8% and an added 7% return is expected, then the MARR is 15%. The overall return on a project must be at least equal to the MARR to consider the project for funding. If a company performs its present-worth and EUAC analyses at the CC rate, obviously the company is satisfied to *just* recover the investment; since this is generally not the case, the economic analysis is usually performed at a MARR > CC. Determination of *actual* CC is virtually impossible. While the quantitative methods available can give an approximate value, when the MARR is established, subjective judgment and experience are vital. Moreover, accuracy in determining CC is justified in terms of neither time nor economics because of a fluctuating economy, inflation, risks, and changing management policies.

It is important to remember that as the demand for capital exceeds its supply, the MARR will far exceed the CC because of the great selectivity that is possible and necessary.

Prob. P18.2

18.3 Variations in MARR

The MARR is not a set, nonvarying value. Rather, it is altered by corporations for different types of projects. For example, a firm may use a MARR of 15% for depreciable assets and a MARR of 20% for diversification investment, that is, purchasing smaller companies, land, etc.

The MARR varies from one project to another and through time because of the following:

1. *Project risk.* The more risk that is *judged* to be associated with a proposed project, the higher the MARR and, for that matter, the higher the CC for the project.
2. *Sensitivity of project area.* If management is determined to diversify (or invest) in a certain area, the MARR may be lowered to encourage investment with the hope of recovering lost profit in other investment areas. This subjective reaction to investment opportunity can create much havoc with an economy study.
3. *Tax structure.* If taxes are increasing because of increased profits, recaptured depreciation and gains from retired assets, and increasing local taxes, the MARR will be increased. An after-tax study would assist in eliminating this reason for a fluctuating MARR.
4. *Capital-financing methods.* As capital becomes limited, the MARR is increased and management begins to look closely at the life of the project. As the demand for the limited capital exceeds supply, the MARR is further increased.
5. *Rates used by other firms.* If the rates of other firms that are used as a standard increase, a company may alter its MARR upward in response. A typical standard may be that of the firm called the government, even though the MARR for government projects varies drastically and is set in a nonquantitative fashion, since profit is not a requisite of government investment.

Prob. P18.3

18.4 Cost of Capital for Debt Financing

Debt financing includes borrowing by bond and mortgages. The interest (or dividend) paid on borrowed money can be used to reduce taxes, as first discussed in Chap. 15; therefore, in computing the CC for debt financing, an after-tax approach should be taken. The two examples below illustrate determination of CC for debt financing.

Example 18.1 A bond face value of $500,000 will be raised by issuing 500 $1000 8%-a-year 10-year bonds. If the effective tax rate is 50% and the bonds are discounted 2% for quick sale, compute the cost of capital (*a*) before taxes and (*b*) after taxes.

SOLUTION

(*a*) The annual dividend payment is $1000(0.08) = $80 and the discount sale price is $980. Using Eq. (7.1), we find the i^* at which

$$0 = -980 + 80(P/A, i^*\%, 10) + 1000(P/F, i^*\%, 10)$$

Trial and error results in a value $i^* = 8.31\%$, which is the before-tax CC for the $500,000.

(*b*) With the allowance to reduce taxes by deducting the interest on borrowed money, a tax savings of $80(0.5) = $40 per year is realized and actual annual dividend outlay is $80 - $40 = $40 in the equation of part *a*. Using Eq. (7.1), the after-tax CC value is now 4.26%.

Example 18.2 A company plans to purchase a certain asset for $20,000 with a zero salvage value and anticipated life of 10 years. Management has decided to put $10,000 down now and borrow $10,000 at an interest rate of 6% on the unpaid balance. The repayment scheme will be $600 interest each year and the $10,000 principal in year 10. (*a*) What is the loan's after-tax CC (ATCC) if the tax rate is 50%? (*b*) If the asset will have a cash flow before taxes (CFBT) of $5000 per year and will be straight-line-depreciated, is the investment justified?

SOLUTION

(*a*) Figure 18.2 presents the cash flow for repayment of the $10,000 loan only with a tax credit for 0.50($600) = $300 on the annual interest. We can set up the following relation for the loan only to find the after-tax CC of 3%.

$$0 = 10{,}000 - 300(P/A, i\%, 10) - 10{,}000(P/F, i\%, 10)$$

(*b*) The CFAT (cash flow after taxes) value for each year of asset ownership is computed using Eqs. (15.12) and (15.13). (See Example 15.8 for a quick review.)

$$\text{CFAT} = \$5000$$

$$\text{Taxable income} = 5000 - (20{,}000/10) - 600 = \$2400$$

$$\text{Taxes} = 0.5(2400) = \$1200$$

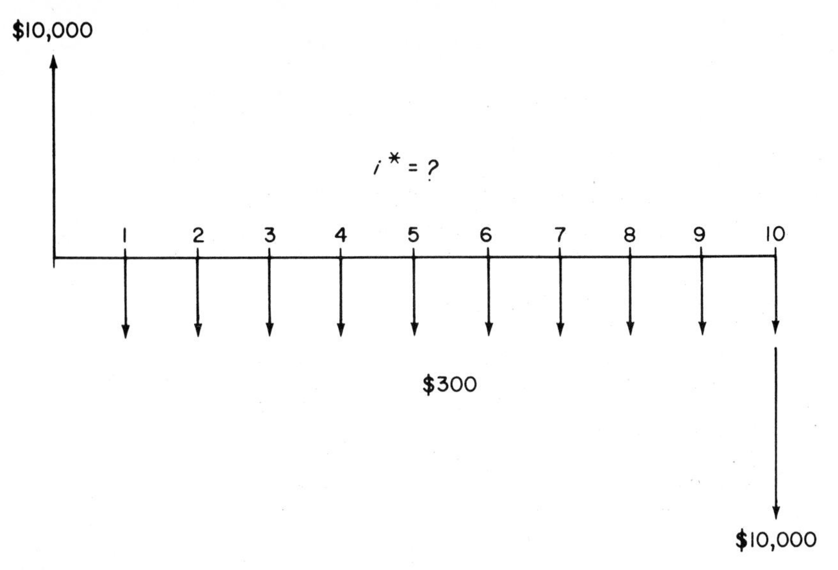

Figure 18.2 Cash flow for a loan of $10,000 at 6%, Example 18.2.

$$\text{CFAT} = 5000 - 1200 - 600 = \$3{,}200 \qquad \text{(years 1-9)}$$

$$= 5000 - 1200 - 600 - 10{,}000 = \$-6800 \qquad \text{(year 10)}$$

The after-tax rate of return is $i^* = 26.5\%$, which satisfies the relation

$$0 = -10{,}000 + 3200(P/A, i^*\%, 9) - 6800(P/F, i^*\%, 10)$$

The investment is justified for all values of MARR $\leqslant 26.5\%$.

COMMENT In part *a*, realize that the CC is *not* the loan rate of 6%, since this is not the rate paid on the entire $20,000 first cost.

Probs. P18.4 to P18.8

18.5 Cost of Capital for Equity Financing

The use of equity funds–stocks or retained earnings–involves *no tax advantage*. If the cost of capital (CC) is computed before taxes, the desired income must include earnings and a dollar amount sufficient to cover the required tax payment. There are many ways to compute CC, some quantitative, some subjective. We briefly look at four methods, the first three quantitative, the last a subjective method that can be easily applied by the corporation analyst or the individual. All methods use the same example to permit a comparison of answers. Realize that since the CC is computed independently for debt or equity financing, the methods presented here have no effect on the CC computation using debt financing.

18.5.1 Dividend method If a company is primarily interested in the dividend paid to stockholders and not in retained earnings as a source of equity financing, the after-tax cost of capital (ATCC) is the stated dividend rate. The before-tax cost of capital (BTCC), with taxes accounted for, is computed as

$$\text{BTCC} = \frac{\text{after-tax dividend rate}}{1 - \text{tax rate}} \tag{18.1}$$

Example 18.3 A total of $50,000 new capital is to be raised by selling 2500 common stocks at the market value of $20 per share. If a dividend rate of 5% is anticipated and the tax rate is 60%, compute (*a*) ATCC and (*b*) BTCC.

SOLUTION

(*a*) ATCC is stated as 5% by management.

(*b*) BTCC may be computed using Eq. (18.1) as

$$\text{BTCC} = \frac{0.05}{1 - 0.60} = 0.125$$

COMMENT In part *b*, the actual earnings needed to cover the dividend will be $50,000(0.125) = $6250 per year.

18.5.2 Earnings/price ratio method A corporation is usually more interested in total earnings per share, which includes dividends *and* retained earnings, that is,

$$\text{Earnings} = \text{dividend} + \text{retained earnings}$$

Here the BTCC is computed as

$$\text{BTCC} = \frac{\text{earnings}}{\text{market price }(1 - \text{tax rate})} \tag{18.2}$$

Example 18.4 Using the details of Example 18.3 and assuming an earning rate of 10% per share of market price, compute (*a*) ATCC and (*b*) BTCC.

SOLUTION

(*a*) ATCC is 10%, as stated.

(*b*) Using Eq. (18.2) with earnings of $20(0.10) = $2 per share,

$$\text{BTCC} = \frac{2}{20(1 - 0.6)} = 0.25$$

COMMENT Of course BTCC can be computed using Eq. (18.1) as

$$\text{BTCC} = 0.10/(1 - 0.6) = 0.25$$

however, dividends are often stated in dollars per share not as a percentage of market price, which justifies the inclusion of Eq. (18.2). Note that doubling the earnings per share has doubled the CC values of Example 18.3.

18.5.3 Gordon-Shapiro method This method is presented without proof here [1]. Both the stock and retained-earning types of equity financing are included in the CC formula. After conversion from a discrete to a continuous time scale, the ATCC is computed as

$$\text{ATCC} = \frac{D}{P} + \frac{Y - D}{\text{BV}} \tag{18.3}$$

where D = current dividend per share
P = current market value per share
Y = current earnings per share
BV = current book value per share

Then,

$$\text{BTCC} = \frac{\text{ATCC}}{1 - \text{tax rate}} \tag{18.4}$$

A simplifying assumption in this formula is that the earnings/book value ratio (Y/BV) remains constant at all times. Therefore, a value of Y/BV = 0.20 now is the same 5 or 7 years from now. Of course, this value actually fluctuates, but it is hoped not chaotically. Also worthy of mention is that the first term in Eq. (18.3), (D/P), is the same as the ATCC that could be computed using the *dividend method* above. For a \$20-per-share stock with a 5% dividend (\$1 per share), ATCC = D/P = 1/20 = 0.05, the same as the ATCC for the dividend method. Thus, with addition of the retained earnings/book value ratio, ($Y - D$)/BV, both types of equity financing are accounted for.

Example 18.5 Using the details of Examples 18.3 and 18.4 and assuming BV is \$15 per share, use the Gordon-Shapiro method to compute (*a*) ATCC and (*b*) BTCC.

SOLUTION
(*a*) Using Eq. (18.3)

$$\text{ATCC} = \frac{1}{20} + \frac{2 - 1}{15} = 0.117$$

(*b*) By Eq. (18.4)

$$\text{BTCC} = \frac{0.117}{1 - 0.6} = 0.292$$

COMMENT A review of these three quantitative, progressively more complex methods shows BTCC to progress from 12.5% (dividend only) to 29.2% (dividend and retained earnings). While all are only approximate, they do give an idea of the cost of capital.

In summary, the selected methods presented here would be utilized under the following conditions:

1. If the consistent payment of a specified dividend rate is of primary concern to company management, then the dividend method should be used.
2. If both the dividend rate and the percent of earnings retained in the company are of primary interest to management, use the earning/price ratio method.
3. When, in addition to the above, the book value of the stock is to be considered, use the Gordon-Shapiro method.
4. If quantitative analysis is not desired and subjective evaluation will suffice, use the opportunity-cost method discussed below.

Other methods usable in obtaining the cost of capital are summarized in Refs. 2 and 3.

18.5.4 Opportunity-cost method If a firm (or individual) considers all present and future investments of about the same risk level as the one presently contemplated, a subjectively determined (and usually experienced) cost of capital can be stated. This rate is the opportunity cost of capital. For example, if a company can make a consistent 15% rate of return before taxes by investing in mining operations, this 15% may well be considered the CC and the MARR without resorting to quantitative methods of determining a cost of capital. Although this method is subjective, it is quite useful when applied in a rational, timely manner.

Probs. P18.9 to P18.14

18.6 Computation of an Average Cost of Capital

Most project capital is not obtained via only debt or equity financing; rather, it is taken from a common pool of capital. Since most firms do not operate on total equity capital and since it is impossible to continue to operate on 100% financing, an average CC value should be computed using the weightings of debt and equity financing of the firm. If a specific project breakdown is definitely known, these weightings should be used. Example 18.6 demonstrates the computation of an average cost of capital.

Example 18.6 A firm must raise a total of $500,000 of new capital for a particular project. If the company raises capital in the proportion 60% equity and 40% debt and if, for this project, equity capital would cost 12% after taxes and debt capital would cost 10% before taxes, find the total earnings necessary (*a*) before taxes and (*b*) after taxes. The tax rate is 60%.

SOLUTION First we must compute the average ATCC and average BTCC values for both equity and debt portions of financing. Refer to Table 18.1 for results of the following computations. For *equity* capital no tax advantage is realizable. Using Eq. (18.1),

$$\text{BTCC} = \frac{0.12}{1 - 0.6} = 0.30$$

Table 18.1 Computation of the average cost of capital before and after taxes for Example 18.6

Source of funds	BTCC	ATCC	Proportion of funds	Average BTCC	Average ATCC
Equity	30%	12%	60%	18%	7.2%
Debt	10	4	40	4	1.6
				22%	8.8%

The average BTCC is found by weighting BTCC with the proportion of funds from each type of financing to obtain

$$\text{Average BTCC} = \text{equity BTCC (equity fraction)} + \text{debt BTCC (debt fraction)}$$

$$= 0.3(0.6) + 0.1(0.4) = 0.22$$

For *debt* financing a tax advantage can be claimed:

$$\text{ATCC} = \text{BTCC}(1 - \text{tax rate}) = 0.10(1 - 0.6) = 0.04$$

Then,

$$\text{Average ATCC} = 0.12(0.6) + 0.04(0.4) = 0.088$$

(*a*) Earnings before taxes = \$500,000(0.22) = \$110,000 per year.
(*b*) Earnings after taxes = \$500,000(0.088) = \$44,000 per year.

Often the MARR is set using a CC value representative of only one type of financing, even though the capital comes from a common fund including both types of capital. If the MARR is determined as though funds were obtained from only one method of financing when in reality both types of capital are used, an unrealistic MARR may result. When the actual type of financing is known, the MARR should be determined for this source; however, if the funds are obtained from a capital pool that includes both types of financing, a compromise MARR may be computed using the overall company split of debt and equity funds.

Solved Problem 18.7
Probs. P18.15 to P18.21

18.7 Effect of the Debt/Equity Ratio on the Cost of Capital

The value of the ratio

$$\text{D/E} = \frac{\text{proportion of debt capital}}{\text{proportion of equity capital}}$$

is called the debt/equity ratio. Of course, this ratio varies with the company and type of industry. The range is $0 < \text{D/E} < \infty$. For example, if debt capital represents 60% of the capital and equity 40%, D/E = 60/40 = 1.5. As the ratio increases, the average CC value decreases, as shown in the Solved Problems. Since CC decreases as D/E goes up, a likely question is, why not use 100% debt financing? You may remember that this question was first posed for *you* in Example 15.9. As debt financing for a firm increases, the lenders take a larger risk and require a greater interest rate on loans. It is impossible to maintain a business and solvency without a healthy share of equity financing. In other words, if the company did not own some part of itself, it would not be able to obtain operating or investment capital.

The advantage that debt financing allows is referred to as *leverage*. More debt capital releases other equity funds for use but in the long run discourages potential stockholders from investing in the company because of the high D/E ratio.

Solved Problem 18.8
Probs. P18.22 to P18.26

SOLVED PROBLEMS

Example 18.7 The Hearty Food chain wants to purchase a fleet of 15 new delivery trucks for $150,000. Each truck has a salvage value of $1,000 after 10 years and will be straight-line-depreciated. There are two methods of capital financing available–equity and debt. Equity financing would involve the selling of stocks at the market value of $15. These stocks would pay a dividend of $0.50 per share and have an anticipated earning rate of 5% of market value. The present book value per share is $12.

Maximum debt financing approved by the bank is 50% of the sum needed. The loan will be for 10 years and repayment will be at 8% interest, to be paid in 10 equal annual installments. If the fleet is expected to produce an annual CFBT of $30,000 and the effective tax rate is 50%, which method of financing is more advantageous? Assume that the company's MARR is two times the cost of capital.

SOLUTION

100% equity financing. We first compute ATCC using the Gordon-Shapiro method, Eq. (18.3):

$$\text{ATCC} = \frac{0.50}{15} + \frac{0.75 - 0.50}{12} = 0.0542$$

Therefore, MARR = 2(5.42) = 10.84%. For 100% equity financing, no interest tax credit is allowed. To find the actual rate of return, we need the CFAT values. For each of the 10 years,

$$\text{CFBT} = \$30{,}000$$

$$\text{Depreciation} = \frac{150{,}000 - 15{,}000}{10} = \$13{,}500$$

$$\text{Taxes} = 0.5(\text{TI}) = 0.5(30{,}000 - 13{,}500) = \$8250$$

$$\text{CFBT} = 30{,}000 - 8250 = \$21{,}750$$

Now, we solve the rate-of-return equation:

$$0 = -150{,}000(A/P, i^*\%, 10) + 15{,}000(A/F, i^*\%, 10) + 21{,}750$$

A return of $i^* = 8.4\%$ satisfies this equation. Since $8.4\% < 10.84\%$, equity financing is not advisable.

50% debt financing. Since two types of financing are involved here, we must find an average ATCC. The ATCC for 50% equity financing is still 5.42%, since all values remain the same. For the \$75,000 loan, we must compute the equivalent annual payment A.

$$A = 75{,}000(A/P, 8\%, 10) = \$11{,}177$$

For ease of calculation, we will assume that the principal is reduced uniformly by an amount of \$75,000/10 = \$7500 per year. Therefore, the annual interest payment is approximated as \$11,177 − \$7500 = \$3677. The annual tax credit for the interest is 0.5(\$3677) = \$1839. Therefore, the CC value for debt financing is approximated as the return which satisfies

$$0 = 75{,}000(A/P, i^*\%, 10) - (11{,}177 - 1839)$$

The value of $(A/P, i^*\%, 10) = 0.12451$ is correct for $i^* = 4.2\%$, which is the ATCC value for debt financing. Note that the 50% debt financing effectively reduced the cost of capital from the loan rate of 8 to 4.2% because of the tax advantage from interest.

Weighting the equity and debt ATCC values by 50% gives

$$\text{Average ATCC} = 0.5(5.42) + 0.5(4.2) = 4.81\%$$

The 50% debt financing has reduced the ATCC from 5.42% for 100% equity financing to 4.81%. Now, MARR = 2(4.81) = 9.62%, compared with the previous value of 10.84%. We are ready to compute the actual rate of return on the \$75,000 put in by the company. The annual CFAT value is computed from Eq. (15.13) using the following data.

$$\text{CFBT} = \$30{,}000$$

$$\text{Depreciation} = \$13{,}500$$

$$\text{Taxes} = 0.5(30{,}000 - 13{,}500 - 3677) = \$6412$$

$$\text{CFAT} = 30{,}000 - 6412 - 3677 - 7500 = \$12{,}411$$

The rate-of-return equation may be written

$$0 = -75{,}000(A/P, i^*\%, 10) + 15{,}000(A/F, i^*\%, 10) + 12{,}411$$

which is satisfied at $i^* = 11.95\%$. Since $11.95\% > \text{MARR} = 9.62\%$, this method of financing is advised.

COMMENT For the 50% debt financing we have used the average annual interest and principal values of \$3677 and \$7500, respectively, to compute $i^* = 11.95\%$. Actually, the changing interest and principal values for each year should be used, since the portion of the $11,177 annual payments applied toward the loan principal increases each year. For example, in the first year

$$\text{Interest} = 75{,}000(0.08) = \$6000$$

$$\text{TI} = 30{,}000 - 13{,}500 - 6000 = \$10{,}500$$

$$\text{Taxes} = 0.5(10{,}500) = \$5250$$

$$\text{CFAT} = 30{,}000 - 5250 - 6000 - 5177 = \$13{,}573$$

However, for year 2

$$\text{Interest} = (75{,}000 - 5177)(0.08) = \$5586$$

$$\text{TI} = 30{,}000 - 13{,}500 - 5586 = \$10{,}914$$

$$\text{Taxes} = 0.5(10{,}914) = \$5457$$

$$\text{CFAT} = 30{,}000 - 5457 - 5586 - 5591 = \$13{,}366$$

Therefore, even though an annual $11,177 payment is made, owing to the 8% interest on the declining balance, the actual rate-of-return equation should be written

$$0 = -75{,}000(A/P, i^*\%, 10) + 15{,}000(A/F, i^*\%, 10) + \left[\sum_{t=1}^{10} (\text{CFAT})_t\,(P/F, i^*\%, t)\right](A/P, i^*\%, 10)$$

However, the additional calculations are not worth the trouble, since the approximation of the actual rate of return is usually quite good.

Secs. 18.4 to 18.6

Example 18.8 Mr. Billy N. Aire is making plans to invest in common stock. He has records of three electronics companies. The total asset value and ownership is given in Table 18.2. For each company, answer the following questions for the particular situation explained:

(*a*) Compute the present debt/equity ratio (D/E).
(*b*) If a decrease of 20% in asset value takes place, compute the new debt/equity ratios.

Table 18.2 Financing profile for three companies*

		Financing	
Firm	Asset value	Debt	Equity
A	$5.0	$1.0	$4.0
B	4.0	2.0	2.0
C	6.0	5.0	1.0

*All $ values in millions.

(*c*) If revenue is \$1.5 million for each company, compute the rate of return on the issued common stock. Assume that interest on debt financing averages 6% for company *A*, 8% for *B*, and 10% for *C*. Compare the returns and comment about their relative magnitudes.

SOLUTION

(*a*) The present D/E ratios (values divided by 1 million) are

$$(D/E)_A = \frac{1}{4} = 0.25$$

$$(D/E)_B = \frac{2}{2} = 1.00$$

$$(D/E)_C = \frac{5}{1} = 5.00$$

(*b*) Table 18.3 presents computations of D/E ratios after a 20% decrease in asset value. The company debt remains the same after the decrease; thus, the equity share must decrease and D/E ratios increase. The zero equity value for company *C* is used, even though the actual loss of \$1.2 million would result in a \$-200,000 stock ownership. In this case, the stockholders of company *C* are completely wiped out. You can see that when a company suffers an asset devaluation, because of obsolescence or some other reason, the stockholder is hurt and the D/E ratio increases.

(*c*) Table 18.4 gives a summary of rate of return on issued stock values using the

Table 18.3 D/E ratios after a 20% decrease in asset value*

	Asset value			Financing		
Firm	Old	Reduction	New	Debt	Equity	D/E ratio
A	\$5.0	\$1.0	\$4.0	\$1.0	\$3.0	0.33
B	4.0	0.8	3.2	2.0	1.2	1.67
C	6.0	1.2	4.8	5.0	0.0	∞

*All \$ values in millions.

Table 18.4 Rate of return on issued stocks*

	(1)	(2)	(3)	(4) = (3) (2)	(5) = (1) – (4)	(6) Stock	(7) = (5)/(6) Rate of
Firm	Revenue	Debt	Rate	Amount	Net income	value	return
A	\$1.5	\$1.0	0.06	\$0.06	\$1.44	\$4.00	36%
B	1.5	2.0	0.08	0.16	1.34	2.00	67
C	1.5	5.0	0.10	0.50	1.00	1.00	100

*All \$ values in millions.

given interest rates. These interest rates increase as the D/E ratio increases, due to the greater *risk* taken by the lender. The return values increase from 36 to 100%. The great increase in return as D/E increases is called *leverage*. However, don't be fooled. This situation is by no means a panacea, since such a high return as 100% is based on a present equity value of zero and a $5 million debt. In short, confidence in a company with a very large D/E ratio is generally bad, no matter how large the return on stock value.

Sec. 18.7

REFERENCES

1. M. J. Gordon and E. Shapiro, Capital Equipment Analysis: The Required Rate of Return, *Management Science*, October, 1959.
2. A Reisman, *Managerial and Engineering Economics*, Allyn and Bacon, Boston, 1971.
3. L. E. Bussey, *The Economic Analysis of Industrial Projects*, Prentice-Hall, Englewood Cliffs, NJ, 1978.

FURTHER INFORMATION

Bussey [2], pp. 152–185; Newnan [11], pp. 329–341; Reisman [12], pp. 283–368; Stevens [15], pp. 85–109.

PROBLEMS

P18.1 State whether each of the following is in the category of debt or equity financing: (*a*) short-term note from the bank; (*b*) $5000 taken from a co-owner's savings account to pay a company bill; (*c*) a $150,000 bond issue; (*d*) an issue of preferred stock worth $55,000; (*e*) Ms. Broke borrows $50,000 from her brother at 3% interest to run her business. The brother is not a co-owner in the company.

P18.2 If the owner of the Dollar Daze Variety Store has computed an overall cost of capital of $5\frac{1}{2}$% and is about to evaluate the purchase of a new cash register using a MARR of 6%, what return is expected on the investment? Does this seem reasonable? Why?

P18.3 Will each of the following cause the MARR to be raised or lowered? State why.

(*a*) Investment in a chain of quick-food stores is contemplated, but the president is very leery of such an undertaking.

(*b*) The Crooked Nail Construction Company built a 250-unit apartment house 3 years ago and still retains ownership. Due to the risk, when the project was undertaken a 12% return was required; however, because of the favorable outcome management feels this is safer than some other types of investments.

P18.4 A large auto-manufacturing firm requires $2 million in new funds. The financial consultants of the company recommend that the firm sell 12-year 8% semiannual bonds to a brokerage firm at a 4% discount.

(*a*) What is the total face value of the bond issue?

(*b*) Compute the after-tax cost of capital obtained in this manner, if the tax rate is 50%.

P18.5 The Barely-Making-It Swimming Suit Company has to raise $50,000 in new capital. Two methods of debt financing are available. The first is to borrow $50,000 from a bank. The company will pay an effective 8% per year for 8 years to the bank. The other method will require issuing 50 \$1000 10-year bonds which will pay a 6% annual dividend. If you assume that the principal on the loan is reduced uniformly for the 8 years, with the remainder going toward interest, which method of financing would you recommend after taxes are accounted for? Assume that the tax rate is 52%.

P18.6 Is the answer the same for Prob. P18.5 if a before-tax analysis is made?

P18.7 Purchase of some new equipment for \$75,000 is contemplated by the Goat and Cow Dairy. The equipment will last 5 years and can be salvaged for \$15,000. The company has \$25,000 in available money and hopes to borrow the remainder from a bank for a 5-year period. The equipment is expected to increase cash flow before taxes by \$18,000 per year. The new equipment will be straight-line-depreciated and a 50% tax rate is applicable. If the economist for the firm estimates that the taxes for this endeavor will be \$1500 per year, what is the (*a*) stated interest rate paid on the loan and (*b*) effective interest rate on the loan after taxes provided that the MARR of 10% is realized? Assume uniform reduction in principal over the life of the loan.

P18.8 Compute the after-tax rate of return for Prob. P18.7 if the stated loan rate is reduced by 10%.

P18.9 The common stock that is outstanding for Bottleneck Contractors earned an average of \$0.75 per share last year. If the market price averaged \$11.50 per share and the tax rate was 47.5%, what was (*a*) the before-tax cost of capital and (*b*) the after-tax cost of capital?

P18.10 The owners of a grocery store plan to construct a laundromat next door. They will use 100% equity financing. It will cost \$22,000 to build the facility, which will be straight-line-depreciated over a 15-year life using a salvage value of \$7000. They have the equity funds invested at the present time and make 10% per year. If the annual CFBT is expected to be \$5000 and the tax rate is 48%, is the venture expected to be profitable?

P18.11 Use the Gordon-Shapiro method to compute the BTCC for the following situation. The AZ Company has \$1.5 million in common stocks outstanding. Last year a dividend of \$1.25 per share was paid. Earnings were reported as \$1.37 per share on a market value of \$13.75. Book value is 75% of market value, and the tax rate was 51%.

P18.12 The AZ Company of Prob. P18.11 would like to reduce its BTCC on equity financing to 15% but pay the same dividend. What earnings per share would be necessary to obtain a 15% BTCC? How much would be left per share for retained earnings?

P18.13 Mr. Snoozer, the president of Westfall Mattresses, sees an opportunity to invest \$100,000 in a new mattress line. He anticipates an annual net income of \$37,800 per year for the next 7 years, which will be taxed at a rate of 50%. No depreciable assets are involved. The president hopes to raise the investment capital by selling stocks at a market value of \$5.80. What percent of market value would have to be earned in order to keep his ATCC at one-half his rate of return on the investment?

P18.14 An investor is interested in a certain company. She finds that stock has a market value of \$28.50 per share and has paid an 8% dividend for the past 5 years. An accountant friend who works for this company tells her that this company retains only 50% of its earnings to keep the ATCC very low. If the accountant also tells the investor that the current book value of stock is 60% of market value, what is the ATCC? Does this ATCC seem high or low?

P18.15 A large company would like to purchase a small firm which has been a supplier for many years. A cost of \$780,000 has been placed on the small firm. The purchasing company does not know exactly how to finance the purchase to obtain an average ATCC as low as that for other ventures. The average BTCC is presently 10%. Two schemes of financing are available. The first requires that the company invest 50% equity funds at 8% and borrow the balance at 11% per year. The second scheme requires only 25% equity funds and the balance can be borrowed at 9%. Which scheme will require the smaller earnings?

P18.16 If ATCC is 10% in Prob. P18.15, what rate of interest could be paid on the debt capital for the two schemes?

P18.17 Rework Example 18.7 by computing the correct interest value each year for the 50% debt financing. Will the decision be altered because of this more exact treatment of debt financing?

P18.18 The Pure Trash Company has always used 100% equity financing in the past. A good opportunity is now offered that will require the raising of $250,000. The owner can supply the money from personal investments, which earn at an ATCC of $8\frac{1}{2}$% per year. The annual CFAT is expected to be $30,000 for the next 15 years at a tax rate of 50%. Alternatively, 60% of the required amount can be borrowed at 5% per year for 15 years. If it is assumed that the principal is uniformly reduced and an average annual interest is paid, use a MARR of 1.2 times ATCC to determine which plan is better.

P18.19 The Sno-Plow Company uses a MARR value that is 1.5 times the cost of capital. Three plans for raising $50,000 are available. These are detailed below.

Type of financing	Plan 1	2	3
Equity	90%	60%	20%
Debt	10	40	80

At present the before-tax cost for equity capital is 10% and for debt capital, 12%. If the project is expected to yield $10,000 for 5 years, do a before-tax analysis to determine the return for each plan and the plans that are acceptable.

P18.20 A conscientious couple devised a plan to buy certain types of groceries now for $600 in order to save a total of 25% in the next 6 months. However, they are not sure how to finance this plan. The couple want to make a return of 50% more than the financing methods will cost over the 6-month period. The financing plans are as follows:

(*a*) Take $600 from a savings account now and put the monthly savings of $125 in as they are received. This account pays 6% per year compounded quarterly.

(*b*) Borrow $600 now from the credit union at an effective 1% per month and repay the loan at $103.54 per month for 6 months and put the difference between the payment and the amount saved each month in the 6% compounded quarterly savings account.

(*c*) Use the extra $300 from this months's budget, borrow $300 at 1% per month, and repay at the rate of $51.77 per month for 6 months. Again, the difference between the payments and the savings would be saved at 6% compounded quarterly.

Perform a before-tax average cost-of-capital analysis to determine which financing plan is the most profitable. Assume that there is no interperiod interest paid on the savings account.

P18.21 The Hack-Away Cough Syrup Company has a total of 153,000 shares of stock outstanding at a market value of $28 per share. Earnings are 12.5% of market and a 48% tax rate is used by the accountant for Hack-Away. Stocks are sufficient to finance only 50% of the company's undertakings. The remaining is financed by bonds and funds borrowed from a bank. Thirty percent of the debt financing is by $1,285,000 worth of $10,000 6%-per-year 15-year bonds, which were sold at a 2% discount for rapid sale. The remaining 70% of debt financing is by loans which are repaid at an effective after-tax rate of 9% per year. Using only the information above, determine (*a*) the average BTCC and (*b*) the average ATCC.

P18.22 Compute the debt-to-equity ratio for all plans in the following problems: (*a*) P18.19, (*b*) P18.20, and (*c*) P18.21.

P18.23 Why is an extremely large D/E value not healthy for a company?

P18.24 Compute the new rate-of-return values for the three firms described in Example 18.8*c* if the D/E ratio for each firm is 1.00.

P18.25 Company *A* has a total asset value of \$2.5 million and a D/E ratio value of 0.40 while company *B* has total assets valued at \$1.6 million and D/E = 2.5.

(*a*) Compute the dollar amount that is from equity and debt sources for each company.

(*b*) If asset value is reduced by 15%, compute the new D/E values for both firms.

P18.26 Assume that the revenue for each company in Prob. P18.25*a* is \$500,000 and that company *A* pays 7% for debt financing and *B* pays 9%, owing to the greater risk taken by the lender because of the larger D/E value. Compute the rate of return on common stock for each company, assuming that common stock is the sole source of equity financing.

CHAPTER

NINETEEN

SENSITIVITY AND RISK ANALYSIS

In this chapter you will learn the use of sensitivity analysis in economy studies and be introduced to the idea of probabilistic engineering economy.

Probabilistic analysis is, in realistic terms, largely an uncharted course for the engineering economist. The computations and decision-making steps are relatively simple, *once the input data are known*. However, to obtain good, real-world probabilistic data is quite difficult and expensive. In this chapter we take only a brief look at risk analysis. This may be considered an optional chapter.

SECTION OBJECTIVES

To complete this chapter you must be able to do the following:

19.1. State the purpose of sensitivity analysis and explain how it differs from break-even analysis.
19.2. Determine the sensitivity of factor(s) for one or more projects, given the factor(s) to be studied, possible variation of each factor, and project details for all other factors.
19.3. Select the most economic of two or more projects using EUAC analysis, given three estimates (pessimistic, reasonable, and optimistic) for important factors of each project and values for the nonvarying factors.
19.4. Compute the expected value of a variable, given the variable values and associated probabilities.
19.5. Determine the desirability of a project using expected value computations, given the time of the project, the cash flows, and/or the interest rate.

STUDY GUIDE

19.1 The Approach of Sensitivity Analysis

Since the workplace of engineering economists is the *future*, the estimates they use can possibly be in error. Sensitivity analysis is a study to see how the economic decision will be altered if certain factors are varied. For example, variation in the minimum attractive rate of return (MARR) may not alter a decision when all compared proposals return far more than the MARR; thus, the decision is relatively insensitive to MARR. However, if a change in the economic life is critical, the decision is sensitive to life estimates.

Usually the variations in values of life, annual costs, or incomes, etc., result from variations in selling price, operation at different levels of capacity, inflation, etc. For example, if 75% capacity is compared with 50% capacity for a contemplated proposal, operating costs and revenue will increase, but anticipated life will probably decrease. Usually several important factors are studied to learn how the uncertainty of estimates will affect the economic study.

Plotting the sensitivity of present worth, EUAC, or rate of return versus the factor studied is quite illustrative. Two projects can be compared with respect to a given factor and the *breakeven point* computed. This is the variable value at which the two proposals are economically equivalent. However, the breakeven chart represents only one factor per chart. Thus, several charts, one for each factor, must be constructed and independence of each factor assumed. In previous uses of breakeven analysis, we computed two values and connected the points with a straight line. However, if a factor generates sensitive results, several intermediate points should be used to be conscious of the sensitivity. This fact is illustrated in this chapter.

If several factors are to be studied, as is the usual case, they may be studied one at a time using manual computation. However, to get an idea of how several factors affect the sensitivity, a computer program should be written using general formulas with the varying factors expressed as unknowns. The computer easily allows more than one basis of comparison to be employed, for example, present-worth and rate-of-return analysis. In addition, the computer can plot the results, giving a rapid visual display of sensitivity.

19.2 Determination of the Sensitivity of Estimates

There is a general procedure that should be followed when conducting a sensitivity analysis. The steps in this procedure are as follows:

1. Determine which factor(s) are most likely to vary from the estimated value.
2. Select the probable range and increment of variation for each factor.
3. Select an evaluation method, such as present-worth, EUAC, or rate-of-return, that will be used to evaluate each factor's sensitivity.
4. Compute and, if desired, plot the results from the evaluation method selected in step 3.

The results of the sensitivity analysis will show the factors that should be carefully estimated by collecting more information when possible. Example 19.1 illustrates sensitivity analysis for one project.

Example 19.1 The ACQ Company is contemplating the purchase of a new piece of automatic machinery for \$80,000 with zero salvage value and an anticipated before-tax cash flow of \$27,000 - 2000k ($k = 1, 2, \ldots, n$) per year. Figure 19.1 is a cash-flow diagram of the asset purchase. The MARR for the company has varied from 10 to 25% for different types of investments. The economic life of similar machinery normally varies from 8 to 12 years. Use EUAC analysis to investigate the sensitivity of varying (*a*) the MARR using $n = 10$ and (*b*) the economic life, assuming a MARR of 15%.

SOLUTION

(*a*) Allowing i to change by a 5% increment should be sufficient for sensitivity purposes. For $i = 10\%$,

$$P = -80{,}000 + 25{,}000(P/A, 10\%, 10) - 2000(P/G, 10\%, 10)$$
$$= \$27{,}830$$
$$\text{EUAC} = P(A/P, 10\%, 10) = \$4529$$

Similarly, other results are

i	P	EUAC
15%	\$ 11,512	\$ 2,294
20	-962	-229
25	-10,711	-3,000

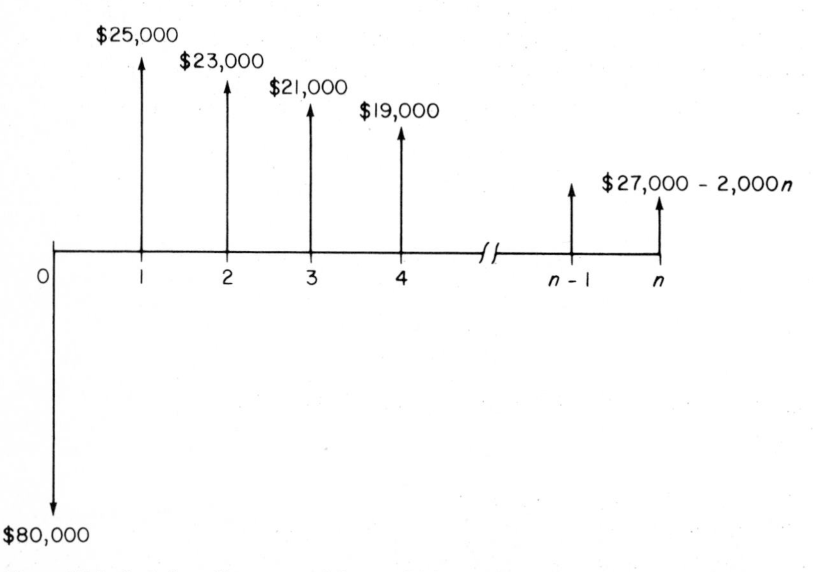

Figure 19.1 Cash-flow diagram used for sensitivity analysis of i and n, Example 19.1.

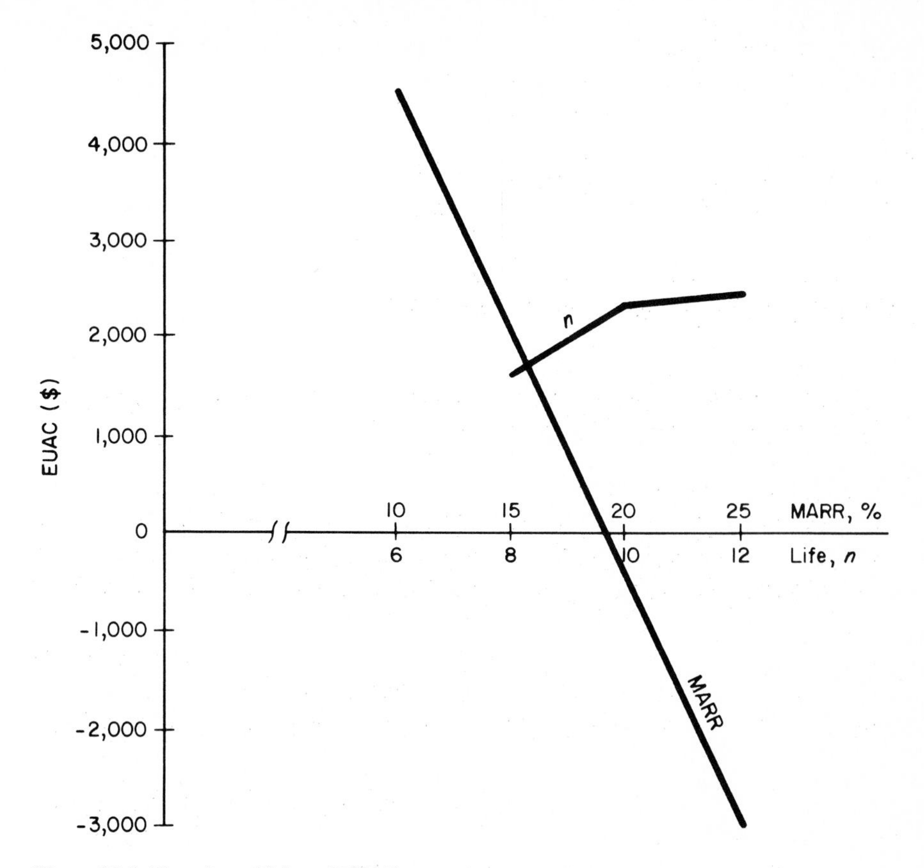

Figure 19.2 Plot of sensitivity of EUAC versus interest and life, Example 19.1.

A plot of MARR versus EUAC is shown in Fig. 19.2. The steep negative slope of EUAC indicates that the decision is quite sensitive to variations in MARR. If it is likely that the MARR will be set toward the upper end of the company's MARR range, the investment will not be attractive.

(*b*) Using an increment of 2 years, the P and EUAC values for $n = 8$, 10, and 12 at $i = 15\%$ are as follows:

n	P	EUAC
8	$ 7,221	$1,609
10	11,511	2,294
12	13,145	2,425

Figure 19.2 presents the characteristic nonlinear relation of EUAC versus n. Since the EUAC is positive for all values of n, the decision to invest would not be affected by the economic life of the machine.

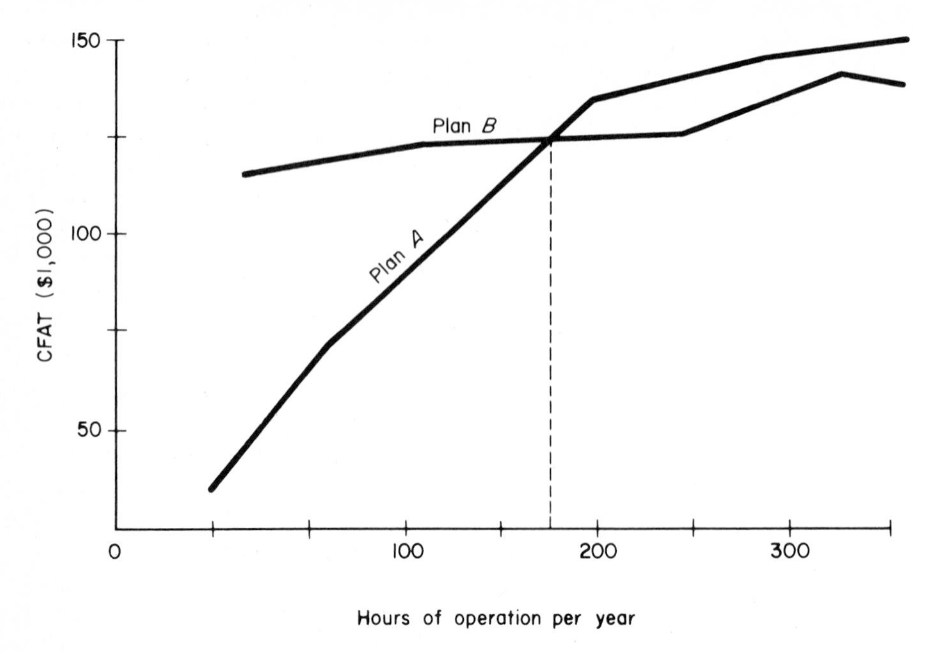

Figure 19.3 Plot of CFAT versus hours for two alternatives.

COMMENT Note that after $n = 10$, the EUAC curve seems to level out and become quite insensitive. This insensitivity to changes in cash flow in the distant future is the expected trait, because when discounted to time 0, the present-worth value gets smaller as n increases.

If two projects are compared and the sensitivity for one factor is to be determined, the actual plot may show quite nonlinear results. Take a look at the general form of the plots in Fig. 19.3. We won't be concerned with the actual computations, but the graph shows that the CFAT of each plan is a nonlinear function of hours of operation. Plan A is extremely sensitive in the range of 50 to 200 hours but is comparatively insensitive above 200 hours. We mentioned in Sec. 19.1 that sensitive plans should have intermediate points plotted for a more accurate sensitivity analysis to be made. If a considerable variation in the hours of operation is expected and if plans A and B are both economically justified, plan B should be selected due to its relative insensitivity to hours of operation. This will provide some assurance of a relatively stable CFAT value.

Solved Problem 19.6
Probs. P19.1 to P19.14

19.3 Sensitivity of Alternatives Using Three Estimates of Factors

We can thoroughly study the economic decision between two or more projects by borrowing from the field of project scheduling the concept of making three estimates for pertinent factors: a pessimistic, a reasonable, and an optimistic estimate. This allows us to study decision sensitivity for each factor, thereby obtaining different decisions depending on the factor considered. After the sensitizing, the analyst should accept the decision representing the economic situation as he or she best understands it. This involves a subjective weighing of the sensitized factors.

Example 19.2 You are an engineering economist attempting to evaluate three alternatives for which pessimistic (P), reasonable (R), and optimistic (O) estimates are made for the life, salvage value, and annual operating costs (Table 19.1). Determine the most economic alternative using EUAC analysis and a before-tax MARR of 12%.

SOLUTION For each alternative description in Table 19.1 we must compute the EUAC. For example, using the pessimistic (P) estimates for alternative A, compute

$$\text{EUAC} = 20{,}000(A/P, 12\%, 3) + 11{,}000 = \$19{,}327$$

Table 19.2 presents EUAC values at 12% for all situations. Figure 19.4 is a plot of EUAC versus the three life estimates for each alternative. Since the EUAC calculated from the "reasonable" estimates for alternative B is less than even the optimistic EUAC values for A and C, alternative B is clearly favored.

Table 19.1 Competing alternatives with three estimates for *n*, SV, and AOC

	First cost, *P*	Salvage, SV	AOC	Life, *n*
Alternative *A*				
P	$20,000	0	$11,000	3
R	20,000	0	9,000	5
O	20,000	0	5,000	8
Alternative *B*				
P	$15,000	$ 500	$ 4,000	2
R	15,000	1,000	3,500	4
O	15,000	2,000	2,000	7
Alternative *C*				
P	$30,000	$3,000	$ 8,000	3
R	30,000	3,000	7,000	7
O	30,000	3,000	3,500	9

Table 19.2 EUAC values for three alternatives with varying factor estimates, Example 19.2

	Alternatives		
Strategy	*A*	*B*	*C*
P	$19,327	$12,640	$19,601
R	14,548	8,229	13,276
O	9,026	5,089	8,927

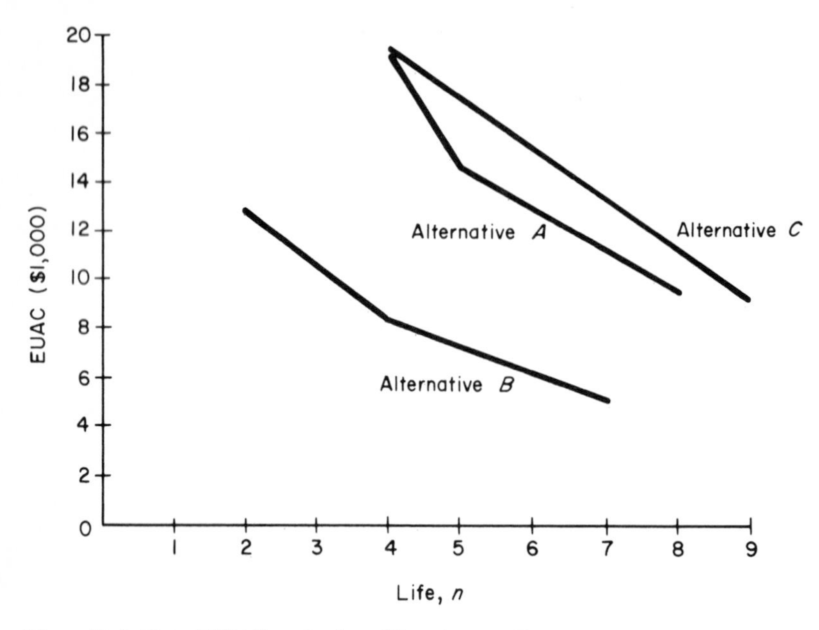

Figure 19.4 Plot of EUAC versus three life estimates, Example 19.2.

COMMENT While the alternative that should be selected in this example is obvious, this may not always be the case. For example, in Table 19.2, if the pessimistic EUAC for alternative *B* were \$21,000 and the optimistic EUAC values for alternatives *A* and *C* were less than \$5089, selection of *B* would not be as apparent. In this case, it would be necessary to decide which strategy (P, R, or O) would control the decision. In determining which strategy should be used, other factors, such as the project MARR, the availability of capital, and the economic stability of the company would have to be taken into consideration.

Probs. P19.15 to P19.20

19.4 Economic Uncertainty and the Expected Value

The use of probability and its computations by the engineering economist are not as common as they should be; the reason for this is not that the computations are difficult to perform or understand but that realistic probability values are difficult to obtain. The economist must deal with *uncertain future* monetary and life values. Often reliance on past data, if any exist, for future cash-flow values is incorrect. However, experience and wise judgment can often be used in conjunction with the expected value to evaluate the desirability of a project. The *expected value* can be interpreted as a long-run average outcome if the project is repeated many times. However, even for a single purchase, the expected value is meaningful.

The expected value $E(X)$ may be computed using the relation

$$E(X) = \sum_{i=1}^{m} XP(X) \qquad (m = 1, 2, \ldots, n) \tag{19.1}$$

where X = specific variable value

$P(X)$ = probability a specific value will occur

A much more detailed explanation of probability, expected values, and statistics can be found in any text on probability and statistics.

Example 19.3 You expect to be mentioned in your favorite uncle's will. You anticipate being willed $5000 with a probability of 0.50, $50,000 with a 0.45 chance, and a value of zero dollars with a 0.05 chance. What is your expected inheritance?

SOLUTION Let

$$X = \text{inheritance value}$$

$$P(X) = \text{associated, subjectively evaluated probability}$$

Your expected inheritance is

$$E(X) = 5000(0.50) + 50{,}000(0.45) + 0(0.05) = \$25{,}000$$

COMMENT As in all probability statements, the sum of all $P(X)$ values must be 1.0. If the $P(X) = 0.05$ for $X = \$0$ were not stated, it would have to be assumed. Of course, it makes no computational difference but is necessary for completeness.

If actual cash flows are the X values, some can be negative, as for the first cost. If the expected value is positive, then the overall outcome is expected to be a cash inflow. Therefore, $E(\text{cash flow}) = \$-1500$ in present worth indicates a losing proposition. The expected-value concept can be used in engineering-economy studies, as discussed below.

Probs. P19.21 to P19.24

19.5 Expected Value of Economy Alternatives

The expected value $E(X)$ can be used in many ways and the probabilities used in its computation can be stated in different fashions. Example 19.4 below is a simple $E(X)$ computation, while Example 19.5 involves the use of PW for an asset with the possibility of different cash-flow sequences.

Example 19.4 A utility company is experiencing a difficult time in obtaining natural gas for electric generation and distribution. Monthly expenses are now averaging $7,750,000. The economist for this city-owned utility has collected average revenue figures for the past 2 years using the categories "fuel plentiful," "less than 30% other fuel purchased," "at least 30% other fuel" (Table 19.3).

Table 19.3 Fuel data for a municipal utility

Fuel situation	Months in past 2 years	Average revenue per month, X
Plentiful	12	$ 5,270,000
< 30% other	6	7,850,000
≥ 30% other	6	12,130,000

"Other" is a fuel other than natural gas, purchased at a premium, the cost of which is transferable directly to the consumer. Can the utility expect to meet future expenses based on the 2 years of data?

SOLUTION Using the 24-month data-collection period, probabilities may be computed as

Fuel	$P(X)$
Plentiful	12/24 = 0.50
< 30%	6/24 = 0.25
≥ 30%	6/24 = 0.25

Probabilities add up to 1.0, so Eq. (19.1) results in

$$E(X) = E(\text{revenue}) = 5{,}270{,}000(0.50) + 7{,}850{,}000(0.25) + 12{,}130{,}000(0.25)$$

$$= \$7{,}630{,}000$$

With average monthly expenses of $7,750,000, the utility will have to raise rates in the near future.

Example 19.5 The Tule Company has had experience with automatic reaming equipment. A certain piece of equipment will cost $5000 and have a life of 3 years. Suspected actual cash flows and probability of each are listed in Table 19.4, depending on a receding, stable, or expanding economy. Using expected-present-

Table 19.4 Equipment cash-flow probabilities for Example 19.5

	Economy		
Year	Receding (Prob. = 0.2)	Stable (Prob. = 0.6)	Expanding (Prob. = 0.2)
0	$-5000	$-5000	$-5000
1	+2500	+2000	+2000
2	+2000	+2000	+3000
3	+1000	+2000	+3500

worth analysis, determine if the equipment should be purchased at a 15% rate of return.

SOLUTION The first step is to find the present worth of the cash flows under each type of anticipated economy and then find the $E(\text{PW})$ using Eq. (19.1). Let the subscript R represent a receding economy, S a stable, and E an expanding economy.

$$\text{PW}_R = 4344 - 5000 = \$ -656$$

$$\text{PW}_S = 4566 - 5000 = -434$$

$$\text{PW}_E = 6309 - 5000 = +1309$$

$$E(\text{PW}) = \sum_{j=R,S}^{E} \text{PW}_j[P(j)]$$

$$= -656(0.2) - 434(0.6) + 1309(0.2)$$

$$= \$-310$$

Since $E(\text{PW}) < 0$, the project is not expected to be a paying proposition at a 15% required rate of return.

The probability analysis presented in this chapter is extremely basic and intended only as an introduction to the field of probabilistic engineering economy. Current research into the area can be found in recent issues of *The Engineering Economist*, *Management Science, The Journal of Finance*, and *The Journal of Business*.

Solved Problem 19.7
Probs. P19.25 to P19.29

SOLVED PROBLEMS

Example 19.6 The city of Booney has a 3-mile stretch of heavily traveled highway to resurface. The Ajax Construction Company offers two methods of resurfacing. The first is a concrete surface for a cost of \$150,000 and an annual maintenance charge of \$1000.

The second method is an asphalt covering with a first cost of \$100,000 and a yearly service charge of \$2000. However, Ajax would also request that every third year the highway be "touched up" at a cost of \$2500 per mile.

The city uses the interest rate on revenue bonds, 6% in this case, as the MARR. (*a*) Determine the breakeven number of years of the two methods. If the city expects an interstate to replace this stretch of highway in 20 years, which method should be selected? (*b*) If touch-up cost increases by \$500 per mile every 3 years, is the decision sensitive to this cost?

SOLUTION

(*a*) We can set up the equation

$$\text{EUAC of concrete} = \text{EUAC of asphalt}$$

and compute the breakeven n value. The breakeven equation is

$$150{,}000(A/P, 6\%, n) + 1000 = 100{,}000(A/P, 6\%, n) + 2000 + 7500\left[\sum_j (P/F, 6\%, j)\right](A/P, 6\%, n) \quad (19.2)$$

where $j = 3, 6, 9, \ldots, n$. This may be rewritten as

$$50{,}000(A/P, 6\%, n) - 1000 - 7500\left[\sum_j (P/F, 6\%, j)\right](A/P, 6\%, n) = 0 \quad (19.3)$$

A value of $n = 39.6$ satisfies the equation, so a life of approximately 40 years is required to break even at 6%. Since the road is required for only 20 years, the asphalt surface should be constructed.

(*b*) The total touch-up cost will increase by \$1500 every 3 years. Equation (19.3) may now be written as

$$50{,}000(A/P, 6\%, n) - 1000 - \sum_j \left[7500 + 1500\left(\frac{j-3}{3}\right)(P/F, 6\%, j)\right](A/P, 6\%, n) = 0$$

where $j = 3, 6, 9, \ldots, n$. The breakeven point is now found to be approximately 21 years, which still favors the asphalt surface. The decision is therefore insensitive to the stated increase in the touch-up cost.

Sec. 19.2

Example 19.7 The Holdup Construction Company plans to build an apartment complex close to the edge of a partially leveled hill. Support of the cliff below the complex will ensure that no damage will occur to the buildings or occupants. The amount of rainfall at any one time can cause varying amounts of damage. Table 19.5 itemizes the probability of certain rainfalls within a period of a few hours

Table 19.5 Rainfall and support-wall cost, Example 19.7

Rainfall, inches	Probability of greater rainfall occurring	Cost of support wall to carry rainfall
2.0	0.300	\$10,000
2.5	0.100	15,000
3.0	0.050	22,000
3.5	0.010	30,000
4.0	0.005	42,000

Table 19.6 EUAC for different support walls

Rainfall, inches	Support-wall cost	Annual loan cost	Expected annual damage	EUAC
2.0	$10,000	$ 973	$6,000	$6,973
2.5	15,000	1,460	2,000	3,460
3.0	22,000	2,141	1,000	3,141
3.5	30,000	2,920	200	3,120
4.0	42,000	4,088	100	4,188

and the first cost to construct a support wall to ensure protection against the corresponding amount of water.

The project of wall construction will be financed by a 30-year 9% loan. Data on record show that an average of $20,000 damage has often occurred when a heavy rain has fallen. Without taking the intangible (and extremely important) fact of human life into account, determine what size support wall is most economic?

SOLUTION We will use EUAC analysis to find the most economic plan. For each rainfall level

$$\text{EUAC} = \text{annual loan cost} + \text{expected annual damage}$$

$$= \text{cost}(A/P,\ 9\%,\ 30) + 20{,}000(\text{probability of greater rainfall})$$

The $20,000 damage figure is used because this has been the previous damage on the *average*. We assume that the probabilities are a yearly value. Computations in Table 19.6 show that a $30,000 wall to protect against a 3.5-inch rainfall is most economic. A wall to protect against a 3.0-inch rain is a close second to the most economic plan.

COMMENT Usually a rather hefty safety factor is automatically added when people are endangered. Pure economic analysis is not used alone as the decision maker. By building to protect to a greater degree than usually needed, the probabilities of damage are lowered–by how much we don't know, but it makes us *feel safe.*

Sec. 19.5

FURTHER INFORMATION

Sensitivity analysis (Secs. 19.1 to 19.3). Canada and White [4], pp. 272-286; Thuesen, Fabrycky, and Thuesen [17], pp. 414-422; White, Agee, and Case [18], pp. 294-303.

Risk analysis (Secs. 19.4 and 19.5). Barish and Kaplan [2], pp. 405-435; Canada and White [4], pp. 251-271, 287-342; Fabrycky and Thuesen [6], pp. 256-284; Mallik [9], pp. 161-180; Morris [10], pp. 235-279; Stevens [15], pp. 248-273.

PROBLEMS

P19.1 The A-1 Salvage Company is contemplating the purchase of a new crane equipped with a magnetic pickup device to be used in moving scrap metal around the yard. The complete crane will cost $62,000 and have an 8-year life and a salvage value of $1500. Annual maintenance, fuel, and overhead costs are estimated at $0.50 per metric ton. Labor cost will be $8 per hour for regular wages and $12 for overtime. A total of 25 tons can be moved in an 8-hour period. The salvage yard has handled in the past anywhere from 10 to 30 tons of scrap per day. If the company uses a MARR of 10%, plot the sensitivity of the present worth of costs versus annual volume moved, assuming that the operator is paid for 200 days of work per year. Use a 5-metric-ton increment for the graph.

P19.2 A new piece of equipment is being economically evaluated by three engineers for the Zoot-Suit Electric Utility. The first cost will be $77,000 and the life is estimated at 6 years with a salvage value of $10,000 at disposal time. The engineers disagree, however, on the annual net income to be credited to this new equipment. Engineer *A* has given an estimate of $14,000 per year. Engineer *B* states that this is too low and estimates $15,000, while *C* estimates $18,000 per year. If the MARR is 8%, use present-worth analysis to determine if these three estimates will change the decision to buy the equipment.

P19.3 Perform the same analysis as in Prob. P19.2, except make it an after-tax consideration using straight-line depreciation with the salvage value considered and a 52% effective tax rate. Assume an annual equipment-operating cost of $1000 and an after-tax rate of return of 5%.

P19.4 In Prob. P5.8 you made a present-worth comparison between building and leasing storage space. Determine the sensitivity of your decision to the following situations: (*a*) construction costs go up 10% and lease costs go down to $1.25 per square meter per month; (*b*) lease costs remain at $1.50 per square meter per month, but construction costs vary from $50 to $90 per square meter.

P19.5 For the situation presented in Prob. P7.6 plot the sensitivity of the rate of return to the amount of the income gradient. Perform the computations for values of the *negative* gradient from $300 to $700 in increments of $100. If the company would like a return of at least 40%, would variation in this income gradient have affected the decision to buy the dump trucks?

P19.6 Consider the two air-conditioning systems detailed below.

	System 1	System 2
First cost	$10,000	$17,000
Annual operating cost	200	150
Salvage value	-100	-300
New compressor and motor cost at midlife	1,750	3,000
Life, years	8	12

Use an EUAC analysis to determine the sensitivity of the economic decision to MARR values of 8, 10, 12, and 15%.

P19.7 Reread Prob. P19.6. If MARR is 10%, plot the EUAC for each system for life values from 4 to 8 for system 1 and 6 to 12 for system 2. Assume that the salvage values and annual operating costs are the same for each life value. Further, assume that the compressor is replaced at midlife. Plot EUAC for even-numbered years only. Which EUAC is more sensitive to a varying-life estimate?

P19.8 A city couple, Joan and John Pollution, would like to buy a small section of land in the woods to be used as a weekend vacation home. Alternatively, they have thought of buying a travel trailer and four-wheel-drive vehicle to pull the trailer for vacations. The Pollutions have found a 5-acre tract with a cabin, well, etc., 25 miles from their home. It will cost them $30,000, but they expect to sell the acreage for $45,000 in 10 years when their children are grown. The insurance, upkeep, etc., costs are estimated at $500 per year, but this weekend site is expected to save the family $50 every day they don't go on a traveling vacation. The Pollutions estimate that even though the cabin is only 25 miles from home, they will travel 50 miles a day when at the cabin while working on it, visiting neighbors, etc. The Pollution car averages 30 miles per gallon of gasoline.

The trailer and van combination would cost $11,000 and could be sold for $2000 in 10 years. Insurance, maintenance, etc., costs will be $750 per year, but the trailer is expected to save $25 per vacation day. On a normal vacation, the Pollutions travel 300 miles each day. Mileage per gallon for the van is estimated at 60% that of the family car.

Gas costs $1.20 per gallon and the Pollution family want a return of fun and 10% from either investment. Plot the sensitivity of EUAC for each plan if the Pollutions' vacation time in the past has been from 6 to 14 days per year. Also compute the breakeven number of days per year for the two plans.

P19.9 Suppose that the Pollution family of Prob. P19.8 plan to purchase the acreage with the cabin and still go on a 4-day traveling vacation in the car at a cost of $65 per day.

(*a*) Was the decision to buy the land still the better decision?

(*b*) Does the breakeven point seem to be sensitive to the types of vacation plans that combine the use of the acreage and traveling?

P19.10 Plot the sensitivity of rate of return versus the life of a 5% $50,000 bond that is offered for $43,500 and has the bond interest paid quarterly. Use life values of 10, 12, 15, 18, and 20.

P19.11 The Charley Horse Company has been offered an investment opportunity that will require a cash outlay of $30,000 now and a cash inflow of $3500 for each year of investment. However, the company must state now the number of years it will retain the investment. If the investment is kept for 6 years, $25,000 can be gotten for the company's share, but after 10 years the resale value will be only $12,000. If money is worth 8%, is the decision sensitive to the retention period?

P19.12 The person who did the analysis in Prob. P16.16 is concerned with the sensitivity of the breakeven point to the expected life of the project. Determine the sensitivity for (*a*) 15 years and (*b*) 25 years.

P19.13 Determine the sensitivity of the most economic life value of Prob. P10.31 to the cost gradient. Investigate the gradient values of $60 to $140 in increments of $20 and plot the results.

P19.14 Rework Prob. P19.13 at an interest rate of $i = 5\%$ and plot the results on the graph used in P19.13. [MINCL (Appendix E) can help solve this problem.]

P19.15 Reread Prob. P6.7. The time of overhaul can vary from 2 to 4 years for the used machine and from 4 to 6 years for the new machine. Plot the EUAC values for these three estimates and determine if they will alter the decision in P6.7.

P19.16 If the spray method of Prob. P6.13 is used, the amount of water used can vary from an optimistic value of 60 liters to a pessimistic value of 120 liters with 80 liters being a reasonable figure. The immersion technique always takes 16 liters per ham. How will this varying use of water for the spray method affect the economic decision?

P19.17 An engineer is trying to decide between two ways to pump concrete up to the top floors of a seven-story office building now under construction. Plan 1 requires the purchase of equipment costing $6000 and costing between $0.40 and $0.75 per metric ton to operate with an expected cost of $0.50 per metric ton. The asset is able to pump 100 metric tons per day. If purchased, the asset will last for 5 years, have no salvage value, and be used from 50 to 100 days per year. Plan 2 is an equipment-leasing option and is expected to cost the company $2500 per year for equipment with an optimistic cost of $1800 and a pessimistic value of $3200 per year. In addition, a $5-per-hour labor cost will be incurred for operation of the leased equipment. Plot the EUAC of each plan versus total annual operating or lease cost at $i = 12\%$. Determine which plan the engineer should select, using the reasonable estimates for a use of (*a*) 50 and (*b*) 100 days per year.

P19.18 When the country's economy is expanding, the AB Investment Company is optimistic and uses a MARR of 8% on all new investments. However, in a receding economy a return of 15% on investments is required. Normally a 10% return is required. Similarly, an expanding economy causes the estimates of asset life to go down about 20% and a receding economy makes the n values increase about 10%. Plot the sensitivity of present worth versus (*a*) MARR and (*b*) life values for the two plans detailed below using the reasonable estimate for other varying factors.

	Plan M	Plan Q
Initial investment	\$-100,000	\$-110,000
Annual CFBT	+15,000	+19,000
Life, years	20	20

P19.19 Rework Prob. P19.18 except use the plans detailed in Prob. P6.15.

P19.20 When is it necessary to select a particular strategy of pessimistic, reasonable, or optimistic and make an economic decision on the basis of the selected strategy?

P19.21 The variable X can take on the values $X = 5$, 10, 15, 20 with a probability of 0.40, 0.30, 0.233, 0.067, respectively. Compute the expected value of X.

P19.22 The AOC value for a plan can take on one of two values. Your office partner told you that the high value is \$2800 per year. If her computations show a probability of 0.75 for the high value and an expected AOC of \$2575, what is the lower AOC value used in the computation of the average AOC value?

P19.23 Find the expected present worth of the following series of payments if each series is expected to be realized with the probability shown at the head of each column (assume $i = 20\%$).

	Annual cash flow		
Year	$P = 0.5$	$P = 0.2$	$P = 0.3$
0	\$-5000	\$-6000	\$-4000
1	+1000	+500	+3000
2	+1000	+1500	+1200
3	+1000	+2000	-800

P19.24 Compute the expected EUAC value for the cash flows of Prob. P19.23.

P19.25 The officers of a resort country club are thinking of constructing an additional 18-hole golf course. Because of the northerly location of the resort there is a 60% chance of a 120-day golf season, a 20% chance of 150 days of golfing weather, and a 20% chance of a 165-day season. The course will be used by an estimated 350 golfers each day of the 4-month season, but by only 100 per day for each extra day in the golfing season. The course will cost \$375,000 to construct and will require a \$25,000 rework cost after 4 years. Annual maintenance cost will be \$36,000 and the green fees will be \$4.25 per person. If a life of 10 years is anticipated before a major rework is required and a 12% return is required, determine if the course should be constructed.

P19.26 The owners of the Dial-A-Hole Roofing Company want to invest \$10,000 in new equipment. A life of 6 years and a salvage value of 12% is anticipated. The annual income will depend upon the state of the housing and construction industry. The income is expected to be \$2000 per year; however, a current slump in the industry is given a 50% chance of lasting 3 years and a 20% chance of continuing for 3 additional years. However, if the outlook of the depressed market does improve, either during the first or second 3-year period, the annual income of the investment is expected to be \$3500. Can the company expect to make 8% on its investment?

P19.27 The High Construction Company is building an apartment complex in the arid southwest on the top of a partially leveled hill. The road that winds around the hill to the apartment complex entrance needs retaining walls for support above and below the road surface. The probability of a rain shower greater than a given amount and the associated damage and retaining wall construction costs are shown in the table. Determine which plan will result in the lowest annual cost over a 25-year period at an interest rate of 10%.

Rainfall, inches	Probability of greater rainfall occurring	Retaining-wall cost to carry rainfall	Expected annual damage for specified rainfall
1.0	0.6	$15,000	$ 1,000
2.0	0.3	16,000	1,500
2.5	0.1	18,000	2,000
3.0	0.02	21,000	5,000
3.5	0.005	28,000	9,000
4.0	0.001	35,000	14,000

P19.28 Rework Prob. P19.27 using an after-tax analysis, assuming that the tax rate is 50% and the retaining-wall construction cost will be secured by an 8% 25-year loan. Assume that the principal amount is reduced an equal amount each year with the remainder of the payment applied to interest.

P19.29 A private citizen has $5000 to invest. If he puts the money in a savings and loan account, he is assured of receiving an effective 6.35% per year on the principal. If he puts the money in stocks he has a 50-50 chance of one of the following cash-flow sequences for the next 5 years.

Year	Stock 1	Stock 2
0	$–5000	$–5000
1–4	+250	+600
5	+6800	+5400

Finally, he can invest his $5000 in improved property for 5 years with the following outcomes and probabilities P.

Year	Cash flow		
	$P = 0.3$	$P = 0.5$	$P = 0.2$
0	$–5000	$–5000	$–5000
1	–425	0	+500
2	–425	0	+600
3	–425	0	+700
4	–425	0	+800
5	+9500	+7200	+5200

Which of the three investments–savings, stocks, or property–is the best?

APPENDIXES

Appendix	Subject
A	Discrete compounding, discrete cash-flow factors
B	Continuous compounding, discrete cash-flow factors
C	Continuous compounding, continuous cash-flow factors
D	Basics of accounting
E	Computer applications
F	Answers to problems

APPENDIX

A

INTEREST FACTORS FOR DISCRETE COMPOUNDING, DISCRETE CASH FLOW

Tabulated here are interest factor values when interest is compounded once each period. (If interest is compounded more or less frequently or if the continuous cash-flow assumption is present, refer to Chap. 3 and Appendix B or C.) The computational forms of the factors are given here.

Factor	Notation	Formula
Single-payment compound amount	$(F/P, i\%, n)$	$(1+i)^n$
Single-payment present worth	$(P/F, i\%, n)$	$\frac{1}{(1+i)^n}$
Sinking fund	$(A/F, i\%, n)$	$\frac{i}{(1+i)^n - 1}$
Uniform-series compound amount	$(F/A, i\%, n)$	$\frac{(1+i)^n - 1}{i}$
Capital recovery	$(A/P, i\%, n)$	$\frac{i(1+i)^n}{(1+i)^n - 1}$
Uniform-series present worth	$(P/A, i\%, n)$	$\frac{(1+i)^n - 1}{i(1+i)^n}$

A few useful computational relations between factors are given below. (The $i\%$ and n are omitted from the notation when possible, simply for the sake of brevity.)

$$(P/F) = \frac{1}{(F/P)} \qquad (F/A) = \frac{1}{(A/F)} \qquad (P/A) = \frac{1}{(A/P)}$$

$$(P/A) = (F/A)(P/F) \qquad (A/P) = (A/F)(F/P)$$

$$(P/A) = \sum_{j=1}^{n} (P/F, i\%, j) \qquad F/A = \sum_{j=1}^{n} (F/P, i\%, j)$$

Tables A-31 to A-38 present factors that convert a uniform gradient (G) of \$1 per period to a present-worth or equivalent uniform annual series, respectively. Computational formulas are as follows:

Factor	Notation	Formula
Uniform-gradient present worth	$(P/G, i\%, n)$	$\frac{1}{i}\left[\frac{(1+i)^n - 1}{i(1+i)^n} - \frac{n}{(1+i)^n}\right]$
Uniform-gradient annual series	$(A/G, i\%, n)$	$\frac{1}{i} - \frac{n}{(1+i)^n - 1}$

Useful gradient relations are

$$(P/G) = (A/G)(P/A) \qquad (A/G) = (P/G)(A/P)$$

TABLE A - 1

DISCRETE CASH FLOW
0.50% DISCRETE COMPOUND INTEREST FACTORS

	SINGLE PAYMENTS		UNIFORM SERIES PAYMENTS				
N	COMPOUND AMOUNT F/P	PRESENT WORTH P/F	SINKING FUND A/F	COMPOUND AMOUNT F/A	CAPITAL RECOVERY A/P	PRESENT WORTH P/A	N
1	1.0050	0.9950	1.00000	1.0000	1.00500	0.9950	1
2	1.0100	0.9901	0.49875	2.0050	0.50375	1.9851	2
3	1.0151	0.9851	0.33167	3.0150	0.33667	2.9702	3
4	1.0202	0.9802	0.24813	4.0301	0.25313	3.9505	4
5	1.0253	0.9754	0.19801	5.0503	0.20301	4.9259	5
6	1.0304	0.9705	0.16460	6.0755	0.16960	5.8964	6
7	1.0355	0.9657	0.14073	7.1059	0.14573	6.8621	7
8	1.0407	0.9609	0.12283	8.1414	0.12783	7.8230	8
9	1.0459	0.9561	0.10891	9.1821	0.11391	8.7791	9
10	1.0511	0.9513	0.09777	10.2280	0.10277	9.7304	10
11	1.0564	0.9466	0.08866	11.2792	0.09366	10.6770	11
12	1.0617	0.9419	0.08107	12.3356	0.08607	11.6189	12
13	1.0670	0.9372	0.07464	13.3972	0.07964	12.5562	13
14	1.0723	0.9326	0.06914	14.4642	0.07414	13.4887	14
15	1.0777	0.9279	0.06436	15.5365	0.06936	14.4166	15
16	1.0831	0.9233	0.06019	16.6142	0.06519	15.3399	16
17	1.0885	0.9187	0.05651	17.6973	0.06151	16.2586	17
18	1.0939	0.9141	0.05323	18.7858	0.05823	17.1728	18
19	1.0994	0.9096	0.05030	19.8797	0.05530	18.0824	19
20	1.1049	0.9051	0.04767	20.9791	0.05267	18.9874	20
22	1.1160	0.8961	0.04311	23.1944	0.04811	20.7841	22
24	1.1272	0.8872	0.03932	25.4320	0.04432	22.5629	24
25	1.1328	0.8828	0.03765	26.5591	0.04265	23.4456	25
26	1.1385	0.8784	0.03611	27.6919	0.04111	24.3240	26
28	1.1499	0.8697	0.03336	29.9745	0.03836	26.0677	28
30	1.1614	0.8610	0.03098	32.2800	0.03598	27.7941	30
32	1.1730	0.8525	0.02889	34.6086	0.03389	29.5033	32
34	1.1848	0.8440	0.02706	36.9606	0.03206	31.1955	34
35	1.1907	0.8398	0.02622	38.1454	0.03122	32.0354	35
36	1.1967	0.8356	0.02542	39.3361	0.03042	32.8710	36
38	1.2087	0.8274	0.02396	41.7354	0.02896	34.5299	38
40	1.2208	0.8191	0.02265	44.1588	0.02765	36.1722	40
45	1.2516	0.7990	0.01987	50.3242	0.02487	40.2072	45
50	1.2832	0.7793	0.01765	56.6452	0.02265	44.1428	50
55	1.3156	0.7601	0.01584	63.1258	0.02084	47.9814	55
60	1.3489	0.7414	0.01433	69.7700	0.01933	51.7256	60
65	1.3829	0.7231	0.01306	76.5821	0.01806	55.3775	65
70	1.4178	0.7053	0.01197	83.5661	0.01697	58.9394	70
75	1.4536	0.6879	0.01102	90.7265	0.01602	62.4136	75
80	1.4903	0.6710	0.01020	98.0677	0.01520	65.8023	80
85	1.5280	0.6545	0.00947	105.5943	0.01447	69.1075	85
90	1.5666	0.6383	0.00883	113.3109	0.01383	72.3313	90
95	1.6061	0.6226	0.00825	121.2224	0.01325	75.4757	95
100	1.6467	0.6073	0.00773	129.3337	0.01273	78.5426	100

TABLE A - 2

DISCRETE CASH FLOW
1.00% DISCRETE COMPOUND INTEREST FACTORS

	SINGLE PAYMENTS		UNIFORM SERIES PAYMENTS				
N	COMPOUND AMOUNT F/P	PRESENT WORTH P/F	SINKING FUND A/F	COMPOUND AMOUNT F/A	CAPITAL RECOVERY A/P	PRESENT WORTH P/A	N
1	1.0100	0.9901	1.00000	1.0000	1.01000	0.9901	1
2	1.0201	0.9803	0.49751	2.0100	0.50751	1.9704	2
3	1.0303	0.9706	0.33002	3.0301	0.34002	2.9410	3
4	1.0406	0.9610	0.24628	4.0604	0.25628	3.9020	4
5	1.0510	0.9515	0.19604	5.1010	0.20604	4.8534	5
6	1.0615	0.9420	0.16255	6.1520	0.17255	5.7955	6
7	1.0721	0.9327	0.13863	7.2135	0.14863	6.7282	7
8	1.0829	0.9235	0.12069	8.2857	0.13069	7.6517	8
9	1.0937	0.9143	0.10674	9.3685	0.11674	8.5660	9
10	1.1046	0.9053	0.09558	10.4622	0.10558	9.4713	10
11	1.1157	0.8963	0.08645	11.5668	0.09645	10.3676	11
12	1.1268	0.8874	0.07885	12.6825	0.08885	11.2551	12
13	1.1381	0.8787	0.07241	13.8093	0.08241	12.1337	13
14	1.1495	0.8700	0.06690	14.9474	0.07690	13.0037	14
15	1.1610	0.8613	0.06212	16.0969	0.07212	13.8651	15
16	1.1726	0.8528	0.05794	17.2579	0.06794	14.7179	16
17	1.1843	0.8444	0.05426	18.4304	0.06426	15.5623	17
18	1.1961	0.8360	0.05098	19.6147	0.06098	16.3983	18
19	1.2081	0.8277	0.04805	20.8109	0.05805	17.2260	19
20	1.2202	0.8195	0.04542	22.0190	0.05542	18.0456	20
22	1.2447	0.8034	0.04086	24.4716	0.05086	19.6604	22
24	1.2697	0.7876	0.03707	26.9735	0.04707	21.2434	24
25	1.2824	0.7798	0.03541	28.2432	0.04541	22.0232	25
26	1.2953	0.7720	0.03387	29.5256	0.04387	22.7952	26
28	1.3213	0.7568	0.03112	32.1291	0.04112	24.3164	28
30	1.3478	0.7419	0.02875	34.7849	0.03875	25.8077	30
32	1.3749	0.7273	0.02667	37.4941	0.03667	27.2696	32
34	1.4026	0.7130	0.02484	40.2577	0.03484	28.7027	34
35	1.4166	0.7059	0.02400	41.6603	0.03400	29.4086	35
36	1.4308	0.6989	0.02321	43.0769	0.03321	30.1075	36
38	1.4595	0.6852	0.02176	45.9527	0.03176	31.4847	38
40	1.4889	0.6717	0.02046	48.8864	0.03046	32.8347	40
45	1.5648	0.6391	0.01771	56.4811	0.02771	36.0945	45
50	1.6446	0.6080	0.01551	64.4632	0.02551	39.1961	50
55	1.7285	0.5785	0.01373	72.8525	0.02373	42.1472	55
60	1.8167	0.5504	0.01224	81.6697	0.02224	44.9550	60
65	1.9094	0.5237	0.01100	90.9366	0.02100	47.6266	65
70	2.0068	0.4983	0.00993	100.6763	0.01993	50.1685	70
75	2.1091	0.4741	0.00902	110.9128	0.01902	52.5871	75
80	2.2167	0.4511	0.00822	121.6715	0.01822	54.8882	80
85	2.3298	0.4292	0.00752	132.9790	0.01752	57.0777	85
90	2.4486	0.4084	0.00690	144.8633	0.01690	59.1609	90
95	2.5735	0.3886	0.00636	157.3538	0.01636	61.1430	95
100	2.7048	0.3697	0.00587	170.4814	0.01587	63.0289	100

TABLE A - 3

DISCRETE CASH FLOW
1.50% DISCRETE COMPOUND INTEREST FACTORS

	SINGLE PAYMENTS		UNIFORM SERIES PAYMENTS				
	COMPOUND AMOUNT	PRESENT WORTH	SINKING FUND	COMPOUND AMOUNT	CAPITAL RECOVERY	PRESENT WORTH	
N	F/P	P/F	A/F	F/A	A/P	P/A	N
1	1.0150	0.9852	1.00000	1.0000	1.01500	0.9852	1
2	1.0302	0.9707	0.49628	2.0150	0.51128	1.9559	2
3	1.0457	0.9563	0.32838	3.0452	0.34338	2.9122	3
4	1.0614	0.9422	0.24444	4.0909	0.25944	3.8544	4
5	1.0773	0.9283	0.19409	5.1523	0.20909	4.7826	5
6	1.0934	0.9145	0.16053	6.2296	0.17553	5.6972	6
7	1.1098	0.9010	0.13656	7.3230	0.15156	6.5982	7
8	1.1265	0.8877	0.11858	8.4328	0.13358	7.4859	8
9	1.1434	0.8746	0.10461	9.5593	0.11961	8.3605	9
10	1.1605	0.8617	0.09343	10.7027	0.10843	9.2222	10
11	1.1779	0.8489	0.08429	11.8633	0.09929	10.0711	11
12	1.1956	0.8364	0.07668	13.0412	0.09168	10.9075	12
13	1.2136	0.8240	0.07024	14.2368	0.08524	11.7315	13
14	1.2318	0.8118	0.06472	15.4504	0.07972	12.5434	14
15	1.2502	0.7999	0.05994	16.6821	0.07494	13.3432	15
16	1.2690	0.7880	0.05577	17.9324	0.07077	14.1313	16
17	1.2880	0.7764	0.05208	19.2014	0.06708	14.9076	17
18	1.3073	0.7649	0.04881	20.4894	0.06381	15.6726	18
19	1.3270	0.7536	0.04588	21.7967	0.06088	16.4262	19
20	1.3469	0.7425	0.04325	23.1237	0.05825	17.1686	20
22	1.3876	0.7207	0.03870	25.8376	0.05370	18.6208	22
24	1.4295	0.6995	0.03492	28.6335	0.04992	20.0304	24
25	1.4509	0.6892	0.03326	30.0630	0.04826	20.7196	25
26	1.4727	0.6790	0.03173	31.5140	0.04673	21.3986	26
28	1.5172	0.6591	0.02900	34.4815	0.04400	22.7267	28
30	1.5631	0.6398	0.02664	37.5387	0.04164	24.0158	30
32	1.6103	0.6210	0.02458	40.6883	0.03958	25.2671	32
34	1.6590	0.6028	0.02276	43.9331	0.03776	26.4817	34
35	1.6839	0.5939	0.02193	45.5921	0.03693	27.0756	35
36	1.7091	0.5851	0.02115	47.2760	0.03615	27.6607	36
38	1.7608	0.5679	0.01972	50.7199	0.03472	28.8051	38
40	1.8140	0.5513	0.01843	54.2679	0.03343	29.9158	40
45	1.9542	0.5117	0.01572	63.6142	0.03072	32.5523	45
50	2.1052	0.4750	0.01357	73.6828	0.02857	34.9997	50
55	2.2679	0.4409	0.01183	84.5296	0.02683	37.2715	55
60	2.4432	0.4093	0.01039	96.2147	0.02539	39.3803	60
65	2.6320	0.3799	0.00919	108.8028	0.02419	41.3378	65
70	2.8355	0.3527	0.00817	122.3638	0.02317	43.1549	70
75	3.0546	0.3274	0.00730	136.9728	0.02230	44.8416	75
80	3.2907	0.3039	0.00655	152.7109	0.02155	46.4073	80
85	3.5450	0.2821	0.00589	169.6652	0.02089	47.8607	85
90	3.8189	0.2619	0.00532	187.9299	0.02032	49.2099	90
95	4.1141	0.2431	0.00482	207.6061	0.01982	50.4622	95
100	4.4320	0.2256	0.00437	228.8030	0.01937	51.6247	100

TABLE A - 4

DISCRETE CASH FLOW
2.00% DISCRETE COMPOUND INTEREST FACTORS

	SINGLE PAYMENTS		UNIFORM SERIES PAYMENTS				
N	COMPOUND AMOUNT F/P	PRESENT WORTH P/F	SINKING FUND A/F	COMPOUND AMOUNT F/A	CAPITAL RECOVERY A/P	PRESENT WORTH P/A	N
1	1.0200	0.9804	1.00000	1.0000	1.02000	0.9804	1
2	1.0404	0.9612	0.49505	2.0200	0.51505	1.9416	2
3	1.0612	0.9423	0.32675	3.0604	0.34675	2.8839	3
4	1.0824	0.9238	0.24262	4.1216	0.26262	3.8077	4
5	1.1041	0.9057	0.19216	5.2040	0.21216	4.7135	5
6	1.1262	0.8880	0.15853	6.3081	0.17853	5.6014	6
7	1.1487	0.8706	0.13451	7.4343	0.15451	6.4720	7
8	1.1717	0.8535	0.11651	8.5830	0.13651	7.3255	8
9	1.1951	0.8368	0.10252	9.7546	0.12252	8.1622	9
10	1.2190	0.8203	0.09133	10.9497	0.11133	8.9826	10
11	1.2434	0.8043	0.08218	12.1687	0.10218	9.7868	11
12	1.2682	0.7885	0.07456	13.4121	0.09456	10.5753	12
13	1.2936	0.7730	0.06812	14.6803	0.08812	11.3484	13
14	1.3195	0.7579	0.06260	15.9739	0.08260	12.1062	14
15	1.3459	0.7430	0.05783	17.2934	0.07783	12.8493	15
16	1.3728	0.7284	0.05365	18.6393	0.07365	13.5777	16
17	1.4002	0.7142	0.04997	20.0121	0.06997	14.2919	17
18	1.4282	0.7002	0.04670	21.4123	0.06670	14.9920	18
19	1.4568	0.6864	0.04378	22.8406	0.06378	15.6785	19
20	1.4859	0.6730	0.04116	24.2974	0.06116	16.3514	20
22	1.5460	0.6468	0.03663	27.2990	0.05663	17.6580	22
24	1.6084	0.6217	0.03287	30.4219	0.05287	18.9139	24
25	1.6406	0.6095	0.03122	32.0303	0.05122	19.5235	25
26	1.6734	0.5976	0.02970	33.6709	0.04970	20.1210	26
28	1.7410	0.5744	0.02699	37.0512	0.04699	21.2813	28
30	1.8114	0.5521	0.02465	40.5681	0.04465	22.3965	30
32	1.8845	0.5306	0.02261	44.2270	0.04261	23.4683	32
34	1.9607	0.5100	0.02082	48.0338	0.04082	24.4986	34
35	1.9999	0.5000	0.02000	49.9945	0.04000	24.9986	35
36	2.0399	0.4902	0.01923	51.9944	0.03923	25.4888	36
38	2.1223	0.4712	0.01782	56.1149	0.03782	26.4406	38
40	2.2080	0.4529	0.01656	60.4020	0.03656	27.3555	40
45	2.4379	0.4102	0.01391	71.8927	0.03391	29.4902	45
50	2.6916	0.3715	0.01182	84.5794	0.03182	31.4236	50
55	2.9717	0.3365	0.01014	98.5865	0.03014	33.1748	55
60	3.2810	0.3048	0.00877	114.0515	0.02877	34.7609	60
65	3.6225	0.2761	0.00763	131.1262	0.02763	36.1975	65
70	3.9996	0.2500	0.00667	149.9779	0.02667	37.4986	70
75	4.4158	0.2265	0.00586	170.7918	0.02586	38.6771	75
80	4.8754	0.2051	0.00516	193.7720	0.02516	39.7445	80
85	5.3829	0.1858	0.00456	219.1439	0.02456	40.7113	85
90	5.9431	0.1683	0.00405	247.1567	0.02405	41.5869	90
95	6.5617	0.1524	0.00360	278.0850	0.02360	42.3800	95
100	7.2446	0.1380	0.00320	312.2323	0.02320	43.0984	100

TABLE A - 5

DISCRETE CASH FLOW
3.00% DISCRETE COMPOUND INTEREST FACTORS

	SINGLE PAYMENTS		UNIFORM SERIES PAYMENTS				
N	COMPOUND AMOUNT F/P	PRESENT WORTH P/F	SINKING FUND A/F	COMPOUND AMOUNT F/A	CAPITAL RECOVERY A/P	PRESENT WORTH P/A	N
1	1.0300	0.9709	1.00000	1.0000	1.03000	0.9709	1
2	1.0609	0.9426	0.49261	2.0300	0.52261	1.9135	2
3	1.0927	0.9151	0.32353	3.0909	0.35353	2.8286	3
4	1.1255	0.8885	0.23903	4.1836	0.26903	3.7171	4
5	1.1593	0.8626	0.18835	5.3091	0.21835	4.5797	5
6	1.1941	0.8375	0.15460	6.4684	0.18460	5.4172	6
7	1.2299	0.8131	0.13051	7.6625	0.16051	6.2303	7
8	1.2668	0.7894	0.11246	8.8923	0.14246	7.0197	8
9	1.3048	0.7664	0.09843	10.1591	0.12843	7.7861	9
10	1.3439	0.7441	0.08723	11.4639	0.11723	8.5302	10
11	1.3842	0.7224	0.07808	12.8078	0.10808	9.2526	11
12	1.4258	0.7014	0.07046	14.1920	0.10046	9.9540	12
13	1.4685	0.6810	0.06403	15.6178	0.09403	10.6350	13
14	1.5126	0.6611	0.05853	17.0863	0.08853	11.2961	14
15	1.5580	0.6419	0.05377	18.5989	0.08377	11.9379	15
16	1.6047	0.6232	0.04961	20.1569	0.07961	12.5611	16
17	1.6528	0.6050	0.04595	21.7616	0.07595	13.1661	17
18	1.7024	0.5874	0.04271	23.4144	0.07271	13.7535	18
19	1.7535	0.5703	0.03981	25.1169	0.06981	14.3238	19
20	1.8061	0.5537	0.03722	26.8704	0.06722	14.8775	20
22	1.9161	0.5219	0.03275	30.5368	0.06275	15.9369	22
24	2.0328	0.4919	0.02905	34.4265	0.05905	16.9355	24
25	2.0938	0.4776	0.02743	36.4593	0.05743	17.4131	25
26	2.1566	0.4637	0.02594	38.5530	0.05594	17.8768	26
28	2.2879	0.4371	0.02329	42.9309	0.05329	18.7641	28
30	2.4273	0.4120	0.02102	47.5754	0.05102	19.6004	30
32	2.5751	0.3883	0.01905	52.5028	0.04905	20.3888	32
34	2.7319	0.3660	0.01732	57.7302	0.04732	21.1318	34
35	2.8139	0.3554	0.01654	60.4621	0.04654	21.4872	35
36	2.8983	0.3450	0.01580	63.2759	0.04580	21.8323	36
38	3.0748	0.3252	0.01446	69.1594	0.04446	22.4925	38
40	3.2620	0.3066	0.01326	75.4013	0.04326	23.1148	40
45	3.7816	0.2644	0.01079	92.7199	0.04079	24.5187	45
50	4.3839	0.2281	0.00887	112.7969	0.03887	25.7298	50
55	5.0821	0.1968	0.00735	136.0716	0.03735	26.7744	55
60	5.8916	0.1697	0.00613	163.0534	0.03613	27.6756	60
65	6.8300	0.1464	0.00515	194.3328	0.03515	28.4529	65
70	7.9178	0.1263	0.00434	230.5941	0.03434	29.1234	70
75	9.1789	0.1089	0.00367	272.6309	0.03367	29.7018	75
80	10.6409	0.0940	0.00311	321.3630	0.03311	30.2008	80
85	12.3357	0.0811	0.00265	377.8570	0.03265	30.6312	85
90	14.3005	0.0699	0.00226	443.3489	0.03226	31.0024	90
95	16.5782	0.0603	0.00193	519.2720	0.03193	31.3227	95
100	19.2186	0.0520	0.00165	607.2877	0.03165	31.5989	100

TABLE A - 6

DISCRETE CASH FLOW
4.00% DISCRETE COMPOUND INTEREST FACTORS

	SINGLE PAYMENTS		UNIFORM SERIES PAYMENTS				
N	COMPOUND AMOUNT F/P	PRESENT WORTH P/F	SINKING FUND A/F	COMPOUND AMOUNT F/A	CAPITAL RECOVERY A/P	PRESENT WORTH P/A	N
1	1.0400	0.9615	1.00000	1.000	1.04000	0.9615	1
2	1.0816	0.9246	0.49020	2.040	0.53020	1.8861	2
3	1.1249	0.8890	0.32035	3.122	0.36035	2.7751	3
4	1.1699	0.8548	0.23549	4.246	0.27549	3.6299	4
5	1.2167	0.8219	0.18463	5.416	0.22463	4.4518	5
6	1.2653	0.7903	0.15076	6.633	0.19076	5.2421	6
7	1.3159	0.7599	0.12661	7.898	0.16661	6.0021	7
8	1.3686	0.7307	0.10853	9.214	0.14853	6.7327	8
9	1.4233	0.7026	0.09449	10.583	0.13449	7.4353	9
10	1.4802	0.6756	0.08329	12.006	0.12329	8.1109	10
11	1.5395	0.6496	0.07415	13.486	0.11415	8.7605	11
12	1.6010	0.6246	0.06655	15.026	0.10655	9.3851	12
13	1.6651	0.6006	0.06014	16.627	0.10014	9.9856	13
14	1.7317	0.5775	0.05467	18.292	0.09467	10.5631	14
15	1.8009	0.5553	0.04994	20.024	0.08994	11.1184	15
16	1.8730	0.5339	0.04582	21.825	0.08582	11.6523	16
17	1.9479	0.5134	0.04220	23.698	0.08220	12.1657	17
18	2.0258	0.4936	0.03899	25.645	0.07899	12.6593	18
19	2.1068	0.4746	0.03614	27.671	0.07614	13.1339	19
20	2.1911	0.4564	0.03358	29.778	0.07358	13.5903	20
22	2.3699	0.4220	0.02920	34.248	0.06920	14.4511	22
24	2.5633	0.3901	0.02559	39.083	0.06559	15.2470	24
25	2.6658	0.3751	0.02401	41.646	0.06401	15.6221	25
26	2.7725	0.3607	0.02257	44.312	0.06257	15.9828	26
28	2.9987	0.3335	0.02001	49.968	0.06001	16.6631	28
30	3.2434	0.3083	0.01783	56.085	0.05783	17.2920	30
32	3.5081	0.2851	0.01595	62.701	0.05595	17.8736	32
34	3.7943	0.2636	0.01431	69.858	0.05431	18.4112	34
35	3.9461	0.2534	0.01358	73.652	0.05358	18.6646	35
36	4.1039	0.2437	0.01289	77.598	0.05289	18.9083	36
38	4.4388	0.2253	0.01163	85.970	0.05163	19.3679	38
40	4.8010	0.2083	0.01052	95.026	0.05052	19.7928	40
45	5.8412	0.1712	0.00826	121.029	0.04826	20.7200	45
50	7.1067	0.1407	0.00655	152.667	0.04655	21.4822	50
55	8.6464	0.1157	0.00523	191.159	0.04523	22.1086	55
60	10.5196	0.0951	0.00420	237.991	0.04420	22.6235	60
65	12.7987	0.0781	0.00339	294.968	0.04339	23.0467	65
70	15.5716	0.0642	0.00275	364.290	0.04275	23.3945	70
75	18.9453	0.0528	0.00223	448.631	0.04223	23.6804	75
80	23.0498	0.0434	0.00181	551.245	0.04181	23.9154	80
85	28.0436	0.0357	0.00148	676.090	0.04148	24.1085	85
90	34.1193	0.0293	0.00121	827.983	0.04121	24.2673	90
95	41.5114	0.0241	0.00099	1012.785	0.04099	24.3978	95
100	50.5049	0.0198	0.00081	1237.624	0.04081	24.5050	100

TABLE A - 7

DISCRETE CASH FLOW
5.00% DISCRETE COMPOUND INTEREST FACTORS

	SINGLE PAYMENTS		UNIFORM SERIES PAYMENTS				
N	COMPOUND AMOUNT F/P	PRESENT WORTH P/F	SINKING FUND A/F	COMPOUND AMOUNT F/A	CAPITAL RECOVERY A/P	PRESENT WORTH P/A	N
1	1.0500	0.9524	1.00000	1.000	1.05000	0.9524	1
2	1.1025	0.9070	0.48780	2.050	0.53780	1.8594	2
3	1.1576	0.8638	0.31721	3.152	0.36721	2.7232	3
4	1.2155	0.8227	0.23201	4.310	0.28201	3.5460	4
5	1.2763	0.7835	0.18097	5.526	0.23097	4.3295	5
6	1.3401	0.7462	0.14702	6.802	0.19702	5.0757	6
7	1.4071	0.7107	0.12282	8.142	0.17282	5.7864	7
8	1.4775	0.6768	0.10472	9.549	0.15472	6.4632	8
9	1.5513	0.6446	0.09069	11.027	0.14069	7.1078	9
10	1.6289	0.6139	0.07950	12.578	0.12950	7.7217	10
11	1.7103	0.5847	0.07039	14.207	0.12039	8.3064	11
12	1.7959	0.5568	0.06283	15.917	0.11283	8.8633	12
13	1.8856	0.5303	0.05646	17.713	0.10646	9.3936	13
14	1.9799	0.5051	0.05102	19.599	0.10102	9.8986	14
15	2.0789	0.4810	0.04634	21.579	0.09634	10.3797	15
16	2.1829	0.4581	0.04227	23.657	0.09227	10.8378	16
17	2.2920	0.4363	0.03870	25.840	0.08870	11.2741	17
18	2.4066	0.4155	0.03555	28.132	0.08555	11.6896	18
19	2.5270	0.3957	0.03275	30.539	0.08275	12.0853	19
20	2.6533	0.3769	0.03024	33.066	0.08024	12.4622	20
22	2.9253	0.3418	0.02597	38.505	0.07597	13.1630	22
24	3.2251	0.3101	0.02247	44.502	0.07247	13.7986	24
25	3.3864	0.2953	0.02095	47.727	0.07095	14.0939	25
26	3.5557	0.2812	0.01956	51.113	0.06956	14.3752	26
28	3.9201	0.2551	0.01712	58.403	0.06712	14.8981	28
30	4.3219	0.2314	0.01505	66.439	0.06505	15.3725	30
32	4.7649	0.2099	0.01328	75.299	0.06328	15.8027	32
34	5.2533	0.1904	0.01176	85.067	0.06176	16.1929	34
35	5.5160	0.1813	0.01107	90.320	0.06107	16.3742	35
36	5.7918	0.1727	0.01043	95.836	0.06043	16.5469	36
38	6.3855	0.1566	0.00928	107.710	0.05928	16.8679	38
40	7.0400	0.1420	0.00828	120.800	0.05828	17.1591	40
45	8.9850	0.1113	0.00626	159.700	0.05626	17.7741	45
50	11.4674	0.0872	0.00478	209.348	0.05478	18.2559	50
55	14.6356	0.0683	0.00367	272.713	0.05367	18.6335	55
60	18.6792	0.0535	0.00283	353.584	0.05283	18.9293	60
65	23.8399	0.0419	0.00219	456.798	0.05219	19.1611	65
70	30.4264	0.0329	0.00170	588.529	0.05170	19.3427	70
75	38.8327	0.0258	0.00132	756.654	0.05132	19.4850	75
80	49.5614	0.0202	0.00103	971.229	0.05103	19.5965	80
85	63.2544	0.0158	0.00080	1245.087	0.05080	19.6838	85
90	80.7304	0.0124	0.00063	1594.607	0.05063	19.7523	90
95	103.035	0.0097	0.00049	2040.694	0.05049	19.8059	95
100	131.501	0.0076	0.00038	2610.025	0.05038	19.8479	100

TABLE A - 8

DISCRETE CASH FLOW
6.00% DISCRETE COMPOUND INTEREST FACTORS

	SINGLE PAYMENTS		UNIFORM SERIES PAYMENTS				
N	COMPOUND AMOUNT F/P	PRESENT WORTH P/F	SINKING FUND A/F	COMPOUND AMOUNT F/A	CAPITAL RECOVERY A/P	PRESENT WORTH P/A	N
1	1.0600	0.9434	1.00000	1.000	1.06000	0.9434	1
2	1.1236	0.8900	0.48544	2.060	0.54544	1.8334	2
3	1.1910	0.8396	0.31411	3.184	0.37411	2.6730	3
4	1.2625	0.7921	0.22859	4.375	0.28859	3.4651	4
5	1.3382	0.7473	0.17740	5.637	0.23740	4.2124	5
6	1.4185	0.7050	0.14336	6.975	0.20336	4.9173	6
7	1.5036	0.6651	0.11914	8.394	0.17914	5.5824	7
8	1.5938	0.6274	0.10104	9.897	0.16104	6.2098	8
9	1.6895	0.5919	0.08702	11.491	0.14702	6.8017	9
10	1.7908	0.5584	0.07587	13.181	0.13587	7.3601	10
11	1.8983	0.5268	0.06679	14.972	0.12679	7.8869	11
12	2.0122	0.4970	0.05928	16.870	0.11928	8.3838	12
13	2.1329	0.4688	0.05296	18.882	0.11296	8.8527	13
14	2.2609	0.4423	0.04758	21.015	0.10758	9.2950	14
15	2.3966	0.4173	0.04296	23.276	0.10296	9.7122	15
16	2.5404	0.3936	0.03895	25.673	0.09895	10.1059	16
17	2.6928	0.3714	0.03544	28.213	0.09544	10.4773	17
18	2.8543	0.3503	0.03236	30.906	0.09236	10.8276	18
19	3.0256	0.3305	0.02962	33.760	0.08962	11.1581	19
20	3.2071	0.3118	0.02718	36.786	0.08718	11.4699	20
22	3.6035	0.2775	0.02305	43.392	0.08305	12.0416	22
24	4.0489	0.2470	0.01968	50.816	0.07968	12.5504	24
25	4.2919	0.2330	0.01823	54.865	0.07823	12.7834	25
26	4.5494	0.2198	0.01690	59.156	0.07690	13.0032	26
28	5.1117	0.1956	0.01459	68.528	0.07459	13.4062	28
30	5.7435	0.1741	0.01265	79.058	0.07265	13.7648	30
32	6.4534	0.1550	0.01100	90.890	0.07100	14.0840	32
34	7.2510	0.1379	0.00960	104.184	0.06960	14.3681	34
35	7.6861	0.1301	0.00897	111.435	0.06897	14.4982	35
36	8.1473	0.1227	0.00839	119.121	0.06839	14.6210	36
38	9.1543	0.1092	0.00736	135.904	0.06736	14.8460	38
40	10.2857	0.0972	0.00646	154.762	0.06646	15.0463	40
45	13.7646	0.0727	0.00470	212.744	0.06470	15.4558	45
50	18.4202	0.0543	0.00344	290.336	0.06344	15.7619	50
55	24.6503	0.0406	0.00254	394.172	0.06254	15.9905	55
60	32.9877	0.0303	0.00188	533.128	0.06188	16.1614	60
65	44.1450	0.0227	0.00139	719.083	0.06139	16.2891	65
70	59.0759	0.0169	0.00103	967.932	0.06103	16.3845	70
75	79.0569	0.0126	0.00077	1300.949	0.06077	16.4558	75
80	105.796	0.0095	0.00057	1746.600	0.06057	16.5091	80
85	141.579	0.0071	0.00043	2342.982	0.06043	16.5489	85
90	189.465	0.0053	0.00032	3141.075	0.06032	16.5787	90
95	253.546	0.0039	0.00024	4209.104	0.06024	16.6009	95
100	339.302	0.0029	0.00018	5638.368	0.06018	16.6175	100

TABLE A - 9

DISCRETE CASH FLOW
7.00% DISCRETE COMPOUND INTEREST FACTORS

	SINGLE PAYMENTS		UNIFORM SERIES PAYMENTS				
N	COMPOUND AMOUNT F/P	PRESENT WORTH P/F	SINKING FUND A/F	COMPOUND AMOUNT F/A	CAPITAL RECOVERY A/P	PRESENT WORTH P/A	N
1	1.0700	0.9346	1.00000	1.000	1.07000	0.9346	1
2	1.1449	0.8734	0.48309	2.070	0.55309	1.8080	2
3	1.2250	0.8163	0.31105	3.215	0.38105	2.6243	3
4	1.3108	0.7629	0.22523	4.440	0.29523	3.3872	4
5	1.4026	0.7130	0.17389	5.751	0.24389	4.1002	5
6	1.5007	0.6663	0.13980	7.153	0.20980	4.7665	6
7	1.6058	0.6227	0.11555	8.654	0.18555	5.3893	7
8	1.7182	0.5820	0.09747	10.260	0.16747	5.9713	8
9	1.8385	0.5439	0.08349	11.978	0.15349	6.5152	9
10	1.9672	0.5083	0.07238	13.816	0.14238	7.0236	10
11	2.1049	0.4751	0.06336	15.784	0.13336	7.4987	11
12	2.2522	0.4440	0.05590	17.888	0.12590	7.9427	12
13	2.4098	0.4150	0.04965	20.141	0.11965	8.3577	13
14	2.5785	0.3878	0.04434	22.550	0.11434	8.7455	14
15	2.7590	0.3624	0.03979	25.129	0.10979	9.1079	15
16	2.9522	0.3387	0.03586	27.888	0.10586	9.4466	16
17	3.1588	0.3166	0.03243	30.840	0.10243	9.7632	17
18	3.3799	0.2959	0.02941	33.999	0.09941	10.0591	18
19	3.6165	0.2765	0.02675	37.379	0.09675	10.3356	19
20	3.8697	0.2584	0.02439	40.995	0.09439	10.5940	20
22	4.4304	0.2257	0.02041	49.006	0.09041	11.0612	22
24	5.0724	0.1971	0.01719	58.177	0.08719	11.4693	24
25	5.4274	0.1842	0.01581	63.249	0.08581	11.6536	25
26	5.8074	0.1722	0.01456	68.676	0.08456	11.8258	26
28	6.6488	0.1504	0.01239	80.698	0.08239	12.1371	28
30	7.6123	0.1314	0.01059	94.461	0.08059	12.4090	30
32	8.7153	0.1147	0.00907	110.218	0.07907	12.6466	32
34	9.9781	0.1002	0.00780	128.259	0.07780	12.8540	34
35	10.6766	0.0937	0.00723	138.237	0.07723	12.9477	35
36	11.4239	0.0875	0.00672	148.913	0.07672	13.0352	36
38	13.0793	0.0765	0.00580	172.561	0.07580	13.1935	38
40	14.9745	0.0668	0.00501	199.635	0.07501	13.3317	40
45	21.0025	0.0476	0.00350	285.749	0.07350	13.6055	45
50	29.4570	0.0339	0.00246	406.529	0.07246	13.8007	50
55	41.3150	0.0242	0.00174	575.929	0.07174	13.9399	55
60	57.9464	0.0173	0.00123	813.520	0.07123	14.0392	60
65	81.2729	0.0123	0.00087	1146.755	0.07087	14.1099	65
70	113.989	0.0088	0.00062	1614.134	0.07062	14.1604	70
75	159.876	0.0063	0.00044	2269.657	0.07044	14.1964	75
80	224.234	0.0045	0.00031	3189.063	0.07031	14.2220	80
85	314.500	0.0032	0.00022	4478.576	0.07022	14.2403	85
90	441.103	0.0023	0.00016	6287.185	0.07016	14.2533	90
95	618.670	0.0016	0.00011	8823.854	0.07011	14.2626	95
100	867.716	0.0012	0.00008	12381.662	0.07008	14.2693	100

TABLE A - 10

DISCRETE CASH FLOW
8.00% DISCRETE COMPOUND INTEREST FACTORS

	SINGLE PAYMENTS		UNIFORM SERIES PAYMENTS				
N	COMPOUND AMOUNT F/P	PRESENT WORTH P/F	SINKING FUND A/F	COMPOUND AMOUNT F/A	CAPITAL RECOVERY A/P	PRESENT WORTH P/A	N
1	1.0800	0.9259	1.00000	1.000	1.08000	0.9259	1
2	1.1664	0.8573	0.48077	2.080	0.56077	1.7833	2
3	1.2597	0.7938	0.30803	3.246	0.38803	2.5771	3
4	1.3605	0.7350	0.22192	4.506	0.30192	3.3121	4
5	1.4693	0.6806	0.17046	5.867	0.25046	3.9927	5
6	1.5869	0.6302	0.13632	7.336	0.21632	4.6229	6
7	1.7138	0.5835	0.11207	8.923	0.19207	5.2064	7
8	1.8509	0.5403	0.09401	10.637	0.17401	5.7466	8
9	1.9990	0.5002	0.08008	12.488	0.16008	6.2469	9
10	2.1589	0.4632	0.06903	14.487	0.14903	6.7101	10
11	2.3316	0.4289	0.06008	16.645	0.14008	7.1390	11
12	2.5182	0.3971	0.05270	18.977	0.13270	7.5361	12
13	2.7196	0.3677	0.04652	21.495	0.12652	7.9038	13
14	2.9372	0.3405	0.04130	24.215	0.12130	8.2442	14
15	3.1722	0.3152	0.03683	27.152	0.11683	8.5595	15
16	3.4259	0.2919	0.03298	30.324	0.11298	8.8514	16
17	3.7000	0.2703	0.02963	33.750	0.10963	9.1216	17
18	3.9960	0.2502	0.02670	37.450	0.10670	9.3719	18
19	4.3157	0.2317	0.02413	41.446	0.10413	9.6036	19
20	4.6610	0.2145	0.02185	45.762	0.10185	9.8181	20
22	5.4365	0.1839	0.01803	55.457	0.09803	10.2007	22
24	6.3412	0.1577	0.01498	66.765	0.09498	10.5288	24
25	6.8485	0.1460	0.01368	73.106	0.09368	10.6748	25
26	7.3964	0.1352	0.01251	79.954	0.09251	10.8100	26
28	8.6271	0.1159	0.01049	95.339	0.09049	11.0511	28
30	10.0627	0.0994	0.00883	113.283	0.08883	11.2578	30
32	11.7371	0.0852	0.00745	134.214	0.08745	11.4350	32
34	13.6901	0.0730	0.00630	158.627	0.08630	11.5869	34
35	14.7853	0.0676	0.00580	172.317	0.08580	11.6546	35
36	15.9682	0.0626	0.00534	187.102	0.08534	11.7172	36
38	18.6253	0.0537	0.00454	220.316	0.08454	11.8289	38
40	21.7245	0.0460	0.00386	259.057	0.08386	11.9246	40
45	31.9204	0.0313	0.00259	386.506	0.08259	12.1084	45
50	46.9016	0.0213	0.00174	573.770	0.08174	12.2335	50
55	68.9139	0.0145	0.00118	848.923	0.08118	12.3186	55
60	101.257	0.0099	0.00080	1253.213	0.08080	12.3766	60
65	148.780	0.0067	0.00054	1847.248	0.08054	12.4160	65
70	218.606	0.0046	0.00037	2720.080	0.08037	12.4428	70
75	321.205	0.0031	0.00025	4002.557	0.08025	12.4611	75
80	471.955	0.0021	0.00017	5886.935	0.08017	12.4735	80
85	693.456	0.0014	0.00012	8655.706	0.08012	12.4820	85
90	1018.915	0.0010	0.00008	12723.939	0.08008	12.4877	90
95	1497.121	0.0007	0.00005	18701.507	0.08005	12.4917	95
100	2199.761	0.0005	0.00004	27484.516	0.08004	12.4943	100

TABLE A - 11

DISCRETE CASH FLOW
9.00% DISCRETE COMPOUND INTEREST FACTORS

	SINGLE PAYMENTS		UNIFORM SERIES PAYMENTS				
N	COMPOUND AMOUNT F/P	PRESENT WORTH P/F	SINKING FUND A/F	COMPOUND AMOUNT F/A	CAPITAL RECOVERY A/P	PRESENT WORTH P/A	N
1	1.0900	0.9174	1.00000	1.000	1.09000	0.9174	1
2	1.1881	0.8417	0.47847	2.090	0.56847	1.7591	2
3	1.2950	0.7722	0.30505	3.278	0.39505	2.5313	3
4	1.4116	0.7084	0.21867	4.573	0.30867	3.2397	4
5	1.5386	0.6499	0.16709	5.985	0.25709	3.8897	5
6	1.6771	0.5963	0.13292	7.523	0.22292	4.4859	6
7	1.8280	0.5470	0.10869	9.200	0.19869	5.0330	7
8	1.9926	0.5019	0.09067	11.028	0.18067	5.5348	8
9	2.1719	0.4604	0.07680	13.021	0.16680	5.9952	9
10	2.3674	0.4224	0.06582	15.193	0.15582	6.4177	10
11	2.5804	0.3875	0.05695	17.560	0.14695	6.8052	11
12	2.8127	0.3555	0.04965	20.141	0.13965	7.1607	12
13	3.0658	0.3262	0.04357	22.953	0.13357	7.4869	13
14	3.3417	0.2992	0.03843	26.019	0.12843	7.7862	14
15	3.6425	0.2745	0.03406	29.361	0.12406	8.0607	15
16	3.9703	0.2519	0.03030	33.003	0.12030	8.3126	16
17	4.3276	0.2311	0.02705	36.974	0.11705	8.5436	17
18	4.7171	0.2120	0.02421	41.301	0.11421	8.7556	18
19	5.1417	0.1945	0.02173	46.018	0.11173	8.9501	19
20	5.6044	0.1784	0.01955	51.160	0.10955	9.1285	20
22	6.6586	0.1502	0.01590	62.873	0.10590	9.4424	22
24	7.9111	0.1264	0.01302	76.790	0.10302	9.7066	24
25	8.6231	0.1160	0.01181	84.701	0.10181	9.8226	25
26	9.3992	0.1064	0.01072	93.324	0.10072	9.9290	26
28	11.1671	0.0895	0.00885	112.968	0.09885	10.1161	28
30	13.2677	0.0754	0.00734	136.308	0.09734	10.2737	30
32	15.7633	0.0634	0.00610	164.037	0.09610	10.4062	32
34	18.7284	0.0534	0.00508	196.982	0.09508	10.5178	34
35	20.4140	0.0490	0.00464	215.711	0.09464	10.5668	35
36	22.2512	0.0449	0.00424	236.125	0.09424	10.6118	36
38	26.4367	0.0378	0.00354	282.630	0.09354	10.6908	38
40	31.4094	0.0318	0.00296	337.882	0.09296	10.7574	40
45	48.3273	0.0207	0.00190	525.859	0.09190	10.8812	45
50	74.3575	0.0134	0.00123	815.084	0.09123	10.9617	50
55	114.408	0.0087	0.00079	1260.092	0.09079	11.0140	55
60	176.031	0.0057	0.00051	1944.792	0.09051	11.0480	60
65	270.846	0.0037	0.00033	2998.288	0.09033	11.0701	65
70	416.730	0.0024	0.00022	4619.223	0.09022	11.0844	70
75	641.191	0.0016	0.00014	7113.232	0.09014	11.0938	75
80	986.552	0.0010	0.00009	10950.574	0.09009	11.0998	80
85	1517.932	0.0007	0.00006	16854.800	0.09006	11.1038	85
90	2335.527	0.0004	0.00004	25939.184	0.09004	11.1064	90
95	3593.497	0.0003	0.00003	39916.635	0.09003	11.1080	95
100	5529.041	0.0002	0.00002	61422.675	0.09002	11.1091	100

TABLE A - 12

DISCRETE CASH FLOW
10.00% DISCRETE COMPOUND INTEREST FACTORS

	SINGLE PAYMENTS		UNIFORM SERIES PAYMENTS				
N	COMPOUND AMOUNT F/P	PRESENT WORTH P/F	SINKING FUND A/F	COMPOUND AMOUNT F/A	CAPITAL RECOVERY A/P	PRESENT WORTH P/A	N
1	1.1000	0.9091	1.00000	1.000	1.10000	0.9091	1
2	1.2100	0.8264	0.47619	2.100	0.57619	1.7355	2
3	1.3310	0.7513	0.30211	3.310	0.40211	2.4869	3
4	1.4641	0.6830	0.21547	4.641	0.31547	3.1699	4
5	1.6105	0.6209	0.16380	6.105	0.26380	3.7908	5
6	1.7716	0.5645	0.12961	7.716	0.22961	4.3553	6
7	1.9487	0.5132	0.10541	9.487	0.20541	4.8684	7
8	2.1436	0.4665	0.08744	11.436	0.18744	5.3349	8
9	2.3579	0.4241	0.07364	13.579	0.17364	5.7590	9
10	2.5937	0.3855	0.06275	15.937	0.16275	6.1446	10
11	2.8531	0.3505	0.05396	18.531	0.15396	6.4951	11
12	3.1384	0.3186	0.04676	21.384	0.14676	6.8137	12
13	3.4523	0.2897	0.04078	24.523	0.14078	7.1034	13
14	3.7975	0.2633	0.03575	27.975	0.13575	7.3667	14
15	4.1772	0.2394	0.03147	31.772	0.13147	7.6061	15
16	4.5950	0.2176	0.02782	35.950	0.12782	7.8237	16
17	5.0545	0.1978	0.02466	40.545	0.12466	8.0216	17
18	5.5599	0.1799	0.02193	45.599	0.12193	8.2014	18
19	6.1159	0.1635	0.01955	51.159	0.11955	8.3649	19
20	6.7275	0.1486	0.01746	57.275	0.11746	8.5136	20
22	8.1403	0.1228	0.01401	71.403	0.11401	8.7715	22
24	9.8497	0.1015	0.01130	88.497	0.11130	8.9847	24
25	10.8347	0.0923	0.01017	98.347	0.11017	9.0770	25
26	11.9182	0.0839	0.00916	109.182	0.10916	9.1609	26
28	14.4210	0.0693	0.00745	134.210	0.10745	9.3066	28
30	17.4494	0.0573	0.00608	164.494	0.10608	9.4269	30
32	21.1138	0.0474	0.00497	201.138	0.10497	9.5264	32
34	25.5477	0.0391	0.00407	245.477	0.10407	9.6086	34
35	28.1024	0.0356	0.00369	271.024	0.10369	9.6442	35
36	30.9127	0.0323	0.00334	299.127	0.10334	9.6765	36
38	37.4043	0.0267	0.00275	364.043	0.10275	9.7327	38
40	45.2593	0.0221	0.00226	442.593	0.10226	9.7791	40
45	72.8905	0.0137	0.00139	718.905	0.10139	9.8628	45
50	117.391	0.0085	0.00086	1163.909	0.10086	9.9148	50
55	189.059	0.0053	0.00053	1880.591	0.10053	9.9471	55
60	304.482	0.0033	0.00033	3034.816	0.10033	9.9672	60
65	490.371	0.0020	0.00020	4893.707	0.10020	9.9796	65
70	789.747	0.0013	0.00013	7887.470	0.10013	9.9873	70
75	1271.895	0.0008	0.00008	12708.954	0.10008	9.9921	75
80	2048.400	0.0005	0.00005	20474.002	0.10005	9.9951	80
85	3298.969	0.0003	0.00003	32979.690	0.10003	9.9970	85
90	5313.023	0.0002	0.00002	53120.226	0.10002	9.9981	90
95	8556.676	0.0001	0.00001	85556.760	0.10001	9.9988	95

TABLE A - 13

DISCRETE CASH FLOW

11.00% DISCRETE COMPOUND INTEREST FACTORS

	SINGLE PAYMENTS		UNIFORM SERIES PAYMENT				
N	COMPOUND AMOUNT F/P	PRESENT WORTH P/F	SINKING FUND A/F	COMPOUND AMOUNT F/A	CAPITAL RECOVERY A/P	PRESENT WORTH P/A	N
1	1.1100	0.9009	1.00000	1.00	1.11000	0.9009	1
2	1.2321	0.8116	0.47393	2.11	0.58393	1.7125	2
3	1.3676	0.7312	0.29921	3.34	0.40921	2.4437	3
4	1.5181	0.6587	0.21233	4.71	0.32233	3.1024	4
5	1.6851	0.5935	0.16057	6.23	0.27057	3.6959	5
6	1.8704	0.5346	0.12638	7.91	0.23638	4.2305	6
7	2.0762	0.4817	0.10222	9.78	0.21222	4.7122	7
8	2.3045	0.4339	0.08432	11.86	0.19432	5.1461	8
9	2.5580	0.3909	0.07060	14.16	0.18060	5.5370	9
10	2.8394	0.3522	0.05980	16.72	0.16980	5.8892	10
11	3.1518	0.3173	0.05112	19.56	0.16112	6.2065	11
12	3.4985	0.2858	0.04403	22.71	0.15403	6.4924	12
13	3.8833	0.2575	0.03815	26.21	0.14815	6.7499	13
14	4.3104	0.2320	0.03323	30.09	0.14323	6.9819	14
15	4.7846	0.2090	0.02907	34.41	0.13907	7.1909	15
16	5.3109	0.1883	0.02552	39.19	0.13552	7.3792	16
17	5.8951	0.1696	0.02247	44.50	0.13247	7.5488	17
18	6.5436	0.1528	0.01984	50.40	0.12984	7.7016	18
19	7.2633	0.1377	0.01756	56.94	0.12756	7.8393	19
20	8.0623	0.1240	0.01558	64.20	0.12558	7.9633	20
22	9.9336	0.1007	0.01231	81.21	0.12231	8.1757	22
24	12.2392	0.0817	0.00979	102.17	0.11979	8.3481	24
25	13.5855	0.0736	0.00874	114.41	0.11874	8.4217	25
26	15.0799	0.0663	0.00781	128.00	0.11781	8.4881	26
28	18.5799	0.0538	0.00626	159.82	0.11626	8.6016	28
30	22.8923	0.0437	0.00502	199.02	0.11502	8.6938	30
32	28.2056	0.0355	0.00404	247.32	0.11404	8.7686	32
34	34.7521	0.0288	0.00326	306.84	0.11326	8.8293	34
35	38.5749	0.0259	0.00293	341.59	0.11293	8.8552	35
36	42.8181	0.0234	0.00263	380.16	0.11263	8.8786	36
38	52.7562	0.0190	0.00213	470.51	0.11213	8.9186	38
40	65.0009	0.0154	0.00172	581.83	0.11172	8.9511	40
45	109.530	0.0091	0.00101	986.64	0.11101	9.0079	45
50	184.565	0.0054	0.00060	1668.77	0.11060	9.0417	50

TABLE A - 14

DISCRETE CASH FLOW
12.00% DISCRETE COMPOUND INTEREST FACTORS

	SINGLE PAYMENTS		UNIFORM SERIES PAYMENTS				
N	COMPOUND AMOUNT F/P	PRESENT WORTH P/F	SINKING FUND A/F	COMPOUND AMOUNT F/A	CAPITAL RECOVERY A/P	PRESENT WORTH P/A	N
1	1.1200	0.8929	1.00000	1.000	1.12000	0.8929	1
2	1.2544	0.7972	0.47170	2.120	0.59170	1.6901	2
3	1.4049	0.7118	0.29635	3.374	0.41635	2.4018	3
4	1.5735	0.6355	0.20923	4.779	0.32923	3.0373	4
5	1.7623	0.5674	0.15741	6.353	0.27741	3.6048	5
6	1.9738	0.5066	0.12323	8.115	0.24323	4.1114	6
7	2.2107	0.4523	0.09912	10.089	0.21912	4.5638	7
8	2.4760	0.4039	0.08130	12.300	0.20130	4.9676	8
9	2.7731	0.3606	0.06768	14.776	0.18768	5.3282	9
10	3.1058	0.3220	0.05698	17.549	0.17698	5.6502	10
11	3.4785	0.2875	0.04842	20.655	0.16842	5.9377	11
12	3.8960	0.2567	0.04144	24.133	0.16144	6.1944	12
13	4.3635	0.2292	0.03568	28.029	0.15568	6.4235	13
14	4.8871	0.2046	0.03087	32.393	0.15087	6.6282	14
15	5.4736	0.1827	0.02682	37.280	0.14682	6.8109	15
16	6.1304	0.1631	0.02339	42.753	0.14339	6.9740	16
17	6.8660	0.1456	0.02046	48.884	0.14046	7.1196	17
18	7.6900	0.1300	0.01794	55.750	0.13794	7.2497	18
19	8.6128	0.1161	0.01576	63.440	0.13576	7.3658	19
20	9.6463	0.1037	0.01388	72.052	0.13388	7.4694	20
22	12.1003	0.0826	0.01081	92.503	0.13081	7.6446	22
24	15.1786	0.0659	0.00846	118.155	0.12846	7.7843	24
25	17.0001	0.0588	0.00750	133.334	0.12750	7.8431	25
26	19.0401	0.0525	0.00665	150.334	0.12665	7.8957	26
28	23.8839	0.0419	0.00524	190.699	0.12524	7.9844	28
30	29.9599	0.0334	0.00414	241.333	0.12414	8.0552	30
32	37.5817	0.0266	0.00328	304.848	0.12328	8.1116	32
34	47.1425	0.0212	0.00260	384.521	0.12260	8.1566	34
35	52.7996	0.0189	0.00232	431.663	0.12232	8.1755	35
36	59.1356	0.0169	0.00206	484.463	0.12206	8.1924	36
38	74.1797	0.0135	0.00164	609.831	0.12164	8.2210	38
40	93.0510	0.0107	0.00130	767.091	0.12130	8.2438	40
45	163.988	0.0061	0.00074	1358.230	0.12074	8.2825	45
50	289.002	0.0035	0.00042	2400.018	0.12042	8.3045	50

TABLE A - 15

DISCRETE CASH FLOW

13.00% DISCRETE COMPOUND INTEREST FACTORS

	SINGLE PAYMENTS		UNIFORM SERIES PAYMENT				
N	COMPOUND AMOUNT F/P	PRESENT WORTH P/F	SINKING FUND A/F	COMPOUND AMOUNT F/A	CAPITAL RECOVERY A/P	PRESENT WORTH P/A	N
1	1.1300	0.8850	1.00000	1.00	1.13000	0.8850	1
2	1.2769	0.7831	0.46948	2.13	0.59948	1.6681	2
3	1.4429	0.6931	0.29352	3.41	0.42352	2.3612	3
4	1.6305	0.6133	0.20619	4.85	0.33619	2.9745	4
5	1.8424	0.5428	0.15431	6.48	0.28431	3.5172	5
6	2.0820	0.4803	0.12015	8.32	0.25015	3.9975	6
7	2.3526	0.4251	0.09611	10.40	0.22611	4.4226	7
8	2.6584	0.3762	0.07839	12.76	0.20839	4.7988	8
9	3.0040	0.3329	0.06487	15.42	0.19487	5.1317	9
10	3.3946	0.2946	0.05429	18.42	0.18429	5.4262	10
11	3.8359	0.2607	0.04584	21.81	0.17584	5.6869	11
12	4.3345	0.2307	0.03899	25.65	0.16899	5.9176	12
13	4.8980	0.2042	0.03335	29.98	0.16335	6.1218	13
14	5.5348	0.1807	0.02867	34.88	0.15867	6.3025	14
15	6.2543	0.1599	0.02474	40.42	0.15474	6.4624	15
16	7.0673	0.1415	0.02143	46.67	0.15143	6.6039	16
17	7.9861	0.1252	0.01861	53.74	0.14861	6.7291	17
18	9.0243	0.1108	0.01620	61.73	0.14620	6.8399	18
19	10.1974	0.0981	0.01413	70.75	0.14413	6.9380	19
20	11.5231	0.0868	0.01235	80.95	0.14235	7.0248	20
22	14.7138	0.0680	0.00948	105.49	0.13948	7.1695	22
24	18.7881	0.0532	0.00731	136.83	0.13731	7.2829	24
25	21.2305	0.0471	0.00643	155.62	0.13643	7.3300	25
26	23.9905	0.0417	0.00565	176.85	0.13565	7.3717	26
28	30.6335	0.0326	0.00439	227.95	0.13439	7.4412	28
30	39.1159	0.0256	0.00341	293.20	0.13341	7.4957	30
32	49.9471	0.0200	0.00266	376.52	0.13266	7.5383	32
34	63.7774	0.0157	0.00207	482.90	0.13207	7.5717	34
35	72.0685	0.0139	0.00183	546.68	0.13183	7.5856	35
36	81.4374	0.0123	0.00162	618.75	0.13162	7.5979	36
38	103.987	0.0096	0.00126	792.21	0.13126	7.6183	38
40	132.782	0.0075	0.00099	1013.70	0.13099	7.6344	40
45	244.641	0.0041	0.00053	1874.16	0.13053	7.6609	45
50	450.736	0.0022	0.00029	3459.51	0.13029	7.6752	50

TABLE A - 16

DISCRETE CASH FLOW

14.00% DISCRETE COMPOUND INTEREST FACTORS

	SINGLE PAYMENTS		UNIFORM SERIES PAYMENT				
N	COMPOUND AMOUNT F/P	PRESENT WORTH P/F	SINKING FUND A/F	COMPOUND AMOUNT F/A	CAPITAL RECOVERY A/P	PRESENT WORTH P/A	N
1	1.1400	0.8772	1.00000	1.00	1.14000	0.8772	1
2	1.2996	0.7695	0.46729	2.14	0.60729	1.6467	2
3	1.4815	0.6750	0.29073	3.44	0.43073	2.3216	3
4	1.6890	0.5921	0.20320	4.92	0.34320	2.9137	4
5	1.9254	0.5194	0.15128	6.61	0.29128	3.4331	5
6	2.1950	0.4556	0.11716	8.54	0.25716	3.8887	6
7	2.5023	0.3996	0.09319	10.73	0.23319	4.2883	7
8	2.8526	0.3506	0.07557	13.23	0.21557	4.6389	8
9	3.2519	0.3075	0.06217	16.09	0.20217	4.9464	9
10	3.7072	0.2697	0.05171	19.34	0.19171	5.2161	10
11	4.2262	0.2366	0.04339	23.04	0.18339	5.4527	11
12	4.8179	0.2076	0.03667	27.27	0.17667	5.6603	12
13	5.4924	0.1821	0.03116	32.09	0.17116	5.8424	13
14	6.2613	0.1597	0.02661	37.58	0.16661	6.0021	14
15	7.1379	0.1401	0.02281	43.84	0.16281	6.1422	15
16	8.1372	0.1229	0.01962	50.98	0.15962	6.2651	16
17	9.2765	0.1078	0.01692	59.12	0.15692	6.3729	17
18	10.5752	0.0946	0.01462	68.39	0.15462	6.4674	18
19	12.0557	0.0829	0.01266	78.97	0.15266	6.5504	19
20	13.7435	0.0728	0.01099	91.02	0.15099	6.6231	20
22	17.8610	0.0560	0.00830	120.44	0.14830	6.7429	22
24	23.2122	0.0431	0.00630	158.66	0.14630	6.8351	24
25	26.4619	0.0378	0.00550	181.87	0.14550	6.8729	25
26	30.1666	0.0331	0.00480	208.33	0.14480	6.9061	26
28	39.2045	0.0255	0.00366	272.89	0.14366	6.9607	28
30	50.9502	0.0196	0.00280	356.79	0.14280	7.0027	30
32	66.2148	0.0151	0.00215	465.82	0.14215	7.0350	32
34	86.0528	0.0116	0.00165	607.52	0.14165	7.0599	34
35	98.1002	0.0102	0.00144	693.57	0.14144	7.0700	35
36	111.834	0.0089	0.00126	791.67	0.14126	7.0790	36
38	145.340	0.0069	0.00097	1031.00	0.14097	7.0937	38
40	188.884	0.0053	0.00075	1342.03	0.14075	7.1050	40
45	363.679	0.0027	0.00039	2590.56	0.14039	7.1232	45
50	700.233	0.0014	0.00020	4994.52	0.14020	7.1327	50

TABLE A - 17

DISCRETE CASH FLOW
15.00% DISCRETE COMPOUND INTEREST FACTORS

N	SINGLE PAYMENTS		UNIFORM SERIES PAYMENTS				N
	COMPOUND AMOUNT F/P	PRESENT WORTH P/F	SINKING FUND A/F	COMPOUND AMOUNT F/A	CAPITAL RECOVERY A/P	PRESENT WORTH P/A	
1	1.1500	0.8696	1.00000	1.000	1.15000	0.8696	1
2	1.3225	0.7561	0.46512	2.150	0.61512	1.6257	2
3	1.5209	0.6575	0.28798	3.472	0.43798	2.2832	3
4	1.7490	0.5718	0.20027	4.993	0.35027	2.8550	4
5	2.0114	0.4972	0.14832	6.742	0.29832	3.3522	5
6	2.3131	0.4323	0.11424	8.754	0.26424	3.7845	6
7	2.6600	0.3759	0.09036	11.067	0.24036	4.1604	7
8	3.0590	0.3269	0.07285	13.727	0.22285	4.4873	8
9	3.5179	0.2843	0.05957	16.786	0.20957	4.7716	9
10	4.0456	0.2472	0.04925	20.304	0.19925	5.0188	10
11	4.6524	0.2149	0.04107	24.349	0.19107	5.2337	11
12	5.3503	0.1869	0.03448	29.002	0.18448	5.4206	12
13	6.1528	0.1625	0.02911	34.352	0.17911	5.5831	13
14	7.0757	0.1413	0.02469	40.505	0.17469	5.7245	14
15	8.1371	0.1229	0.02102	47.580	0.17102	5.8474	15
16	9.3576	0.1069	0.01795	55.717	0.16795	5.9542	16
17	10.7613	0.0929	0.01537	65.075	0.16537	6.0472	17
18	12.3755	0.0808	0.01319	75.836	0.16319	6.1280	18
19	14.2318	0.0703	0.01134	88.212	0.16134	6.1982	19
20	16.3665	0.0611	0.00976	102.444	0.15976	6.2593	20
22	21.6447	0.0462	0.00727	137.632	0.15727	6.3587	22
24	28.6252	0.0349	0.00543	184.168	0.15543	6.4338	24
25	32.9190	0.0304	0.00470	212.793	0.15470	6.4641	25
26	37.8568	0.0264	0.00407	245.712	0.15407	6.4906	26
28	50.0656	0.0200	0.00306	327.104	0.15306	6.5335	28
30	66.2118	0.0151	0.00230	434.745	0.15230	6.5660	30
32	87.5651	0.0114	0.00173	577.100	0.15173	6.5905	32
34	115.805	0.0086	0.00131	765.365	0.15131	6.6091	34
35	133.176	0.0075	0.00113	881.170	0.15113	6.6166	35
36	153.152	0.0065	0.00099	1014.346	0.15099	6.6231	36
38	202.543	0.0049	0.00074	1343.622	0.15074	6.6338	38
40	267.864	0.0037	0.00056	1779.090	0.15056	6.6418	40
45	538.769	0.0019	0.00028	3585.128	0.15028	6.6543	45
50	1083.657	0.0009	0.00014	7217.716	0.15014	6.6605	50

TABLE A - 18

DISCRETE CASH FLOW

16.00% DISCRETE COMPOUND INTEREST FACTORS

	SINGLE PAYMENTS		UNIFORM SERIES PAYMENT				
N	COMPOUND AMOUNT F/P	PRESENT WORTH P/F	SINKING FUND A/F	COMPOUND AMOUNT F/A	CAPITAL RECOVERY A/P	PRESENT WORTH P/A	N
1	1.1600	0.8621	1.00000	1.00	1.16000	0.8621	1
2	1.3456	0.7432	0.46296	2.16	0.62296	1.6052	2
3	1.5609	0.6407	0.28526	3.51	0.44526	2.2459	3
4	1.8106	0.5523	0.19738	5.07	0.35738	2.7982	4
5	2.1003	0.4761	0.14541	6.88	0.30541	3.2743	5
6	2.4364	0.4104	0.11139	8.98	0.27139	3.6847	6
7	2.8262	0.3538	0.08761	11.41	0.24761	4.0386	7
8	3.2784	0.3050	0.07022	14.24	0.23022	4.3436	8
9	3.8030	0.2630	0.05708	17.52	0.21708	4.6065	9
10	4.4114	0.2267	0.04690	21.32	0.20690	4.8332	10
11	5.1173	0.1954	0.03886	25.73	0.19886	5.0286	11
12	5.9360	0.1685	0.03241	30.85	0.19241	5.1971	12
13	6.8858	0.1452	0.02718	36.79	0.18718	5.3423	13
14	7.9875	0.1252	0.02290	43.67	0.18290	5.4675	14
15	9.2655	0.1079	0.01936	51.66	0.17936	5.5755	15
16	10.7480	0.0930	0.01641	60.93	0.17641	5.6685	16
17	12.4677	0.0802	0.01395	71.67	0.17395	5.7487	17
18	14.4625	0.0691	0.01188	84.14	0.17188	5.8178	18
19	16.7765	0.0596	0.01014	98.60	0.17014	5.8775	19
20	19.4608	0.0514	0.00867	115.38	0.16867	5.9288	20
22	26.1864	0.0382	0.00635	157.41	0.16635	6.0113	22
24	35.2364	0.0284	0.00467	213.98	0.16467	6.0726	24
25	40.8742	0.0245	0.00401	249.21	0.16401	6.0971	25
26	47.4141	0.0211	0.00345	290.09	0.16345	6.1182	26
28	63.8004	0.0157	0.00255	392.50	0.16255	6.1520	28
30	85.8499	0.0116	0.00189	530.31	0.16189	6.1772	30
32	115.520	0.0087	0.00140	715.75	0.16140	6.1959	32
34	155.443	0.0064	0.00104	965.27	0.16104	6.2098	34
35	180.314	0.0055	0.00089	1120.71	0.16089	6.2153	35
36	209.164	0.0048	0.00077	1301.03	0.16077	6.2201	36
38	281.452	0.0036	0.00057	1752.82	0.16057	6.2278	38
40	378.721	0.0026	0.00042	2360.76	0.16042	6.2335	40
45	795.444	0.0013	0.00020	4965.27	0.16020	6.2421	45
50	1670.704	0.0006	0.00010	10435.65	0.16010	6.2463	50

TABLE A - 19

DISCRETE CASH FLOW

17.00% DISCRETE COMPOUND INTEREST FACTORS

	SINGLE PAYMENTS		UNIFORM SERIES PAYMENT				
N	COMPOUND AMOUNT F/P	PRESENT WORTH P/F	SINKING FUND A/F	COMPOUND AMOUNT F/A	CAPITAL RECOVERY A/P	PRESENT WORTH P/A	N
1	1.1700	0.8547	1.00000	1.00	1.17000	0.8547	1
2	1.3689	0.7305	0.46083	2.17	0.63083	1.5852	2
3	1.6016	0.6244	0.28257	3.54	0.45257	2.2096	3
4	1.8739	0.5337	0.19453	5.14	0.36453	2.7432	4
5	2.1924	0.4561	0.14256	7.01	0.31256	3.1993	5
6	2.5652	0.3898	0.10861	9.21	0.27861	3.5892	6
7	3.0012	0.3332	0.08495	11.77	0.25495	3.9224	7
8	3.5115	0.2848	0.06769	14.77	0.23769	4.2072	8
9	4.1084	0.2434	0.05469	18.28	0.22469	4.4506	9
10	4.8068	0.2080	0.04466	22.39	0.21466	4.6586	10
11	5.6240	0.1778	0.03676	27.20	0.20676	4.8364	11
12	6.5801	0.1520	0.03047	32.82	0.20047	4.9884	12
13	7.6987	0.1299	0.02538	39.40	0.19538	5.1183	13
14	9.0075	0.1110	0.02123	47.10	0.19123	5.2293	14
15	10.5387	0.0949	0.01782	56.11	0.18782	5.3242	15
16	12.3303	0.0811	0.01500	66.65	0.18500	5.4053	16
17	14.4265	0.0693	0.01266	78.98	0.18266	5.4746	17
18	16.8790	0.0592	0.01071	93.41	0.18071	5.5339	18
19	19.7484	0.0506	0.00907	110.28	0.17907	5.5845	19
20	23.1056	0.0433	0.00769	130.03	0.17769	5.6278	20
22	31.6293	0.0316	0.00555	180.17	0.17555	5.6964	22
24	43.2973	0.0231	0.00402	248.81	0.17402	5.7465	24
25	50.6578	0.0197	0.00342	292.10	0.17342	5.7662	25
26	59.2697	0.0169	0.00292	342.76	0.17292	5.7831	26
28	81.1342	0.0123	0.00212	471.38	0.17212	5.8099	28
30	111.065	0.0090	0.00154	647.44	0.17154	5.8294	30
32	152.036	0.0066	0.00113	888.45	0.17113	5.8437	32
34	208.123	0.0048	0.00082	1218.37	0.17082	5.8541	34
35	243.503	0.0041	0.00070	1426.49	0.17070	5.8582	35
36	284.899	0.0035	0.00060	1669.99	0.17060	5.8617	36
38	389.998	0.0026	0.00044	2288.23	0.17044	5.8673	38
40	533.869	0.0019	0.00032	3134.52	0.17032	5.8713	40
45	1170.479	0.0009	0.00015	6879.29	0.17015	5.8773	45
50	2566.215	0.0004	0.00007	15089.50	0.17007	5.8801	50

TABLE A - 20

DISCRETE CASH FLOW
18.00% DISCRETE COMPOUND INTEREST FACTORS

	SINGLE PAYMENTS		UNIFORM SERIES PAYMENTS				
N	COMPOUND AMOUNT F/P	PRESENT WORTH P/F	SINKING FUND A/F	COMPOUND AMOUNT F/A	CAPITAL RECOVERY A/P	PRESENT WORTH P/A	N
1	1.1800	0.8475	1.00000	1.000	1.18000	0.8475	1
2	1.3924	0.7182	0.45872	2.180	0.63872	1.5656	2
3	1.6430	0.6086	0.27992	3.572	0.45992	2.1743	3
4	1.9388	0.5158	0.19174	5.215	0.37174	2.6901	4
5	2.2878	0.4371	0.13978	7.154	0.31978	3.1272	5
6	2.6996	0.3704	0.10591	9.442	0.28591	3.4976	6
7	3.1855	0.3139	0.08236	12.142	0.26236	3.8115	7
8	3.7589	0.2660	0.06524	15.327	0.24524	4.0776	8
9	4.4355	0.2255	0.05239	19.086	0.23239	4.3030	9
10	5.2338	0.1911	0.04251	23.521	0.22251	4.4941	10
11	6.1759	0.1619	0.03478	28.755	0.21478	4.6560	11
12	7.2876	0.1372	0.02863	34.931	0.20863	4.7932	12
13	8.5994	0.1163	0.02369	42.219	0.20369	4.9095	13
14	10.1472	0.0985	0.01968	50.818	0.19968	5.0081	14
15	11.9737	0.0835	0.01640	60.965	0.19640	5.0916	15
16	14.1290	0.0708	0.01371	72.939	0.19371	5.1624	16
17	16.6722	0.0600	0.01149	87.068	0.19149	5.2223	17
18	19.6733	0.0508	0.00964	103.740	0.18964	5.2732	18
19	23.2144	0.0431	0.00810	123.414	0.18810	5.3162	19
20	27.3930	0.0365	0.00682	146.628	0.18682	5.3527	20
22	38.1421	0.0262	0.00485	206.345	0.18485	5.4099	22
24	53.1090	0.0188	0.00345	289.494	0.18345	5.4509	24
25	62.6686	0.0160	0.00292	342.603	0.18292	5.4669	25
26	73.9490	0.0135	0.00247	405.272	0.18247	5.4804	26
28	102.9666	0.0097	0.00177	566.481	0.18177	5.5016	28
30	143.3706	0.0070	0.00126	790.948	0.18126	5.5168	30
32	199.6293	0.0050	0.00091	1103.496	0.18091	5.5277	32
34	277.9638	0.0036	0.00065	1538.688	0.18065	5.5356	34
35	327.9973	0.0030	0.00055	1816.652	0.18055	5.5386	35
36	387.0368	0.0026	0.00047	2144.649	0.18047	5.5412	36
38	538.9100	0.0019	0.00033	2988.389	0.18033	5.5452	38
40	750.3783	0.0013	0.00024	4163.213	0.18024	5.5482	40
45	1716.684	0.0006	0.18010	9531.577	0.18010	5.5523	45
50	3927.357	0.0003	0.18005	21813.094	0.18005	5.5541	50

TABLE A - 21

DISCRETE CASH FLOW

19.00% DISCRETE COMPCUND INTEREST FACTORS

	SINGLE PAYMENTS		UNIFORM SERIES PAYMENT				
N	COMPOUND AMOUNT F/P	PRESENT WORTH P/F	SINKING FUND A/F	COMPOUND AMOUNT F/A	CAPITAL RECOVERY A/P	PRESENT WORTH P/A	N
1	1.1900	0.8403	1.00000	1.00	1.19000	0.8403	1
2	1.4161	0.7062	0.45662	2.19	0.64662	1.5465	2
3	1.6852	0.5934	0.27731	3.61	0.46731	2.1399	3
4	2.0053	0.4987	0.18899	5.29	0.37899	2.6386	4
5	2.3864	0.4190	0.13705	7.30	0.32705	3.0576	5
6	2.8398	0.3521	0.10327	9.68	0.29327	3.4098	6
7	3.3793	0.2959	0.07985	12.52	0.26985	3.7057	7
8	4.0214	0.2487	0.06289	15.90	0.25289	3.9544	8
9	4.7854	0.2090	0.05019	19.92	0.24019	4.1633	9
10	5.6947	0.1756	0.04047	24.71	0.23047	4.3389	10
11	6.7767	0.1476	0.03289	30.40	0.22289	4.4865	11
12	8.0642	0.1240	0.02690	37.18	0.21690	4.6105	12
13	9.5964	0.1042	0.02210	45.24	0.21210	4.7147	13
14	11.4198	0.0876	0.01823	54.84	0.20823	4.8023	14
15	13.5895	0.0736	0.01509	66.26	0.20509	4.8759	15
16	16.1715	0.0618	0.01252	79.85	0.20252	4.9377	16
17	19.2441	0.0520	0.01041	96.02	0.20041	4.9897	17
18	22.9005	0.0437	0.00868	115.27	0.19868	5.0333	18
19	27.2516	0.0367	0.00724	138.17	0.19724	5.0700	19
20	32.4294	0.0308	0.00605	165.42	0.19605	5.1009	20
22	45.9233	0.0218	0.00423	236.44	0.19423	5.1486	22
24	65.0320	0.0154	0.00297	337.01	0.19297	5.1822	24
25	77.3881	0.0129	0.00249	402.04	0.19249	5.1951	25
26	92.0918	0.0109	0.00209	479.43	0.19209	5.2060	26
28	130.4112	0.0077	0.00147	681.11	0.19147	5.2228	28
30	184.6753	0.0054	0.00103	966.71	0.19103	5.2347	30
32	261.5187	0.0038	0.00073	1371.15	0.19073	5.2430	32
34	370.3366	0.0027	0.00051	1943.88	0.19051	5.2489	34
35	440.7006	0.0023	0.00043	2314.21	0.19043	5.2512	35
36	524.4337	0.0019	0.00036	2754.91	0.19036	5.2531	36
38	742.6506	0.0013	0.00026	3903.42	0.19026	5.2561	38
40	1051.668	0.0010	0.19018	5529.83	0.19018	5.2582	40
45	2509.651	0.0004	0.19008	13203.42	0.19008	5.2611	45
50	5988.914	0.0002	0.19003	31515.34	0.19003	5.2623	50

TABLE A - 22

DISCRETE CASH FLOW
20.00% DISCRETE COMPOUND INTEREST FACTORS

	SINGLE PAYMENTS		UNIFORM SERIES PAYMENTS				
N	COMPOUND AMOUNT F/P	PRESENT WORTH P/F	SINKING FUND A/F	COMPOUND AMOUNT F/A	CAPITAL RECOVERY A/P	PRESENT WORTH P/A	N
1	1.2000	0.8333	1.00000	1.000	1.20000	0.8333	1
2	1.4400	0.6944	0.45455	2.200	0.65455	1.5278	2
3	1.7280	0.5787	0.27473	3.640	0.47473	2.1065	3
4	2.0736	0.4823	0.18629	5.368	0.38629	2.5887	4
5	2.4883	0.4019	0.13438	7.442	0.33438	2.9906	5
6	2.9860	0.3349	0.10071	9.930	0.30071	3.3255	6
7	3.5832	0.2791	0.07742	12.916	0.27742	3.6046	7
8	4.2998	0.2326	0.06061	16.499	0.26061	3.8372	8
9	5.1598	0.1938	0.04808	20.799	0.24808	4.0310	9
10	6.1917	0.1615	0.03852	25.959	0.23852	4.1925	10
11	7.4301	0.1346	0.03110	32.150	0.23110	4.3271	11
12	8.9161	0.1122	0.02526	39.581	0.22526	4.4392	12
13	10.6993	0.0935	0.02062	48.497	0.22062	4.5327	13
14	12.8392	0.0779	0.01689	59.196	0.21689	4.6106	14
15	15.4070	0.0649	0.01388	72.035	0.21388	4.6755	15
16	18.4884	0.0541	0.01144	87.442	0.21144	4.7296	16
17	22.1861	0.0451	0.00944	105.931	0.20944	4.7746	17
18	26.6233	0.0376	0.00781	128.117	0.20781	4.8122	18
19	31.9480	0.0313	0.00646	154.740	0.20646	4.8435	19
20	38.3376	0.0261	0.00536	186.688	0.20536	4.8696	20
22	55.2061	0.0181	0.00369	271.031	0.20369	4.9094	22
24	79.4968	0.0126	0.00255	392.484	0.20255	4.9371	24
25	95.3962	0.0105	0.00212	471.981	0.20212	4.9476	25
26	114.4755	0.0087	0.00176	567.377	0.20176	4.9563	26
28	164.8447	0.0061	0.00122	819.223	0.20122	4.9697	28
30	237.3763	0.0042	0.00085	1181.882	0.20085	4.9789	30
32	341.8219	0.0029	0.00059	1704.109	0.20059	4.9854	32
34	492.2235	0.0020	0.00041	2456.118	0.20041	4.9898	34
35	590.6682	0.0017	0.00034	2948.341	0.20034	4.9915	35
36	708.8019	0.0014	0.00028	3539.009	0.20028	4.9929	36
38	1020.675	0.0010	0.20020	5098.373	0.20020	4.9951	38
40	1469.772	0.0007	0.20014	7343.858	0.20014	4.9966	40
45	3657.262	0.0003	0.20005	18281.310	0.20005	4.9986	45
50	9100.438	0.0001	0.20002	45497.191	0.20002	4.9995	50

TABLE A - 23

DISCRETE CASH FLOW

22.00% DISCRETE COMPOUND INTEREST FACTORS

	SINGLE PAYMENTS		UNIFORM SERIES PAYMENT				
N	COMPOUND AMOUNT F/P	PRESENT WORTH P/F	SINKING FUND A/F	COMPOUND AMOUNT F/A	CAPITAL RECOVERY A/P	PRESENT WORTH P/A	N
1	1.2200	0.8197	1.00000	1.00	1.22000	0.8197	1
2	1.4884	0.6719	0.45045	2.22	0.67045	1.4915	2
3	1.8158	0.5507	0.26966	3.71	0.48966	2.0422	3
4	2.2153	0.4514	0.18102	5.52	0.40102	2.4936	4
5	2.7027	0.3700	0.12921	7.74	0.34921	2.8636	5
6	3.2973	0.3033	0.09576	10.44	0.31576	3.1669	6
7	4.0227	0.2486	0.07278	13.74	0.29278	3.4155	7
8	4.9077	0.2038	0.05630	17.76	0.27630	3.6193	8
9	5.9874	0.1670	0.04411	22.67	0.26411	3.7863	9
10	7.3046	0.1369	0.03489	28.66	0.25489	3.9232	10
11	8.9117	0.1122	0.02781	35.96	0.24781	4.0354	11
12	10.8722	0.0920	0.02228	44.87	0.24228	4.1274	12
13	13.2641	0.0754	0.01794	55.75	0.23794	4.2028	13
14	16.1822	0.0618	0.01449	69.01	0.23449	4.2646	14
15	19.7423	0.0507	0.01174	85.19	0.23174	4.3152	15
16	24.0856	0.0415	0.00953	104.93	0.22953	4.3567	16
17	29.3844	0.0340	0.00775	129.02	0.22775	4.3908	17
18	35.8490	0.0279	0.00631	158.40	0.22631	4.4187	18
19	43.7358	0.0229	0.00515	194.25	0.22515	4.4415	19
20	53.3576	0.0187	0.00420	237.99	0.22420	4.4603	20
22	79.4175	0.0126	0.00281	356.44	0.22281	4.4882	22
24	118.2050	0.0085	0.00188	532.75	0.22188	4.5070	24
25	144.2101	0.0069	0.00154	650.96	0.22154	4.5139	25
26	175.9364	0.0057	0.00126	795.17	0.22126	4.5196	26
28	261.8637	0.0038	0.00084	1185.74	0.22084	4.5281	28
30	389.7579	0.0026	0.00057	1767.08	0.22057	4.5338	30
32	580.1156	0.0017	0.00038	2632.34	0.22038	4.5376	32
34	863.4441	0.0012	0.00026	3920.20	0.22026	4.5402	34
35	1053.402	0.0009	0.00021	4783.64	0.22021	4.5411	35
36	1285.150	0.0008	0.00017	5837.05	0.22017	4.5419	36
38	1912.818	0.0005	0.00012	8690.08	0.22012	4.5431	38
40	2847.038	0.0004	0.00008	12936.54	0.22008	4.5439	40
45	7694.712	0.0001	0.00003	34971.42	0.22003	4.5449	45
50	20796.56	0.0000	0.00001	94525.28	0.22001	4.5452	50

TABLE A - 24

DISCRETE CASH FLOW

24.00% DISCRETE COMPOUND INTEREST FACTORS

	SINGLE PAYMENTS		UNIFORM SERIES PAYMENT				
N	COMPOUND AMOUNT F/P	PRESENT WORTH P/F	SINKING FUND A/F	COMPOUND AMOUNT F/A	CAPITAL RECOVERY A/P	PRESENT WORTH P/A	N
1	1.2400	0.8065	1.00000	1.00	1.24000	0.8065	1
2	1.5376	0.6504	0.44643	2.24	0.68643	1.4568	2
3	1.9066	0.5245	0.26472	3.78	0.50472	1.9813	3
4	2.3642	0.4230	0.17593	5.68	0.41593	2.4043	4
5	2.9316	0.3411	0.12425	8.05	0.36425	2.7454	5
6	3.6352	0.2751	0.09107	10.98	0.33107	3.0205	6
7	4.5077	0.2218	0.06842	14.62	0.30842	3.2423	7
8	5.5895	0.1789	0.05229	19.12	0.29229	3.4212	8
9	6.9310	0.1443	0.04047	24.71	0.28047	3.5655	9
10	8.5944	0.1164	0.03160	31.64	0.27160	3.6819	10
11	10.6571	0.0938	0.02485	40.24	0.26485	3.7757	11
12	13.2148	0.0757	0.01965	50.89	0.25965	3.8514	12
13	16.3863	0.0610	0.01560	64.11	0.25560	3.9124	13
14	20.3191	0.0492	0.01242	80.50	0.25242	3.9616	14
15	25.1956	0.0397	0.00992	100.82	0.24992	4.0013	15
16	31.2426	0.0320	0.00794	126.01	0.24794	4.0333	16
17	38.7408	0.0258	0.00636	157.25	0.24636	4.0591	17
18	48.0386	0.0208	0.00510	195.99	0.24510	4.0799	18
19	59.5679	0.0168	0.00410	244.03	0.24410	4.0967	19
20	73.8641	0.0135	0.00329	303.60	0.24329	4.1103	20
22	113.5735	0.0088	0.00213	469.06	0.24213	4.1300	22
24	174.6306	0.0057	0.00138	723.46	0.24138	4.1428	24
25	216.5420	0.0046	0.00111	898.09	0.24111	4.1474	25
26	268.5121	0.0037	0.00090	1114.63	0.24090	4.1511	26
28	412.8642	0.0024	0.00058	1716.10	0.24058	4.1566	28
30	634.8199	0.0016	0.00038	2640.92	0.24038	4.1601	30
32	976.0991	0.0010	0.00025	4062.91	0.24025	4.1624	32
34	1500.850	0.0007	0.00016	6249.38	0.24016	4.1639	34
35	1861.054	0.0005	0.00013	7750.23	0.24013	4.1644	35
36	2307.707	0.0004	0.00010	9611.28	0.24010	4.1649	36
38	3548.330	0.0003	0.00007	14780.54	0.24007	4.1655	38
40	5455.913	0.0002	0.00004	22728.80	0.24004	4.1659	40
45	15994.69	0.0001	0.00002	66640.38	0.24002	4.1664	45
50	46890.43	0.0000	0.00001	195372.64	0.24001	4.1666	50

TABLE A - 25

DISCRETE CASH FLOW
25.00% DISCRETE COMPOUND INTEREST FACTORS

	SINGLE PAYMENTS		UNIFORM SERIES PAYMENTS				
N	COMPOUND AMOUNT F/P	PRESENT WORTH P/F	SINKING FUND A/F	COMPOUND AMOUNT F/A	CAPITAL RECOVERY A/P	PRESENT WORTH P/A	N
1	1.2500	0.8000	1.00000	1.000	1.25000	0.8000	1
2	1.5625	0.6400	0.44444	2.250	0.69444	1.4400	2
3	1.9531	0.5120	0.26230	3.813	0.51230	1.9520	3
4	2.4414	0.4096	0.17344	5.766	0.42344	2.3616	4
5	3.0518	0.3277	0.12185	8.207	0.37185	2.6893	5
6	3.8147	0.2621	0.08882	11.259	0.33882	2.9514	6
7	4.7684	0.2097	0.06634	15.073	0.31634	3.1611	7
8	5.9605	0.1678	0.05040	19.842	0.30040	3.3289	8
9	7.4506	0.1342	0.03876	25.802	0.28876	3.4631	9
10	9.3132	0.1074	0.03007	33.253	0.28007	3.5705	10
11	11.6415	0.0859	0.02349	42.566	0.27349	3.6564	11
12	14.5519	0.0687	0.01845	54.208	0.26845	3.7251	12
13	18.1899	0.0550	0.01454	68.760	0.26454	3.7801	13
14	22.7374	0.0440	0.01150	86.949	0.26150	3.8241	14
15	28.4217	0.0352	0.00912	109.687	0.25912	3.8593	15
16	35.5271	0.0281	0.00724	138.109	0.25724	3.8874	16
17	44.4089	0.0225	0.00576	173.636	0.25576	3.9099	17
18	55.5112	0.0180	0.00459	218.045	0.25459	3.9279	18
19	69.3889	0.0144	0.00366	273.556	0.25366	3.9424	19
20	86.7362	0.0115	0.00292	342.945	0.25292	3.9539	20
22	135.5253	0.0074	0.00186	538.101	0.25186	3.9705	22
24	211.7582	0.0047	0.00119	843.033	0.25119	3.9811	24
25	264.6978	0.0038	0.00095	1054.791	0.25095	3.9849	25
26	330.8722	0.0030	0.00076	1319.489	0.25076	3.9879	26
28	516.9879	0.0019	0.00048	2063.952	0.25048	3.9923	28
30	807.7936	0.0012	0.00031	3227.174	0.25031	3.9950	30
32	1262.177	0.0008	0.00020	5044.710	0.25020	3.9968	32
34	1972.152	0.0005	0.00013	7884.609	0.25013	3.9980	34
35	2465.190	0.0004	0.00010	9856.761	0.25010	3.9984	35
36	3081.488	0.0003	0.00008	12321.952	0.25008	3.9987	36
38	4814.825	0.0002	0.00005	19255.299	0.25005	3.9992	38
40	7523.164	0.0001	0.00003	30088.655	0.25003	3.9995	40
45	22958.87	0.0000	0.00001	91831.496	0.25001	3.9998	45

TABLE A - 26

DISCRETE CASH FLOW
30.00% DISCRETE COMPOUND INTEREST FACTORS

	SINGLE PAYMENTS		UNIFORM SERIES PAYMENTS				
N	COMPOUND AMOUNT F/P	PRESENT WORTH P/F	SINKING FUND A/F	COMPOUND AMOUNT F/A	CAPITAL RECOVERY A/P	PRESENT WORTH P/A	N
1	1.3000	0.7692	1.00000	1.000	1.30000	0.7692	1
2	1.6900	0.5917	0.43478	2.300	0.73478	1.3609	2
3	2.1970	0.4552	0.25063	3.990	0.55063	1.8161	3
4	2.8561	0.3501	0.16163	6.187	0.46163	2.1662	4
5	3.7129	0.2693	0.11058	9.043	0.41058	2.4356	5
6	4.8268	0.2072	0.07839	12.756	0.37839	2.6427	6
7	6.2749	0.1594	0.05687	17.583	0.35687	2.8021	7
8	8.1573	0.1226	0.04192	23.858	0.34192	2.9247	8
9	10.6045	0.0943	0.03124	32.015	0.33124	3.0190	9
10	13.7858	0.0725	0.02346	42.619	0.32346	3.0915	10
11	17.9216	0.0558	0.01773	56.405	0.31773	3.1473	11
12	23.2981	0.0429	0.01345	74.327	0.31345	3.1903	12
13	30.2875	0.0330	0.01024	97.625	0.31024	3.2233	13
14	39.3738	0.0254	0.00782	127.913	0.30782	3.2487	14
15	51.1859	0.0195	0.00598	167.286	0.30598	3.2682	15
16	66.5417	0.0150	0.00458	218.472	0.30458	3.2832	16
17	86.5042	0.0116	0.00351	285.014	0.30351	3.2948	17
18	112.4554	0.0089	0.00269	371.518	0.30269	3.3037	18
19	146.1920	0.0068	0.00207	483.973	0.30207	3.3105	19
20	190.0496	0.0053	0.00159	630.165	0.30159	3.3158	20
22	321.1839	0.0031	0.00094	1067.280	0.30094	3.3230	22
24	542.8008	0.0018	0.00055	1806.003	0.30055	3.3272	24
25	705.6410	0.0014	0.00043	2348.803	0.30043	3.3286	25
26	917.3333	0.0011	0.00033	3054.444	0.30033	3.3297	26
28	1550.293	0.0006	0.00019	5164.311	0.30019	3.3312	28
30	2619.996	0.0004	0.00011	8729.985	0.30011	3.3321	30
32	4427.793	0.0002	0.00007	14755.975	0.30007	3.3326	32
34	7482.970	0.0001	0.00004	24939.899	0.30004	3.3329	34
35	9727.860	0.0001	0.00003	32422.868	0.30003	3.3330	35

TABLE A - 27

DISCRETE CASH FLOW
35.00% DISCRETE COMPOUND INTEREST FACTORS

	SINGLE PAYMENTS		UNIFORM SERIES PAYMENTS				
N	COMPOUND AMOUNT F/P	PRESENT WORTH P/F	SINKING FUND A/F	COMPOUND AMOUNT F/A	CAPITAL RECOVERY A/P	PRESENT WORTH P/A	N
1	1.3500	0.7407	1.00000	1.000	1.35000	0.7407	1
2	1.8225	0.5487	0.42553	2.350	0.77553	1.2894	2
3	2.4604	0.4064	0.23966	4.172	0.58966	1.6959	3
4	3.3215	0.3011	0.15076	6.633	0.50076	1.9969	4
5	4.4840	0.2230	0.10046	9.954	0.45046	2.2200	5
6	6.0534	0.1652	0.06926	14.438	0.41926	2.3852	6
7	8.1722	0.1224	0.04880	20.492	0.39880	2.5075	7
8	11.0324	0.0906	0.03489	28.664	0.38489	2.5982	8
9	14.8937	0.0671	0.02519	39.696	0.37519	2.6653	9
10	20.1066	0.0497	0.01832	54.590	0.36832	2.7150	10
11	27.1439	0.0368	0.01339	74.697	0.36339	2.7519	11
12	36.6442	0.0273	0.00982	101.841	0.35982	2.7792	12
13	49.4697	0.0202	0.00722	138.485	0.35722	2.7994	13
14	66.7841	0.0150	0.00532	187.954	0.35532	2.8144	14
15	90.1585	0.0111	0.00393	254.738	0.35393	2.8255	15
16	121.7139	0.0082	0.00290	344.897	0.35290	2.8337	16
17	164.3138	0.0061	0.00214	466.611	0.35214	2.8398	17
18	221.8236	0.0045	0.00158	630.925	0.35158	2.8443	18
19	299.4619	0.0033	0.00117	852.748	0.35117	2.8476	19
20	404.2736	0.0025	0.00087	1152.210	0.35087	2.8501	20
22	736.7886	0.0014	0.00048	2102.253	0.35048	2.8533	22
24	1342.797	0.0007	0.00026	3833.706	0.35026	2.8550	24
25	1812.776	0.0006	0.00019	5176.504	0.35019	2.8556	25
26	2447.248	0.0004	0.00014	6989.280	0.35014	2.8560	26
28	4460.109	0.0002	0.00008	12740.313	0.35008	2.8565	28
30	8128.550	0.0001	0.00004	23221.570	0.35004	2.8568	30
32	14814.28	0.0001	0.00002	42323.661	0.35002	2.8569	32
34	26999.03	0.0000	0.00001	77137.223	0.35001	2.8570	34
35	36448.69	0.0000	0.00001	104136.25	0.35001	2.8571	35

TABLE A - 28

DISCRETE CASH FLOW
40.00% DISCRETE COMPOUND INTEREST FACTORS

	SINGLE PAYMENTS		UNIFORM SERIES PAYMENTS				
N	COMPOUND AMOUNT F/P	PRESENT WORTH P/F	SINKING FUND A/F	COMPOUND AMOUNT F/A	CAPITAL RECOVERY A/P	PRESENT WORTH P/A	N
1	1.4000	0.7143	1.00000	1.000	1.40000	0.7143	1
2	1.9600	0.5102	0.41667	2.400	0.81667	1.2245	2
3	2.7440	0.3644	0.22936	4.360	0.62936	1.5889	3
4	3.8416	0.2603	0.14077	7.104	0.54077	1.8492	4
5	5.3782	0.1859	0.09136	10.946	0.49136	2.0352	5
6	7.5295	0.1328	0.06126	16.324	0.46126	2.1680	6
7	10.5414	0.0949	0.04192	23.853	0.44192	2.2628	7
8	14.7579	0.0678	0.02907	34.395	0.42907	2.3306	8
9	20.6610	0.0484	0.02034	49.153	0.42034	2.3790	9
10	28.9255	0.0346	0.01432	69.814	0.41432	2.4136	10
11	40.4957	0.0247	0.01013	98.739	0.41013	2.4383	11
12	56.6939	0.0176	0.00718	139.235	0.40718	2.4559	12
13	79.3715	0.0126	0.00510	195.929	0.40510	2.4685	13
14	111.1201	0.0090	0.00363	275.300	0.40363	2.4775	14
15	155.5681	0.0064	0.00259	386.420	0.40259	2.4839	15
16	217.7953	0.0046	0.00185	541.988	0.40185	2.4885	16
17	304.9135	0.0033	0.00132	759.784	0.40132	2.4918	17
18	426.8789	0.0023	0.00094	1064.697	0.40094	2.4941	18
19	597.6304	0.0017	0.00067	1491.576	0.40067	2.4958	19
20	836.6826	0.0012	0.00048	2089.206	0.40048	2.4970	20
22	1639.898	0.0006	0.00024	4097.245	0.40024	2.4985	22
24	3214.200	0.0003	0.00012	8032.999	0.40012	2.4992	24
25	4499.880	0.0002	0.00009	11247.199	0.40009	2.4994	25
26	6299.831	0.0002	0.00006	15747.079	0.40006	2.4996	26
28	12347.67	0.0001	0.00003	30866.674	0.40003	2.4998	28
30	24201.43	0.0000	0.00002	60501.081	0.40002	2.4999	30
32	47434.81	0.0000	0.00001	118584.52	0.40001	2.4999	32
34	92972.22	0.0000	0.00000	232428.06	0.40000	2.5000	34
35	130161.1	0.0000	0.00000	325400.28	0.40000	2.5000	35

TABLE A - 29

DISCRETE CASH FLOW
45.00% DISCRETE COMPOUND INTEREST FACTORS

N	SINGLE PAYMENTS COMPOUND AMOUNT F/P	PRESENT WORTH P/F	UNIFORM SERIES PAYMENTS SINKING FUND A/F	COMPOUND AMOUNT F/A	CAPITAL RECOVERY A/P	PRESENT WORTH P/A	N
1	1.4500	0.6897	1.00000	1.000	1.45000	0.6897	1
2	2.1025	0.4756	0.40816	2.450	0.85816	1.1653	2
3	3.0486	0.3280	0.21966	4.552	0.66966	1.4933	3
4	4.4205	0.2262	0.13156	7.601	0.58156	1.7195	4
5	6.4097	0.1560	0.08318	12.022	0.53318	1.8755	5
6	9.2941	0.1076	0.05426	18.431	0.50426	1.9831	6
7	13.4765	0.0742	0.03607	27.725	0.48607	2.0573	7
8	19.5409	0.0512	0.02427	41.202	0.47427	2.1085	8
9	28.3343	0.0353	0.01646	60.743	0.46646	2.1438	9
10	41.0847	0.0243	0.01123	89.077	0.46123	2.1681	10
11	59.5728	0.0168	0.00768	130.162	0.45768	2.1849	11
12	86.3806	0.0116	0.00527	189.735	0.45527	2.1965	12
13	125.2518	0.0080	0.00362	276.115	0.45362	2.2045	13
14	181.6151	0.0055	0.00249	401.367	0.45249	2.2100	14
15	263.3419	0.0038	0.00172	582.982	0.45172	2.2138	15
16	381.8458	0.0026	0.00118	846.324	0.45118	2.2164	16
17	553.6764	0.0018	0.00081	1228.170	0.45081	2.2182	17
18	802.8308	0.0012	0.00056	1781.846	0.45056	2.2195	18
19	1164.105	0.0009	0.00039	2584.677	0.45039	2.2203	19
20	1687.952	0.0006	0.00027	3748.782	0.45027	2.2209	20
22	3548.919	0.0003	0.00013	7884.264	0.45013	2.2216	22
24	7461.602	0.0001	0.00006	16579.115	0.45006	2.2219	24
25	10819.32	0.0001	0.00004	24040.716	0.45004	2.2220	25
26	15688.02	0.0001	0.00003	34860.038	0.45003	2.2221	26
28	32984.06	0.0000	0.00001	73295.681	0.45001	2.2222	28
30	69348.98	0.0000	0.00001	154106.62	0.45001	2.2222	30
32	145806.2	0.0000	0.00000	324011.62	0.45000	2.2222	32
34	306557.6	0.0000	0.00000	681236.87	0.45000	2.2222	34
35	444508.5	0.0000	0.00000	987794.46	0.45000	2.2222	35

TABLE A - 30

DISCRETE CASH FLOW
50.00% DISCRETE COMPOUND INTEREST FACTORS

	SINGLE PAYMENTS		UNIFORM SERIES PAYMENTS				
N	COMPOUND AMOUNT F/P	PRESENT WORTH P/F	SINKING FUND A/F	COMPOUND AMOUNT F/A	CAPITAL RECOVERY A/P	PRESENT WORTH P/A	N
1	1.5000	0.6667	1.00000	1.000	1.50000	0.6667	1
2	2.2500	0.4444	0.40000	2.500	0.90000	1.1111	2
3	3.3750	0.2963	0.21053	4.750	0.71053	1.4074	3
4	5.0625	0.1975	0.12308	8.125	0.62308	1.6049	4
5	7.5938	0.1317	0.07583	13.188	0.57583	1.7366	5
6	11.3906	0.0878	0.04812	20.781	0.54812	1.8244	6
7	17.0859	0.0585	0.03108	32.172	0.53108	1.8829	7
8	25.6289	0.0390	0.02030	49.258	0.52030	1.9220	8
9	38.4434	0.0260	0.01335	74.887	0.51335	1.9480	9
10	57.6650	0.0173	0.00882	113.330	0.50882	1.9653	10
11	86.4976	0.0116	0.00585	170.995	0.50585	1.9769	11
12	129.7463	0.0077	0.00388	257.493	0.50388	1.9846	12
13	194.6195	0.0051	0.00258	387.239	0.50258	1.9897	13
14	291.9293	0.0034	0.00172	581.859	0.50172	1.9931	14
15	437.8939	0.0023	0.00114	873.788	0.50114	1.9954	15
16	656.8408	0.0015	0.00076	1311.682	0.50076	1.9970	16
17	985.2613	0.0010	0.00051	1968.523	0.50051	1.9980	17
18	1477.892	0.0007	0.00034	2953.784	0.50034	1.9986	18
19	2216.838	0.0005	0.00023	4431.676	0.50023	1.9991	19
20	3325.257	0.0003	0.00015	6648.513	0.50015	1.9994	20
22	7481.828	0.0001	0.00007	14961.655	0.50007	1.9997	22
24	16834.11	0.0001	0.00003	33666.224	0.50003	1.9999	24
25	25251.17	0.0000	0.00002	50500.337	0.50002	1.9999	25
26	37876.75	0.0000	0.00001	75751.505	0.50001	1.9999	26
28	85222.69	0.0000	0.00001	170443.39	0.50001	2.0000	28
30	191751.1	0.0000	0.00000	383500.12	0.50000	2.0000	30
32	431439.9	0.0000	0.00000	862877.77	0.50000	2.0000	32
34	970739.7	0.0000	0.00000	1941477.5	0.50000	2.0000	34

TABLE A - 31 "Only Gradient."

PRESENT WORTH GRADIENT FACTORS (P/G)
DISCRETE CASH FLOW, DISCRETE COMPOUNDING

N	1%	2%	3%	4%	5%	6%	N
2	0.980	0.961	0.943	0.925	0.907	0.890	2
3	2.921	2.846	2.773	2.703	2.635	2.569	3
4	5.804	5.617	5.438	5.267	5.103	4.946	4
5	9.610	9.240	8.889	8.555	8.237	7.935	5
6	14.321	13.680	13.076	12.506	11.968	11.459	6
7	19.917	18.903	17.955	17.066	16.232	15.450	7
8	26.381	24.878	23.481	22.181	20.970	19.842	8
9	33.696	31.572	29.612	27.801	26.127	24.577	9
10	41.843	38.955	36.309	33.881	31.652	29.602	10
11	50.807	46.998	43.533	40.377	37.499	34.870	11
12	60.569	55.671	51.248	47.248	43.624	40.337	12
13	71.113	64.948	59.420	54.455	49.988	45.963	13
14	82.422	74.800	68.014	61.962	56.554	51.713	14
15	94.481	85.202	77.000	69.735	63.288	57.555	15
16	107.273	96.129	86.348	77.744	70.160	63.459	16
17	120.783	107.555	96.028	85.958	77.140	69.401	17
18	134.996	119.458	106.014	94.350	84.204	75.357	18
19	149.895	131.814	116.279	102.893	91.328	81.306	19
20	165.466	144.600	126.799	111.565	98.488	87.230	20
21	181.695	157.796	137.550	120.341	105.667	93.114	21
22	198.566	171.379	148.509	129.202	112.846	98.941	22
23	216.066	185.331	159.657	138.128	120.009	104.701	23
24	234.180	199.630	170.971	147.101	127.140	110.381	24
25	252.894	214.259	182.434	156.104	134.228	115.973	25
26	272.196	229.199	194.026	165.121	141.259	121.468	26
27	292.070	244.431	205.731	174.138	148.223	126.860	27
28	312.505	259.939	217.532	183.142	155.110	132.142	28
29	333.486	275.706	229.414	192.121	161.913	137.310	29
30	355.002	291.716	241.361	201.062	168.623	142.359	30
31	377.039	307.954	253.361	209.956	175.233	147.286	31
32	399.586	324.403	265.399	218.792	181.739	152.090	32
33	422.629	341.051	277.464	227.563	188.135	156.768	33
34	446.157	357.882	289.544	236.261	194.417	161.319	34
35	470.158	374.883	301.627	244.877	200.581	165.743	35
36	494.621	392.040	313.703	253.405	206.624	170.039	36
37	519.533	409.342	325.762	261.840	212.543	174.207	37
38	544.884	426.776	337.796	270.175	218.338	178.249	38
39	570.662	444.330	349.794	278.407	224.005	182.165	39
40	596.856	461.993	361.750	286.530	229.545	185.957	40
42	650.451	497.601	385.502	302.437	240.239	193.173	42
44	705.585	533.517	408.997	317.870	250.417	199.913	44
46	762.176	569.662	432.186	332.810	260.084	206.194	46
48	820.146	605.966	455.025	347.245	269.247	212.035	48
50	879.418	642.361	477.480	361.164	277.915	217.457	50

TABLE A - 32

PRESENT WORTH GRADIENT FACTORS (P/G)
DISCRETE CASH FLOW, DISCRETE COMPOUNDING

N	7%	8%	9%	10%	11%	12%	N
2	0.873	0.857	0.842	0.826	0.812	0.797	2
3	2.506	2.445	2.386	2.329	2.274	2.221	3
4	4.795	4.650	4.511	4.378	4.250	4.127	4
5	7.647	7.372	7.111	6.862	6.624	6.397	5
6	10.978	10.523	10.092	9.684	9.297	8.930	6
7	14.715	14.024	13.375	12.763	12.187	11.644	7
8	18.789	17.806	16.888	16.029	15.225	14.471	8
9	23.140	21.808	20.571	19.421	18.352	17.356	9
10	27.716	25.977	24.373	22.891	21.522	20.254	10
11	32.466	30.266	28.248	26.396	24.695	23.129	11
12	37.351	34.634	32.159	29.901	27.839	25.952	12
13	42.330	39.046	36.073	33.377	30.929	28.702	13
14	47.372	43.472	39.963	36.800	33.945	31.362	14
15	52.446	47.886	43.807	40.152	36.871	33.920	15
16	57.527	52.264	47.585	43.416	39.695	36.367	16
17	62.592	56.588	51.282	46.582	42.409	38.697	17
18	67.622	60.843	54.886	49.640	45.007	40.908	18
19	72.599	65.013	58.387	52.583	47.486	42.998	19
20	77.509	69.090	61.777	55.407	49.842	44.968	20
21	82.339	73.063	65.051	58.110	52.077	46.819	21
22	87.079	76.926	68.205	60.689	54.191	48.554	22
23	91.720	80.673	71.236	63.146	56.186	50.178	23
24	96.255	84.300	74.143	65.481	58.066	51.693	24
25	100.676	87.804	76.926	67.696	59.832	53.105	25
26	104.981	91.184	79.586	69.794	61.490	54.418	26
27	109.166	94.439	82.124	71.777	63.043	55.637	27
28	113.226	97.569	84.542	73.650	64.497	56.767	28
29	117.162	100.574	86.842	75.415	65.854	57.814	29
30	120.972	103.456	89.028	77.077	67.121	58.782	30
31	124.655	106.216	91.102	78.640	68.302	59.676	31
32	128.212	108.857	93.069	80.108	69.401	60.501	32
33	131.643	111.382	94.931	81.486	70.423	61.261	33
34	134.951	113.792	96.693	82.777	71.372	61.961	34
35	138.135	116.092	98.359	83.987	72.254	62.605	35
36	141.199	118.284	99.932	85.119	73.071	63.197	36
37	144.144	120.371	101.416	86.178	73.829	63.741	37
38	146.973	122.358	102.816	87.167	74.530	64.239	38
39	149.688	124.247	104.135	88.091	75.179	64.697	39
40	152.293	126.042	105.376	88.953	75.779	65.116	40
42	157.181	129.365	107.643	90.505	76.845	65.851	42
44	161.661	132.355	109.646	91.851	77.753	66.466	44
46	165.758	135.038	111.410	93.016	78.525	66.979	46
48	169.498	137.443	112.962	94.022	79.180	67.407	48
50	172.905	139.593	114.325	94.889	79.734	67.762	50

TABLE A - 33

PRESENT WORTH GRADIENT FACTORS (P/G)
DISCRETE CASH FLOW, DISCRETE COMPOUNDING

N	13%	14%	15%	18%	20%	22%	N
2	0.783	0.769	0.756	0.718	0.694	0.672	2
3	2.169	2.119	2.071	1.935	1.852	1.773	3
4	4.009	3.896	3.786	3.483	3.299	3.127	4
5	6.180	5.973	5.775	5.231	4.906	4.607	5
6	8.582	8.251	7.937	7.083	6.581	6.124	6
7	11.132	10.649	10.192	8.967	8.255	7.615	7
8	13.765	13.103	12.481	10.829	9.883	9.042	8
9	16.428	15.563	14.755	12.633	11.434	10.378	9
10	19.080	17.991	16.979	14.352	12.887	11.610	10
11	21.687	20.357	19.129	15.972	14.233	12.732	11
12	24.224	22.640	21.185	17.481	15.467	13.744	12
13	26.574	24.825	23.135	18.877	16.588	14.649	13
14	29.023	26.901	24.972	20.158	17.601	15.452	14
15	31.262	28.862	26.693	21.327	18.509	16.161	15
16	33.384	30.706	28.296	22.389	19.321	16.784	16
17	35.388	32.430	29.783	23.348	20.042	17.328	17
18	37.271	34.038	31.156	24.212	20.680	17.803	18
19	39.037	35.531	32.421	24.988	21.244	18.214	19
20	40.685	36.914	33.582	25.681	21.739	18.570	20
21	42.221	38.190	34.645	26.300	22.174	18.877	21
22	43.649	39.366	35.615	26.851	22.555	19.142	22
23	44.972	40.446	36.499	27.339	22.887	19.369	23
24	46.196	41.437	37.302	27.772	23.176	19.563	24
25	47.326	42.344	38.031	28.155	23.428	19.730	25
26	48.369	43.173	38.692	28.494	23.646	19.872	26
27	49.328	43.929	39.289	28.791	23.835	19.993	27
28	50.209	44.618	39.828	29.054	23.999	20.096	28
29	51.018	45.244	40.315	29.284	24.141	20.184	29
30	51.759	45.813	40.753	29.486	24.263	20.258	30
31	52.438	46.330	41.147	29.664	24.368	20.321	31
32	53.059	46.798	41.501	29.819	24.459	20.375	32
33	53.626	47.222	41.818	29.955	24.537	20.420	33
34	54.143	47.605	42.103	30.074	24.604	20.458	34
35	54.615	47.952	42.359	30.177	24.661	20.491	35
36	55.045	48.265	42.587	30.268	24.711	20.518	36
37	55.436	48.547	42.792	30.347	24.753	20.541	37
38	55.792	48.802	42.974	30.415	24.789	20.560	38
39	56.115	49.031	43.137	30.475	24.820	20.576	39
40	56.409	49.238	43.283	30.527	24.847	20.590	40
42	56.917	49.590	43.529	30.611	24.889	20.611	42
44	57.335	49.875	43.723	30.675	24.920	20.626	44
46	57.678	50.105	43.878	30.723	24.942	20.637	45
48	57.958	50.289	44.000	30.759	24.958	20.644	48
50	58.187	50.438	44.096	30.786	24.970	20.649	50

TABLE A - 34

PRESENT WORTH GRADIENT FACTORS (P/G)
DISCRETE CASH FLOW, DISCRETE COMPOUNDING

N	25%	30%	35%	40%	45%	50%	N
2	0.640	0.592	0.549	0.510	0.476	0.444	2
3	1.664	1.502	1.362	1.239	1.132	1.037	3
4	2.893	2.552	2.265	2.020	1.810	1.630	4
5	4.204	3.630	3.157	2.764	2.434	2.156	5
6	5.514	4.666	3.983	3.428	2.972	2.595	6
7	6.773	5.622	4.717	3.997	3.418	2.947	7
8	7.947	6.480	5.352	4.471	3.776	3.220	8
9	9.021	7.234	5.889	4.858	4.058	3.428	9
10	9.987	7.887	6.336	5.170	4.277	3.584	10
11	10.846	8.445	6.705	5.417	4.445	3.699	11
12	11.602	8.917	7.005	5.611	4.572	3.784	12
13	12.262	9.314	7.247	5.762	4.668	3.846	13
14	12.833	9.644	7.442	5.879	4.740	3.890	14
15	13.326	9.917	7.597	5.969	4.793	3.922	15
16	13.748	10.143	7.721	6.038	4.832	3.945	16
17	14.108	10.328	7.818	6.090	4.861	3.961	17
18	14.415	10.479	7.895	6.130	4.882	3.973	18
19	14.674	10.602	7.955	6.160	4.898	3.981	19
20	14.893	10.702	8.002	6.183	4.909	3.987	20
21	15.078	10.783	8.038	6.200	4.917	3.991	21
22	15.233	10.848	8.067	6.213	4.923	3.994	22
23	15.362	10.901	8.089	6.222	4.927	3.996	23
24	15.471	10.943	8.106	6.229	4.930	3.997	24
25	15.562	10.977	8.119	6.235	4.933	3.998	25
26	15.637	11.005	8.130	6.239	4.934	3.999	26
27	15.700	11.026	8.137	6.242	4.935	3.999	27
28	15.752	11.044	8.143	6.244	4.936	3.999	28
29	15.796	11.058	8.148	6.245	4.937	4.000	29
30	15.832	11.069	8.152	6.247	4.937	4.000	30
31	15.861	11.078	8.154	6.248	4.938	4.000	31
32	15.886	11.085	8.157	6.248	4.938	4.000	32
33	15.906	11.090	8.158	6.249	4.938	4.000	33
34	15.923	11.094	8.159	6.249	4.938	4.000	34
35	15.937	11.098	8.160	6.249	4.938	4.000	35
36	15.948	11.101	8.161	6.249	4.938	4.000	36
37	15.957	11.103	8.162	6.250	4.938	4.000	37
38	15.965	11.105	8.162	6.250	4.938	4.000	38
39	15.971	11.106	8.162	6.250	4.938	4.000	39
40	15.977	11.107	8.163	6.250	4.938	4.000	40
42	15.984	11.109	8.163	6.250	4.938	4.000	42
44	15.990	11.110	8.163	6.250	4.938	4.000	44
46	15.993	11.110	8.163	6.250	4.938	4.000	46
48	15.995	11.111	8.163	6.250	4.938	4.000	48
50	15.997	11.111	8.163	6.250	4.938	4.000	50

TABLE A - 35

ANNUAL COST GRADIENT FACTORS(A/G)
DISCRETE CASH FLOW, DISCRETE COMPOUNDING

N	1/2%	1%	1.5%	2%	3%	4%	5%	N
2	0.499	0.498	0.496	0.495	0.493	0.490	0.488	2
3	0.997	0.993	0.990	0.987	0.980	0.974	0.967	3
4	1.494	1.488	1.481	1.475	1.463	1.451	1.439	4
5	1.990	1.980	1.970	1.960	1.941	1.922	1.903	5
6	2.485	2.471	2.457	2.442	2.414	2.386	2.358	6
7	2.980	2.960	2.940	2.921	2.882	2.843	2.805	7
8	3.474	3.448	3.422	3.396	3.345	3.294	3.245	8
9	3.967	3.934	3.901	3.868	3.803	3.739	3.676	9
10	4.459	4.418	4.377	4.337	4.256	4.177	4.099	10
11	4.950	4.901	4.851	4.802	4.705	4.609	4.514	11
12	5.441	5.381	5.323	5.264	5.148	5.034	4.922	12
13	5.930	5.861	5.792	5.723	5.587	5.453	5.322	13
14	6.419	6.338	6.258	6.179	6.021	5.866	5.713	14
15	6.907	6.814	6.722	6.631	6.450	6.272	6.097	15
16	7.394	7.289	7.184	7.080	6.874	6.672	6.474	16
17	7.880	7.761	7.643	7.526	7.294	7.066	6.842	17
18	8.366	8.232	8.100	7.968	7.708	7.453	7.203	18
19	8.850	8.702	8.554	8.407	8.118	7.834	7.557	19
20	9.334	9.169	9.006	8.843	8.523	8.209	7.903	20
22	10.299	10.100	9.902	9.705	9.319	8.941	8.573	22
24	11.261	11.024	10.788	10.555	10.095	9.648	9.214	24
25	11.741	11.483	11.228	10.974	10.477	9.993	9.524	25
26	12.220	11.941	11.665	11.391	10.853	10.331	9.827	26
28	13.175	12.852	12.531	12.214	11.593	10.991	10.411	28
30	14.126	13.756	13.388	13.025	12.314	11.627	10.969	30
32	15.075	14.653	14.236	13.823	13.017	12.241	11.501	32
34	16.020	15.544	15.073	14.608	13.702	12.832	12.006	34
35	16.492	15.987	15.488	14.996	14.037	13.120	12.250	35
36	16.962	16.428	15.901	15.381	14.369	13.402	12.487	36
38	17.901	17.306	16.719	16.141	15.018	13.950	12.944	38
40	18.836	18.178	17.528	16.889	15.650	14.477	13.377	40
45	21.159	20.327	19.507	18.703	17.156	15.705	14.364	45
50	23.462	22.436	21.428	20.442	18.558	16.812	15.223	50
55	25.745	24.505	23.289	22.106	19.860	17.807	15.966	55
60	28.006	26.533	25.093	23.696	21.067	18.697	16.606	60
65	30.247	28.522	26.839	25.215	22.184	19.491	17.154	65
70	32.468	30.470	28.529	26.663	23.215	20.196	17.621	70
75	34.668	32.379	30.163	28.043	24.163	20.821	18.018	75
80	36.847	34.249	31.742	29.357	25.035	21.372	18.353	80
85	39.006	36.080	33.268	30.606	25.835	21.857	18.635	85
90	41.145	37.872	34.740	31.793	26.567	22.283	18.871	90
95	43.263	39.626	36.160	32.919	27.235	22.655	19.069	95
100	45.361	41.343	37.530	33.986	27.844	22.980	19.234	100

TABLE A - 36

ANNUAL COST GRADIENT FACTORS(A/G)
DISCRETE CASH FLOW, DISCRETE COMPOUNDING

N	6%	7%	8%	9%	10%	11%	12%	N
2	0.485	0.483	0.481	0.478	0.476	0.474	0.472	2
3	0.961	0.955	0.949	0.943	0.937	0.931	0.925	3
4	1.427	1.416	1.404	1.393	1.381	1.370	1.359	4
5	1.884	1.865	1.846	1.828	1.810	1.792	1.775	5
6	2.330	2.303	2.276	2.250	2.224	2.198	2.172	6
7	2.768	2.730	2.694	2.657	2.622	2.586	2.551	7
8	3.195	3.147	3.099	3.051	3.004	2.958	2.913	8
9	3.613	3.552	3.491	3.431	3.372	3.314	3.257	9
10	4.022	3.946	3.871	3.798	3.725	3.654	3.585	10
11	4.421	4.330	4.240	4.151	4.064	3.979	3.895	11
12	4.811	4.703	4.596	4.491	4.388	4.288	4.190	12
13	5.192	5.065	4.940	4.818	4.699	4.582	4.468	13
14	5.564	5.417	5.273	5.133	4.996	4.862	4.732	14
15	5.926	5.758	5.594	5.435	5.279	5.127	4.980	15
16	6.279	6.090	5.905	5.724	5.549	5.379	5.215	16
17	6.624	6.411	6.204	6.002	5.807	5.618	5.435	17
18	6.960	6.722	6.492	6.269	6.053	5.844	5.643	18
19	7.287	7.024	6.770	6.524	6.286	6.057	5.838	19
20	7.605	7.316	7.037	6.767	6.508	6.259	6.020	20
22	8.217	7.872	7.541	7.223	6.919	6.628	6.351	22
24	8.795	8.392	8.007	7.638	7.288	6.956	6.641	24
25	9.072	8.639	8.225	7.832	7.458	7.104	6.771	25
26	9.341	8.877	8.435	8.016	7.619	7.244	6.892	26
28	9.857	9.329	8.829	8.357	7.914	7.498	7.110	28
30	10.342	9.749	9.190	8.666	8.176	7.721	7.297	30
32	10.799	10.138	9.520	8.944	8.409	7.915	7.459	32
34	11.228	10.499	9.821	9.193	8.615	8.084	7.596	34
35	11.432	10.669	9.961	9.308	8.709	8.159	7.658	35
36	11.630	10.832	10.095	9.417	8.796	8.230	7.714	36
38	12.007	11.140	10.344	9.617	8.956	8.357	7.814	38
40	12.359	11.423	10.570	9.796	9.096	8.466	7.899	40
45	13.141	12.036	11.045	10.160	9.374	8.676	8.057	45
50	13.796	12.529	11.411	10.430	9.570	8.819	8.160	50
55	14.341	12.921	11.690	10.626	9.708	8.913	8.225	55
60	14.791	13.232	11.902	10.768	9.802	8.976	8.266	60
65	15.160	13.476	12.060	10.870	9.867	9.017	8.292	65
70	15.461	13.666	12.178	10.943	9.911	9.044	8.308	70
75	15.706	13.814	12.266	10.994	9.941	9.061	8.318	75
80	15.903	13.927	12.330	11.030	9.961	9.072	8.324	80
85	16.062	14.015	12.377	11.055	9.974	9.079	8.328	85
90	16.189	14.081	12.412	11.073	9.983	9.083	8.330	90
95	16.290	14.132	12.437	11.085	9.989	9.086	8.331	95
100	16.371	14.170	12.455	11.093	9.993	9.088	8.332	100

TABLE A - 37

ANNUAL COST GRADIENT FACTORS(A/G)
DISCRETE CASH FLOW, DISCRETE COMPOUNDING

N	13%	14%	15%	16%	17%	18%	19%	N
2	0.469	0.467	0.465	0.463	0.461	0.459	0.457	2
3	0.919	0.913	0.907	0.901	0.896	0.890	0.885	3
4	1.348	1.337	1.326	1.316	1.305	1.295	1.284	4
5	1.757	1.740	1.723	1.706	1.689	1.673	1.657	5
6	2.147	2.122	2.097	2.073	2.049	2.025	2.002	6
7	2.517	2.483	2.450	2.417	2.385	2.353	2.321	7
8	2.869	2.825	2.781	2.739	2.697	2.656	2.615	8
9	3.201	3.146	3.092	3.039	2.987	2.936	2.886	9
10	3.516	3.449	3.383	3.319	3.255	3.194	3.133	10
11	3.813	3.733	3.655	3.578	3.503	3.430	3.359	11
12	4.094	4.000	3.908	3.819	3.732	3.647	3.564	12
13	4.357	4.249	4.144	4.041	3.942	3.845	3.751	13
14	4.605	4.482	4.362	4.246	4.134	4.025	3.920	14
15	4.837	4.699	4.565	4.435	4.310	4.189	4.072	15
16	5.055	4.901	4.752	4.609	4.470	4.337	4.209	16
17	5.259	5.089	4.925	4.768	4.616	4.471	4.331	17
18	5.449	5.263	5.084	4.913	4.749	4.592	4.441	18
19	5.627	5.424	5.231	5.046	4.869	4.700	4.539	19
20	5.792	5.573	5.365	5.167	4.978	4.798	4.627	20
22	6.088	5.838	5.601	5.377	5.164	4.963	4.773	22
24	6.343	6.062	5.798	5.549	5.315	5.095	4.888	24
25	6.457	6.161	5.883	5.623	5.379	5.150	4.936	25
26	6.561	6.251	5.961	5.690	5.436	5.199	4.978	26
28	6.747	6.410	6.096	5.804	5.533	5.281	5.047	28
30	6.905	6.542	6.207	5.896	5.610	5.345	5.100	30
32	7.039	6.652	6.297	5.971	5.670	5.394	5.140	32
34	7.151	6.743	6.371	6.030	5.718	5.433	5.171	34
35	7.200	6.782	6.402	6.055	5.738	5.449	5.184	35
36	7.245	6.818	6.430	6.077	5.756	5.462	5.194	36
38	7.323	6.880	6.478	6.115	5.785	5.485	5.212	38
40	7.389	6.930	6.517	6.144	5.807	5.502	5.225	40
45	7.508	7.019	6.583	6.193	5.844	5.529	5.245	45
50	7.581	7.071	6.620	6.220	5.863	5.543	5.255	50
55	7.626	7.102	6.641	6.234	5.873	5.549	5.259	55
60	7.653	7.120	6.653	6.242	5.877	5.553	5.261	60
65	7.669	7.130	6.659	6.246	5.880	5.554	5.262	65
70	7.679	7.136	6.663	6.248	5.881	5.555	5.263	70
75	7.684	7.139	6.665	6.249	5.882	5.555	5.263	75
80	7.688	7.141	6.666	6.249	5.882	5.555	5.263	80
85	7.690	7.142	6.666	6.250	5.882	5.555	5.263	85
90	7.691	7.142	6.666	6.250	5.882	5.556	5.263	90
95	7.691	7.142	6.667	6.250	5.882	5.556	5.263	95
100	7.692	7.143	6.667	6.250	5.882	5.556	5.263	100

TABLE A - 38

ANNUAL CCST GRADIENT FACTORS (A/G)
DISCRETE CASH FLOW, DISCRETE COMPOUNDING

N	20%	25%	30%	35%	40%	45%	50%	N
2	0.455	0.444	0.435	0.426	0.417	0.408	0.400	2
3	0.879	0.852	0.827	0.803	0.780	0.758	0.737	3
4	1.274	1.225	1.178	1.134	1.092	1.053	1.015	4
5	1.641	1.563	1.490	1.422	1.358	1.298	1.242	5
6	1.979	1.868	1.765	1.670	1.581	1.499	1.423	6
7	2.290	2.142	2.006	1.881	1.766	1.661	1.565	7
8	2.576	2.387	2.216	2.060	1.919	1.791	1.675	8
9	2.836	2.605	2.396	2.209	2.042	1.893	1.760	9
10	3.074	2.797	2.551	2.334	2.142	1.973	1.824	10
11	3.289	2.966	2.683	2.436	2.221	2.034	1.871	11
12	3.484	3.115	2.795	2.520	2.285	2.082	1.907	12
13	3.660	3.244	2.889	2.589	2.334	2.118	1.933	13
14	3.817	3.356	2.969	2.644	2.373	2.145	1.952	14
15	3.959	3.453	3.034	2.689	2.403	2.165	1.966	15
16	4.085	3.537	3.089	2.725	2.426	2.180	1.976	16
17	4.198	3.608	3.135	2.753	2.444	2.191	1.983	17
18	4.298	3.670	3.172	2.776	2.458	2.200	1.988	18
19	4.386	3.722	3.202	2.793	2.468	2.206	1.991	19
20	4.464	3.767	3.228	2.808	2.476	2.210	1.994	20
22	4.594	3.836	3.265	2.827	2.487	2.216	1.997	22
24	4.694	3.886	3.289	2.839	2.493	2.219	1.999	24
25	4.735	3.905	3.298	2.843	2.494	2.220	1.999	25
26	4.771	3.921	3.305	2.847	2.496	2.221	1.999	26
28	4.829	3.946	3.315	2.851	2.498	2.221	2.000	28
30	4.873	3.963	3.322	2.853	2.499	2.222	2.000	30
32	4.906	3.975	3.326	2.855	2.499	2.222	2.000	32
34	4.931	3.983	3.329	2.856	2.500	2.222	2.000	34
35	4.941	3.986	3.330	2.856	2.500	2.222	2.000	35
36	4.949	3.988	3.330	2.856	2.500	2.222	2.000	36
38	4.963	3.992	3.332	2.857	2.500	2.222	2.000	38
40	4.973	3.995	3.332	2.857	2.500	2.222	2.000	40
45	4.988	3.998	3.333	2.857	2.500	2.222	2.000	45
50	4.995	3.999	3.333	2.857	2.500	2.222	2.000	50
55	4.998	4.000	3.333	2.857	2.500	2.222	2.000	55
60	4.999	4.000	3.333	2.857	2.500	2.222	2.000	60
65	5.000	4.000	3.333	2.857	2.500	2.222	2.000	65
70	5.000	4.000	3.333	2.857	2.500	2.222	2.000	70
75	5.000	4.000	3.333	2.857	2.500	2.222	2.000	75
80	5.000	4.000	3.333	2.857	2.500	2.222	2.000	80
85	5.000	4.000	3.333	2.857	2.500	2.222	2.000	85
90	5.000	4.000	3.333	2.857	2.500	2.222	2.000	90
95	5.000	4.000	3.333	2.857	2.500	2.222	2.000	95
100	5.000	4.000	3.333	2.857	2.500	2.222	2.000	100

APPENDIX

B

INTEREST FACTORS FOR CONTINUOUS COMPOUNDING, DISCRETE CASH FLOW

Tabulated here are the interest factors and gradient conversion factors with the nominal interest rate r compounded continuously and the cash flows occurring at the end of each period (Sec. 3.7 of the text). The computational forms of the factors are given here.

Factor	Notation	Formula
Single-payment compound amount	$(F/P, r\%, n)$	e^{rn}
Single-payment present worth	$(P/F, r\%, n)$	e^{-rn}
Sinking fund	$(A/F, r\%, n)$	$\dfrac{e^r - 1}{e^{rn} - 1}$
Uniform-series compound amount	$(F/A, r\%, n)$	$\dfrac{e^{rn} - 1}{e^r - 1}$
Capital recovery	$(A/P, r\%, n)$	$\dfrac{e^r - 1}{1 - e^{-rn}}$
Uniform-series present worth	$(P/A, r\%, n)$	$\dfrac{1 - e^{-rn}}{e^r - 1}$
Uniform-gradient present worth	$(P/G, r\%, n)$	$\dfrac{1 - e^{-rn}}{(e^r - 1)^2} - \dfrac{n}{e^{rn}(e^r - 1)}$
Uniform-gradient capital recovery	$(A/G, r\%, n)$	$\dfrac{1}{e^r - 1} - \dfrac{n}{e^{rn} - 1}$

The same computational relations given at the beginning of Appendix A may be used for the factors in this appendix.

TABLE B - 1

DISCRETE CASH FLOW
1.00% CONTINUOUS COMPOUND INTEREST FACTORS
(EFFECTIVE RATE IS 1.0050%)

	SINGLE PAYMENTS		UNIFORM SERIES PAYMENTS				
N	COMPOUND AMOUNT F/P	PRESENT WORTH P/F	SINKING FUND A/F	COMPOUND AMOUNT F/A	CAPITAL RECOVERY A/P	PRESENT WORTH P/A	N
1	1.0101	0.9900	1.00000	1.0000	1.01005	0.9900	1
2	1.0202	0.9802	0.49750	2.0101	0.50755	1.9702	2
3	1.0305	0.9704	0.33001	3.0303	0.34006	2.9407	3
4	1.0408	0.9608	0.24626	4.0607	0.25631	3.9015	4
5	1.0513	0.9512	0.19602	5.1015	0.20607	4.8527	5
6	1.0618	0.9418	0.16253	6.1528	0.17258	5.7945	6
7	1.0725	0.9324	0.13861	7.2146	0.14866	6.7269	7
8	1.0833	0.9231	0.12067	8.2871	0.13072	7.6500	8
9	1.0942	0.9139	0.10672	9.3704	0.11677	8.5639	9
10	1.1052	0.9048	0.09556	10.4646	0.10561	9.4688	10
11	1.1163	0.8958	0.08643	11.5698	0.09648	10.3646	11
12	1.1275	0.8869	0.07883	12.6860	0.08888	11.2515	12
13	1.1388	0.8781	0.07239	13.8135	0.08244	12.1296	13
14	1.1503	0.8694	0.06688	14.9524	0.07693	12.9990	14
15	1.1618	0.8607	0.06210	16.1026	0.07215	13.8597	15
16	1.1735	0.8521	0.05792	17.2645	0.06797	14.7118	16
17	1.1853	0.8437	0.05424	18.4380	0.06429	15.5555	17
18	1.1972	0.8353	0.05096	19.6233	0.06101	16.3908	18
19	1.2092	0.8270	0.04803	20.8205	0.05808	17.2177	19
20	1.2214	0.8187	0.04539	22.0298	0.05544	18.0364	20
22	1.2461	0.8025	0.04084	24.4848	0.05089	19.6495	22
24	1.2712	0.7866	0.03705	26.9895	0.04710	21.2307	24
25	1.2840	0.7788	0.03538	28.2608	0.04543	22.0095	25
26	1.2969	0.7711	0.03385	29.5448	0.04390	22.7806	26
28	1.3231	0.7558	0.03110	32.1517	0.04115	24.2997	28
30	1.3499	0.7408	0.02873	34.8112	0.03878	25.7888	30
32	1.3771	0.7261	0.02665	37.5245	0.03670	27.2484	32
34	1.4049	0.7118	0.02482	40.2926	0.03487	28.6791	34
35	1.4191	0.7047	0.02398	41.6976	0.03403	29.3838	35
36	1.4333	0.6977	0.02319	43.1166	0.03324	30.0815	36
38	1.4623	0.6839	0.02174	45.9977	0.03179	31.4561	38
40	1.4918	0.6703	0.02043	48.9370	0.03048	32.8034	40
45	1.5683	0.6376	0.01768	56.5475	0.02773	36.0563	45
50	1.6487	0.6065	0.01549	64.5483	0.02554	39.1505	50
45	1.5683	0.6376	0.01768	56.5475	0.02773	36.0563	45
50	1.6487	0.6065	0.01549	64.5483	0.02554	39.1505	50
55	1.7333	0.5769	0.01371	72.9593	0.02376	42.0938	55
60	1.8221	0.5488	0.01222	81.8015	0.02227	44.8936	60
65	1.9155	0.5220	0.01098	91.0971	0.02103	47.5568	65
70	2.0138	0.4966	0.00991	100.8692	0.01996	50.0902	70
75	2.1170	0.4724	0.00900	111.1424	0.01905	52.5000	75
80	2.2255	0.4493	0.00820	121.9423	0.01825	54.7922	80
85	2.3396	0.4274	0.00750	133.2960	0.01755	56.9727	85
90	2.4596	0.4066	0.00689	145.2317	0.01694	59.0468	90
95	2.5857	0.3867	0.00634	157.7794	0.01639	61.0198	95
100	2.7183	0.3679	0.00585	170.9705	0.01590	62.8965	100

TABLE B - 2

DISCRETE CASH FLOW
2.00% CONTINUOUS COMPOUND INTEREST FACTORS
(EFFECTIVE RATE IS 2.0201%)

	SINGLE PAYMENTS		UNIFORM SERIES PAYMENTS				
	COMPOUND AMOUNT	PRESENT WORTH	SINKING FUND	COMPOUND AMOUNT	CAPITAL RECOVERY	PRESENT WORTH	
N	F/P	P/F	A/F	F/A	A/P	P/A	N
1	1.0202	0.9802	1.00000	1.0000	1.02020	0.9802	1
2	1.0408	0.9608	0.49500	2.0202	0.51520	1.9410	2
3	1.0618	0.9418	0.32669	3.0610	0.34689	2.8828	3
4	1.0833	0.9231	0.24255	4.1228	0.26275	3.8059	4
5	1.1052	0.9048	0.19208	5.2061	0.21228	4.7107	5
6	1.1275	0.8869	0.15845	6.3113	0.17865	5.5976	6
7	1.1503	0.8694	0.13443	7.4388	0.15463	6.4670	7
8	1.1735	0.8521	0.11643	8.5891	0.13663	7.3191	8
9	1.1972	0.8353	0.10243	9.7626	0.12263	8.1544	9
10	1.2214	0.8187	0.09124	10.9598	0.11144	8.9731	10
11	1.2461	0.8025	0.08209	12.1812	0.10230	9.7756	11
12	1.2712	0.7866	0.07448	13.4273	0.09468	10.5623	12
13	1.2969	0.7711	0.06803	14.6985	0.08824	11.3333	13
14	1.3231	0.7558	0.06252	15.9955	0.08272	12.0891	14
15	1.3499	0.7408	0.05774	17.3186	0.07794	12.8299	15
16	1.3771	0.7261	0.05357	18.6685	0.07377	13.5561	16
17	1.4049	0.7118	0.04989	20.0456	0.07009	14.2678	17
18	1.4333	0.6977	0.04662	21.4505	0.06682	14.9655	18
19	1.4623	0.6839	0.04370	22.8839	0.06390	15.6494	19
20	1.4918	0.6703	0.04107	24.3461	0.06128	16.3197	20
22	1.5527	0.6440	0.03655	27.3599	0.05675	17.6208	22
24	1.6161	0.6188	0.03279	30.4967	0.05299	18.8709	24
25	1.6487	0.6065	0.03114	32.1128	0.05134	19.4774	25
26	1.6820	0.5945	0.02962	33.7615	0.04982	20.0719	26
28	1.7507	0.5712	0.02691	37.1595	0.04711	21.2259	28
30	1.8221	0.5488	0.02457	40.6963	0.04477	22.3346	30
32	1.8965	0.5273	0.02253	44.3773	0.04274	23.3998	32
34	1.9739	0.5066	0.02074	48.2086	0.04094	24.4233	34
35	2.0138	0.4966	0.01993	50.1824	0.04013	24.9199	35
36	2.0544	0.4868	0.01916	52.1962	0.03936	25.4066	36
38	2.1383	0.4677	0.01775	56.3466	0.03795	26.3514	38
40	2.2255	0.4493	0.01648	60.6663	0.03668	27.2591	40
45	2.4596	0.4066	0.01384	72.2528	0.03404	29.3758	45
50	2.7183	0.3679	0.01176	85.0578	0.03196	31.2910	50
45	2.4596	0.4066	0.01384	72.2528	0.03404	29.3758	45
50	2.7183	0.3679	0.01176	85.0578	0.03196	31.2910	50
55	3.0042	0.3329	0.01008	99.2096	0.03028	33.0240	55
60	3.3201	0.3012	0.00871	114.8497	0.02891	34.5921	60
65	3.6693	0.2725	0.00757	132.1346	0.02777	36.0109	65
70	4.0552	0.2466	0.00661	151.2375	0.02681	37.2947	70
75	4.4817	0.2231	0.00580	172.3494	0.02600	38.4564	75
80	4.9530	0.2019	0.00511	195.6817	0.02531	39.5075	80
85	5.4739	0.1827	0.00452	221.4679	0.02472	40.4585	85
90	6.0496	0.1653	0.00400	249.9660	0.02420	41.3191	90
95	6.6859	0.1496	0.00355	281.4613	0.02375	42.0978	95
100	7.3891	0.1353	0.00316	316.2689	0.02336	42.8023	100

TABLE B - 3

DISCRETE CASH FLOW
4.00% CONTINUOUS COMPOUND INTEREST FACTORS
(EFFECTIVE RATE IS 4.0811%)

	SINGLE PAYMENTS		UNIFORM SERIES PAYMENTS				
N	COMPOUND AMOUNT F/P	PRESENT WORTH P/F	SINKING FUND A/F	COMPOUND AMOUNT F/A	CAPITAL RECOVERY A/P	PRESENT WORTH P/A	N
1	1.0408	0.9608	1.00000	1.000	1.04081	0.9608	1
2	1.0833	0.9231	0.49000	2.041	0.53081	1.8839	2
3	1.1275	0.8869	0.32009	3.124	0.36090	2.7708	3
4	1.1735	0.8521	0.23521	4.252	0.27602	3.6230	4
5	1.2214	0.8187	0.18433	5.425	0.22514	4.4417	5
6	1.2712	0.7866	0.15045	6.647	0.19127	5.2283	6
7	1.3231	0.7558	0.12630	7.918	0.16711	5.9841	7
8	1.3771	0.7261	0.10821	9.241	0.14903	6.7103	8
9	1.4333	0.6977	0.09418	10.618	0.13499	7.4079	9
10	1.4918	0.6703	0.08298	12.051	0.12379	8.0783	10
11	1.5527	0.6440	0.07384	13.543	0.11465	8.7223	11
12	1.6161	0.6188	0.06624	15.096	0.10705	9.3411	12
13	1.6820	0.5945	0.05984	16.712	0.10065	9.9356	13
14	1.7507	0.5712	0.05437	18.394	0.09518	10.5068	14
15	1.8221	0.5488	0.04964	20.145	0.09045	11.0556	15
16	1.8965	0.5273	0.04552	21.967	0.08633	11.5829	16
17	1.9739	0.5066	0.04191	23.863	0.08272	12.0895	17
18	2.0544	0.4868	0.03870	25.837	0.07951	12.5763	18
19	2.1383	0.4677	0.03585	27.892	0.07666	13.0439	19
20	2.2255	0.4493	0.03330	30.030	0.07411	13.4933	20
22	2.4109	0.4148	0.02893	34.572	0.06974	14.3398	22
24	2.6117	0.3829	0.02532	39.492	0.06613	15.1212	24
25	2.7183	0.3679	0.02375	42.104	0.06456	15.4891	25
26	2.8292	0.3535	0.02231	44.822	0.06312	15.8425	26
28	3.0649	0.3263	0.01976	50.596	0.06058	16.5084	28
30	3.3201	0.3012	0.01759	56.851	0.05840	17.1231	30
32	3.5966	0.2780	0.01572	63.626	0.05653	17.6905	32
34	3.8962	0.2567	0.01409	70.966	0.05490	18.2143	34
35	4.0552	0.2466	0.01336	74.863	0.05417	18.4609	35
36	4.2207	0.2369	0.01267	78.918	0.05348	18.6978	36
38	4.5722	0.2187	0.01142	87.531	0.05224	19.1442	38
40	4.9530	0.2019	0.01032	96.862	0.05113	19.5562	40
45	6.0496	0.1653	0.00808	123.733	0.04889	20.4530	45
50	7.3891	0.1353	0.00639	156.553	0.04720	21.1872	50
45	6.0496	0.1653	0.00808	123.733	0.04889	20.4530	45
50	7.3891	0.1353	0.00639	156.553	0.04720	21.1872	50
55	9.0250	0.1108	0.00509	196.640	0.04590	21.7883	55
60	11.0232	0.0907	0.00407	245.601	0.04488	22.2804	60
65	13.4637	0.0743	0.00327	305.403	0.04409	22.6834	65
70	16.4446	0.0608	0.00264	378.445	0.04345	23.0133	70
75	20.0855	0.0498	0.00214	467.659	0.04295	23.2834	75
80	24.5325	0.0408	0.00173	576.625	0.04255	23.5045	80
85	29.9641	0.0334	0.00141	709.717	0.04222	23.6856	85
90	36.5982	0.0273	0.00115	872.275	0.04196	23.8338	90
95	44.7012	0.0224	0.00093	1070.825	0.04174	23.9552	95
100	54.5982	0.0183	0.00076	1313.333	0.04157	24.0545	100

TABLE B - 4

DISCRETE CASH FLOW
5.00% CONTINUOUS COMPOUND INTEREST FACTORS
(EFFECTIVE RATE IS 5.1271%)

	SINGLE PAYMENTS		UNIFORM SERIES PAYMENTS				
N	COMPOUND AMOUNT F/P	PRESENT WORTH P/F	SINKING FUND A/F	COMPOUND AMOUNT F/A	CAPITAL RECOVERY A/P	PRESENT WORTH P/A	N
1	1.0513	0.9512	1.00000	1.000	1.05127	0.9512	1
2	1.1052	0.9048	0.48750	2.051	0.53877	1.8561	2
3	1.1618	0.8607	0.31681	3.156	0.36808	2.7168	3
4	1.2214	0.8187	0.23157	4.318	0.28284	3.5355	4
5	1.2840	0.7788	0.18052	5.540	0.23179	4.3143	5
6	1.3499	0.7408	0.14655	6.824	0.19782	5.0551	6
7	1.4191	0.7047	0.12235	8.174	0.17362	5.7598	7
8	1.4918	0.6703	0.10425	9.593	0.15552	6.4301	8
9	1.5683	0.6376	0.09022	11.084	0.14149	7.0678	9
10	1.6487	0.6065	0.07903	12.653	0.13031	7.6743	10
11	1.7333	0.5769	0.06992	14.301	0.12119	8.2512	11
12	1.8221	0.5488	0.06236	16.035	0.11364	8.8001	12
13	1.9155	0.5220	0.05600	17.857	0.10727	9.3221	13
14	2.0138	0.4966	0.05058	19.772	0.10185	9.8187	14
15	2.1170	0.4724	0.04590	21.786	0.09717	10.2911	15
16	2.2255	0.4493	0.04184	23.903	0.09311	10.7404	16
17	2.3396	0.4274	0.03827	26.129	0.08954	11.1678	17
18	2.4596	0.4066	0.03513	28.468	0.08640	11.5744	18
19	2.5857	0.3867	0.03233	30.928	0.08360	11.9611	19
20	2.7183	0.3679	0.02984	33.514	0.08111	12.3290	20
22	3.0042	0.3329	0.02558	39.090	0.07685	13.0118	22
24	3.3201	0.3012	0.02210	45.252	0.07337	13.6296	24
25	3.4903	0.2865	0.02059	48.572	0.07186	13.9161	25
26	3.6693	0.2725	0.01921	52.062	0.07048	14.1887	26
28	4.0552	0.2466	0.01678	59.589	0.06805	14.6945	28
30	4.4817	0.2231	0.01473	67.907	0.06600	15.1522	30
32	4.9530	0.2019	0.01297	77.101	0.06424	15.5663	32
34	5.4739	0.1827	0.01146	87.261	0.06273	15.9411	34
35	5.7546	0.1738	0.01078	92.735	0.06205	16.1149	35
36	6.0496	0.1653	0.01015	98.489	0.06142	16.2801	36
38	6.6859	0.1496	0.00902	110.899	0.06029	16.5870	38
40	7.3891	0.1353	0.00802	124.613	0.05930	16.8646	40
45	9.4877	0.1054	0.00604	165.546	0.05731	17.4484	45
50	12.1825	0.0821	0.00458	218.105	0.05586	17.9032	50
45	9.4877	0.1054	0.00604	165.546	0.05731	17.4484	45
50	12.1825	0.0821	0.00458	218.105	0.05586	17.9032	50
55	15.6426	0.0639	0.00350	285.592	0.05477	18.2573	55
60	20.0855	0.0498	0.00269	372.247	0.05396	18.5331	60
65	25.7903	0.0388	0.00207	483.515	0.05334	18.7479	65
70	33.1155	0.0302	0.00160	626.385	0.05287	18.9152	70
75	42.5211	0.0235	0.00123	809.834	0.05251	19.0455	75
80	54.5982	0.0183	0.00096	1045.387	0.05223	19.1469	80
85	70.1054	0.0143	0.00074	1347.843	0.05201	19.2260	85
90	90.0171	0.0111	0.00058	1736.205	0.05185	19.2875	90
95	115.584	0.0087	0.00045	2234.871	0.05172	19.3354	95
100	148.413	0.0067	0.00035	2875.171	0.05162	19.3727	100

TABLE B - 5

DISCRETE CASH FLOW
6.00% CONTINUOUS COMPOUND INTEREST FACTORS
(EFFECTIVE RATE IS 6.1837%)

	SINGLE PAYMENTS		UNIFORM SERIES PAYMENTS				
N	COMPOUND AMOUNT F/P	PRESENT WORTH P/F	SINKING FUND A/F	COMPOUND AMOUNT F/A	CAPITAL RECOVERY A/P	PRESENT WORTH P/A	N
1	1.0618	0.9418	1.00000	1.000	1.06184	0.9418	1
2	1.1275	0.8869	0.48500	2.062	0.54684	1.8287	2
3	1.1972	0.8353	0.31355	3.189	0.37538	2.6640	3
4	1.2712	0.7866	0.22797	4.387	0.28981	3.4506	4
5	1.3499	0.7408	0.17675	5.658	0.23858	4.1914	5
6	1.4333	0.6977	0.14270	7.008	0.20454	4.8891	6
7	1.5220	0.6570	0.11847	8.441	0.18031	5.5461	7
8	1.6161	0.6188	0.10037	9.963	0.16221	6.1649	8
9	1.7160	0.5827	0.08636	11.579	0.14820	6.7477	9
10	1.8221	0.5488	0.07522	13.295	0.13705	7.2965	10
11	1.9348	0.5169	0.06615	15.117	0.12799	7.8133	11
12	2.0544	0.4868	0.05864	17.052	0.12048	8.3001	12
13	2.1815	0.4584	0.05234	19.106	0.11418	8.7585	13
14	2.3164	0.4317	0.04698	21.288	0.10881	9.1902	14
15	2.4596	0.4066	0.04237	23.604	0.10420	9.5968	15
16	2.6117	0.3829	0.03837	26.064	0.10020	9.9797	16
17	2.7732	0.3606	0.03487	28.676	0.09671	10.3402	17
18	2.9447	0.3396	0.03180	31.449	0.09363	10.6798	18
19	3.1268	0.3198	0.02908	34.393	0.09091	10.9997	19
20	3.3201	0.3012	0.02665	37.520	0.08849	11.3009	20
22	3.7434	0.2671	0.02254	44.366	0.08438	11.8516	22
24	4.2207	0.2369	0.01920	52.084	0.08104	12.3401	24
25	4.4817	0.2231	0.01776	56.305	0.07960	12.5633	25
26	4.7588	0.2101	0.01645	60.786	0.07829	12.7734	26
28	5.3656	0.1864	0.01416	70.598	0.07600	13.1577	28
30	6.0496	0.1653	0.01225	81.661	0.07408	13.4985	30
32	6.8210	0.1466	0.01062	94.135	0.07246	13.8008	32
34	7.6906	0.1300	0.00924	108.198	0.07108	14.0689	34
35	8.1662	0.1225	0.00863	115.889	0.07047	14.1913	35
36	8.6711	0.1153	0.00806	124.055	0.06990	14.3067	36
38	9.7767	0.1023	0.00705	141.934	0.06888	14.5176	38
40	11.0232	0.0907	0.00617	162.091	0.06801	14.7046	40
45	14.8797	0.0672	0.00446	224.458	0.06629	15.0848	45
50	20.0855	0.0498	0.00324	308.645	0.06508	15.3665	50
45	14.8797	0.0672	0.00446	224.458	0.06629	15.0848	45
50	20.0855	0.0498	0.00324	308.645	0.06508	15.3665	50
55	27.1126	0.0369	0.00237	422.285	0.06420	15.5752	55
60	36.5982	0.0273	0.00174	575.683	0.06357	15.7298	60
65	49.4024	0.0202	0.00128	782.748	0.06311	15.8443	65
70	66.6863	0.0150	0.00094	1062.257	0.06278	15.9292	70
75	90.0171	0.0111	0.00069	1439.555	0.06253	15.9920	75
80	121.510	0.0082	0.00051	1948.854	0.06235	16.0386	80
85	164.022	0.0061	0.00038	2636.336	0.06222	16.0731	85
90	221.406	0.0045	0.00028	3564.339	0.06212	16.0986	90
95	298.867	0.0033	0.00021	4817.012	0.06204	16.1176	95
100	403.429	0.0025	0.00015	6507.944	0.06199	16.1316	100

TABLE B - 6

DISCRETE CASH FLOW
8.00% CONTINUOUS COMPOUND INTEREST FACTORS
(EFFECTIVE RATE IS 8.3287%)

	SINGLE PAYMENTS		UNIFORM SERIES PAYMENTS				
N	COMPOUND AMOUNT F/P	PRESENT WORTH P/F	SINKING FUND A/F	COMPOUND AMOUNT F/A	CAPITAL RECOVERY A/P	PRESENT WORTH P/A	N
1	1.0833	0.9231	1.00000	1.000	1.08329	0.9231	1
2	1.1735	0.8521	0.48001	2.083	0.56330	1.7753	2
3	1.2712	0.7866	0.30705	3.257	0.39034	2.5619	3
4	1.3771	0.7261	0.22085	4.528	0.30413	3.2880	4
5	1.4918	0.6703	0.16934	5.905	0.25263	3.9584	5
6	1.6161	0.6188	0.13519	7.397	0.21848	4.5771	6
7	1.7507	0.5712	0.11095	9.013	0.19424	5.1483	7
8	1.8965	0.5273	0.09290	10.764	0.17619	5.6756	8
9	2.0544	0.4868	0.07899	12.660	0.16227	6.1624	9
10	2.2255	0.4493	0.06796	14.715	0.15125	6.6117	10
11	2.4109	0.4148	0.05903	16.940	0.14232	7.0265	11
12	2.6117	0.3829	0.05168	19.351	0.13496	7.4094	12
13	2.8292	0.3535	0.04553	21.963	0.12882	7.7629	13
14	3.0649	0.3263	0.04034	24.792	0.12362	8.0891	14
15	3.3201	0.3012	0.03590	27.857	0.11918	8.3903	15
16	3.5966	0.2780	0.03207	31.177	0.11536	8.6684	16
17	3.8962	0.2567	0.02876	34.774	0.11204	8.9250	17
18	4.2207	0.2369	0.02586	38.670	0.10915	9.1620	18
19	4.5722	0.2187	0.02332	42.891	0.10660	9.3807	19
20	4.9530	0.2019	0.02107	47.463	0.10436	9.5826	20
22	5.8124	0.1720	0.01731	57.781	0.10059	9.9410	22
24	6.8210	0.1466	0.01431	69.890	0.09760	10.2464	24
25	7.3891	0.1353	0.01304	76.711	0.09632	10.3817	25
26	8.0045	0.1249	0.01189	84.100	0.09518	10.5067	26
28	9.3933	0.1065	0.00992	100.776	0.09321	10.7285	28
30	11.0232	0.0907	0.00831	120.345	0.09160	10.9174	30
32	12.9358	0.0773	0.00698	143.309	0.09026	11.0785	32
34	15.1803	0.0659	0.00587	170.258	0.08916	11.2157	34
35	16.4446	0.0608	0.00539	185.439	0.08868	11.2765	35
36	17.8143	0.0561	0.00495	201.883	0.08824	11.3327	36
38	20.9052	0.0478	0.00418	238.996	0.08747	11.4323	38
40	24.5325	0.0408	0.00354	282.547	0.08683	11.5172	40
45	36.5982	0.0273	0.00234	427.416	0.08563	11.6786	45
50	54.5982	0.0183	0.00155	643.535	0.08484	11.7868	50
45	36.5982	0.0273	0.00234	427.416	0.08563	11.6786	45
50	54.5982	0.0183	0.00155	643.535	0.08484	11.7868	50
55	81.4509	0.0123	0.00104	965.947	0.08432	11.8593	55
60	121.510	0.0082	0.00069	1446.928	0.08398	11.9079	60
65	181.272	0.0055	0.00046	2164.469	0.08375	11.9404	65
70	270.426	0.0037	0.00031	3234.913	0.08360	11.9623	70
75	403.429	0.0025	0.00021	4831.828	0.08349	11.9769	75
80	601.845	0.0017	0.00014	7214.146	0.08343	11.9867	80
85	897.847	0.0011	0.00009	10768.146	0.08338	11.9933	85
90	1339.431	0.0007	0.00006	16070.091	0.08335	11.9977	90
95	1998.196	0.0005	0.00004	23979.664	0.08333	12.0007	95
100	2980.958	0.0003	0.00003	35779.360	0.08332	12.0026	100

TABLE B - 7

DISCRETE CASH FLOW
10.00% CONTINUOUS COMPOUND INTEREST FACTORS
(EFFECTIVE RATE IS 10.5171%)

	SINGLE PAYMENTS		UNIFORM SERIES PAYMENTS				
N	COMPOUND AMOUNT F/P	PRESENT WORTH P/F	SINKING FUND A/F	COMPOUND AMOUNT F/A	CAPITAL RECOVERY A/P	PRESENT WORTH P/A	N
1	1.1052	0.9048	1.00000	1.000	1.10517	0.9048	1
2	1.2214	0.8187	0.47502	2.105	0.58019	1.7236	2
3	1.3499	0.7408	0.30061	3.327	0.40578	2.4644	3
4	1.4918	0.6703	0.21384	4.676	0.31901	3.1347	4
5	1.6487	0.6065	0.16212	6.168	0.26729	3.7412	5
6	1.8221	0.5488	0.12793	7.817	0.23310	4.2900	6
7	2.0138	0.4966	0.10374	9.639	0.20892	4.7866	7
8	2.2255	0.4493	0.08582	11.653	0.19099	5.2360	8
9	2.4596	0.4066	0.07205	13.878	0.17723	5.6425	9
10	2.7183	0.3679	0.06121	16.338	0.16638	6.0104	10
11	3.0042	0.3329	0.05248	19.056	0.15765	6.3433	11
12	3.3201	0.3012	0.04533	22.060	0.15050	6.6445	12
13	3.6693	0.2725	0.03940	25.381	0.14457	6.9170	13
14	4.0552	0.2466	0.03442	29.050	0.13959	7.1636	14
15	4.4817	0.2231	0.03021	33.105	0.13538	7.3867	15
16	4.9530	0.2019	0.02661	37.587	0.13178	7.5886	16
17	5.4739	0.1827	0.02351	42.540	0.12868	7.7713	17
18	6.0496	0.1653	0.02083	48.014	0.12600	7.9366	18
19	6.6859	0.1496	0.01850	54.063	0.12367	8.0862	19
20	7.3891	0.1353	0.01646	60.749	0.12163	8.2215	20
22	9.0250	0.1108	0.01311	76.304	0.11828	8.4548	22
24	11.0232	0.0907	0.01049	95.304	0.11566	8.6458	24
25	12.1825	0.0821	0.00940	106.327	0.11458	8.7278	25
26	13.4637	0.0743	0.00844	118.509	0.11361	8.8021	26
28	16.4446	0.0608	0.00681	146.853	0.11198	8.9301	28
30	20.0855	0.0498	0.00551	181.472	0.11068	9.0349	30
32	24.5325	0.0408	0.00447	223.755	0.10964	9.1208	32
34	29.9641	0.0334	0.00363	275.400	0.10880	9.1910	34
35	33.1155	0.0302	0.00327	305.364	0.10845	9.2212	35
36	36.5982	0.0273	0.00295	338.480	0.10813	9.2485	36
38	44.7012	0.0224	0.00241	415.525	0.10758	9.2956	38
40	54.5982	0.0183	0.00196	509.629	0.10713	9.3342	40
45	90.0171	0.0111	0.00118	846.404	0.10635	9.4027	45
50	148.413	0.0067	0.00071	1401.653	0.10588	9.4443	50
45	90.0171	0.0111	0.00118	846.404	0.10635	9.4027	45
50	148.413	0.0067	0.00071	1401.653	0.10588	9.4443	50
55	244.692	0.0041	0.00043	2317.104	0.10560	9.4695	55
60	403.429	0.0025	0.00026	3826.427	0.10543	9.4848	60
65	665.142	0.0015	0.00016	6314.879	0.10533	9.4940	65
70	1096.633	0.0009	0.00010	10417.644	0.10527	9.4997	70
75	1808.042	0.0006	0.00006	17181.959	0.10523	9.5031	75
80	2980.958	0.0003	0.00004	28334.430	0.10521	9.5051	80
85	4914.769	0.0002	0.00002	46721.745	0.10519	9.5064	85
90	8103.084	0.0001	0.00001	77037.303	0.10518	9.5072	90

TABLE B - 8

DISCRETE CASH FLOW
12.00% CONTINUOUS COMPOUND INTEREST FACTORS
(EFFECTIVE RATE IS 12.7497%)

	SINGLE PAYMENTS		UNIFORM SERIES PAYMENTS				
N	COMPOUND AMOUNT F/P	PRESENT WORTH P/F	SINKING FUND A/F	COMPOUND AMOUNT F/A	CAPITAL RECOVERY A/P	PRESENT WORTH P/A	N
1	1.1275	0.8869	1.00000	1.000	1.12750	0.8869	1
2	1.2712	0.7866	0.47004	2.127	0.59753	1.6735	2
3	1.4333	0.6977	0.29423	3.399	0.42172	2.3712	3
4	1.6161	0.6188	0.20695	4.832	0.33445	2.9900	4
5	1.8221	0.5488	0.15508	6.448	0.28258	3.5388	5
6	2.0544	0.4868	0.12092	8.270	0.24841	4.0256	6
7	2.3164	0.4317	0.09686	10.325	0.22435	4.4573	7
8	2.6117	0.3829	0.07911	12.641	0.20660	4.8402	8
9	2.9447	0.3396	0.06556	15.253	0.19306	5.1798	9
10	3.3201	0.3012	0.05495	18.197	0.18245	5.4810	10
11	3.7434	0.2671	0.04647	21.518	0.17397	5.7481	11
12	4.2207	0.2369	0.03959	25.261	0.16708	5.9850	12
13	4.7588	0.2101	0.03392	29.482	0.16142	6.1952	13
14	5.3656	0.1864	0.02921	34.241	0.15670	6.3815	14
15	6.0496	0.1653	0.02525	39.606	0.15275	6.5468	15
16	6.8210	0.1466	0.02190	45.656	0.14940	6.6934	16
17	7.6906	0.1300	0.01906	52.477	0.14655	6.8235	17
18	8.6711	0.1153	0.01662	60.167	0.14412	6.9388	18
19	9.7767	0.1023	0.01453	68.838	0.14202	7.0411	19
20	11.0232	0.0907	0.01272	78.615	0.14022	7.1318	20
22	14.0132	0.0714	0.00980	102.067	0.13729	7.2836	22
24	17.8143	0.0561	0.00758	131.880	0.13508	7.4030	24
25	20.0855	0.0498	0.00668	149.694	0.13418	7.4528	25
26	22.6464	0.0442	0.00589	169.780	0.13339	7.4970	26
28	28.7892	0.0347	0.00459	217.960	0.13208	7.5709	28
30	36.5982	0.0273	0.00358	279.209	0.13108	7.6290	30
32	46.5255	0.0215	0.00280	357.071	0.13030	7.6747	32
34	59.1455	0.0169	0.00219	456.054	0.12969	7.7107	34
35	66.6863	0.0150	0.00194	515.200	0.12944	7.7257	35
36	75.1886	0.0133	0.00172	581.886	0.12922	7.7390	36
38	95.5835	0.0105	0.00135	741.850	0.12884	7.7613	38
40	121.510	0.0082	0.00106	945.203	0.12855	7.7788	40

TABLE B - 9

DISCRETE CASH FLOW
15.00% CONTINUOUS COMPOUND INTEREST FACTORS
(EFFECTIVE RATE IS 16.1834%)

	SINGLE PAYMENTS		UNIFORM SERIES PAYMENTS				
N	COMPOUND AMOUNT F/P	PRESENT WORTH P/F	SINKING FUND A/F	COMPOUND AMOUNT F/A	CAPITAL RECOVERY A/P	PRESENT WORTH P/A	N
1	1.1618	0.8607	1.00000	1.000	1.16183	0.8607	1
2	1.3499	0.7408	0.46257	2.162	0.62440	1.6015	2
3	1.5683	0.6376	0.28476	3.512	0.44660	2.2392	3
4	1.8221	0.5488	0.19685	5.080	0.35868	2.7880	4
5	2.1170	0.4724	0.14488	6.902	0.30672	3.2603	5
6	2.4596	0.4066	0.11088	9.019	0.27271	3.6669	6
7	2.8577	0.3499	0.08712	11.479	0.24895	4.0168	7
8	3.3201	0.3012	0.06975	14.336	0.23159	4.3180	8
9	3.8574	0.2592	0.05664	17.656	0.21847	4.5773	9
10	4.4817	0.2231	0.04648	21.514	0.20832	4.8004	10
11	5.2070	0.1920	0.03847	25.996	0.20030	4.9925	11
12	6.0496	0.1653	0.03205	31.203	0.19388	5.1578	12
13	7.0287	0.1423	0.02684	37.252	0.18868	5.3000	13
14	8.1662	0.1225	0.02258	44.281	0.18442	5.4225	14
15	9.4877	0.1054	0.01907	52.447	0.18090	5.5279	15
16	11.0232	0.0907	0.01615	61.935	0.17798	5.6186	16
17	12.8071	0.0781	0.01371	72.958	0.17554	5.6967	17
18	14.8797	0.0672	0.01166	85.765	0.17349	5.7639	18
19	17.2878	0.0578	0.00994	100.645	0.17177	5.8217	19
20	20.0855	0.0498	0.00848	117.933	0.17031	5.8715	20
22	27.1126	0.0369	0.00620	161.354	0.16803	5.9513	22
24	36.5982	0.0273	0.00455	219.967	0.16638	6.0103	24
25	42.5211	0.0235	0.00390	256.565	0.16573	6.0338	25
26	49.4024	0.0202	0.00334	299.087	0.16518	6.0541	26
28	66.6863	0.0150	0.00246	405.886	0.16430	6.0865	28
30	90.0171	0.0111	0.00182	550.051	0.16365	6.1105	30
32	121.510	0.0082	0.00134	744.653	0.16318	6.1283	32
34	164.022	0.0061	0.00099	1007.339	0.16283	6.1415	34
35	190.566	0.0052	0.00085	1171.361	0.16269	6.1467	35
36	221.406	0.0045	0.00073	1361.927	0.16257	6.1513	36
38	298.867	0.0033	0.00054	1840.571	0.16238	6.1585	38
40	403.429	0.0025	0.00040	2486.673	0.16224	6.1638	40

TABLE B - 10

DISCRETE CASH FLOW
18.00% CONTINUOUS COMPOUND INTEREST FACTORS
(EFFECTIVE RATE IS 19.7217%)

	SINGLE PAYMENTS		UNIFORM SERIES PAYMENTS				
	COMPOUND AMOUNT	PRESENT WORTH	SINKING FUND	COMPOUND AMOUNT	CAPITAL RECOVERY	PRESENT WORTH	
N	F/P	P/F	A/F	F/A	A/P	P/A	N
1	1.1972	0.8353	1.00000	1.000	1.19722	0.8353	1
2	1.4333	0.6977	0.45512	2.197	0.65234	1.5329	2
3	1.7160	0.5827	0.27544	3.631	0.47266	2.1157	3
4	2.0544	0.4868	0.18704	5.347	0.38425	2.6024	4
5	2.4596	0.4066	0.13512	7.401	0.33233	3.0090	5
6	2.9447	0.3396	0.10141	9.861	0.29863	3.3486	6
7	3.5254	0.2837	0.07809	12.805	0.27531	3.6323	7
8	4.2207	0.2369	0.06123	16.331	0.25845	3.8692	8
9	5.0531	0.1979	0.04866	20.551	0.24588	4.0671	9
10	6.0496	0.1653	0.03906	25.604	0.23627	4.2324	10
11	7.2427	0.1381	0.03159	31.654	0.22881	4.3705	11
12	8.6711	0.1153	0.02571	38.897	0.22293	4.4858	12
13	10.3812	0.0963	0.02102	47.568	0.21824	4.5821	13
14	12.4286	0.0805	0.01726	57.949	0.21447	4.6626	14
15	14.8797	0.0672	0.01421	70.378	0.21143	4.7298	15
16	17.8143	0.0561	0.01173	85.258	0.20895	4.7859	16
17	21.3276	0.0469	0.00970	103.072	0.20692	4.8328	17
18	25.5337	0.0392	0.00804	124.399	0.20526	4.8720	18
19	30.5694	0.0327	0.00667	149.933	0.20389	4.9047	19
20	36.5982	0.0273	0.00554	180.503	0.20276	4.9320	20
22	52.4573	0.0191	0.00383	260.917	0.20105	4.9739	22
24	75.1886	0.0133	0.00266	376.177	0.19988	5.0031	24
25	90.0171	0.0111	0.00222	451.366	0.19943	5.0142	25
26	107.7701	0.0093	0.00185	541.383	0.19906	5.0235	26
28	154.4700	0.0065	0.00129	778.177	0.19850	5.0377	28
30	221.4064	0.0045	0.00089	1117.581	0.19811	5.0476	30
32	317.3483	0.0032	0.00062	1604.059	0.19784	5.0546	32
34	454.8647	0.0022	0.00043	2301.342	0.19765	5.0594	34
35	544.5719	0.0018	0.00036	2756.207	0.19758	5.0612	35
36	651.9709	0.0015	0.00030	3300.779	0.19752	5.0628	36
38	934.4891	0.0011	0.00021	4733.301	0.19743	5.0651	38
40	1339.431	0.0007	0.19736	6786.577	0.19736	5.0668	40

TABLE B - 11

DISCRETE CASH FLOW
20.00% CONTINUOUS COMPOUND INTEREST FACTORS
(EFFECTIVE RATE IS 22.1403%)

	SINGLE PAYMENTS		UNIFORM SERIES PAYMENTS				
N	COMPOUND AMOUNT F/P	PRESENT WORTH P/F	SINKING FUND A/F	COMPOUND AMOUNT F/A	CAPITAL RECOVERY A/P	PRESENT WORTH P/A	N
1	1.2214	0.8187	1.00000	1.000	1.22140	0.8187	1
2	1.4918	0.6703	0.45017	2.221	0.67157	1.4891	2
3	1.8221	0.5488	0.26931	3.713	0.49071	2.0379	3
4	2.2255	0.4493	0.18066	5.535	0.40206	2.4872	4
5	2.7183	0.3679	0.12885	7.761	0.35025	2.8551	5
6	3.3201	0.3012	0.09543	10.479	0.31683	3.1563	6
7	4.0552	0.2466	0.07247	13.799	0.29387	3.4029	7
8	4.9530	0.2019	0.05601	17.854	0.27741	3.6048	8
9	6.0496	0.1653	0.04385	22.808	0.26525	3.7701	9
10	7.3891	0.1353	0.03465	28.857	0.25606	3.9054	10
11	9.0250	0.1108	0.02759	36.246	0.24899	4.0162	11
12	11.0232	0.0907	0.02209	45.271	0.24349	4.1069	12
13	13.4637	0.0743	0.01776	56.294	0.23917	4.1812	13
14	16.4446	0.0608	0.01434	69.758	0.23574	4.2420	14
15	20.0855	0.0498	0.01160	86.203	0.23300	4.2918	15
16	24.5325	0.0408	0.00941	106.288	0.23081	4.3325	16
17	29.9641	0.0334	0.00764	130.821	0.22905	4.3659	17
18	36.5982	0.0273	0.00622	160.785	0.22762	4.3932	18
19	44.7012	0.0224	0.00507	197.383	0.22647	4.4156	19
20	54.5982	0.0183	0.00413	242.084	0.22553	4.4339	20
22	81.4509	0.0123	0.00275	363.369	0.22415	4.4612	22
24	121.5104	0.0082	0.00184	544.304	0.22324	4.4795	24
25	148.4132	0.0067	0.00150	665.814	0.22290	4.4862	25
26	181.2722	0.0055	0.00123	814.228	0.22263	4.4917	26
28	270.4264	0.0037	0.00082	1216.906	0.22222	4.5000	28
30	403.4288	0.0025	0.00055	1817.632	0.22195	4.5055	30
32	601.8450	0.0017	0.00037	2713.810	0.22177	4.5092	32
34	897.8473	0.0011	0.00025	4050.750	0.22165	4.5116	34
35	1096.633	0.0009	0.22160	4948.598	0.22160	4.5125	35

TABLE B - 12

DISCRETE CASH FLOW
25.00% CONTINUOUS COMPOUND INTEREST FACTORS
(EFFECTIVE RATE IS 28.4025%)

	SINGLE PAYMENTS		UNIFORM SERIES PAYMENTS				
N	COMPOUND AMOUNT F/P	PRESENT WORTH P/F	SINKING FUND A/F	COMPOUND AMOUNT F/A	CAPITAL RECOVERY A/P	PRESENT WORTH P/A	N
1	1.2840	0.7788	1.00000	1.000	1.28403	0.7788	1
2	1.6487	0.6065	0.43782	2.284	0.72185	1.3853	2
3	2.1170	0.4724	0.25428	3.933	0.53830	1.8577	3
4	2.7183	0.3679	0.16530	6.050	0.44932	2.2256	4
5	3.4903	0.2865	0.11405	8.768	0.39808	2.5121	5
6	4.4817	0.2231	0.08158	12.258	0.36560	2.7352	6
7	5.7546	0.1738	0.05974	16.740	0.34376	2.9090	7
8	7.3891	0.1353	0.04445	22.495	0.32848	3.0443	8
9	9.4877	0.1054	0.03346	29.884	0.31749	3.1497	9
10	12.1825	0.0821	0.02540	39.371	0.30942	3.2318	10
11	15.6426	0.0639	0.01940	51.554	0.30342	3.2957	11
12	20.0855	0.0498	0.01488	67.197	0.29891	3.3455	12
13	25.7903	0.0388	0.01146	87.282	0.29548	3.3843	13
14	33.1155	0.0302	0.00884	113.072	0.29287	3.4145	14
15	42.5211	0.0235	0.00684	146.188	0.29087	3.4380	15
16	54.5982	0.0183	0.00530	188.709	0.28932	3.4563	16
17	70.1054	0.0143	0.00411	243.307	0.28814	3.4706	17
18	90.0171	0.0111	0.00319	313.413	0.28722	3.4817	18
19	115.5843	0.0087	0.00248	403.430	0.28650	3.4904	19
20	148.4132	0.0067	0.00193	519.014	0.28595	3.4971	20
22	244.6919	0.0041	0.00117	857.993	0.28519	3.5064	22
24	403.4288	0.0025	0.00071	1416.876	0.28473	3.5121	24
25	518.0128	0.0019	0.00055	1820.305	0.28457	3.5140	25
26	665.1416	0.0015	0.00043	2338.318	0.28445	3.5155	26
28	1096.633	0.0009	0.00026	3857.518	0.28428	3.5176	28
30	1808.042	0.0006	0.00016	6362.256	0.28418	3.5189	30
32	2980.958	0.0003	0.00010	10491.871	0.28412	3.5196	32
34	4914.769	0.0002	0.00006	17300.455	0.28408	3.5201	34
35	6310.688	0.0002	0.00005	22215.223	0.28407	3.5203	35

TABLE B - 13

DISCRETE CASH FLOW
30.00% CONTINUOUS COMPOUND INTEREST FACTORS
(EFFECTIVE RATE IS 34.9859%)

	SINGLE PAYMENTS		UNIFORM SERIES PAYMENTS				
N	COMPOUND AMOUNT F/P	PRESENT WORTH P/F	SINKING FUND A/F	COMPOUND AMOUNT F/A	CAPITAL RECOVERY A/P	PRESENT WORTH P/A	N
1	1.3499	0.7408	1.00000	1.000	1.34986	0.7408	1
2	1.8221	0.5488	0.42556	2.350	0.77542	1.2896	2
3	2.4596	0.4066	0.23969	4.172	0.58955	1.6962	3
4	3.3201	0.3012	0.15079	6.632	0.50065	1.9974	4
5	4.4817	0.2231	0.10049	9.952	0.45034	2.2205	5
6	6.0496	0.1653	0.06928	14.433	0.41914	2.3858	6
7	8.1662	0.1225	0.04882	20.483	0.39868	2.5083	7
8	11.0232	0.0907	0.03490	28.649	0.38476	2.5990	8
9	14.8797	0.0672	0.02521	39.672	0.37507	2.6662	9
10	20.0855	0.0498	0.01833	54.552	0.36819	2.7160	10
11	27.1126	0.0369	0.01340	74.638	0.36326	2.7529	11
12	36.5982	0.0273	0.00983	101.750	0.35969	2.7802	12
13	49.4024	0.0202	0.00723	138.349	0.35709	2.8004	13
14	66.6863	0.0150	0.00533	187.751	0.35519	2.8154	14
15	90.0171	0.0111	0.00393	254.437	0.35379	2.8265	15
16	121.5104	0.0082	0.00290	344.454	0.35276	2.8348	16
17	164.0219	0.0061	0.00215	465.965	0.35200	2.8409	17
18	221.4064	0.0045	0.00159	629.987	0.35145	2.8454	18
19	298.8674	0.0033	0.00117	851.393	0.35103	2.8487	19
20	403.4288	0.0025	0.00087	1150.261	0.35073	2.8512	20
22	735.0952	0.0014	0.00048	2098.261	0.35034	2.8544	22
24	1339.431	0.0007	0.00026	3825.631	0.35012	2.8562	24
25	1808.042	0.0006	0.00019	5165.062	0.35005	2.8567	25

TABLE B - 14

DISCRETE CASH FLOW
35.00% CONTINUOUS COMPOUND INTEREST FACTORS
(EFFECTIVE RATE IS 41.9068%)

	SINGLE PAYMENTS		UNIFORM SERIES PAYMENTS				
N	COMPOUND AMOUNT F/P	PRESENT WORTH P/F	SINKING FUND A/F	COMPOUND AMOUNT F/A	CAPITAL RECOVERY A/P	PRESENT WORTH P/A	N
1	1.4191	0.7047	1.00000	1.000	1.41907	0.7047	1
2	2.0138	0.4966	0.41338	2.419	0.83245	1.2013	2
3	2.8577	0.3499	0.22559	4.433	0.64466	1.5512	3
4	4.0552	0.2466	0.13717	7.290	0.55623	1.7978	4
5	5.7546	0.1738	0.08814	11.346	0.50721	1.9716	5
6	8.1662	0.1225	0.05848	17.100	0.47755	2.0940	6
7	11.5883	0.0863	0.03958	25.266	0.45865	2.1803	7
8	16.4446	0.0608	0.02713	36.855	0.44620	2.2411	8
9	23.3361	0.0429	0.01876	53.299	0.43783	2.2840	9
10	33.1155	0.0302	0.01305	76.636	0.43212	2.3142	10
11	46.9931	0.0213	0.00911	109.751	0.42818	2.3355	11
12	66.6863	0.0150	0.00638	156.744	0.42545	2.3505	12
13	94.6324	0.0106	0.00448	223.430	0.42354	2.3610	13
14	134.2898	0.0074	0.00314	318.063	0.42221	2.3685	14
15	190.5663	0.0052	0.00221	452.353	0.42128	2.3737	15
16	270.4264	0.0037	0.00156	642.919	0.42062	2.3774	16
17	383.7533	0.0026	0.00109	913.345	0.42016	2.3800	17
18	544.5719	0.0018	0.00077	1297.099	0.41984	2.3819	18
19	772.7843	0.0013	0.00054	1841.670	0.41961	2.3832	19
20	1096.633	0.0009	0.00038	2614.455	0.41945	2.3841	20
22	2208.348	0.0005	0.00019	5267.284	0.41926	2.3852	22
24	4447.067	0.0002	0.00009	10609.427	0.41916	2.3857	24
25	6310.688	0.0002	0.00007	15056.494	0.41913	2.3859	25

TABLE B - 15

DISCRETE CASH FLOW
40.00% CONTINUOUS COMPOUND INTEREST FACTORS
(EFFECTIVE RATE IS 49.1825%)

	SINGLE PAYMENTS		UNIFORM SERIES PAYMENTS				
N	COMPOUND AMOUNT F/P	PRESENT WORTH P/F	SINKING FUND A/F	COMPOUND AMOUNT F/A	CAPITAL RECOVERY A/P	PRESENT WORTH P/A	N
1	1.4918	0.6703	1.00000	1.000	1.49182	0.6703	1
2	2.2255	0.4493	0.40131	2.492	0.89314	1.1196	2
3	3.3201	0.3012	0.21198	4.717	0.70381	1.4208	3
4	4.9530	0.2019	0.12442	8.037	0.61624	1.6227	4
5	7.3891	0.1353	0.07698	12.991	0.56880	1.7581	5
6	11.0232	0.0907	0.04907	20.380	0.54089	1.8488	6
7	16.4446	0.0608	0.03184	31.403	0.52367	1.9096	7
8	24.5325	0.0408	0.02090	47.847	0.51272	1.9504	8
9	36.5982	0.0273	0.01382	72.380	0.50564	1.9777	9
10	54.5982	0.0183	0.00918	108.978	0.50100	1.9960	10
11	81.4509	0.0123	0.00611	163.576	0.49794	2.0083	11
12	121.5104	0.0082	0.00408	245.027	0.49591	2.0165	12
13	181.2722	0.0055	0.00273	366.538	0.49455	2.0220	13
14	270.4264	0.0037	0.00183	547.810	0.49365	2.0257	14
15	403.4288	0.0025	0.00122	818.236	0.49305	2.0282	15
16	601.8450	0.0017	0.00082	1221.665	0.49264	2.0299	16
17	897.8473	0.0011	0.00055	1823.510	0.49237	2.0310	17
18	1339.431	0.0007	0.00037	2721.357	0.49219	2.0317	18
19	1998.196	0.0005	0.00025	4060.788	0.49207	2.0322	19
20	2980.958	0.0003	0.00017	6058.984	0.49199	2.0326	20
22	6634.244	0.0002	0.00007	13487.009	0.49190	2.0329	22
24	14764.78	0.0001	0.00003	30018.382	0.49186	2.0331	24
25	22026.47	0.0000	0.00002	44783.163	0.49185	2.0332	25

TABLE B - 16

DISCRETE CASH FLOW
45.00% CONTINUOUS COMPOUND INTEREST FACTORS
(EFFECTIVE RATE IS 56.8312%)

	SINGLE PAYMENTS		UNIFORM SERIES PAYMENTS				
N	COMPOUND AMOUNT F/P	PRESENT WORTH P/F	SINKING FUND A/F	COMPOUND AMOUNT F/A	CAPITAL RECOVERY A/P	PRESENT WORTH P/A	N
1	1.5683	0.6376	1.00000	1.000	1.56831	0.6376	1
2	2.4596	0.4066	0.38936	2.568	0.95767	1.0442	2
3	3.8574	0.2592	0.19889	5.028	0.76720	1.3034	3
4	6.0496	0.1653	0.11254	8.885	0.68086	1.4687	4
5	9.4877	0.1054	0.06696	14.935	0.63527	1.5741	5
6	14.8797	0.0672	0.04095	24.423	0.60926	1.6413	6
7	23.3361	0.0429	0.02544	39.302	0.59376	1.6842	7
8	36.5982	0.0273	0.01596	62.639	0.58428	1.7115	8
9	57.3975	0.0174	0.01008	99.237	0.57839	1.7289	9
10	90.0171	0.0111	0.00638	156.634	0.57470	1.7400	10
11	141.1750	0.0071	0.00405	246.651	0.57237	1.7471	11
12	221.4064	0.0045	0.00258	387.826	0.57089	1.7516	12
13	347.2344	0.0029	0.00164	609.233	0.56995	1.7545	13
14	544.5719	0.0018	0.00105	956.467	0.56936	1.7564	14
15	854.0588	0.0012	0.00067	1501.039	0.56898	1.7575	15
16	1339.431	0.0007	0.00042	2355.098	0.56874	1.7583	16
17	2100.646	0.0005	0.00027	3694.529	0.56858	1.7588	17
18	3294.468	0.0003	0.00017	5795.174	0.56848	1.7591	18
19	5166.754	0.0002	0.00011	9089.642	0.56842	1.7593	19
20	8103.084	0.0001	0.00007	14256.397	0.56838	1.7594	20
22	19930.37	0.0001	0.00003	35067.646	0.56834	1.7595	22
24	49020.80	0.0000	0.00001	86255.059	0.56832	1.7596	24

TABLE B - 17

DISCRETE CASH FLOW
50.00% CONTINUOUS COMPOUND INTEREST FACTORS
(EFFECTIVE RATE IS 64.8721%)

	SINGLE PAYMENTS		UNIFORM SERIES PAYMENTS				
N	COMPOUND AMOUNT F/P	PRESENT WORTH P/F	SINKING FUND A/F	COMPOUND AMOUNT F/A	CAPITAL RECOVERY A/P	PRESENT WORTH P/A	N
1	1.6487	0.6065	1.00000	1.000	1.64872	0.6065	1
2	2.7183	0.3679	0.37754	2.649	1.02626	0.9744	2
3	4.4817	0.2231	0.18632	5.367	0.83504	1.1975	3
4	7.3891	0.1353	0.10154	9.849	0.75026	1.3329	4
5	12.1825	0.0821	0.05801	17.238	0.70673	1.4150	5
6	20.0855	0.0498	0.03399	29.420	0.68271	1.4647	6
7	33.1155	0.0302	0.02020	49.506	0.66892	1.4949	7
8	54.5982	0.0183	0.01210	82.621	0.66082	1.5133	8
9	90.0171	0.0111	0.00729	137.219	0.65601	1.5244	9
10	148.4132	0.0067	0.00440	227.237	0.65312	1.5311	10
11	244.6919	0.0041	0.00266	375.650	0.65138	1.5352	11
12	403.4288	0.0025	0.00161	620.342	0.65033	1.5377	12
13	665.1416	0.0015	0.00098	1023.770	0.64970	1.5392	13
14	1096.633	0.0009	0.00059	1688.912	0.64931	1.5401	14
15	1808.042	0.0006	0.00036	2785.545	0.64908	1.5406	15
16	2980.958	0.0003	0.00022	4593.588	0.64894	1.5410	16
17	4914.769	0.0002	0.00013	7574.546	0.64885	1.5412	17
18	8103.084	0.0001	0.00008	12489.314	0.64880	1.5413	18
19	13359.73	0.0001	0.00005	20592.398	0.64877	1.5414	19
20	22026.47	0.0000	0.00003	33952.125	0.64875	1.5414	20

TABLE B - 18

PRESENT WORTH GRADIENT FACTORS (P/G)
DISCRETE CASH FLOW, CONTINUOUS COMPOUNDING

N	2% (2.020)	5% (5.127)	8% (8.329)	10% (10.517)	12% (12.750)	15% (16.183)	N
2	0.961	0.905	0.852	0.819	0.787	0.741	2
3	2.844	2.626	2.425	2.300	2.182	2.016	3
4	5.614	5.082	4.604	4.311	4.038	3.663	4
5	9.233	8.198	7.285	6.737	6.234	5.552	5
6	13.668	11.902	10.379	9.482	8.667	7.585	6
7	18.884	16.130	13.806	12.461	11.258	9.684	7
8	24.849	20.822	17.457	15.606	13.938	11.793	8
9	31.531	25.923	21.391	18.859	16.655	13.867	9
10	38.900	31.382	25.435	22.170	19.365	15.875	10
11	46.925	37.151	29.583	25.499	22.037	17.795	11
12	55.578	43.188	33.795	28.812	24.643	19.614	12
13	64.830	49.453	38.036	32.082	27.165	21.321	13
14	74.655	55.908	42.278	35.288	29.587	22.913	14
15	85.027	62.522	46.495	38.412	31.902	24.389	15
16	95.919	69.262	50.665	41.440	34.101	25.749	16
17	107.307	76.100	54.772	44.363	36.181	26.999	17
18	119.168	83.012	58.800	47.173	38.142	28.141	18
19	131.477	89.973	62.737	49.865	39.983	29.182	19
20	144.214	96.963	66.573	52.437	41.706	30.128	20
21	157.354	103.962	70.300	54.886	43.316	30.985	21
22	170.879	110.952	73.913	57.213	44.814	31.760	22
23	184.767	117.918	77.407	59.418	46.207	32.458	23
24	198.999	124.845	80.779	61.505	47.498	33.087	24
25	213.556	131.722	84.027	63.475	48.693	33.651	25
26	228.419	138.535	87.150	65.332	49.797	34.157	26
27	243.571	145.275	90.149	67.079	50.815	34.610	27
28	258.993	151.933	93.023	68.721	51.753	35.015	28
29	274.670	158.501	95.775	70.262	52.615	35.376	29
30	290.586	164.972	98.406	71.705	53.408	35.699	30
31	306.724	171.339	100.918	73.057	54.135	35.985	31
32	323.070	177.598	103.314	74.321	54.801	36.241	32
33	339.610	183.744	105.598	75.501	55.411	36.467	33
34	356.328	189.772	107.772	76.602	55.969	36.668	34
35	373.212	195.681	109.839	77.629	56.479	36.847	35
36	390.248	201.466	111.804	78.585	56.944	37.005	36
37	407.424	207.127	113.669	79.475	57.369	37.145	37
38	424.728	212.661	115.439	80.303	57.756	37.269	38
39	442.147	218.067	117.117	81.072	58.109	37.378	39
40	459.671	223.345	118.707	81.786	58.430	37.475	40
42	494.989	233.515	121.636	83.064	58.987	37.635	42
44	530.597	243.172	124.256	84.162	59.447	37.760	44
46	566.420	252.321	126.593	85.103	59.826	37.857	46
48	602.384	260.972	128.674	85.908	60.138	37.932	48
50	638.425	269.136	130.524	86.596	60.393	37.990	50

TABLE B - 19

PRESENT WORTH GRADIENT FACTORS (P/G)
DISCRETE CASH FLOW, CONTINUOUS COMPOUNDING

N	18% (19.722)	20% (22.140)	25% (28.403)	30% (34.986)	40% (49.182)	50% (64.872)	N
2	0.698	0.670	0.607	0.549	0.449	0.368	2
3	1.863	1.768	1.551	1.362	1.052	0.814	3
4	3.323	3.116	2.655	2.266	1.657	1.220	4
5	4.950	4.587	3.801	3.158	2.199	1.548	5
6	6.648	6.093	4.917	3.985	2.652	1.797	6
7	8.350	7.573	5.959	4.719	3.017	1.979	7
8	10.008	8.986	6.907	5.354	3.303	2.107	8
9	11.591	10.309	7.750	5.892	3.521	2.196	9
10	13.079	11.527	8.489	6.340	3.686	2.256	10
11	14.460	12.635	9.128	6.709	3.809	2.297	11
12	15.728	13.633	9.675	7.009	3.899	2.324	12
13	16.884	14.524	10.141	7.252	3.965	2.343	13
14	17.930	15.314	10.533	7.447	4.014	2.354	14
15	18.871	16.011	10.863	7.603	4.048	2.362	15
16	19.713	16.623	11.137	7.726	4.073	2.367	16
17	20.463	17.157	11.366	7.824	4.091	2.370	17
18	21.129	17.621	11.554	7.901	4.104	2.372	18
19	21.718	18.024	11.710	7.961	4.113	2.374	19
20	22.237	18.372	11.838	8.008	4.119	2.375	20
21	22.693	18.672	11.943	8.045	4.124	2.375	21
22	23.094	18.930	12.029	8.073	4.127	2.376	22
23	23.444	19.151	12.099	8.095	4.129	2.376	23
24	23.750	19.340	12.156	8.113	4.130	2.376	24
25	24.017	19.502	12.202	8.126	4.132	2.376	25
26	24.249	19.640	12.240	8.136	4.132	2.376	26
27	24.450	19.757	12.270	8.144	4.133	2.376	27
28	24.625	19.857	12.295	8.150	4.133	2.376	28
29	24.776	19.942	12.315	8.155	4.134	2.376	29
30	24.907	20.014	12.331	8.158	4.134	2.376	30
31	25.020	20.075	12.344	8.161	4.134	2.376	31
32	25.118	20.126	12.354	8.163	4.134	2.376	32
33	25.202	20.170	12.363	8.165	4.134	2.376	33
34	25.275	20.206	12.369	8.166	4.134	2.376	34
35	25.337	20.237	12.375	8.167	4.134	2.376	35
36	25.391	20.264	12.379	8.168	4.134	2.376	36
37	25.437	20.286	12.382	8.168	4.134	2.376	37
38	25.477	20.304	12.385	8.169	4.134	2.376	38
39	25.511	20.320	12.387	8.169	4.134	2.376	39
40	25.540	20.333	12.389	8.169	4.134	2.376	40
42	25.586	20.353	12.392	8.169	4.134	2.376	42
44	25.620	20.367	12.393	8.170	4.134	2.376	44
46	25.645	20.377	12.394	8.170	4.134	2.376	46
48	25.663	20.384	12.395	8.170	4.134	2.376	48
50	25.676	20.389	12.395	8.170	4.134	2.376	50

TABLE B - 20

ANNUAL COST GRADIENT FACTORS (A/G)
DISCRETE CASH FLOW, CONTINUOUS COMPOUNDING

N	2% (2.020)	5% (5.127)	8% (8.329)	10% (10.517)	12% (12.750)	15% (16.183)	N
2	0.495	0.488	0.480	0.475	0.470	0.463	2
3	0.987	0.967	0.947	0.933	0.920	0.900	3
4	1.475	1.438	1.400	1.375	1.351	1.314	4
5	1.960	1.900	1.840	1.801	1.761	1.703	5
6	2.442	2.354	2.268	2.210	2.153	2.068	6
7	2.920	2.800	2.682	2.603	2.526	2.411	7
8	3.395	3.238	3.083	2.981	2.880	2.731	8
9	3.867	3.668	3.471	3.342	3.215	3.029	9
10	4.335	4.089	3.847	3.689	3.533	3.307	10
11	4.800	4.503	4.210	4.020	3.834	3.564	11
12	5.262	4.908	4.561	4.336	4.117	3.803	12
13	5.720	5.305	4.900	4.638	4.385	4.023	13
14	6.175	5.694	5.227	4.926	4.636	4.226	14
15	6.627	6.075	5.541	5.200	4.873	4.412	15
16	7.076	6.449	5.845	5.461	5.095	4.583	16
17	7.521	6.814	6.137	5.709	5.302	4.739	17
18	7.963	7.172	6.418	5.944	5.497	4.882	18
19	8.401	7.522	6.688	6.167	5.679	5.013	19
20	8.837	7.865	6.947	6.378	5.848	5.131	20
21	9.269	8.200	7.196	6.578	6.006	5.239	21
22	9.698	8.527	7.435	6.767	6.153	5.337	22
23	10.123	8.847	7.664	6.945	6.289	5.425	23
24	10.545	9.160	7.884	7.114	6.416	5.505	24
25	10.964	9.465	8.094	7.273	6.533	5.577	25
26	11.380	9.764	8.295	7.422	6.642	5.642	26
27	11.793	10.055	8.487	7.563	6.743	5.700	27
28	12.202	10.339	8.671	7.695	6.836	5.753	28
29	12.608	10.617	8.846	7.820	6.922	5.800	29
30	13.011	10.888	9.014	7.936	7.001	5.842	30
31	13.410	11.152	9.173	8.046	7.073	5.880	31
32	13.807	11.409	9.326	8.149	7.140	5.914	32
33	14.200	11.660	9.471	8.245	7.202	5.944	33
34	14.590	11.905	9.609	8.334	7.259	5.971	34
35	14.976	12.143	9.741	8.419	7.310	5.995	35
36	15.360	12.375	9.866	8.497	7.358	6.016	36
37	15.741	12.601	9.985	8.570	7.402	6.035	37
38	16.118	12.821	10.098	8.639	7.442	6.052	38
39	16.492	13.035	10.205	8.703	7.478	6.067	39
40	16.863	13.243	10.307	8.762	7.511	6.080	40
42	17.596	13.643	10.495	8.869	7.570	6.102	42
44	18.316	14.021	10.665	8.961	7.618	6.119	44
46	19.024	14.378	10.816	9.041	7.658	6.133	46
48	19.719	14.715	10.952	9.110	7.692	6.143	48
50	20.403	15.033	11.074	9.169	7.719	6.151	50

TABLE B - 21

ANNUAL CCST GRADIENT FACTORS (A/G)
DISCRETE CASH FLOW, CONTINUOUS COMPOUNDING

N	18% (19.722)	20% (22.140)	25% (28.403)	30% (34.986)	40% (49.182)	50% (64.872)	N
2	0.455	0.450	0.438	0.426	0.401	0.378	2
3	0.881	0.868	0.835	0.803	0.740	0.680	3
4	1.277	1.253	1.193	1.134	1.021	0.915	4
5	1.645	1.607	1.513	1.422	1.251	1.094	5
6	1.985	1.931	1.798	1.670	1.435	1.227	6
7	2.299	2.225	2.049	1.881	1.580	1.324	7
8	2.587	2.493	2.269	2.060	1.693	1.392	8
9	2.850	2.734	2.460	2.210	1.780	1.440	9
10	3.090	2.951	2.627	2.334	1.847	1.474	10
11	3.309	3.146	2.770	2.437	1.897	1.496	11
12	3.506	3.319	2.892	2.521	1.934	1.512	12
13	3.685	3.474	2.996	2.590	1.961	1.522	13
14	3.846	3.610	3.085	2.645	1.981	1.529	14
15	3.990	3.731	3.160	2.690	1.996	1.533	15
16	4.119	3.837	3.222	2.726	2.007	1.536	16
17	4.234	3.930	3.275	2.754	2.014	1.538	17
18	4.337	4.011	3.319	2.777	2.020	1.539	18
19	4.428	4.082	3.355	2.795	2.024	1.540	19
20	4.509	4.144	3.385	2.809	2.027	1.541	20
21	4.580	4.197	3.410	2.820	2.029	1.541	21
22	4.643	4.243	3.431	2.828	2.030	1.541	22
23	4.698	4.283	3.447	2.835	2.031	1.541	23
24	4.747	4.318	3.461	2.840	2.032	1.541	24
25	4.790	4.347	3.472	2.844	2.032	1.541	25
26	4.827	4.372	3.482	2.848	2.032	1.541	26
27	4.860	4.394	3.489	2.850	2.033	1.541	27
28	4.888	4.413	3.495	2.852	2.033	1.541	28
29	4.913	4.429	3.500	2.853	2.033	1.541	29
30	4.934	4.442	3.504	2.855	2.033	1.541	30
31	4.953	4.454	3.507	2.855	2.033	1.541	31
32	4.969	4.463	3.510	2.856	2.033	1.541	32
33	4.983	4.472	3.512	2.857	2.033	1.541	33
34	4.996	4.479	3.514	2.857	2.033	1.541	34
35	5.006	4.485	3.515	2.857	2.033	1.541	35
36	5.015	4.490	3.516	2.858	2.033	1.541	36
37	5.023	4.494	3.517	2.858	2.033	1.541	37
38	5.030	4.498	3.518	2.858	2.033	1.541	38
39	5.036	4.501	3.519	2.858	2.033	1.541	39
40	5.041	4.503	3.519	2.858	2.033	1.541	40
42	5.049	4.507	3.520	2.858	2.033	1.541	42
44	5.055	4.510	3.520	2.858	2.033	1.541	44
46	5.059	4.512	3.520	2.858	2.033	1.541	46
48	5.062	4.513	3.521	2.858	2.033	1.541	48
50	5.064	4.514	3.521	2.858	2.033	1.541	50

APPENDIX

C

INTEREST FACTORS FOR CONTINUOUS COMPOUNDING, CONTINUOUS CASH FLOW

Tabulated here are the interest factors with the nominal interest rate r compounded continuously and cash flow taking place continuously throughout each period (Sec. 3.8 of text). The symbol $\bar{A}$ is used to indicate the total flow amount accumulated in one period with a portion of $\bar{A}$ occurring in each time increment during the period. The computational forms of the factors are given here.

Factor	Notation	Formula
Single-payment compound amount	$(F/P, r\%, n)$	e^{rn}
Single-payment present worth	$(P/F, r\%, n)$	e^{-rn}
Sinking fund	$(\bar{A}/F, r\%, n)$	$\dfrac{r}{e^{rn}-1}$
Uniform-series compound amount	$(F/\bar{A}, r\%, n)$	$\dfrac{e^{rn}-1}{r}$
Capital recovery	$(\bar{A}/P, r\%, n)$	$\dfrac{re^{rn}}{e^{rn}-1}$
Uniform-series present worth	$(P/\bar{A}, r\%, n)$	$\dfrac{e^{rn}-1}{re^{rn}}$

TABLE C - 1

CONTINUOUS CASH FLOW
1.00% CONTINUOUS COMPOUND INTEREST FACTORS
(EFFECTIVE RATE IS 1.0050%)

	SINGLE PAYMENTS		UNIFORM SERIES PAYMENTS				
	COMPOUND AMOUNT	PRESENT WORTH	SINKING FUND	COMPOUND AMOUNT	CAPITAL RECOVERY	PRESENT WORTH	
N	F/P	P/F	$\bar{A}$/F	F/$\bar{A}$	$\bar{A}$/P	P/$\bar{A}$	N
1	1.0101	0.9900	0.99501	1.0050	1.00501	0.9950	1
2	1.0202	0.9802	0.49502	2.0201	0.50502	1.9801	2
3	1.0305	0.9704	0.32836	3.0455	0.33836	2.9554	3
4	1.0408	0.9608	0.24503	4.0811	0.25503	3.9211	4
5	1.0513	0.9512	0.19504	5.1271	0.20504	4.8771	5
6	1.0618	0.9418	0.16172	6.1837	0.17172	5.8235	6
7	1.0725	0.9324	0.13792	7.2508	0.14792	6.7606	7
8	1.0833	0.9231	0.12007	8.3287	0.13007	7.6884	8
9	1.0942	0.9139	0.10619	9.4174	0.11619	8.6069	9
10	1.1052	0.9048	0.09508	10.5171	0.10508	9.5163	10
11	1.1163	0.8958	0.08600	11.6278	0.09600	10.4166	11
12	1.1275	0.8869	0.07843	12.7497	0.08843	11.3080	12
13	1.1388	0.8781	0.07203	13.8828	0.08203	12.1905	13
14	1.1503	0.8694	0.06655	15.0274	0.07655	13.0642	14
15	1.1618	0.8607	0.06179	16.1834	0.07179	13.9292	15
16	1.1735	0.8521	0.05763	17.3511	0.06763	14.7856	16
17	1.1853	0.8437	0.05397	18.5305	0.06397	15.6335	17
18	1.1972	0.8353	0.05071	19.7217	0.06071	16.4730	18
19	1.2092	0.8270	0.04779	20.9250	0.05779	17.3041	19
20	1.2214	0.8187	0.04517	22.1403	0.05517	18.1269	20
22	1.2461	0.8025	0.04064	24.6077	0.05064	19.7481	22
24	1.2712	0.7866	0.03687	27.1249	0.04687	21.3372	24
25	1.2840	0.7788	0.03521	28.4025	0.04521	22.1199	25
26	1.2969	0.7711	0.03368	29.6930	0.04368	22.8948	26
28	1.3231	0.7558	0.03095	32.3130	0.04095	24.4216	28
30	1.3499	0.7408	0.02858	34.9859	0.03858	25.9182	30
32	1.3771	0.7261	0.02652	37.7128	0.03652	27.3851	32
34	1.4049	0.7118	0.02469	40.4948	0.03469	28.8230	34
35	1.4191	0.7047	0.02386	41.9068	0.03386	29.5312	35
36	1.4333	0.6977	0.02308	43.3329	0.03308	30.2324	36
38	1.4623	0.6839	0.02163	46.2285	0.03163	31.6139	38
40	1.4918	0.6703	0.02033	49.1825	0.03033	32.9680	40
45	1.5683	0.6376	0.01760	56.8312	0.02760	36.2372	45
50	1.6487	0.6065	0.01541	64.8721	0.02541	39.3469	50
45	1.5683	0.6376	0.01760	56.8312	0.02760	36.2372	45
50	1.6487	0.6065	0.01541	64.8721	0.02541	39.3469	50
55	1.7333	0.5769	0.01364	73.3253	0.02364	42.3050	55
60	1.8221	0.5488	0.01216	82.2119	0.02216	45.1188	60
65	1.9155	0.5220	0.01092	91.5541	0.02092	47.7954	65
70	2.0138	0.4966	0.00986	101.3753	0.01986	50.3415	70
75	2.1170	0.4724	0.00895	111.7000	0.01895	52.7633	75
80	2.2255	0.4493	0.00816	122.5541	0.01816	55.0671	80
85	2.3396	0.4274	0.00746	133.9647	0.01746	57.2585	85
90	2.4596	0.4066	0.00685	145.9603	0.01685	59.3430	90
95	2.5857	0.3867	0.00631	158.5710	0.01631	61.3259	95
100	2.7183	0.3679	0.00582	171.8282	0.01582	63.2121	100

TABLE C - 2

CONTINUOUS CASH FLOW
2.00% CONTINUOUS COMPOUND INTEREST FACTORS
(EFFECTIVE RATE IS 2.0201%)

	SINGLE PAYMENTS		UNIFORM SERIES PAYMENTS				
N	COMPOUND AMOUNT F/P	PRESENT WORTH P/F	SINKING FUND $\bar{A}$/F	COMPOUND AMOUNT F/$\bar{A}$	CAPITAL RECOVERY $\bar{A}$/P	PRESENT WORTH P/$\bar{A}$	N
1	1.0202	0.9802	0.99003	1.0101	1.01003	0.9901	1
2	1.0408	0.9608	0.49007	2.0405	0.51007	1.9605	2
3	1.0618	0.9418	0.32343	3.0918	0.34343	2.9118	3
4	1.0833	0.9231	0.24013	4.1644	0.26013	3.8442	4
5	1.1052	0.9048	0.19017	5.2585	0.21017	4.7581	5
6	1.1275	0.8869	0.15687	6.3748	0.17687	5.6540	6
7	1.1503	0.8694	0.13309	7.5137	0.15309	6.5321	7
8	1.1735	0.8521	0.11527	8.6755	0.13527	7.3928	8
9	1.1972	0.8353	0.10141	9.8609	0.12141	8.2365	9
10	1.2214	0.8187	0.09033	11.0701	0.11033	9.0635	10
11	1.2461	0.8025	0.08128	12.3038	0.10128	9.8741	11
12	1.2712	0.7866	0.07373	13.5625	0.09373	10.6686	12
13	1.2969	0.7711	0.06736	14.8465	0.08736	11.4474	13
14	1.3231	0.7558	0.06189	16.1565	0.08189	12.2108	14
15	1.3499	0.7408	0.05717	17.4929	0.07717	12.9591	15
16	1.3771	0.7261	0.05303	18.8564	0.07303	13.6925	16
17	1.4049	0.7118	0.04939	20.2474	0.06939	14.4115	17
18	1.4333	0.6977	0.04615	21.6665	0.06615	15.1162	18
19	1.4623	0.6839	0.04326	23.1142	0.06326	15.8069	19
20	1.4918	0.6703	0.04066	24.5912	0.06066	16.4840	20
22	1.5527	0.6440	0.03619	27.6354	0.05619	17.7982	22
24	1.6161	0.6188	0.03246	30.8037	0.05246	19.0608	24
25	1.6487	0.6065	0.03083	32.4361	0.05083	19.6735	25
26	1.6820	0.5945	0.02932	34.1014	0.04932	20.2740	26
28	1.7507	0.5712	0.02664	37.5336	0.04664	21.4395	28
30	1.8221	0.5488	0.02433	41.1059	0.04433	22.5594	30
32	1.8965	0.5273	0.02231	44.8240	0.04231	23.6354	32
34	1.9739	0.5066	0.02054	48.6939	0.04054	24.6692	34
35	2.0138	0.4966	0.01973	50.6876	0.03973	25.1707	35
36	2.0544	0.4868	0.01897	52.7217	0.03897	25.6624	36
38	2.1383	0.4677	0.01757	56.9138	0.03757	26.6167	38
40	2.2255	0.4493	0.01632	61.2770	0.03632	27.5336	40
45	2.4596	0.4066	0.01370	72.9802	0.03370	29.6715	45
50	2.7183	0.3679	0.01164	85.9141	0.03164	31.6060	50
45	2.4596	0.4066	0.01370	72.9802	0.03370	29.6715	45
50	2.7183	0.3679	0.01164	85.9141	0.03164	31.6060	50
55	3.0042	0.3329	0.00998	100.2083	0.02998	33.3564	55
60	3.3201	0.3012	0.00862	116.0058	0.02862	34.9403	60
65	3.6693	0.2725	0.00749	133.4648	0.02749	36.3734	65
70	4.0552	0.2466	0.00655	152.7600	0.02655	37.6702	70
75	4.4817	0.2231	0.00574	174.0845	0.02574	38.8435	75
80	4.9530	0.2019	0.00506	197.6516	0.02506	39.9052	80
85	5.4739	0.1827	0.00447	223.6974	0.02447	40.8658	85
90	6.0496	0.1653	0.00396	252.4824	0.02396	41.7351	90
95	6.6859	0.1496	0.00352	284.2947	0.02352	42.5216	95
100	7.3891	0.1353	0.00313	319.4528	0.02313	43.2332	100

TABLE C - 3

CONTINUOUS CASH FLOW
4.00% CONTINUOUS COMPOUND INTEREST FACTORS
(EFFECTIVE RATE IS 4.0811%)

	SINGLE PAYMENTS		UNIFORM SERIES PAYMENTS				
N	COMPOUND AMOUNT F/P	PRESENT WORTH P/F	SINKING FUND $\bar{A}/F$	COMPOUND AMOUNT $F/\bar{A}$	CAPITAL RECOVERY $\bar{A}/P$	PRESENT WORTH $P/\bar{A}$	N
1	1.0408	0.9608	C.98013	1.020	1.02013	0.9803	1
2	1.0833	0.9231	C.48027	2.082	0.52027	1.9221	2
3	1.1275	0.8869	0.31373	3.187	0.35373	2.8270	3
4	1.1735	0.8521	0.23053	4.338	0.27053	3.6964	4
5	1.2214	0.8187	0.18067	5.535	0.22067	4.5317	5
6	1.2712	0.7866	0.14747	6.781	0.18747	5.3343	6
7	1.3231	0.7558	0.12379	8.078	0.16379	6.1054	7
8	1.3771	0.7261	0.10606	9.428	0.14606	6.8463	8
9	1.4333	0.6977	0.09231	10.833	0.13231	7.5581	9
10	1.4918	0.6703	0.08133	12.296	0.12133	8.2420	10
11	1.5527	0.6440	0.07237	13.818	0.11237	8.8991	11
12	1.6161	0.6188	0.06493	15.402	0.10493	9.5304	12
13	1.6820	0.5945	0.05865	17.051	0.09865	10.1370	13
14	1.7507	0.5712	0.05329	18.767	0.09329	10.7198	14
15	1.8221	0.5488	0.04865	20.553	0.08865	11.2797	15
16	1.8965	0.5273	0.04462	22.412	0.08462	11.8177	16
17	1.9739	0.5066	0.04107	24.347	0.08107	12.3346	17
18	2.0544	0.4868	0.03794	26.361	0.07794	12.8312	18
19	2.1383	0.4677	0.03514	28.457	0.07514	13.3083	19
20	2.2255	0.4493	0.03264	30.639	0.07264	13.7668	20
22	2.4109	0.4148	0.02835	35.272	0.06835	14.6304	22
24	2.6117	0.3829	0.02482	40.292	0.06482	15.4277	24
25	2.7183	0.3679	0.02328	42.957	0.06328	15.8030	25
26	2.8292	0.3535	0.02187	45.730	0.06187	16.1636	26
28	3.0649	0.3263	0.01937	51.621	0.05937	16.8430	28
30	3.3201	0.3012	0.01724	58.003	0.05724	17.4701	30
32	3.5966	0.2780	0.01540	64.916	0.05540	18.0491	32
34	3.8962	0.2567	0.01381	72.405	0.05381	18.5835	34
35	4.0552	0.2466	0.01309	76.380	0.05309	18.8351	35
36	4.2207	0.2369	0.01242	80.517	0.05242	19.0768	36
38	4.5722	0.2187	0.01120	89.306	0.05120	19.5322	38
40	4.9530	0.2019	0.01012	98.826	0.05012	19.9526	40
45	6.0496	0.1653	0.00792	126.241	0.04792	20.8675	45
50	7.3891	0.1353	0.00626	159.726	0.04626	21.6166	50
45	6.0496	0.1653	0.00792	126.241	0.04792	20.8675	45
50	7.3891	0.1353	0.00626	159.726	0.04626	21.6166	50
55	9.0250	0.1108	0.00498	200.625	0.04498	22.2299	55
60	11.0232	0.0907	0.00399	250.579	0.04399	22.7321	60
65	13.4637	0.0743	0.00321	311.593	0.04321	23.1432	65
70	16.4446	0.0608	0.00259	386.116	0.04259	23.4797	70
75	20.0855	0.0498	0.00210	477.138	0.04210	23.7553	75
80	24.5325	0.0408	0.00170	588.313	0.04170	23.9809	80
85	29.9641	0.0334	0.00138	724.103	0.04138	24.1657	85
90	36.5982	0.0273	0.00112	889.956	0.04112	24.3169	90
95	44.7012	0.0224	0.00092	1092.530	0.04092	24.4407	95
100	54.5982	0.0183	0.00075	1339.954	0.04075	24.5421	100

TABLE C - 4

CONTINUOUS CASH FLOW
5.00% CONTINUOUS COMPOUND INTEREST FACTORS
(EFFECTIVE RATE IS 5.1271%)

	SINGLE PAYMENTS		UNIFORM SERIES PAYMENTS				
N	COMPOUND AMOUNT F/P	PRESENT WORTH P/F	SINKING FUND $\bar{A}/F$	COMPOUND AMOUNT $F/\bar{A}$	CAPITAL RECOVERY $\bar{A}/P$	PRESENT WORTH $P/\bar{A}$	N
1	1.0513	0.9512	0.97521	1.025	1.02521	0.9754	1
2	1.1052	0.9048	0.47542	2.103	0.52542	1.9033	2
3	1.1618	0.8607	0.30896	3.237	0.35896	2.7858	3
4	1.2214	0.8187	0.22583	4.428	0.27583	3.6254	4
5	1.2840	0.7788	0.17604	5.681	0.22604	4.4240	5
6	1.3499	0.7408	0.14291	6.997	0.19291	5.1836	6
7	1.4191	0.7047	0.11931	8.381	0.16931	5.9062	7
8	1.4918	0.6703	0.10166	9.836	0.15166	6.5936	8
9	1.5683	0.6376	0.08798	11.366	0.13798	7.2474	9
10	1.6487	0.6065	0.07707	12.974	0.12707	7.8694	10
11	1.7333	0.5769	0.06819	14.665	0.11819	8.4610	11
12	1.8221	0.5488	0.06082	16.442	0.11082	9.0238	12
13	1.9155	0.5220	0.05461	18.311	0.10461	9.5591	13
14	2.0138	0.4966	0.04932	20.275	0.09932	10.0683	14
15	2.1170	0.4724	0.04476	22.340	0.09476	10.5527	15
16	2.2255	0.4493	0.04080	24.511	0.09080	11.0134	16
17	2.3396	0.4274	0.03732	26.793	0.08732	11.4517	17
18	2.4596	0.4066	0.03426	29.192	0.08426	11.8686	18
19	2.5857	0.3867	0.03153	31.714	0.08153	12.2652	19
20	2.7183	0.3679	0.02910	34.366	0.07910	12.6424	20
22	3.0042	0.3329	0.02495	40.083	0.07495	13.3426	22
24	3.3201	0.3012	0.02155	46.402	0.07155	13.9761	24
25	3.4903	0.2865	0.02008	49.807	0.07008	14.2699	25
26	3.6693	0.2725	0.01873	53.386	0.06873	14.5494	26
28	4.0552	0.2466	0.01637	61.104	0.06637	15.0681	28
30	4.4817	0.2231	0.01436	69.634	0.06436	15.5374	30
32	4.9530	0.2019	0.01265	79.061	0.06265	15.9621	32
34	5.4739	0.1827	0.01118	89.479	0.06118	16.3463	34
35	5.7546	0.1738	0.01052	95.092	0.06052	16.5245	35
36	6.0496	0.1653	0.00990	100.993	0.05990	16.6940	36
38	6.6859	0.1496	0.00879	113.718	0.05879	17.0086	38
40	7.3891	0.1353	0.00783	127.781	0.05783	17.2933	40
45	9.4877	0.1054	0.00589	169.755	0.05589	17.8920	45
50	12.1825	0.0821	0.00447	223.650	0.05447	18.3583	50
45	9.4877	0.1054	0.00589	169.755	0.05589	17.8920	45
50	12.1825	0.0821	0.00447	223.650	0.05447	18.3583	50
55	15.6426	0.0639	0.00341	292.853	0.05341	18.7214	55
60	20.0855	0.0498	0.00262	381.711	0.05262	19.0043	60
65	25.7903	0.0388	0.00202	495.807	0.05202	19.2245	65
70	33.1155	0.0302	0.00156	642.309	0.05156	19.3961	70
75	42.5211	0.0235	0.00120	830.422	0.05120	19.5296	75
80	54.5982	0.0183	0.00093	1071.963	0.05093	19.6337	80
85	70.1054	0.0143	0.00072	1382.108	0.05072	19.7147	85
90	90.0171	0.0111	0.00056	1780.343	0.05056	19.7778	90
95	115.584	0.0087	0.00044	2291.686	0.05044	19.8270	95
100	148.413	0.0067	0.00034	2948.263	0.05034	19.8652	100

TABLE C - 5

CONTINUOUS CASH FLOW
6.00% CONTINUOUS COMPOUND INTEREST FACTORS
(EFFECTIVE RATE IS 6.1837%)

	SINGLE PAYMENTS		UNIFORM SERIES PAYMENTS				
N	COMPOUND AMOUNT F/P	PRESENT WORTH P/F	SINKING FUND $\bar{A}/F$	COMPOUND AMOUNT $F/\bar{A}$	CAPITAL RECOVERY $\bar{A}/P$	PRESENT WORTH $P/\bar{A}$	N
1	1.0618	0.9418	0.97030	1.031	1.03030	0.9706	1
2	1.1275	0.8869	0.47060	2.125	0.53060	1.8847	2
3	1.1972	0.8353	0.30423	3.287	0.36423	2.7455	3
4	1.2712	0.7866	0.22120	4.521	0.28120	3.5562	4
5	1.3499	0.7408	0.17150	5.831	0.23150	4.3197	5
6	1.4333	0.6977	0.13846	7.222	0.19846	5.0387	6
7	1.5220	0.6570	0.11495	8.699	0.17495	5.7159	7
8	1.6161	0.6188	0.09739	10.268	0.15739	6.3536	8
9	1.7160	0.5827	0.08380	11.933	0.14380	6.9542	9
10	1.8221	0.5488	0.07298	13.702	0.13298	7.5198	10
11	1.9348	0.5169	0.06419	15.580	0.12419	8.0525	11
12	2.0544	0.4868	0.05690	17.574	0.11690	8.5541	12
13	2.1815	0.4584	0.05078	19.691	0.11078	9.0266	13
14	2.3164	0.4317	0.04558	21.939	0.10558	9.4715	14
15	2.4596	0.4066	0.04111	24.327	0.10111	9.8905	15
16	2.6117	0.3829	0.03723	26.862	0.09723	10.2851	16
17	2.7732	0.3606	0.03384	29.553	0.09384	10.6568	17
18	2.9447	0.3396	0.03085	32.411	0.09085	11.0067	18
19	3.1268	0.3198	0.02821	35.446	0.08821	11.3363	19
20	3.3201	0.3012	0.02586	38.669	0.08586	11.6468	20
22	3.7434	0.2671	0.02187	45.724	0.08187	12.2144	22
24	4.2207	0.2369	0.01863	53.678	0.07863	12.7179	24
25	4.4817	0.2231	0.01723	58.028	0.07723	12.9478	25
26	4.7588	0.2101	0.01596	62.647	0.07596	13.1644	26
28	5.3656	0.1864	0.01374	72.759	0.07374	13.5604	28
30	6.0496	0.1653	0.01188	84.161	0.07188	13.9117	30
32	6.8210	0.1466	0.01031	97.016	0.07031	14.2232	32
34	7.6906	0.1300	0.00897	111.510	0.06897	14.4995	34
35	8.1662	0.1225	0.00837	119.436	0.06837	14.6257	35
36	8.6711	0.1153	0.00782	127.852	0.06782	14.7446	36
38	9.7767	0.1023	0.00684	146.278	0.06684	14.9619	38
40	11.0232	0.0907	0.00599	167.053	0.06599	15.1547	40
45	14.8797	0.0672	0.00432	231.329	0.06432	15.5466	45
50	20.0855	0.0498	0.00314	318.092	0.06314	15.8369	50
45	14.8797	0.0672	0.00432	231.329	0.06432	15.5466	45
50	20.0855	0.0498	0.00314	318.092	0.06314	15.8369	50
55	27.1126	0.0369	0.00230	435.211	0.06230	16.0519	55
60	36.5982	0.0273	0.00169	593.304	0.06169	16.2113	60
65	49.4024	0.0202	0.00124	806.707	0.06124	16.3293	65
70	66.6863	0.0150	0.00091	1094.772	0.06091	16.4167	70
75	90.0171	0.0111	0.00067	1483.619	0.06067	16.4815	75
80	121.510	0.0082	0.00050	2008.507	0.06050	16.5295	80
85	164.022	0.0061	0.00037	2717.032	0.06037	16.5651	85
90	221.406	0.0045	0.00027	3673.440	0.06027	16.5914	90
95	298.867	0.0033	0.00020	4964.457	0.06020	16.6109	95
100	403.429	0.0025	0.00015	6707.147	0.06015	16.6254	100

TABLE C - 6

CONTINUOUS CASH FLOW
8.00% CONTINUOUS COMPOUND INTEREST FACTORS
(EFFECTIVE RATE IS 8.3287%)

	SINGLE PAYMENTS		UNIFORM SERIES PAYMENTS				
N	COMPOUND AMOUNT F/P	PRESENT WORTH P/F	SINKING FUND $\bar{A}$/F	COMPOUND AMOUNT F/$\bar{A}$	CAPITAL RECOVERY $\bar{A}$/P	PRESENT WORTH P/$\bar{A}$	N
1	1.0833	0.9231	0.96053	1.041	1.04053	0.9610	1
2	1.1735	0.8521	0.46107	2.169	0.54107	1.8482	2
3	1.2712	0.7866	0.29493	3.391	0.37493	2.6672	3
4	1.3771	0.7261	0.21213	4.714	0.29213	3.4231	4
5	1.4918	0.6703	0.16266	6.148	0.24266	4.1210	5
6	1.6161	0.6188	0.12985	7.701	0.20985	4.7652	6
7	1.7507	0.5712	0.10657	9.383	0.18657	5.3599	7
8	1.8965	0.5273	0.08924	11.206	0.16924	5.9088	8
9	2.0544	0.4868	0.07587	13.180	0.15587	6.4156	9
10	2.2255	0.4493	0.06528	15.319	0.14528	6.8834	10
11	2.4109	0.4148	0.05670	17.636	0.13670	7.3152	11
12	2.6117	0.3829	0.04964	20.146	0.12964	7.7138	12
13	2.8292	0.3535	0.04373	22.865	0.12373	8.0818	13
14	3.0649	0.3263	0.03874	25.811	0.11874	8.4215	14
15	3.3201	0.3012	0.03448	29.001	0.11448	8.7351	15
16	3.5966	0.2780	0.03081	32.458	0.11081	9.0245	16
17	3.8962	0.2567	0.02762	36.202	0.10762	9.2917	17
18	4.2207	0.2369	0.02484	40.259	0.10484	9.5384	18
19	4.5722	0.2187	0.02240	44.653	0.10240	9.7661	19
20	4.9530	0.2019	0.02024	49.413	0.10024	9.9763	20
22	5.8124	0.1720	0.01662	60.155	0.09662	10.3494	22
24	6.8210	0.1466	0.01374	72.762	0.09374	10.6674	24
25	7.3891	0.1353	0.01252	79.863	0.09252	10.8083	25
26	8.0045	0.1249	0.01142	87.556	0.09142	10.9384	26
28	9.3933	0.1065	0.00953	104.917	0.08953	11.1693	28
30	11.0232	0.0907	0.00798	125.290	0.08798	11.3660	30
32	12.9358	0.0773	0.00670	149.198	0.08670	11.5337	32
34	15.1803	0.0659	0.00564	177.254	0.08564	11.6766	34
35	16.4446	0.0608	0.00518	193.058	0.08518	11.7399	35
36	17.8143	0.0561	0.00476	210.178	0.08476	11.7983	36
38	20.9052	0.0478	0.00402	248.816	0.08402	11.9021	38
40	24.5325	0.0408	0.00340	294.157	0.08340	11.9905	40
45	36.5982	0.0273	0.00225	444.978	0.08225	12.1585	45
50	54.5982	0.0183	0.00149	669.977	0.08149	12.2711	50
45	36.5982	0.0273	0.00225	444.978	0.08225	12.1585	45
50	54.5982	0.0183	0.00149	669.977	0.08149	12.2711	50
55	81.4509	0.0123	0.00099	1005.636	0.08099	12.3465	55
60	121.510	0.0082	0.00066	1506.380	0.08066	12.3971	60
65	181.272	0.0055	0.00044	2253.403	0.08044	12.4310	65
70	270.426	0.0037	0.00030	3367.830	0.08030	12.4538	70
75	403.429	0.0025	0.00020	5030.360	0.08020	12.4690	75
80	601.845	0.0017	0.00013	7510.563	0.08013	12.4792	80
85	897.847	0.0011	0.00009	11210.591	0.08009	12.4861	85
90	1339.431	0.0007	0.00006	16730.385	0.08006	12.4907	90
95	1998.196	0.0005	0.00004	24964.949	0.08004	12.4937	95
100	2980.958	0.0003	0.00003	37249.475	0.08003	12.4958	100

TABLE C - 7

CONTINUOUS CASH FLOW
10.00% CONTINUOUS COMPOUND INTEREST FACTORS
(EFFECTIVE RATE IS 10.5171%)

	SINGLE PAYMENTS		UNIFORM SERIES PAYMENTS				
N	COMPOUND AMOUNT F/P	PRESENT WORTH P/F	SINKING FUND $\bar{A}/F$	COMPOUND AMOUNT $F/\bar{A}$	CAPITAL RECOVERY $\bar{A}/P$	PRESENT WORTH $P/\bar{A}$	N
1	1.1052	0.9048	0.95083	1.052	1.05083	0.9516	1
2	1.2214	0.8187	0.45167	2.214	0.55167	1.8127	2
3	1.3499	0.7408	0.28583	3.499	0.38583	2.5918	3
4	1.4918	0.6703	0.20332	4.918	0.30332	3.2968	4
5	1.6487	0.6065	0.15415	6.487	0.25415	3.9347	5
6	1.8221	0.5488	0.12164	8.221	0.22164	4.5119	6
7	2.0138	0.4966	0.09864	10.138	0.19864	5.0341	7
8	2.2255	0.4493	0.08160	12.255	0.18160	5.5067	8
9	2.4596	0.4066	0.06851	14.596	0.16851	5.9343	9
10	2.7183	0.3679	0.05820	17.183	0.15820	6.3212	10
11	3.0042	0.3329	0.04990	20.042	0.14990	6.6713	11
12	3.3201	0.3012	0.04310	23.201	0.14310	6.9881	12
13	3.6693	0.2725	0.03746	26.693	0.13746	7.2747	13
14	4.0552	0.2466	0.03273	30.552	0.13273	7.5340	14
15	4.4817	0.2231	0.02872	34.817	0.12872	7.7687	15
16	4.9530	0.2019	0.02530	39.530	0.12530	7.9810	16
17	5.4739	0.1827	0.02235	44.739	0.12235	8.1732	17
18	6.0496	0.1653	0.01980	50.496	0.11980	8.3470	18
19	6.6859	0.1496	0.01759	56.859	0.11759	8.5043	19
20	7.3891	0.1353	0.01565	63.891	0.11565	8.6466	20
22	9.0250	0.1108	0.01246	80.250	0.11246	8.8920	22
24	11.0232	0.0907	0.00998	100.232	0.10998	9.0928	24
25	12.1825	0.0821	0.00894	111.825	0.10894	9.1792	25
26	13.4637	0.0743	0.00802	124.637	0.10802	9.2573	26
28	16.4446	0.0608	0.00647	154.446	0.10647	9.3919	28
30	20.0855	0.0498	0.00524	190.855	0.10524	9.5021	30
32	24.5325	0.0408	0.00425	235.325	0.10425	9.5924	32
34	29.9641	0.0334	0.00345	289.641	0.10345	9.6663	34
35	33.1155	0.0302	0.00311	321.155	0.10311	9.6980	35
36	36.5982	0.0273	0.00281	355.982	0.10281	9.7268	36
38	44.7012	0.0224	0.00229	437.012	0.10229	9.7763	38
40	54.5982	0.0183	0.00187	535.982	0.10187	9.8168	40
45	90.0171	0.0111	0.00112	890.171	0.10112	9.8889	45
50	148.413	0.0067	0.00068	1474.132	0.10068	9.9326	50
45	90.0171	0.0111	0.00112	890.171	0.10112	9.8889	45
50	148.413	0.0067	0.00068	1474.132	0.10068	9.9326	50
55	244.692	0.0041	0.00041	2436.919	0.10041	9.9591	55
60	403.429	0.0025	0.00025	4024.288	0.10025	9.9752	60
65	665.142	0.0015	0.00015	6641.416	0.10015	9.9850	65
70	1096.633	0.0009	0.00009	10956.332	0.10009	9.9909	70
75	1808.042	0.0006	0.00006	18070.424	0.10006	9.9945	75
80	2980.958	0.0003	0.00003	29799.580	0.10003	9.9966	80
85	4914.769	0.0002	0.00002	49137.688	0.10002	9.9980	85
90	8103.084	0.0001	0.00001	81020.839	0.10001	9.9988	90

TABLE C - 8

CONTINUOUS CASH FLOW
12.00% CONTINUOUS COMPOUND INTEREST FACTORS
(EFFECTIVE RATE IS 12.7497%)

	SINGLE PAYMENTS		UNIFORM SERIES PAYMENTS				
N	COMPOUND AMOUNT F/P	PRESENT WORTH P/F	SINKING FUND $\bar{A}/F$	COMPOUND AMOUNT $F/\bar{A}$	CAPITAL RECOVERY $\bar{A}/P$	PRESENT WORTH $P/\bar{A}$	N
1	1.1275	0.8869	0.94120	1.062	1.06120	0.9423	1
2	1.2712	0.7866	0.44240	2.260	0.56240	1.7781	2
3	1.4333	0.6977	0.27693	3.611	0.39693	2.5194	3
4	1.6161	0.6188	0.19478	5.134	0.31478	3.1768	4
5	1.8221	0.5488	0.14596	6.851	0.26596	3.7599	5
6	2.0544	0.4868	0.11381	8.787	0.23381	4.2771	6
7	2.3164	0.4317	0.09116	10.970	0.21116	4.7357	7
8	2.6117	0.3829	0.07446	13.431	0.19446	5.1426	8
9	2.9447	0.3396	0.06171	16.206	0.18171	5.5034	9
10	3.3201	0.3012	0.05172	19.334	0.17172	5.8234	10
11	3.7434	0.2671	0.04374	22.862	0.16374	6.1072	11
12	4.2207	0.2369	0.03726	26.839	0.15726	6.3589	12
13	4.7588	0.2101	0.03192	31.324	0.15192	6.5822	13
14	5.3656	0.1864	0.02749	36.380	0.14749	6.7802	14
15	6.0496	0.1653	0.02376	42.080	0.14376	6.9558	15
16	6.8210	0.1466	0.02062	48.508	0.14062	7.1116	16
17	7.6906	0.1300	0.01794	55.755	0.13794	7.2498	17
18	8.6711	0.1153	0.01564	63.926	0.13564	7.3723	18
19	9.7767	0.1023	0.01367	73.139	0.13367	7.4810	19
20	11.0232	0.0907	0.01197	83.526	0.13197	7.5774	20
22	14.0132	0.0714	0.00922	108.443	0.12922	7.7387	22
24	17.8143	0.0561	0.00714	140.119	0.12714	7.8655	24
25	20.0855	0.0498	0.00629	159.046	0.12629	7.9184	25
26	22.6464	0.0442	0.00554	180.386	0.12554	7.9654	26
28	28.7892	0.0347	0.00432	231.577	0.12432	8.0439	28
30	36.5982	0.0273	0.00337	296.652	0.12337	8.1056	30
32	46.5255	0.0215	0.00264	379.379	0.12264	8.1542	32
34	59.1455	0.0169	0.00206	484.546	0.12206	8.1924	34
35	66.6863	0.0150	0.00183	547.386	0.12183	8.2084	35
36	75.1886	0.0133	0.00162	618.239	0.12162	8.2225	36
38	95.5835	0.0105	0.00127	788.196	0.12127	8.2461	38
40	121.510	0.0082	0.00100	1004.253	0.12100	8.2648	40

TABLE C - 9

CONTINUOUS CASH FLOW
15.00% CONTINUOUS COMPOUND INTEREST FACTORS
(EFFECTIVE RATE IS 16.1834%)

	SINGLE PAYMENTS		UNIFORM SERIES PAYMENTS				
N	COMPOUND AMOUNT F/P	PRESENT WORTH P/F	SINKING FUND $\bar{A}/F$	COMPOUND AMOUNT $F/\bar{A}$	CAPITAL RECOVERY $\bar{A}/P$	PRESENT WORTH $P/\bar{A}$	N
1	1.1618	0.8607	0.92687	1.079	1.07687	0.9286	1
2	1.3499	0.7408	0.42874	2.332	0.57874	1.7279	2
3	1.5683	0.6376	0.26394	3.789	0.41394	2.4158	3
4	1.8221	0.5488	0.18246	5.481	0.33246	3.0079	4
5	2.1170	0.4724	0.13429	7.447	0.28429	3.5176	5
6	2.4596	0.4066	0.10277	9.731	0.25277	3.9562	6
7	2.8577	0.3499	0.08075	12.384	0.23075	4.3337	7
8	3.3201	0.3012	0.06465	15.467	0.21465	4.6587	8
9	3.8574	0.2592	0.05249	19.050	0.20249	4.9384	9
10	4.4817	0.2231	0.04308	23.211	0.19308	5.1791	10
11	5.2070	0.1920	0.03566	28.047	0.18566	5.3863	11
12	6.0496	0.1653	0.02971	33.664	0.17971	5.5647	12
13	7.0287	0.1423	0.02488	40.191	0.17488	5.7182	13
14	8.1662	0.1225	0.02093	47.774	0.17093	5.8503	14
15	9.4877	0.1054	0.01767	56.585	0.16767	5.9640	15
16	11.0232	0.0907	0.01497	66.821	0.16497	6.0619	16
17	12.8071	0.0781	0.01270	78.714	0.16270	6.1461	17
18	14.8797	0.0672	0.01081	92.532	0.16081	6.2186	18
19	17.2878	0.0578	0.00921	108.585	0.15921	6.2810	19
20	20.0855	0.0498	0.00786	127.237	0.15786	6.3348	20
22	27.1126	0.0369	0.00574	174.084	0.15574	6.4208	22
24	36.5982	0.0273	0.00421	237.322	0.15421	6.4845	24
25	42.5211	0.0235	0.00361	276.807	0.15361	6.5099	25
26	49.4024	0.0202	0.00310	322.683	0.15310	6.5317	26
28	66.6863	0.0150	0.00228	437.909	0.15228	6.5667	28
30	90.0171	0.0111	0.00169	593.448	0.15169	6.5926	30
32	121.510	0.0082	0.00124	803.403	0.15124	6.6118	32
34	164.022	0.0061	0.00092	1086.813	0.15092	6.6260	34
35	190.566	0.0052	0.00079	1263.775	0.15079	6.6317	35
36	221.406	0.0045	0.00068	1469.376	0.15068	6.6366	36
38	298.867	0.0033	0.00050	1985.783	0.15050	6.6444	38
40	403.429	0.0025	0.00037	2682.859	0.15037	6.6501	40

TABLE C - 10

CONTINUOUS CASH FLOW
18.00% CONTINUOUS COMPOUND INTEREST FACTORS
(EFFECTIVE RATE IS 19.7217%)

	SINGLE PAYMENTS		UNIFORM SERIES PAYMENTS				
N	COMPOUND AMOUNT F/P	PRESENT WORTH P/F	SINKING FUND $\bar{A}/F$	COMPOUND AMOUNT $F/\bar{A}$	CAPITAL RECOVERY $\bar{A}/P$	PRESENT WORTH $P/\bar{A}$	N
1	1.1972	0.8353	0.91270	1.096	1.09270	0.9152	1
2	1.4333	0.6977	0.41539	2.407	0.59539	1.6796	2
3	1.7160	0.5827	0.25139	3.978	0.43139	2.3181	3
4	2.0544	0.4868	0.17071	5.858	0.35071	2.8514	4
5	2.4596	0.4066	0.12332	8.109	0.30332	3.2968	5
6	2.9447	0.3396	0.09256	10.804	0.27256	3.6689	6
7	3.5254	0.2837	0.07128	14.030	0.25128	3.9797	7
8	4.2207	0.2369	0.05589	17.893	0.23589	4.2393	8
9	5.0531	0.1979	0.04441	22.517	0.22441	4.4561	9
10	6.0496	0.1653	0.03565	28.054	0.21565	4.6372	10
11	7.2427	0.1381	0.02883	34.682	0.20883	4.7885	11
12	8.6711	0.1153	0.02346	42.617	0.20346	4.9149	12
13	10.3812	0.0963	0.01919	52.118	0.19919	5.0204	13
14	12.4286	0.0805	0.01575	63.492	0.19575	5.1086	14
15	14.8797	0.0672	0.01297	77.110	0.19297	5.1822	15
16	17.8143	0.0561	0.01071	93.413	0.19071	5.2437	16
17	21.3276	0.0469	0.00885	112.931	0.18885	5.2951	17
18	25.5337	0.0392	0.00734	136.298	0.18734	5.3380	18
19	30.5694	0.0327	0.00609	164.275	0.18609	5.3738	19
20	36.5982	0.0273	0.00506	197.768	0.18506	5.4038	20
22	52.4573	0.0191	0.00350	285.874	0.18350	5.4496	22
24	75.1886	0.0133	0.00243	412.159	0.18243	5.4817	24
25	90.0171	0.0111	0.00202	494.540	0.18202	5.4938	25
26	107.7701	0.0093	0.00169	593.167	0.18169	5.5040	26
28	154.4700	0.0065	0.00117	852.611	0.18117	5.5196	28
30	221.4064	0.0045	0.00082	1224.480	0.18082	5.5305	30
32	317.3483	0.0032	0.00057	1757.491	0.18057	5.5380	32
34	454.8647	0.0022	0.00040	2521.471	0.18040	5.5433	34
35	544.5719	0.0018	0.00033	3019.844	0.18033	5.5454	35
36	651.9709	0.0015	0.00028	3616.505	0.18028	5.5470	36
38	934.4991	0.0011	0.00019	5186.051	0.18019	5.5496	38
40	1339.431	0.0007	0.18013	7435.726	0.18013	5.5514	40

TABLE C - 11

CONTINUOUS CASH FLOW
20.00% CONTINUOUS COMPOUND INTEREST FACTORS
(EFFECTIVE RATE IS 22.1403%)

	SINGLE PAYMENTS		UNIFORM SERIES PAYMENTS				
N	COMPOUND AMOUNT F/P	PRESENT WORTH P/F	SINKING FUND $\bar{A}$/F	COMPOUND AMOUNT F/$\bar{A}$	CAPITAL RECOVERY $\bar{A}$/P	PRESENT WORTH P/$\bar{A}$	N
1	1.2214	0.8187	0.90333	1.107	1.10333	0.9063	1
2	1.4918	0.6703	0.40665	2.459	0.60665	1.6484	2
3	1.8221	0.5488	0.24327	4.111	0.44327	2.2559	3
4	2.2255	0.4493	0.16319	6.128	0.36319	2.7534	4
5	2.7183	0.3679	0.11640	8.591	0.31640	3.1606	5
6	3.3201	0.3012	0.08620	11.601	0.28620	3.4940	6
7	4.0552	0.2466	0.06546	15.276	0.26546	3.7670	7
8	4.9530	0.2019	0.05059	19.765	0.25059	3.9905	8
9	6.0496	0.1653	0.03961	25.248	0.23961	4.1735	9
10	7.3891	0.1353	0.03130	31.945	0.23130	4.3233	10
11	9.0250	0.1108	0.02492	40.125	0.22492	4.4460	11
12	11.0232	0.0907	0.01995	50.116	0.21995	4.5464	12
13	13.4637	0.0743	0.01605	62.319	0.21605	4.6286	13
14	16.4446	0.0608	0.01295	77.223	0.21295	4.6959	14
15	20.0855	0.0498	0.01048	95.428	0.21048	4.7511	15
16	24.5325	0.0408	0.00850	117.663	0.20850	4.7962	16
17	29.9641	0.0334	0.00691	144.821	0.20691	4.8331	17
18	36.5982	0.0273	0.00562	177.991	0.20562	4.8634	18
19	44.7012	0.0224	0.00458	218.506	0.20458	4.8881	19
20	54.5982	0.0183	0.00373	267.991	0.20373	4.9084	20
22	81.4509	0.0123	0.00249	402.254	0.20249	4.9386	22
24	121.5104	0.0082	0.00166	602.552	0.20166	4.9589	24
25	148.4132	0.0067	0.00136	737.066	0.20136	4.9663	25
26	181.2722	0.0055	0.00111	901.361	0.20111	4.9724	26
28	270.4264	0.0037	0.00074	1347.132	0.20074	4.9815	28
30	403.4288	0.0025	0.00050	2012.144	0.20050	4.9876	30
32	601.8450	0.0017	0.00033	3004.225	0.20033	4.9917	32
34	897.8473	0.0011	0.00022	4484.236	0.20022	4.9944	34
35	1096.633	0.0009	0.20018	5478.166	0.20018	4.9954	35

TABLE C - 12

CONTINUOUS CASH FLOW
25.00% CONTINUOUS COMPOUND INTEREST FACTORS
(EFFECTIVE RATE IS 28.4025%)

	SINGLE PAYMENTS		UNIFORM SERIES PAYMENTS				
N	COMPOUND AMOUNT F/P	PRESENT WORTH P/F	SINKING FUND $\bar{A}/F$	COMPOUND AMOUNT $F/\bar{A}$	CAPITAL RECOVERY $\bar{A}/P$	PRESENT WORTH $P/\bar{A}$	N
1	1.2840	0.7788	0.88020	1.136	1.13020	0.8848	1
2	1.6487	0.6065	0.38537	2.595	0.63537	1.5739	2
3	2.1170	0.4724	0.22381	4.468	0.47381	2.1105	3
4	2.7183	0.3679	0.14549	6.873	0.39549	2.5285	4
5	3.4903	0.2865	0.10039	9.961	0.35039	2.8540	5
6	4.4817	0.2231	0.07180	13.927	0.32180	3.1075	6
7	5.7546	0.1738	0.05258	19.018	0.30258	3.3049	7
8	7.3891	0.1353	0.03913	25.556	0.28913	3.4587	8
9	9.4877	0.1054	0.02945	33.951	0.27945	3.5784	9
10	12.1825	0.0821	0.02236	44.730	0.27236	3.6717	10
11	15.6426	0.0639	0.01707	58.571	0.26707	3.7443	11
12	20.0855	0.0498	0.01310	76.342	0.26310	3.8009	12
13	25.7903	0.0388	0.01008	99.161	0.26008	3.8449	13
14	33.1155	0.0302	0.00778	128.462	0.25778	3.8792	14
15	42.5211	0.0235	0.00602	166.084	0.25602	3.9059	15
16	54.5982	0.0183	0.00466	214.393	0.25466	3.9267	16
17	70.1054	0.0143	0.00362	276.422	0.25362	3.9429	17
18	90.0171	0.0111	0.00281	356.069	0.25281	3.9556	18
19	115.5843	0.0087	0.00218	458.337	0.25218	3.9654	19
20	148.4132	0.0067	0.00170	589.653	0.25170	3.9730	20
22	244.6919	0.0041	0.00103	974.768	0.25103	3.9837	22
24	403.4288	0.0025	0.00062	1609.715	0.25062	3.9901	24
25	518.0128	0.0019	0.00048	2068.051	0.25048	3.9923	25
26	665.1416	0.0015	0.00038	2656.567	0.25038	3.9940	26
28	1096.633	0.0009	0.00023	4382.533	0.25023	3.9964	28
30	1808.042	0.0006	0.00014	7228.170	0.25014	3.9978	30
32	2980.958	0.0003	0.00008	11919.832	0.25008	3.9987	32
34	4914.769	0.0002	0.00005	19655.075	0.25005	3.9992	34
35	6310.688	0.0002	0.00004	25238.752	0.25004	3.9994	35

TABLE C - 13

CONTINUOUS CASH FLOW
30.00% CONTINUOUS COMPOUND INTEREST FACTORS
(EFFECTIVE RATE IS 34.9859%)

	SINGLE PAYMENTS		UNIFORM SERIES PAYMENTS				
N	COMPOUND AMOUNT F/P	PRESENT WORTH P/F	SINKING FUND $\bar{A}/F$	COMPOUND AMOUNT $F/\bar{A}$	CAPITAL RECOVERY $\bar{A}/P$	PRESENT WORTH $P/\bar{A}$	N
1	1.3499	0.7408	0.85749	1.166	1.15749	0.8639	1
2	1.8221	0.5488	0.36491	2.740	0.66491	1.5040	2
3	2.4596	0.4066	0.20554	4.865	0.50554	1.9781	3
4	3.3201	0.3012	0.12930	7.734	0.42930	2.3294	4
5	4.4817	0.2231	0.08617	11.606	0.38617	2.5896	5
6	6.0496	0.1653	0.05941	16.832	0.35941	2.7823	6
7	8.1662	0.1225	0.04186	23.887	0.34186	2.9251	7
8	11.0232	0.0907	0.02993	33.411	0.32993	3.0309	8
9	14.8797	0.0672	0.02161	46.266	0.32161	3.1093	9
10	20.0855	0.0498	0.01572	63.618	0.31572	3.1674	10
11	27.1126	0.0369	0.01149	87.042	0.31149	3.2104	11
12	36.5982	0.0273	0.00843	118.661	0.30843	3.2423	12
13	49.4024	0.0202	0.00620	161.341	0.30620	3.2659	13
14	66.6863	0.0150	0.00457	218.954	0.30457	3.2833	14
15	90.0171	0.0111	0.00337	296.724	0.30337	3.2963	15
16	121.5104	0.0082	0.00249	401.701	0.30249	3.3059	16
17	164.0219	0.0061	0.00184	543.406	0.30184	3.3130	17
18	221.4064	0.0045	0.00136	734.688	0.30136	3.3183	18
19	298.8674	0.0033	0.00101	992.891	0.30101	3.3222	19
20	403.4288	0.0025	0.00075	1341.429	0.30075	3.3251	20
22	735.0952	0.0014	0.00041	2446.984	0.30041	3.3288	22
24	1339.431	0.0007	0.00022	4461.436	0.30022	3.3308	24
25	1808.042	0.0006	0.00017	6023.475	0.30017	3.3315	25

TABLE C - 14

CONTINUOUS CASH FLOW
35.00% CONTINUOUS COMPOUND INTEREST FACTORS
(EFFECTIVE RATE IS 41.9068%)

	SINGLE PAYMENTS		UNIFORM SERIES PAYMENTS				
N	COMPOUND AMOUNT F/P	PRESENT WORTH P/F	SINKING FUND $\bar{A}/F$	COMPOUND AMOUNT $F/\bar{A}$	CAPITAL RECOVERY $\bar{A}/P$	PRESENT WORTH $P/\bar{A}$	N
1	1.4191	0.7047	0.83519	1.197	1.18519	0.8437	1
2	2.0138	0.4966	0.34525	2.896	0.69525	1.4383	2
3	2.8577	0.3499	0.18841	5.308	0.53841	1.8573	3
4	4.0552	0.2466	0.11456	8.729	0.46456	2.1526	4
5	5.7546	0.1738	0.07361	13.585	0.42361	2.3606	5
6	8.1662	0.1225	0.04884	20.475	0.39884	2.5073	6
7	11.5883	0.0863	0.03306	30.252	0.38306	2.6106	7
8	16.4446	0.0608	0.02266	44.128	0.37266	2.6834	8
9	23.3361	0.0429	0.01567	63.817	0.36567	2.7347	9
10	33.1155	0.0302	0.01090	91.758	0.36090	2.7709	10
11	46.9931	0.0213	0.00761	131.409	0.35761	2.7963	11
12	66.6863	0.0150	0.00533	187.675	0.35533	2.8143	12
13	94.6324	0.0106	0.00374	267.521	0.35374	2.8270	13
14	134.2898	0.0074	0.00263	380.828	0.35263	2.8359	14
15	190.5663	0.0052	0.00185	541.618	0.35185	2.8421	15
16	270.4264	0.0037	0.00130	769.790	0.35130	2.8466	16
17	383.7533	0.0026	0.00091	1093.581	0.35091	2.8497	17
18	544.5719	0.0018	0.00064	1553.063	0.35064	2.8519	18
19	772.7843	0.0013	0.00045	2205.098	0.35045	2.8534	19
20	1096.633	0.0009	0.00032	3130.380	0.35032	2.8545	20
22	2208.348	0.0005	0.00016	6306.709	0.35016	2.8558	22
24	4447.067	0.0002	0.00008	12703.048	0.35008	2.8565	24
25	6310.688	0.0002	0.00006	18027.680	0.35006	2.8567	25

TABLE C - 15

CONTINUOUS CASH FLOW
40.00% CONTINUOUS COMPOUND INTEREST FACTORS
(EFFECTIVE RATE IS 49.1825%)

	SINGLE PAYMENTS		UNIFORM SERIES PAYMENTS				
N	COMPOUND AMOUNT F/P	PRESENT WORTH P/F	SINKING FUND $\bar{A}/F$	COMPOUND AMOUNT $F/\bar{A}$	CAPITAL RECOVERY $\bar{A}/P$	PRESENT WORTH $P/\bar{A}$	N
1	1.4918	0.6703	0.81330	1.230	1.21330	0.8242	1
2	2.2255	0.4493	0.32639	3.064	0.72639	1.3767	2
3	3.3201	0.3012	0.17241	5.800	0.57241	1.7470	3
4	4.9530	0.2019	0.10119	9.883	0.50119	1.9953	4
5	7.3891	0.1353	0.06261	15.973	0.46261	2.1617	5
6	11.0232	0.0907	0.03991	25.058	0.43991	2.2732	6
7	16.4446	0.0608	0.02590	38.612	0.42590	2.3480	7
8	24.5325	0.0408	0.01700	58.831	0.41700	2.3981	8
9	36.5982	0.0273	0.01124	88.996	0.41124	2.4317	9
10	54.5982	0.0183	0.00746	133.995	0.40746	2.4542	10
11	81.4509	0.0123	0.00497	201.127	0.40497	2.4693	11
12	121.5104	0.0082	0.00332	301.276	0.40332	2.4794	12
13	181.2722	0.0055	0.00222	450.681	0.40222	2.4862	13
14	270.4264	0.0037	0.00148	673.566	0.40148	2.4908	14
15	403.4288	0.0025	0.00099	1006.072	0.40099	2.4938	15
16	601.8450	0.0017	0.00067	1502.113	0.40067	2.4958	16
17	897.8473	0.0011	0.00045	2242.118	0.40045	2.4972	17
18	1339.431	0.0007	0.00030	3346.077	0.40030	2.4981	18
19	1998.196	0.0005	0.00020	4992.990	0.40020	2.4987	19
20	2980.958	0.0003	0.00013	7449.895	0.40013	2.4992	20
22	6634.244	0.0002	0.00006	16583.110	0.40006	2.4996	22
24	14764.78	0.0001	0.00003	36909.454	0.40003	2.4998	24
25	22026.47	0.0000	0.00002	55063.664	0.40002	2.4999	25

TABLE C - 16

CONTINUOUS CASH FLOW
45.00% CONTINUOUS COMPOUND INTEREST FACTORS
(EFFECTIVE RATE IS 56.8312%)

	SINGLE PAYMENTS		UNIFORM SERIES PAYMENTS				
N	COMPOUND AMOUNT F/P	PRESENT WORTH P/F	SINKING FUND $\bar{A}/F$	COMPOUND AMOUNT $F/\bar{A}$	CAPITAL RECOVERY $\bar{A}/P$	PRESENT WORTH $P/\bar{A}$	N
1	1.5683	0.6376	0.79182	1.263	1.24182	0.8053	1
2	2.4596	0.4066	0.30830	3.244	0.75830	1.3187	2
3	3.8574	0.2592	0.15748	6.350	0.60748	1.6461	3
4	6.0496	0.1653	0.08912	11.221	0.53912	1.8549	4
5	9.4877	0.1054	0.05302	18.862	0.50302	1.9880	5
6	14.8797	0.0672	0.03242	30.844	0.48242	2.0729	6
7	23.3361	0.0429	0.02015	49.636	0.47015	2.1270	7
8	36.5982	0.0273	0.01264	79.107	0.46264	2.1615	8
9	57.3975	0.0174	0.00798	125.328	0.45798	2.1835	9
10	90.0171	0.0111	0.00506	197.816	0.45506	2.1975	10
11	141.1750	0.0071	0.00321	311.500	0.45321	2.2065	11
12	221.4064	0.0045	0.00204	489.792	0.45204	2.2122	12
13	347.2344	0.0029	0.00130	769.410	0.45130	2.2158	13
14	544.5719	0.0018	0.00083	1207.938	0.45083	2.2181	14
15	854.0588	0.0012	0.00053	1895.686	0.45053	2.2196	15
16	1339.431	0.0007	0.00034	2974.291	0.45034	2.2206	16
17	2100.646	0.0005	0.00021	4665.879	0.45021	2.2212	17
18	3294.468	0.0003	0.00014	7318.818	0.45014	2.2215	18
19	5166.754	0.0002	0.00009	11479.454	0.45009	2.2218	19
20	8103.084	0.0001	0.00006	18004.631	0.45006	2.2219	20
22	19930.37	0.0001	0.00002	44287.490	0.45002	2.2221	22
24	49020.80	0.0000	0.00001	108932.89	0.45001	2.2222	24

TABLE C - 17

CONTINUOUS CASH FLOW
50.00% CONTINUOUS COMPOUND INTEREST FACTORS
(EFFECTIVE RATE IS 64.8721%)

	SINGLE PAYMENTS		UNIFORM SERIES PAYMENTS				
N	COMPOUND AMOUNT F/P	PRESENT WORTH P/F	SINKING FUND $\bar{A}/F$	COMPOUND AMOUNT $F/\bar{A}$	CAPITAL RECOVERY $\bar{A}/P$	PRESENT WORTH $P/\bar{A}$	N
1	1.6487	0.6065	0.77075	1.297	1.27075	0.7869	1
2	2.7183	0.3679	0.29099	3.437	0.79099	1.2642	2
3	4.4817	0.2231	0.14361	6.963	0.64361	1.5537	3
4	7.3891	0.1353	0.07826	12.778	0.57826	1.7293	4
5	12.1825	0.0821	0.04471	22.365	0.54471	1.8358	5
6	20.0855	0.0498	0.02620	38.171	0.52620	1.9004	6
7	33.1155	0.0302	0.01557	64.231	0.51557	1.9396	7
8	54.5982	0.0183	0.00933	107.196	0.50933	1.9634	8
9	90.0171	0.0111	0.00562	178.034	0.50562	1.9778	9
10	148.4132	0.0067	0.00339	294.826	0.50339	1.9865	10
11	244.6919	0.0041	0.00205	487.384	0.50205	1.9918	11
12	403.4288	0.0025	0.00124	804.858	0.50124	1.9950	12
13	665.1416	0.0015	0.00075	1328.283	0.50075	1.9970	13
14	1096.633	0.0009	0.00046	2191.266	0.50046	1.9982	14
15	1808.042	0.0006	0.00028	3614.085	0.50028	1.9989	15
16	2980.958	0.0003	0.00017	5959.916	0.50017	1.9993	16
17	4914.769	0.0002	0.00010	9827.538	0.50010	1.9996	17
18	8103.084	0.0001	0.00006	16204.168	0.50006	1.9998	18
19	13359.73	0.0001	0.00004	26717.454	0.50004	1.9999	19
20	22026.47	0.0000	0.00002	44050.932	0.50002	1.9999	20

APPENDIX

D

BASICS OF ACCOUNTING

The overall objective of this appendix is to give the engineering economist an idea about the basics of financial statements and cost-accounting records, the information they contain, and how that information may be used in an economic study.

By no means is this material thorough enough to permit the engineering economist who has mastered it to practice accounting. However, some of the basic approaches of cost accounting discussed here will assist in gathering data for the study to be performed.

SECTION OBJECTIVES

To understand this material you must be able to do the following:

D.1. For a *balance sheet*, name and explain the major accounting categories and state the basic equations used in its preparation. Given the financial data, prepare a balance sheet.

D.2. For an *income statement* and a *cost-of-goods-sold statement*, name and explain the major categories and state the basic equations used in their preparation. Given the financial data, prepare each of these statements.

D.3. Compute and give a meaning for each of the following business ratios, given a balance sheet, income statement, and cost-of-goods-sold statement: current ratio, acid-test ratio, equity ratio, income ratio, and inventory-turnover ratio.

D.4. Compute overhead rates, given the allocation basis, total overhead expenses, and estimated activity level.

D.5. Compute the actual cost of overhead and the variance, given the allocation basis, overhead rate, and observed activity level.

STUDY GUIDE

D.1 The Balance Sheet

At the end of each fiscal year a company publishes a *balance sheet*. A sample balance sheet for the Wimble Corporation is presented in Table D.1. This is a yearly presentation of the state of the firm at a particular time—for example, December 31, 19XX; however, a balance sheet is also usually prepared quarterly and monthly. Note that three main categories are used. These divisions and their meanings are the following:

Assets. A summary of all resources owned by or owed to the firm. Two classes of assets are distinguishable. *Current* assets represent the short-lived working capital of the company (cash, accounts receivable, etc.), which can be converted to cash in approximately 1 year. Long-lived assets are referred to as *fixed*. Conversion of these holdings (land, equipment, etc.) to cash in a short period of time would require a major company reorientation.

Liabilities. A summary of financial obligations of the company.

Net worth (equity). A summary of the worth of ownership, including outstanding stock issues and earnings retained for expansion.

Table D.1 Sample balance sheet

Wimble Corporation
Balance Sheet
December 31, 19XX

Assets				Liabilities		
Current						
Cash		$ 10,500		Accounts payable	$ 19,700	
Accounts receivable		18,700		Dividends payable	7,000	
Interest accrued receivable		500		Long-term notes payable	16,000	
Inventories		52,000		Bonds payable	20,000	
Total current assets			$ 81,700	Total liabilities		$ 62,700
Fixed				**Net worth**		
Land		$ 25,000				
Building and equipment	$438,000			Common stock	$275,000	
				Preferred stock	100,000	
Less: Depreciation allowance	82,000	356,000		Retained earnings	25,000	
Total fixed assets			381,000	Total equity		400,000
Total assets			$462,700	Total liabilities and equity		$462,700

The balance sheet is constructed using the basic equation

$$\text{Assets} = \text{liabilities} + \text{net worth}$$

Note in Table D.1 that each major category is subdivided into particular titles. For example, accounts receivable represents all money owed to the firm by its customers. Further utilization of some of these subcategories will be made when business ratios are discussed.

Prob. D.1

D.2 The Income Statement and the Cost-of-Goods-Sold Statement

A second important financial statement utilized to present the relations between the major categories in the balance sheet is the *income statement* (Table D.2). The income statement shows this relation by summarizing the profit or loss situation of the firm for a preceding, stated amount of time. The major categories of an income statement are the following:

Revenues. All sales and interest revenue that the company has received in the past accounting period.

Expenses. A summary of all expenses for the accounting period. Some expense values (for example, income taxes and cost of goods sold) are itemized in other statements.

Table D.2 Sample income statement

Wimble Corporation
Income Statement
Year Ended December 31, 19XX

Revenues		
Sales	$505,000	
Interest revenue	3,500	
Total revenues		$508,500
Expenses		
Cost of goods sold (Table D.3)	$290,000	
Selling	28,000	
Administrative	35,000	
Other	12,000	
Total expenses		365,000
Income before taxes		143,500
Taxes for year		71,750
Net profit for year		$ 71,750

Table D.3 Sample cost-of-goods-sold statement

Wimble Corporation Statement of Cost of Goods Sold Year Ended December 31, 19XX		
Materials		
Inventory, January 1, 19XX	$ 54,000	
Purchases during year	174,500	
Total	$228,500	
Less: Inventory, December 31, 19XX	50,000	
Cost of materials		$178,500
Direct labor		110,000
Prime cost		288,500
Factory expense		7,000
Factory cost		295,500
Less: Increase in finished goods inventory during year		5,500
Cost of goods sold (Table D.2)		$290,000

The income statement is published at the same time as the balance sheet. The income statement uses the equation

$$\text{Revenues} - \text{expenses} = \text{profit (or loss)}$$

The *cost of goods sold* is an important accounting term to the engineering economist. It represents the net cost of producing the product marketed by the firm. The economist will find a statement of the cost of goods sold, such as that shown in Table D.3, useful in determining exactly how much it costs to make a particular product over a stated time period, usually a year. Note that the total of the cost-of-goods-sold statement is entered as an expense item on the income statement. This total is determined using the relations

$$\begin{aligned} \text{Direct materials} + \text{direct labor} &= \text{prime cost} \\ \text{Prime cost} + \text{factory expense} &= \text{cost of goods sold} \end{aligned} \tag{D.1}$$

The item factory expense includes all overhead charged to produce a product. Overhead accumulation systems are discussed later in this chapter. Cost of goods sold is also called *factory cost*.

Probs. D.2 and D.3

D.3 Business Ratios

Accountants frequently utilize ratio analysis in discussing the state of the company over time and in relation to industry norms. Because the engineering economist must communicate with the accountant, the economist should have a general idea of the types of ratios commonly employed by accountants. Information for these ratios is extracted from the balance sheet and income statement. Tables D.1 and D.2 are used for the numerical examples of this section.

Current ratio This ratio, utilized to analyze the company's working-capital conditions, is defined as

$$\text{Current ratio} = \frac{\text{current assets}}{\text{current liabilities}}$$

Current liabilities include all short-term debts, such as accounts and dividends payable. Note that only balance-sheet data are utilized in the current ratio; thus, no association with revenues or expenses is made. For the balance sheet of Table D.1 current liabilities amount to \$19,700 + \$7000 = \$26,700 and

$$\text{Current ratio} = \frac{81{,}700}{26{,}700} = 3.06$$

Since current liabilities are those debts payable in the next year, the current-ratio value of 3.06 means that the current assets would "cover" short-term debts approximately three times. Current-ratio values of 2 to 3 are common. For comparison purposes, it is necessary to compute the current ratio, and all other ratios, for several companies in the same industry. Industry-wide median ratio values are periodically published by firms such as Dun and Bradstreet [1]. You should realize that the current ratio assumes that the working capital invested in inventory can be converted to cash quite rapidly. Often, however, a better idea of a company's *immediate* financial position can be obtained by using the acid-test ratio.

Acid-test ratio (quick ratio) This ratio, defined by the relation

$$\text{Acid-test ratio} = \frac{\text{quick assets}}{\text{current liabilities}} = \frac{\text{current assets} - \text{inventories}}{\text{current liabilities}}$$

is meaningful for the emergency situation when the firm must cover short-term debts using its readily convertible assets. For the Wimble Corporation,

$$\text{Acid-test ratio} = \frac{81{,}700 - 52{,}000}{26{,}700} = 1.11$$

Comparison of this and the current ratio shows that approximately two times the current debts of the company are invested in inventories. However, an acid-test ratio of approximately 1.0 is generally regarded as a strong *current* position, regardless of the amount of assets in inventories.

Equity ratio This ratio has historically been a measure of financial strength since it is defined as

$$\text{Equity ratio} = \frac{\text{stockholder's equity}}{\text{total assets}}$$

For the Wimble Corporation,

$$\text{Equity ratio} = \frac{400{,}000}{462{,}700} = 0.865$$

that is, 86.5% of Wimble is stockholder-owned. The $25,000 retained earnings is also called *equity* since it is actually owned by the stockholders, not the corporation. An equity ratio in the range 0.80 to 1.0 usually indicates sound financial condition, with little fear of forced reorganization because of unpaid liabilities. However, a company with virtually no debts, that is, one with a very high equity ratio, may not have a promising future, because of its inexperience in dealing with short- and long-term debt financing.

Income ratio This often quoted, and often misused, ratio is defined as

$$\text{Income ratio} = \frac{\text{net profit}}{\text{total revenue}} (100\%)$$

Net profit is the after-tax value from the income statement. For the Wimble Corporation,

$$\text{Income ratio} = \frac{71{,}750}{508{,}500} (100\%) = 14.1\%$$

Such a high percentage as this for the Wimble Corporation may, at first glance, look highly favorable. However, often a large income ratio reflects an inability to earn a high rate of return on resources owned and employed by the firm. In recent years, corporations have pointed to small income ratios, say 2.5 to 4.0%, as indications of sagging economic conditions. In truth, for a relatively large-volume, high-turnover business, an income ratio of 3% is quite healthy. Of course, a steadily decreasing ratio indicates rising company expenses, which absorb the potentially higher net profit after taxes.

Inventory turnover ratio This ratio indicates the number of times the average inventory value passes through the operations of the company. If turnover of inventory to net sales is desired the formula is

$$\text{Net sales to inventory} = \frac{\text{net sales}}{\text{average inventory}}$$

where average inventory is the figure recorded in the balance sheet. For the Wimble Corporation this ratio is

$$\text{Net sales to inventory} = \frac{505{,}000}{52{,}000} = 9.71$$

which means that the average value of the inventory has been sold 9.71 times during the year. Values of this ratio vary greatly from one industry to another.

If inventory turnover is related to cost of goods sold, the ratio to use is

$$\text{Cost of goods sold to inventory} = \frac{\text{cost of goods sold}}{\text{average inventory}}$$

where average inventory is computed as the average of the beginning and ending inventory values in the statement of cost of goods sold. This ratio is commonly used as a measure of the inventory turnover rate in manufacturing companies. It varies with industries, but management likes to see it remain relatively constant as business increases. For Wimble, using the values in Table D.3

$$\text{Cost of goods sold to inventory} = \frac{290{,}000}{1/2(54{,}000 + 50{,}000)} = 5.58$$

There are naturally many other ratios a company's personnel can use in various circumstances; however, the ones presented here are commonly used by both accountants and engineering economists.

Example D.1 The median value for three ratios of nationally surveyed companies are presented hère. They are extracted from Ref. 1 for three different industries [Standard Industrial Codes (SIC) are also shown]. Compare the Wimble Corporation ratios with these norms and comment on differences and similarities.

Ratio	House furnishings (SIC 2392)	Manufactured ice (SIC 2097)	Soaps and detergents (SIC 2841)
Current	1.89	1.37	2.71
Net sales to inventory	7.7	37.3	8.9
Income	2.85%	6.71%	3.37%

SOLUTION The corresponding ratio values for Wimble computed above are

$$\text{Current ratio} = 3.06$$

$$\text{Net sales to inventory} = 9.71$$

$$\text{Income ratio} = 14.1\%$$

If Wimble makes furniture or ice, its current ratio is higher than expected, since it can cover current liabilities over three times compared with the averages of 1.89 and 1.37 found in the survey. If Wimble makes ice, its inventory turnover is too low compared with the 37.3 times published for other companies.

If Wimble makes soaps and detergents, it is probably better off using these few ratios as criteria, since the current and inventory turnover are comparable and the income ratio is much higher than that of other companies in the same SIC classification.

Probs. D4 and D5

Table D.4 Factory-expense allocation bases

Overhead category	Allocation basis
Taxes	Space occupied
Heat, light	Space, usage, number of outlets
Power	Space, direct labor hours, direct labor cost, machine hours
Receiving, purchasing	Cost of materials, number of orders, number of items
Personnel, machine shop	Direct labor hours, direct labor cost
Building maintenance	Space occupied, direct labor cost

D.4 Allocation of Factory Expense

All costs incurred in the production of an item are accounted for by the cost-accounting system. It can be generally stated that the statement of cost of goods sold (Table D.3) is the end product of this system. Cost accounting accumulates the material, direct labor, and factory-expense costs by using *cost centers*. For example, a department or a machine may be a cost center. All costs incurred in the department or in the utilization of the machine are thus collected under one cost-center title, such as machine 300. Since direct materials and labor are assignable to a cost center, the accountant has only to keep track of these costs. Of course, this is no easy chore and often the cost of the cost-accounting system prohibits collection of direct costing data in the detail that the accountant or the engineering economist might desire.

By far the most burdensome chore of cost accounting is the allocation of factory expense, or overhead. The costs associated with property taxes, service and maintenance departments, personnel, supervision, utilities, etc., must be allocated to the using cost center. Detailed collection of these data is cost prohibitive and often impossible; thus, allocation schemes are utilized to distribute the expenses on a reasonable basis. A description of some of the bases is given in Table D.4. The most common bases used are direct labor cost and direct labor hours.

Most allocation is accomplished utilizing a predetermined factory expense rate, computed using the relation

$$\text{Overhead rate} = \frac{\text{estimated overhead costs}}{\text{estimated basis level}} \tag{D.2}$$

The estimated overhead cost is the amount allocated to the particular cost center. For example, if a company has two producing departments, the overhead allocated to *each department* is used in Eq. (D.2) when determining the departmental overhead rate. Example D.2 illustrates the use of Eq. (D.2) when the cost center is a machine.

Example D.2 An engineering economist is attempting to compute overhead rates for the production of glass bookends. The following information is obtained from last year's budget for the three machines used to produce this product.

Cost source	Allocation basis	Estimated activity level
Machine 1	Labor cost	\$10,000
Machine 2	Labor hours	2,000 hours
Machine 3	Material cost	\$12,000

Determine overhead rates for each machine if the estimated factory expense budget for producing the bookends is \$5000 per machine.

SOLUTION Applying Eq. (D.2) for each machine, rates for the year are

$$\text{Machine 1 rate} = \frac{\text{overhead budget}}{\text{direct labor cost}} = \frac{5000}{10{,}000}$$

$$= \$0.50 \text{ per dollar labor}$$

$$\text{Machine 2 rate} = \frac{\text{overhead budget}}{\text{direct labor hours}} = \frac{5000}{2000}$$

$$= \$2.50 \text{ per labor hour}$$

$$\text{Machine 3 rate} = \frac{\text{overhead budget}}{\text{material cost}} = \frac{5000}{12{,}000}$$

$$= \$0.42 \text{ per material dollar}$$

COMMENT Once the product has been manufactured and actual direct labor costs, hours, and material costs are computed, each dollar of direct labor cost spent on machine 1 implies that \$0.50 in factory expense will be added to the cost of the product. Similar expense values will be added for machines 2 and 3.

The amount of overhead allocated to an asset is a component of the annual operating cost (AOC). Other elements of AOC are directly assignable expenses such as insurance and direct labor cost.

Solved Problem D.4
Probs. D.6 to D.8

D.5 Factory-Expense Computation and Variance

Once the overhead rates are computed and data collected, it is possible to actually determine the cost of production. This cost may be to operate a department or a machine or to manufacture a product, depending upon the manner in which the cost centers are set up. A *cost center* is simply a term used for the specific overhead account; for example, department *A* may be one cost center.

Assuming that the estimated overhead budget was correct, we know that total allocated overhead should equal estimated overhead. However, since some error in budgeting always exists, there will be some over- or underallocation of overhead, termed *variance*. Experience in overhead estimation assists in reducing the variance at the end of the accounting period. Example D.3 illustrates overhead allocation and variance computation.

Example D.3 The economist who collected the data and determined rates in Example D.2 is now ready to compute the actual cost of producing the product. Assume that you are the economist and do the computation, if the actual cost data of Table D.5 are available from cost-accounting records for the year.

SOLUTION To determine actual production costs, the economist should use the factory cost relation given in Eq. (D.1). That is,

$$\text{Factory cost} = \text{materials} + \text{labor} + \text{factory expense}$$

To determine factory expense, which is the overhead expense, the rates from Example D.2 are utilized as follows:

$$\begin{aligned}\text{Machine 1 overhead} &= (\text{labor cost})(\text{rate}) = 2500(0.50)\\ &= \$1250\end{aligned}$$

$$\begin{aligned}\text{Machine 2 overhead} &= (\text{labor hours})(\text{rate}) = 750(2.50)\\ &= \$1875\end{aligned}$$

$$\begin{aligned}\text{Machine 3 overhead} &= (\text{material cost})(\text{rate}) = 19{,}550(0.42)\\ &= \$8211\end{aligned}$$

$$\text{Total factory overhead expense} = \$11{,}336$$

Thus, factory cost is the sum of material, labor, and overhead charges, or a total of \$43,186 for the year.

COMMENT The variance for factory expense in this example is 3(\$5000) − \$11,336 = \$3664 under the budget. The \$15,000 budget for the three machines represents a 32.3% allocation over actual overhead. This analysis should allow a more realistic overhead budget for the production in the future years.

Actually, once estimates of overhead costs are determined, it is possible to

Table D.5 Actual data used for overhead allocation for Example D.3

Cost source	Machine number	Actual cost	Actual hours
Material	1	\$ 3,800	
	3	19,550	
Labor	1	2,500	650
	2	3,200	750
	3	2,800	720

make an economic analysis of the present operation versus a proposed or anticipated operation. Such a study is explained in the Solved Problems.

Solved Problems D.5 and D.6
Probs. D.9 to D.12

SOLVED PROBLEMS

Example D.4 The ABC Company produces many products, two of which are stereo cabinets and Formica-topped dinette sets. A total of \$33,000 is allocated to overhead for next year. Management wants to determine overhead rates on the basis of direct labor hours for the two processing and one finishing departments (*a*) individually and (*b*) using an overall (blanket) basis. Table D.6 presents departmental overhead allocation and processing time. Develop the rates for management.

SOLUTION

(*a*) The rate for each department is computed using Eq. (D.2):

$$\text{Stereo processing} = \frac{145{,}000}{20{,}000} = \$7.25 \text{ per hour}$$

$$\text{Dinette processing} = \frac{145{,}000}{10{,}000} = \$14.50 \text{ per hour}$$

$$\text{Finishing} = \frac{10{,}000}{6000} = \$1.67 \text{ per hour}$$

(*b*) A blanket rate is found by computing

$$\text{Overhead rate} = \frac{\text{total overhead allocation}}{\text{total labor hours}}$$

$$= \frac{300{,}000}{36{,}000}$$

$$= \$8.33 \text{ per hour}$$

COMMENT Of course, an overall rate is easier to compute and use; however, it

Table D.6 Overhead allocation for 1 year for Example D.4

Allocation	Stereo processing	Dinette processing	Finishing
Overhead dollars	\$145,000	\$145,000	\$10,000
Direct labor hours	20,000	10,000	6,000

cannot account for differences in the nature of the work of departments "blanketed" under the rate. See Example D.5.

Sec. D.4

Example D.5 Use the two sets of rates determined in Example D.4 to compute factory expense if the following processing times per unit are correct:

Stereo: Processing = 2 hours
Finishing = 8 hours

Dinette: Processing = 7 hours
Finishing = 3 hours

SOLUTION

(*a*) For the stereo, total overhead per unit is computed as 2($7.25) + 8($1.67) = $27.86; for the dinette sets 7($14.50) + 3($1.67) = $106.51. The larger expense is allocated to the dinette because of its large processing time and high overhead rate compared with low costs in finishing. Factory expense is the sum of $134.37 per unit.

(*b*) For the blanket rate we use a total of 10 hours to manufacture either item, so overhead for both is 10($8.33) = $83.33. Factory expense is $166.66 per unit.

COMMENT Factory expense doesn't change much, but by using the blanket rate, the stereo is overcharged and the dinette undercharged in overhead. For this reason, blanket rates can be inaccurate and misleading.

Sec. D.5

Example D.6 For several years a certain company has purchased the motor and frame assembly of its major product line at an annual cost of $1.5 million. The thought now is to make the components in-house, using existing departmental facilities. For the three departments involved the overhead rates, estimated material, labor, and hours are quoted in Table D.7. The allocated hours column is the time necessary to produce the motor and frame only.

Table D.7 Production cost estimates for Example D.6

	Overhead				
Department	Basis, hours	Rate, $/hour	Allocated hours	Material cost	Direct labor cost
A	Labor	$10	25,000	$200,000	$200,000
B	Machine	5	25,000	50,000	200,000
C	Labor	15	10,000	50,000	100,000
				$300,000	$500,000

To produce the products, equipment must be purchased. The machinery has a first cost of \$2 million, a salvage value of \$50,000, and a life of 10 years. Perform an economic analysis of the suggestion to make the components, assuming a 15% return is required.

SOLUTION For making the components in-house, the annual operating costs (AOC) are composed of labor, material, and overhead costs. Using the data of Table D.7, we find that overhead is

Department A:	25,000(10) =	\$250,000
Department B:	25,000(5) =	125,000
Department C:	10,000(15) =	150,000
		\$525,000

Thus,

$$\text{AOC} = 500{,}000 + 300{,}000 + 525{,}000 = \$1{,}325{,}000$$

The EUAC is

$$\begin{aligned}\text{EUAC}_{\text{make}} &= P(A/P, i\%, n) - \text{SV}(A/F, i\%, n) + \text{AOC}\\ &= 2{,}000{,}000(A/P, 15\%, 10) - 50{,}000(A/F, 15\%, 10) + 1{,}325{,}000\\ &= \$1{,}721{,}037\end{aligned}$$

Currently,

$$\text{EUAC}_{\text{buy}} = \$1{,}500{,}000$$

Therefore, it is cheaper to continue to buy the motor and frames assembly.

Sec. D.5

REFERENCE

1. Dun and Bradstreet, *Key Business Ratios*, Dun and Bradstreet, New York, 1979.

PROBLEMS

The following financial data have been gathered for the month of July 19XX for the Stop-Gap Tool and Die Company. You will use this information in Probs. D.1 to D.5.

Present Situation, July 31, 19XX

Account	Balance
Accounts payable	$ 35,000
Accounts receivable	29,000
Bonds payable (20 year)	110,000
Buildings (net value)	605,000
Cash on hand	17,000
Dividends payable	8,000
Inventory value (all inventories)	31,000
Land value	450,000
Long-term mortgage payable	450,000
Retained earnings	154,000
Stock value outstanding	375,000

Transaction for July, 19XX

Category		Amount
Direct labor		$ 50,000
Expenses		
Insurance	$ 20,000	
Selling	62,000	
Rent and lease	40,000	
Salaries	110,000	
Other	62,000	
Total		294,000
Income taxes		20,000
Increase in finished goods inventory		25,000
Materials inventory, July 1, 19XX		46,000
Materials inventory, July 31, 19XX		25,000
Materials purchases		20,000
Overhead charges		75,000
Revenue from sales		500,000

D.1 Using the data summary in the tables (*a*) construct a balance sheet for the Stop-Gap Company as of July 31, 19XX and (*b*) determine the value of each term in the basic equation of the balance sheet.

D.2 (*a*) Prepare the cost-of-goods-sold statement for July 19XX.
(*b*) What was the net change in materials inventory value during the month?

D.3 Use the account summary above to develop (*a*) an income statement for July 19XX and (*b*) the basic equation of the income statement. (*c*) What percent of revenue is reported as after-tax income?

D.4 (*a*) Compute the value of each accounting ratio that uses only balance-sheet information from the statement you constructed in Prob. D.1.
(*b*) What percent of the company's current debt is "tied up" in inventory?

D.5 (*a*) Compute the turnover of inventory (based on net sales) and the income ratio for the Stop-Gap Company.
(*b*) What percent of each sales dollar can the company rely upon as profit?

D.6 A company has a processing department with 25 machines. Owing to the nature and use of three of these machines, each is considered a separate account for factory overhead. The remaining 22 machines are grouped under one account number, M104. Machine hours are used as an overhead allocation basis for all machines. A total of $25,000 in overhead is allocated to the department for next year. Using the data below, determine (*a*) the overhead rate for each account and (*b*) the blanket rate for the department.

Account number	Overhead allocated	Estimated machine hours
M101	$ 5,000	600
M102	5,000	200
M103	5,000	800
M104	10,000	1,600

D.7 A machining department's overhead is allocated by accounting. The department manager has just obtained records of allocation rates and actual charges for the prior months and estimates for this month (May) and next month. However, the basis of allocation is not indicated and accounting has no record of the basis used. You, an engineer with the company, are asked to investigate the allocation for each month. You are given the following information by the department manager as it comes from accounting.

	Overhead		
Month	Rate	Allocated	Charged
February	$1.40	$2800	$2600
March	1.33	3400	3800
April	1.37	3500	3500
May	1.03	3600	
June	0.92	6000	

You collect the following data from departmental and accounting records.

	Direct labor		Material	Space,
Month	Hours	Costs	costs	square feet
February	640	$2560	$5400	2000
March	640	2560	4600	2000
April	640	2560	5700	3500
May	640	2720	6300	3500
June	800	3320	6500	3500

(*a*) Determine the actual allocation basis used each month and (*b*) comment on the decreasing overhead rate published each month by accounting.

D.8 An electronics manufacturing company has five separate departments. Departmental charges for a certain month are given below. Also detailed are the space allocations, direct labor hours, and direct labor costs for each producing department.

Department	Overhead costs	Space, square feet	Direct labor	
			Hours	Costs
Preparation	$20,000	10,000	480	$1,680
Subassemblies	15,000	18,000	1,000	3,250
Final assembly	10,000	6,000	600	2,460
Quality control	5,000	1,200		
Engineering	9,000	2,000		

Determine the departmental overhead rates to be used in making a redistribution of the overhead cost for quality control and engineering ($14,000) to the other three departments on the basis of (*a*) space, (*b*) direct labor hours, and (*c*) direct labor costs.

D.9 Compute the actual overhead allocation and department variance for Prob. D.6 using (*a*) the individual account rates and (*b*) the blanket rate. The actual hours credited to each account are as follows: M101, 700 hours; M102, 350 hours; M103, 650 hours; M104, 1300 hours.

D.10 Overhead allocation rates and the allocation bases for the six producing departments of the E-Z-Duz-It Calculator Manufacturers are listed below. (*a*) Use the reported data to distribute overhead to the departments. (*b*) Compute the variance of the allocation if a total of $750,000 had been allocated.

Department	Allocation basis*	Rate	Direct labor hours	Direct labor cost	Machine hours
1	DLH	$2.50	5,000	$20,000	3,500
2	MH	0.95	5,000	35,000	25,000
3	DLH	1.25	10,500	44,100	5,000
4	DLC	5.75	12,000	84,000	40,000
5	DLC	3.45	10,200	54,700	10,200
6	DLH	1.00	29,000	89,000	60,500

*DLH = direct labor hours; MH = machine hours; DLC = direct labor costs.

D.11 Perform the overhead allocation of Prob. D.8 using the rates you determined. Use a basis of direct labor hours for the preparation and subassembly departments and direct labor cost for final assembly.

D.12 (*a*) The electronics firm in Prob. D.8 presently makes all the components required by the preparation department. The firm is considering the possibility of purchasing rather than making all these components. The firm has a quote of $67,500 a month from an outside contractor to make the items. If the costs for the month stated in Prob. D.8 are representative and if $41,000 worth of materials is charged to preparation, make an economic comparison of the present versus proposed situation. Assume that the preparation department's share of the quality control and engineering costs is a total of $3230 per month.

(*b*) Another alternative for the company is to purchase new equipment in the preparation department. The machinery will cost $375,000 and have a 5-year life, no salvage value, and a monthly operating cost of $475. This purchase is expected to reduce the quality control and engineering costs by $2000 and $3000, respectively, and reduce direct labor hours to 200 and direct labor cost to $850 for the preparation department. The redistribution of the overhead costs in quality control and engineering to the three production departments is on the basis of direct labor hours. If other costs remain the same, compare the present cost of making the components with the cost of the equipment purchase alternative. A return of a nominal 12% per year compounded monthly is required on all capital investments.

APPENDIX

E

COMPUTER APPLICATIONS

This appendix briefly introduces several computer programs available to solve specific problems in engineering economy. The programs are:

1. *R*eturn *o*n *i*nvestment *d*etermination *s*ystem (ROIDS)
2. *Min*imum *c*ost *l*ife determination (MINCL)
3. *Dep*reciation techniques and *strat*egies (DEPSTRAT)
4. *A*ccelerated *C*ost *R*ecovery *S*ystem (ACRS)

The programs are all interactive and written in BASIC for easy implementation on microprocessor systems such as the Apple series. No expertise in the technique is necessary to operate the programs, but a familiarity with the material in this text will allow the user to understand more easily the program output.

Code for all programs is available in the instructors manual or from the authors. The instructors manual includes a users guide explaining the use of each system.

ROIDS—Return on Investment Determination System

This program will find the rate-of-return value i^* given any cash-flow sequence for a single project (Chap. 7) or the incremental cash-flow sequence for two projects (Chap. 8). Both fixed amount and fixed percent gradients are allowed in the cash-flow sequence. The program will find all real i^* values between −12.5 and 500% to an accuracy of 0.1% If there is no i^* found in this range, the user is so informed.

If the conditions for multiple rate-of-return values are present in the cash-flow sequence, all real roots within the specified range are determined. As discussed in Sec. 7.5, these rates each assume that any net positive cash flow in a particular year is reinvested at the rate of return which balances the rate-of-return equation. If a specific reinvestment rate applies (Sec. 7.5), the user simply inputs this figure and ROIDS proceeds by converting the cash-flow sequence to a conventional sequence and finds the correct i^* value to balance the equation.

It is possible to see the effect of inflation on the rate of return using ROIDS. The user can input an inflation factor for each year of the sequence—for example, 10% per year. When inflation is accounted for, the user is given the i^* value with inflation considered. Once i^* is computed ROIDS allows the user to input a new cash-flow sequence or edit the current sequence to solve for i^* under slightly different conditions.

MINCL—Minimum Cost Life Determination

This program will calculate the EUAC value for an asset, assuming it is retained for 1 year, 2 years, and so on until the EUAC value reaches a minimum and begins to increase thereafter (Sec. 10.6). The year value (n) and the minimum EUAC are given to the user for use in economic analysis. Additionally, the EUAC for all years are displayed for user information.

The user inputs the first cost, salvage value, and annual operating cost for each year of possible ownership. The salvage value and operating costs may be input as a constant for all years, with a positive or negative gradient (uniform or percentage) or as different values for each year.

If the minimum EUAC life cannot be found in the range of years input, the user is given the opportunity to input salvage values and annual operating costs for additional years.

DEPSTRAT—Depreciation Techniques and Strategies

The DEPSTRAT program package is capable of displaying the annual depreciation amounts, additional 20% first-year depreciation (where applicable), book values, and present worth of the tax savings due to depreciation cash flow for the following pre-1981 (classical) depreciation methods and switching between methods as indicated by the user:

Straight line (SL)
Declining balance (DB)
Sum-of-years-digits (SYD)
Switch from SYD to SL
Switch from DB to SL

Switching takes place to the SL method from the SYD or DB method in the user-specified year. Additionally, the user may request optimal switching between all three methods so that the best depreciation strategy is determined by the software with the switch taking place when the annual SL depreciation exceeds the SYD or DB depreciation. This request will also produce the information for all methods without switching.

The depreciation rate for declining balance is input as a percent—for instance, 200% is interpreted as double-declining-balance depreciation (Sec. 13.4) and the rate of depreciation in Eq. (13.8) would be $d_M = 200/100n = 2/n$, whereas an input of 150% causes the rate to be $1.5/n$.

Income taxes are accounted for by inputting a value from 0 (no taxes considered) to 100 percent. The software will compute two present-worth values, one of the tax savings and one of the actual depreciation. The present worth of tax savings due to depreciation is computed as

$$\text{PW}_{\text{save}} = \sum_{t=1}^{n} \text{(Depreciation for year } t\text{) (tax rate) (single payment present-worth factor)}$$

$$= \sum_{t=1}^{n} D_t T(1+i)^{-t}$$

where n is the depreciable life and i is the expected rate of return. The depreciation schedule with the largest PW_{save} value indicates the selected method because it offers the greatest tax advantage in present worth dollars.

ACRS–Accelerated Cost Recovery System[1]

This program computes the depreciation schedule and present-worth values for any personal or real property utilizing the inputs of year of purchase, recovery period, type of property, and appropriate ACRS depreciation rates. The rates are permanently stored in the system and applied to the unadjusted basis (first cost) of the property.

It is possible to perform a depreciation analysis utilizing the standard ACRS recovery periods or the optional (longer) recovery periods with the required half-year convention. Included in the software is the ability to neglect or accept the Section 179 capital expense and the investment tax credit as discussed in Sec. 13.7.

[1]We are indebted to Don Smith for his contribution of the ACRS program to this collection.

APPENDIX

F

ANSWERS TO PROBLEMS

CHAPTER 1

P1.1 See Sec. 1.1.
P1.2 See Sec. 1.1.
P1.3 See Sec. 1.1.
P1.4 6 months.
P1.5 \$600.
P1.6 \$1666.67.
P1.7 10.55%.
P1.8 See Sec. 1.3.
P1.9 15%.
P1.10 There is some risk associated with investments other than those considered safe. Higher risk requires higher potential return.
P1.11 Invest at 7% simple interest for extra interest of \$18.98.
P1.12 (*a*) \$27.41; (*b*) 4.57% of *P*.
P1.13 (*a*) \$27.00; (*b*) 4.50% of *P*.
P1.14 \$1280.
P1.15 \$620.
P1.16 \$756.50.
P1.17 9.54%.
P1.18 Invest for 4 years.
P1.19 \$150.
P1.20 5.36 years.

P1.21 5.67%.
P1.22 $P = \$1000$ every 2 years, $F = ?, i = 10\%, n = 9$ years.
P1.23 $P = \$709.90, A_1 = \$100, n_1 = 5, A_2 = \$200, n = 2, i = 6\%$.
P1.24 $P = \$1400, F = \$4200, i = 6\%, n = ?$.

Note: In P1.26 to P1.41, we tabulate or explain the cash flows rather than draw the cash-flow diagrams, but the numbers should be the same as yours.

P1.26

Year	Payment	Income	Net cash flow
0	0	\$2000	\$+2000
1–4	0	0	0
5	$F = ?$	0	$-F = ?$

P1.27

Year	Deposit	Withdrawal	Net cash flow
May 1, 1984	\$500	0	\$–500
May 1, 1985–1994	0	$A = ?$	$+A = ?$

P1.28

Year	0, 2, 4	1, 3	5, 7, 9, 11, 12, 13	6, 8, 10
Net cash flow	\$–500	0	+300	–200

P1.29 Plan 1: +\$5000 in year 0, –\$10,056.80 in year 5; plan 2: +\$5000 in year 0, –\$750 in years 1 to 4, –\$5750 in year 5; plan 3: +\$5000 in year 0, –\$1750 in year 1, –\$1600 in year 2, –\$1450 in year 3, –\$1300 in year 4, –\$1150 in year 5; plan 4: +\$5000 in year 0, –\$1491.58 in years 1 to 5.
P1.30 $P = -?$ in year 0, $F = \$+3000$ in year 5, $i = 8\%/\text{year}$.

P1.31

Year	Deposit	Withdrawal	Net cash flow
0–4	\$700	\$ 0	\$–700
5–8	0	0	0
9	0	3000	3000
10–12	0	$A_2 = +?$	$A_2 = +?$

P1.32

Year	Deposit	Withdrawal	Net cash flow
0–1	\$0	\$ 0	\$ 0
2	$P_1 = -?$	0	$P_1 = -?$
3	0	0	0
4	$P_2 = -?$	0	$P_2 = -?$
4–8	0	100	100
9	0	500	500

P1.33 Plan 1: P_1 = \$-351.80 in year 0, P_2 = \$-351.80 in year 3, F = +? in year 6; plan 2: A = \$-136.32 in years 1 to 6, F = +? in year 6.
P1.34 P = -?, F = \$+580 in year 8.

P1.35

Year	Deposit	Withdrawal	Net cash flow
0	\$ 0	\$0	\$ 0
1–5	100	0	−100
6–14	0	0	0
15	0	F = +?	F = +?

P1.36 P = +? in year 0, F_1 = \$-1200 in year 5, F_2 = \$-2200 in year 8, i = 10%/year.
P1.37 P = +? in year 0, A = \$-85 in years 3 to 8, i = 20% per year.
P1.38 P = \$-10,000 in year 0, F = +? in year 10, n = 10, i = 12%.
P1.39 P = \$-4100 in year 0, F = \$+7500 in year 5, n = 5, i = ?%.

P1.40

Year, k	Deposit	Withdrawal	Net cash flow
0	\$500	\$0	\$-500
1–6	500 + 50(k)	0	−500 − 50(k)
6	0	F = +?	F = +?

P1.41 P_1 = -\$4500 in year 0, P_2 = -\$3300 in year 3, P_3 = -\$6800 in year 5, A = +? in years 1 to 8, i = 8%/year.

CHAPTER 2

P2.1 SSPWF: Move F to year $n - 1$ to get a factor of $1/(1+i)^{n-1}$.
USPWF: Move all A values back 1 year to get a P/A factor in year 0 of $(1+i)^{n-1}/[i(1+i)^{n-1}]$.
USCAF: Move all A values back 1 year to get an F/A factor in year n of $[(1+i)/i]\ [(1+i)^n - 1]$.
P2.2 (*a*) 0.02718; (*b*) 1.4693; (*c*) 3.8372; (*d*) 0.11017; (*e*) 33.582.
P2.3

Year	1	2	3	4	5	6	7
Cash flow, \$	60	60	60	100	110	120	130

P2.5 3.484
P2.6 (*a*) 7.6954, 5513.3834, 0.5133, 0.00001; (*b*) 7.6910, 5162.5537, 0.5129, 0.00001.
P2.7 (*a*) 3.1684, 0.16493, 0.00308, 0.00225; (*b*) 3.1670, 0.16473, 0.00305, 0.00219.

P2.8 (*a*) 0.7509, 10.3126, 44.374, 0.00488; (*b*) 0.7504, 10.2903, 43.6445, 0.00472.
P2.9 16.029, 1.723, 5.2395, 13.057.
P2.10 F = \$13,636.
P2.11 A = \$725.36.
P2.12 P = \$2369.68.
P2.13 F = \$52,251.45.
P2.14 A = \$1073.34.
P2.15 P = \$6810.
P2.16 P = \$8233.77.
P2.17 P = \$2353.44.
P2.18 P = \$3331.50.
P2.19 A = \$2374.20.
P2.20 F = \$153,220.02.
P2.21 F = \$15,016.50.
P2.22 P = \$5956.63.
P2.23 F = \$12,084,737 (from equation).
P2.24 P = \$1357.16.
P2.25 n = 3.6 years.
P2.26 P = \$4105.60.
P2.27 \$29.08.
P2.28 A = \$5239.25.
P2.29 A = \$115.50.
P2.30 F = \$164,494.
P2.31 F = \$14,586.12.
P2.32 A = \$990.
P2.33 \$11,183.38.
P2.34 \$10,497.23.
P2.35 Year 1, \$8700; year 2, \$7706.11.
P2.36 P_1 = \$2500, P_2 = \$2380.50.
P2.37 F simple = \$69,300; F compounded = \$182,648.40.
P2.38 P later = \$22,569.30, buy later.
P2.39 (*b*) G = \$22; (*c*) P_G is in year 0; (*d*) $n = 8$.
P2.40 (*a*) A = \$5710; (*b*) P = \$20,582.80.
P2.41 x = \$818.70.
P2.42 P = \$12,019.14.
P2.43 A = \$9693.50.
P2.44 P = \$17,545.72.
P2.45 A = \$24,230.20.
P2.46 \$3147.42.
P2.47 G = \$4964.69.
P2.48 P = \$2755.65.
P2.49 G = \$1163.50.

P2.50 $G = \$-6391.97$, reduction: year 2 = \$33,608.03, year 3 = \$27,216.06, year 4 = \$20,824.09.
P2.51 $i = 4.36\%$.
P2.52 $i = 3.62\%$.
P2.53 $i = 4.0\%$.
P2.54 $i = 12.05\%$.
P2.55 $i = 10.78\%$.
P2.56 $i = 6.87\%$.
P2.57 (*a*) $i = 0\%$; (*b*) $i = 5.18\%$.
P2.58 $i = 8.88\%$.
P2.59 $n = 9.68$ years.
P2.60 $n = 20.95$ years.
P2.61 $n = 15$ years.
P2.62 $n = 6.88$, or 7 years.
P2.63 $n = 6.82$ years.
P2.64 $n = 7.11$ years.
P2.65 $n = 22.3$ years.
P2.66 $n = 17.7$ years.
P2.67 $n = 16.9$ years.

CHAPTER 3

P3.1 (*a*) 13% per year; (*b*) 4% per year.
P3.2 Year.
P3.3 Percent per compounding period.
P3.4 $r = 18\%$, $i = 19.56\%$ per year.
P3.5 $r = 8\%$, $i = 8.16\%$ per year.
P3.6 12.36%.
P3.7 16.99%.
P3.8 $r = 21\%$, $i = 23.14\%$.
P3.9 1.47% per quarter.
P3.10 $F = \$4682.50$.
P3.11 $r = 18.44\%$, $i = 20.08\%$ per year.
P3.12 $P = \$5080.65$.
P3.13 $F = \$17{,}252.90$.
P3.14 $A = \$141.21$ per month.
P3.15 $P = \$5906.15$.
P3.16 $P = \$18{,}130.60$.
P3.17 $F = \$80{,}353.90$.
P3.18 $r = 43.676\%$, $i = 53.57\%$.
P3.19 $r = 0.656\%$.
P3.20 $A = \$542.50$ per month.
P3.21 $P = \$3531.30$.
P3.22 $A = \$1433.88$ semiannually.

P3.23 $F = \$2253.68$.
P3.24 $A = \$71.64$.
P3.26 $F = \$6797.10$.
P3.27 (*a*) $F = \$6801.30$; (*b*) $F = \$6802.30$.
P3.28 $F = \$332.74$.
P3.29 $F = \$338.64$.
P3.30 $D = \$258.46$ per month to obtain, $A = \$1570.14$ semiannually.
P3.31 $A = \$199.02$ per month.
P3.32 $F = \$2472.96$.
P3.33 $n = 140$.
P3.34 $i = 15.604\%$.
P3.35 Yes.
P3.36 (*a*) $r = 19.89\%$; (*b*) $r = 12.88\%$.
P3.38 $A = \$2672.91$.
P3.39 \$371.54 is difference.
P3.40 $A = \$937.44$.
P3.41 (*a*) $r = 5.055\%$; (*b*) $i = 5.18\%$.
P3.42 (*a*) $i = 8.15\%$; (*b*) $r = 7.84\%$.
P3.43 (*a*) $\overline{A} = \$46{,}260$; (*b*) $A = \$47{,}704$.
P3.44 $P = \$22{,}996.50$.
P3.45 (*a*) 1629.84; (*b*) $A = \$1590$ per year; (*c*) $A = \$1621.20$; (*d*) $\overline{A} = \$1541.50$, *d* is smallest.
P3.46 (*a*) $r = 6.97\%$; (*b*) $i = 7.22\%$.

CHAPTER 4

P4.1 (*a*) $A = \$-600$ for years 4 to 10, $P = ?$ in year 3, $F = ?$ in year 10, (*b*) $A_1 = \$-2000$ for months 1 to 6 and $A_2 = \$+1500$ for months 3 to 6, $P_1 = ?$ in month 0 and $F_1 = ?$ in month 6, $P_2 = ?$ in month 2 and $F_2 = ?$ in month 6,
(*c*)

Month	Deposit	Withdrawal
December	$P_1 = ?$	\$ 0
January	$A_1 = \$20$	0
February	$A_1 = 20$	$P_4 = ?, 0$
March	$A_1 = 20$	$A_4 = 10$
April	$F_1 = ?, A_1 = 20, P_2 = ?$	$A_4 = 10$
May	$A_2 = 75$	$F_4 = ?, A_4 = 10, P_5 = ?$
June	$A_2 = 75$	$A_5 = 25$
July	$A_2 = 75$	$A_5 = 25$
August	$F_2 = ?, A_2 = 75, P_3 = ?$	$F_5 = ?, A_5 = 25, P_6 = ?$
September	$A_3 = 25$	$A_6 = 10$
October	$A_3 = 25$	$A_6 = 10$
November	$F_3 = ?, A_3 = 25$	$F_6 = ?, A_6 = 10$
December	0	250

P4.2 P in year 19 = \$18,777.96, P = \$1556.69.
P4.3 F in year 16 = \$9895.25, A = 113.20.
P4.4 A = \$4109.60.
P4.5 F = \$27,418.
P4.6 F = \$295,051.
P4.7 P = \$4456.
P4.8 F in year 24 = \$59,078, A = \$7334 for years 25 to 54.
P4.9 P = \$452.92.
P4.10 P = \$1342.
P4.11 A = \$172.
P4.12 F = \$57,907 in year 28.
P4.13 F = \$28,880.
P4.14 P = \$82,228.
P4.15 A = \$1014.93.
P4.16 P in year −1 = \$6312.68, A = 1322.95 in years 0 to 8.
P4.17 x = \$1361.62.
P4.18 F = \$55,920.
P4.19 F = \$63,719.
P4.20 x = \$2277.24.
P4.21 x = \$91,373.
P4.22 \$13,795 at effective i of 13.4% (by interpolation).
P4.23 P = \$25,297.
P4.24 P = \$22,424,631.
P4.25 (*a*) P = \$25,959; (*b*) F = \$120,575 at effective i = 16.6%.
P4.26 A = \$8831.
P4.27 A = \$2555.33.
P4.28 A = \$5492.
P4.29 A = \$3339.
P4.30 n = 9.8 (or 10) for F_4 = \$69,334.30.
P4.31 (*a*) A = \$615.20, (*b*) P = \$3601.01.
P4.32 A = \$931.74.
P4.33 A = \$684.11.
P4.34 A = \$4161.
P4.35 (*a*) P = \$2820 (P for gradient is in year 1); (*b*) A = \$436.32.
P4.36 PW_A = \$14,493, PW_B = \$15,498.
P4.37 P in period −1 is \$7862.40 and A = \$560.06 at effective semiannual i of 3.53%.
P4.38 F in year 7 = \$7149.60, G = \$1457.32.
P4.39 n = 14, last payment is \$240.
P4.40 x = \$234.58.
P4.41 G = \$877.52.
P4.42 x = \$789.29.
P4.43 P = \$24,394.
P4.44 P = \$26,459.
P4.45 x = \$818.70.
P4.46 x = \$853.19.

P4.47 F = \$5142.61.
P4.48 i per year = 8.16%, P = \$223,271.
P4.49 P = \$918.83.
P4.50 P = \$1847.10.
P4.51 n = 7 months with F = \$2368.60 in month 17.
P4.52 P = \$9821.38.
P4.53 P = \$6092.71, A = \$992.
P4.54 P = \$949.97.
P4.55 P = \$31,830, A = \$7,650.
P4.56 P = \$4501.60, F = \$14,758.

CHAPTER 5

P5.1 PW_A = \$30,985, PW_B = \$28,865; select B.
P5.2 PW_{250} = \$53,778, PW_{300} = \$55,000; select 250-mm pipe.
P5.3 PW buy = \$53,529, PW rent = \$43,367; select rent.
P5.4 PW manual = \$6460, PW auto = \$4367; select automatic.
P5.5 PW purchase = \$9573.78, PW lease = \$13,238.10; select purchase.
P5.6 i = 5% per quarter, PW no drains = \$8863.30, PW drains = \$11,330; select no drains.
P5.7 PW none = \$61,679, PW auto = \$24,416; select automatic system.
P5.8 PW purchase = \$65,847, PW lease = \$45,500; select lease.
P5.9 PW_G = \$190,344, PW_H = \$217,928; select G.
P5.10 PW_G = \$204,700, PW_H = \$217,928; select G.
P5.11 PW manual = \$9981.23, PW auto = \$6287.02; select automatically cleaned screen.
P5.12 PW_P = \$83,706, PW_Q = \$78,607; select Q.
P5.13 PW buy = \$100,560, PW lease = \$132,465; purchase clamshell.
P5.14 PW_A = \$35,767, PW_B = \$42,842; select A.
P5.15 PW high = \$16,195, PW low = \$43,186; select high-pressure system.
P5.16 PW pave = \$38,053, PW gravel = \$32,353; select gravel.
P5.17 PW radial = \$1909.81, PW recap = \$1970.06; purchase radial.
P5.18 PW hot-patch = \$55,860, PW resurface = \$69,598; select hot-patch.
P5.19 PW regular = \$66,969, PW auto = \$88,475; select regular.
P5.20 PW manual = \$92,592, PW computer = \$143,851; select manual.
P5.21 PW disposable = \$14,330, PW reusable = \$28,203; use disposable utensils.
P5.22 Expense is \$375,790, income is \$462,441; total capitalized cost is \$86,652 (net profit).
P5.23 A = \$5199.
P5.24 Capitalized cost = \$120,212.
P5.25 Capitalized cost = \$417,391.
P5.26 Capitalized cost = \$42,417.
P5.27 Capitalized cost = \$50,505.

P5.28 EUAC Y = \$27,857.02, capitalized cost Y = \$185,713; EUAC NOT = \$29,562, capitalized cost NOT = \$197,084; select Y.
P5.29 EUAC X = \$75,096.90, capitalized cost X = \$375,484, capitalized cost Z = \$332,014; select machine Z.
P5.30 EUAC X = \$72,920.20, capitalized cost X = \$503,245; capitalized cost Z = \$437,841; select Z.
P5.31 Capitalized cost F = \$11,333,997, capitalized cost G = \$16,207,732; select F.

CHAPTER 6

P6.1 EUAC is cost for infinite number of renewals and PW is cost for one life cycle.
P6.2 A = \$2801.92.
P6.3 A = \$7167.34.
P6.4 (*a*) A = \$2801.94; (*b*) A = \$7167.30.
P6.5 (*a*) A = \$2801.92; (*b*) A = \$7167.34.
P6.6 (*a*) $EUAC_A$ = \$4589.44, $EUAC_B$ = \$4482.62; select B; (*b*) 70 years.
P6.7 EUAC new = \$16,094.45, EUAC used = \$15,542.38; select used.
P6.8 $EUAC_M$ = –\$6624.30 (net income), $EUAC_R$ = –\$10,349.20; select R.
P6.9 $EUAC_1$ = \$11,857.44, $EUAC_2$ = \$15,218.91; select proposal 1.
P6.10 EUAC gaseous = \$789.92, EUAC dry = \$1300; use gaseous chlorine.
P6.11 EUAC R-11 = \$93.86, EUAC R-19 = \$96.40; use R-11.
P6.12 \$27.54.
P6.13 EUAC spray = \$128,000, EUAC immersion = \$26,098.50; use immersion.
P6.14 (*a*) $EUAC_A$ = \$1769.05, $EUAC_B$ = \$1826.42; select A; (*b*) \$157.37.
P6.15 $EUAC_A$ = \$2005.04, $EUAC_B$ = \$5614.75; select plan A.
P6.16 EUAC scram = \$3419.04, EUAC catch = \$5311.87; purchase scram-um.
P6.17 EUAC spray = \$163,109, EUAC truck = \$175,956; select spray.
P6.18 $EUAC_1$ = \$15,978.90, $EUAC_4$ = \$14,955.52; select method 4.
P6.19 n diesel = 9, n gasoline = 6, EUAC diesel = \$996.45, EUAC gasoline = \$1369.07; purchase diesel trucks.
P6.20 n = 8, EUAC asphalt = \$4991.49, EUAC fiberglass = \$5959.26; select asphalt.
P6.21 EUAC metal = \$38,992.82, EUAC concrete = \$26,127.59; select concrete.
P6.22 EUAC foam = \$146.43 (net annual cost), EUAC batts = \$– 634.84 (net annual savings); use batts.
P6.23 $EUAC_A$ = \$10,900, $EUAC_B$ = \$12,545; select plan A.
P6.24 EUAC = \$7043.88.
P6.25 Effective i = 18.81%, EUAC = \$8088.73.
P6.26 EUAC = \$375,509.
P6.27 EUAC = \$1,483,779.
P6.28 P = \$165,722, A per student = \$2651.55.
P6.29 P in year 14 is \$108,333.33, A = \$3344.25.
P6.30 First deposit = \$5734.59, second deposit = \$6734.59.
P6.31 F in year 12 = \$19,930.15, A = \$2989.52.
P6.32 Time between last deposit and first withdrawal is 5.3 years.
P6.33 F in year 11 = \$23,225.42, A = \$3019.30.

CHAPTER 7

P7.1 See Sec. 7.1.
P7.3 $i^* = 6.26\%$.
P7.4 $i^* = 5.97\%$.
P7.5 $i^* = 1.91\%$.
P7.6 $i^* = 40.95\%$.
P7.7 $i^* = 10.78\%$.
P7.8 $r = 16.5\%$, $i = 1.37\%$ effective monthly (maintenance cost occurs in year 12 also).
P7.9 $i^* = 3.63\%$.
P7.10 $i_A^* = 8.56\%$, $i_B^* = 17.26\%$, $i_C^* = 9.56\%$, overall $i^* = 10.4\%$.
P7.11 $i^* = 6.22\%$ per month nominal or 74.64% per year.
P7.12 $i^* = 37.19\%$.
P7.13 $i^* = 10.06\%$.
P7.14 (*a*)

i, %	0	10	20	30	40
P, $	0	711	745	116	-625

;

(*b*) 0 and 32%; (*c*) 0 and 31.6%.
P7.15 $i_1^* = 39.9\%$, $i_2^* = 53.77\%$ (ROIDS).
P7.17 (*a*) $i_1^* = 25\%$, $i_2^* = 400\%$.
P7.18 (*a*) $i_1^* = 0\%$, $i_2^* = 256.15\%$, (*b*) $i' = 3.13\%$.
P7.19 (*a*) $i^* = 400\%$; (*b*) $i_1^* = 21.55\%$, $i_2^* = 278.45\%$; (*c*) $i' = 46.15\%$; (*d*) $i' = 0\%$ (ROIDS).
P7.20 See Sec. 7.5.
P7.21 (*a*)

c, %	15	25	35	45	50
i', %	0	3.20	59.26	82.76	93.33

CHAPTER 8

P8.1

Year	0	1–4	5	6–9	10
S-R	$–4000	–500	6500	–500	500

P8.2

Year	0	1	2	3	4	5	6
A-B	$–4000	1200	10,700	–11,800	10,700	1200	1700

P8.3

Year	0	1–3	4	5–7	8	9–11	12
Y-X	$–9000	1800	–700	1800	–700	1800	3300

P8.4

Year	0	1, 2, 4, 5, 7, 8, 10, 11, 13, 14, 16, 17, 18	3, 9, 15	6, 12
Des.-Gr.	$–2340	110	130	–220

P8.6 $i_{S\text{-}R}^* < 0\%$; select *R*.

P8.7 $i^*_{A\text{-}B} = 58.2\%$; select A.
P8.8 $i^*_{Y\text{-}X} = 11.85\%$; select X.
P8.9 $i^*_{D\text{-}G} = -4.87\%$; select grass.
P8.10 $i^* = 7.07\%$; not justified.
P8.11 $i^* = 11.85\%$; select X.
P8.12 $i^* = 1.27\%$; select grass.
P8.13 $i^*_{A\text{-}M} = 37.85\%$, no difference; select A.
P8.14 Select M, different answer now.
P8.15 (*a*) $i^* = 11.6\%$; select gravel; (*b*) $i^* = 6.71\%$; select gravel.
P8.16 See Sec. 8.5.
P8.17 See Sec. 8.5.
P8.18 (*a*) Select method 2, EUAC = \$2512; (*b*) incremental $i^* = 31.9\%$.
P8.19 Select method 3, incremental $i^* > 20\%$.
P8.20 Select proposal 6, incremental $i^* > 500\%$.
P8.21 (*a*) Select machine 4; incremental $i^* = 38.45\%$; (*b*) $P = \$-104{,}924$ at 18%.
P8.22 (*a*) Select 200 mm, $P = \$-59{,}089$ at 8%; (*b*) incremental $i^* = 16.49\%$.
P8.23 (*a*) Select 10-square-meter truck, incremental $i^* = 20.54\%$; (*b*) EUAC = \$2222 at 18%.
P8.24 (*a*) Select process 4, $P = \$-60{,}242$, at 15%; (*b*) incremental $i^* = 28.3\%$.
P8.25 (*a*) Select pressing 2, slicing 2, weighing 2, and wrapping 1; (*b*) investment = \$45,000, AOC = \$37,000.
P8.26 (*a*) Select pressing-slicing 3 and weighing-wrapping 3; (*b*) investment = \$55,000, AOC = \$27,000; (*c*) selection same as *a*.

CHAPTER 9

P9.1 Disbenefits are losses to the people and should therefore be considered in the numerator.
P9.2 (*a*) Disbenefit; (*b*) cost; (*c*) benefit; (*d*) disbenefit; (*e*) cost.
P9.3 $B/C = 0.29$ by EUAC; do not construct dam.
P9.4 (*a*) $B - C = \$-139{,}380$; no; (*b*) $B/C = 0.715$; no.
P9.5 EUAC disbenefits = \$24,072, B = \$513,452, G = \$18,017.
P9.6 EUAC = \$155,770, $B/C = 1.12$; extend canals.
P9.7 EUAC = \$173,560, $B/C = 0.92$; no.
P9.8 (*a*) B = user cost savings = \$576,000 per year, cost is extra EUAC = \$2,117,320 for transmountain, $B/C = 0.27$; select long route; (*b*) $B - C = \$-1{,}541{,}320$; select long route.
P9.9 $(B/C)_A = 0.77$, $(B/C)_B = 1.28$, $(B/C)_C = 1.20$; build at sites B and C.
P9.10 Resurface EUAC = \$296,850, new EUAC = \$918,603, annual user costs: resurface \$720,000, new \$600,000, $B/C = 0.99$; select resurface.
P9.11 (*a*) Location E only, $(B - D)/C = 1.15$, W versus E, $(B - D)/C = 0.40$, W not justified; select E; (*b*) $(B - D) - C = \$60{,}000$ for E only, W versus E, $(B - D) - C = \$-280{,}000$; select E.
P9.12 (*a*) Lined EUAC = \$337,920, not lined EUAC = \$122,905, $B/C = 0.56$; do not line; (*b*) $B - C = \$-95{,}015$; do not line.

P9.13 EUAC difference in costs = \$20,732,036, EUAC difference in benefits = \$2,123,504, $B/C = 0.10$; construct conventional stadium.

P9.14 Only the best alternative should be selected.

P9.15 Select C with $B/C = 1.07$ for B versus C.

P9.16 Do nothing versus 1 $B/C = 1.67$–select 1; method 2 versus method 1 $B/C = 3.76$–eliminate method 1; method 3 versus method 2 $B/C = 1.67$–eliminate method 2; method 4 versus method 3 $B/C = 0.63$–eliminate method 3; method 5 versus method 3 $B/C = 0.87$–eliminate method 5, select method 3.

P9.17 Method 1 $B/C = 1.67$, method 2 $B/C = 2.11$, method 3 $B/C = 2.01$, method 4 $B/C = 1.67$, method 5 $B/C = 1.46$; select all methods.

P9.18 Rank alternatives: DN, 2, 3, 6, 1, 5, 4, 7; DN versus 2 $B/C = 1.39$, eliminate DN; 2 versus 3 $BC = 3.73$, eliminate 2; 3 versus 6 $B/C = 63.55$, eliminate 3; 6 versus 1 $B/C = -15.11$, eliminate 1; 6 versus 5 $B/C < 1.0$, eliminate 5; 6 versus 4 $B/C < 1.0$, eliminate 4; 6 versus 7 $B/C < 1.0$; select proposal 6.

P9.19 140 mm versus 160 mm $B/C = 200/178.75 = 1.12$–eliminate 140 mm; 160 mm versus 200 mm $B/C = 1.57$–eliminate 160 mm; 200 mm versus 240 mm $B/C = 0.93$–eliminate 240 mm; 200 mm versus 300 mm $B/C = 0.19$–eliminate 300 mm; select 200-mm pipeline.

P9.20 200 mm versus 240 mm: benefits same but costs lower for 240-mm pipe; therefore, eliminate 200 mm; 240 mm versus 300 mm $B/C = 0.07$–eliminate 300-mm pipe; select 240-mm pipe.

CHAPTER 10

P10.3 (*a*) P = \$18,000, $n = 8$, SV = \$1000, AOC = \$150; (*b*) sunk cost = \$-8000.

P10.4 (*a*)

	P	n	SV	AOC
Downtown	\$25,000	10	0	\$1080
Suburban	750	10	\$750	6600

(*b*) Sunk cost = \$-3424.

P10.5 (*a*) \$2925; (*b*) \$5300.

P10.6 Select old with EUAC_{old} = \$79,731.

P10.7 Select old.

P10.8 Select old with EUAC_{old} = \$87,117.

P10.9 Select new with EUAC_{new} = \$2646.

P10.10 Select suburban with EUAC_s = \$2606.

P10.11 Select conveyor plus old mover with EUAC = \$23,955.50.

P10.12 Select two movers with EUAC = \$35,344.56.

P10.13 Select plan III with EUAC_c = \$6348.

P10.14 Select combination electric-microwave range with EUAC = \$294.86.

P10.15 Select old with EUAC_{old} = \$64,736.

P10.16 Select two mover plan with EUAC = \$21,727.86.

P10.17 Select plan I with EUAC_A = \$5275.27.

P10.18 Select B-2 with $EUAC_{B-2}$ = \$33,971.65.
P10.19 Purchase challenger now.
P10.20 Keep old compressor 1 year, then trade on new one.
P10.21 Keep pressing equipment all 5 years.
P10.22 Keep old car all 3 years.
P10.23 RV = \$11,527.
P10.24 RV = \$51,607.
P10.25 (*a*) RV = \$54,622; (*b*) increase of 5.84%.
P10.26 RV = \$20.65.
P10.27 RV = \$-26,620 (pay less than RV for disposal).
P10.28 Retain ownership 9 years at EUAC = \$2256.
P10.29 Retain ownership 8 or 9 years at EUAC = \$2148.
P10.30 Retain full 10 years at EUAC = \$10,837.
P10.31 n^* = 12.6 years.
P10.32 Retain for total of 14 years at EUAC = \$1879.

CHAPTER 11

P11.1 See Sec. 11.1.
P11.2 I = \$125, 6 months.
P11.3 Three months (quarterly), I = \$150.
P11.4 I = \$18.75, monthly.
P11.5 P = \$10,680.
P11.6 P = \$16,357.
P11.7 P = \$6611.
P11.8 P = \$16,524, no.
P11.9 Semiannual i = 6.77%, P = \$10,372 (interpolation).
P11.10 (*a*) P = \$17,900 (interpolation); (*b*) semiannual i = 9.34%, P = \$11,716.
P11.11 Conventional bonds, i = 3.5% per quarter, I = \$300 per quarter, P = \$8804; put bonds, i = 3.125% per quarter, P = \$9702; difference = \$898 per \$10,000 bond.
P11.12 I = \$120,000 per 6 months, P = \$2,485,092.
P11.13 Semiannual i = 5.06%, P = \$2,461,776.
P11.14 \$2,628,598.
P11.15 I = \$400 per 6 months, sales price = \$13,810.
P11.16 F in year 9 = \$2032.53, F in year 17 = \$8070.
P11.17 Original i = 8.38%, F = \$1380.
P11.18 n = 8.07 semiannual periods, new maturity date is 4.04 years from now, years moved forward is 21.
P11.19 n = 9.32 semiannual periods or 4.7 years.
P11.20 n = 17.4 semiannual periods or 8.7 years.
P11.21 P = \$5,680,164, face value of convertible bonds = \$14,668,273; for continuous compounding, face value = \$14,775,482.

P11.22 Interest on conventional = \$1,387,500, interest on industrial bonds = \$975,000, P = \$4,111,470.
P11.23 P = \$9,403,450, loss = \$96,550.
P11.24 Nominal i = 15.1% per year compounded semiannually.
P11.25 Nominal i = 12.1% per year compounded semiannually.
P11.26 Nominal i = 11.9% per year compounded quarterly.
P11.27 Nominal i = 5.8% per year compounded semiannually.
P11.28 Effective i = 5.86%.
P11.29 b = 10.78% per year.
P11.30 b = 6.8% per year payable semiannually.
P11.31 b = 11.8% per year payable semiannually.
P11.32 Effective = 16.99%, A = \$10,313.
P11.33 Nominal i = 14.82% per year.
P11.34 Nominal i = 14.22% per year, keep bonds 5 years.
P11.35 Total semiannual expense = \$265,000, nominal i = 11.88% per year.
P11.36 I = \$20 per quarter, nominal i = 11.12% per year.
P11.37 (*a*) Nominal i = 12.46% per year; (*b*) call bonds.

CHAPTER 12

P12.1 i_f = 26.5%, P = \$5612.87 (from equation).
P12.2 i_f = 23.2%, P = \$2826.22.
P12.3 (*a*) P = \$2287.71; (*b*) P = \$1575.22.
P12.4 P = \$36,772.80.
P12.5 i = 2.515% per month, P = \$9515.36.
P12.6 F = \$84,387.
P12.7 (*a*) i = 20%, F = \$82,413.60, (*b*) i_f = 32%, F = \$161,000.
P12.8 F = \$426,275.30.
P12.9 (*a*) i = 15%, F = \$133,504.80; (*b*) i_f = 25.6%, F = \$346,273.
P12.10 F = \$58,007.
P12.11 F(accumulated) = \$37,052.60; F(today's) = \$13,173.49.
P12.12 (*a*) P = \$11,142.67; (*b*) P = \$11,142.54.
P12.13 P = \$14,556.26.
P12.14 F = \$48,470.59.
P12.15 P_A = \$44,418.26, P_C = 32,036.27; select machine C.
P12.16 $P_{used} = P_{new}$, P = \$627,479.
P12.17 P_A = \$215,302 (i = 15%), P_B = \$217,725 (i_f = 25.6%); purchase from A.
P12.18 i_f per quarter = 6.08%, P = \$6343.73.
P12.19 A = \$10,461.
P12.20 A = \$16,712.
P12.21 F in year 16 = \$422,535 (i_f = 23.2%), A = \$9883 (i = 12%).
P12.22 P today's = \$533,115.50 (i = 14%), A = \$136,892 (i_f = 25.4%).
P12.23 A = \$30,585.07.
P12.24 Fixed dollar increase versus fixed percentage increase.

P12.25 $P_E = \$18,435.52$.
P12.26 $P_E = \$48,435.60, P = \$65,132$.
P12.27 $P = \$75,103, F = \$108,535$.
P12.28 $P = \$3425.10, D = \$239.76$.
P12.29 $P = \$2580.90, D = \$209.30$.
P12.30 $n = 7.8$ years.
P12.31 $P = \$94,520$.
P12.32 $P = \$90,041$.
P12.33 $i_f = 24.2\%, P = \$96,995$.
P12.34 (*a*) $A = \$41,381$; (*b*) $A = \$49,605$.

CHAPTER 13

P13.1 $P = \$55,000$, $n = 10$, SV $= \$5200$, actual life $= 5$, market value $= \$8700$, book value $= \$13,750$.
P13.2 (*a*) $P = \$350,000$; (*b*) SV $= \$27,500$; (*c*) $D_t = \$10,750$; (*d*) $BV_{20} = \$135,000$.
P13.3 (*a*) $D_t = \$1250$; (*b*) $BV_1 = \$10,750$, $BV_2 = \$9500, \ldots, BV_8 = \$2000$; (*c*) $d_t = 1/8$.
P13.5 Selected answers are as follows.

t	D_t	BV_t	d_t
1	\$2222.22	\$9777.78	8/36
4	1388.89	4777.78	5/36
8	277.78	2000.00	1/36

P13.6

t	D_t	BV_t
2	\$6660.82	\$68,286.55
7	4701.75	40,859.65
12	2742.69	23,228.07
18	391.81	15,000.00

P13.7 (*a*) $BV_3 = \$4714.19$; (*b*) $d_4 = 1/7$; (*c*) $D_4 = \$1457.14$.
P13.8 In DDB, $d = 2/n$.

P13.9

t	D_t	BV_t	d_t
1	\$3000.00	\$9000.00	0.2500
2	2250.00	6750.00	0.1875
3	1687.50	5062.50	0.1406
4	1265.63	3796.88	0.1055
5	949.22	2847.66	0.0791
6	711.91	2135.74	0.0593
7	135.74	2000.00	0.0113
8	0	2000.00	0

P13.10

t	D_t SL	SYD	DDB
4	\$8000	\$13,935.48	\$17,345
9	8000	11,354.84	12,285
18	8000	6,709.68	6,602
26	8000	2,580.65	0

P13.11 (*a*) BV_{13} = \$216,000; (*b*) BV_{13} = \$158,967.74; (*c*) BV_{13} = \$130,499.22.

P13.12 (*a*)

t	D_t DDB	D_t DB	BV_t DDB	BV_t DB
2	\$8099	\$6719	\$64,790	\$67,897
7	4494	4192	35,954	42,358
12	2494	2615	19,952	26,426
18	0	1485	15,000	15,000

P13.13 (*a*)

t	D_t	BV_t
1	\$21,875.00	\$153,125.00
12	5,035.43	35,248.02

(*b*) SV for DB is larger.

P13.14 Switch to SL in year 5, P_D = \$30,198.

P13.15 Switch to SYD in year 3, P_D = \$30,259.

P13.16 (*a*) P_D = \$65,067; (*b*) no switch for P_D = \$65,906.

P13.17 (*a*) No; (*b*) $d < 0.04425$.

P13.18 (*a*) Switch to SYD in year 4 for P_D = \$5936.06; (*b*) switch to SYD in year 1 for P_D = \$5790.97.

P13.19 (*a*) P_D = \$6511; (*b*) switch in year 7 for P_D = \$6611.

P13.20

Method	P_D
SL	\$12,977
SYD	13,952
125% DB	11,694
DDB	14,454

No switching strategy will increase P_D.

P13.21 $t = 7$.

P13.22 (*b*) SL gives P_D = \$12,577.50 and 1985 ACRS gives P_D = \$14,880.93.

P13.23 Switch from DDB to SYD in year 3 for P_D = \$30,259, 1985 ACRS gives P_D = \$29,829.78.

P13.24 (*a*) P_D = \$14,099; (*b*) \$11,125 for DDB.

P13.25 (*a*) Switch is okay; (*b*) pre-1981 switch gives P_D = \$7358, ACRS gives P_D = \$7017 for 5-year recovery on both.

P13.26 179 expense is \$5000, ACRS write off in \$3225.

P13.27 (*a*) \$7500; (*b*) D_1 = \$747; D_2 = \$1577, . . . , D_{10} = \$166.

P13.28

t	D_t (*a*) Group	D_t (*b*) Composite
1–8	\$111,300	\$134,500
9, 10	55,650	14,500
11, 12	55,650	4,000

(*c*) 8.28 years.

P13.29 (*a*)

t	1	2	3–5
D_t	\$191,700	281,160	268,380

(*b*) Same total depreciation as in *a*.

P13.30 (*a*)

t	1	2	3	4	5
Depletion	\$70,000	91,000	104,400	108,000	100,800

(*b*) 13.55%.

P13.31 (*a*)

t	1	2	3	4	5
Depletion	\$70,000	91,000	104,400	129,600	120,960

(*b*) 14.74%.

P13.32

t	D_t	BV_t
1	\$5000	\$77,000
2	6250	70,750
3	4375	66,375

CHAPTER 14

P14.1 (*a*) Recaptured depreciation; (*b*) taxable income; (*c*) operating loss; (*d*) taxable income; (*e*) capital loss; (*f*) taxable income.

P14.2 Dough: \$315,090; broke: \$71,750.

P14.3 Dough: \$335,340 (−6.04%); broke: \$92,000 (−22.01%).

P14.4 (*a*) 37.9%; (*b*) 46%; (*c*) 39.01%; (*d*) \$21,600.

P14.5 \$717,895.

P14.6 (*a*) 22.19%; (*b*) 28.41%; (*c*) $22,728.
P14.7 $1725 increase with new machinery.
P14.8 $20,924.
P14.9 (*a*) LTL = $1000; (*b*) STL = $200; (*c*) STG = $800; (*d*) recaptured depreciation = $300; (*e*) STL = $1750, taxes = $6468.
P14.10 STL = $1750 to offset STG in tax computations.

P14.11

t	Taxes (*a*) Group	Taxes (*b*) Individually*
1976	$76,200	$76,200
1977	86,200	86,200
1978	78,700	79,340
1979	94,800	100,220
1980	60,600	60,600
1981	40,600	44,150
1982	72,800	75,100
	$513,500	$521,810

*Recaptured depreciation taxed as capital gain using the 20% rate.

P14.12 $1775 less in taxes.
P14.13 $115,838.
P14.14 (*a*) $1000; (*b*) $550; (*c*) $420; (*d*) $570; (*e*) $180.
P14.15 (*a*) $104,070; (*b*) $105,206; (*c*) $106,240; (*d*) $104,638; (*e*) $105,155; (*f*) $105,723; (*g*) $105,172, recovery in 1980 is best.
P14.16 (*a*) SYD is less by $3017, or 3.78%; (*b*) SYD is less by $20,198, or 20.8%.
P14.17 *A*: P_{tax} = $23,208; *B*: P_{tax} = $23,431; ACRS is less for *A*.
P14.18 *A*: P_{tax} = $26,820; *B*: P_{tax} = $26,077; SYD is less for *B*.
P14.19 (*a*) $10,239.

(*b*)

Method	P_{tax}
SL	$10,320
DDB	8,371
DDB to SL	8,355
1984 ACRS	10,239

P14.20 (*a*)

t	1	2	3-5
Taxes, $	3460	3208	3244

(*b*) $11,858.
P14.21 $25,000-asset P_{tax} = $2192.90, $40,000-asset P_{tax} = $3958.18; select $25,000-asset.

P14.22

Method	P_{TS}
1984 ACRS	\$13,794
SL	13,134
DDB	14,990
DDB to SL	15,099

P14.23

Method	(*a*) P_{tax}	(*b*) P_{TS}
SL	\$1822.97	\$2664.33
DDB	1464.58	3022.75
DDB to SL	1376.64	3110.70

P14.24 Select 1986 ACRS with P_{tax} = \$2572.15.

P14.25

Method	(*a*) P_D	(*b*) P_{TS}	(*c*) P_{tax}
SL	\$12,997.28	\$5969.55	\$10,612.53
SYD	13,952.24	6418.03	10,163.92
125% DB	11,693.88	5379.18	10,486.04
DDB to SL (or SYD)	14,453.41	6648.57	9,933.30

(*d*) DDB switch to SL (or SYD) is best.

P14.26 $n = 4.97$.

P14.27 Select 1986 ACRS with P_{tax} = \$43,931.

P14.28 Select 3-year period with P_{tax} = \$1999.30.

P14.29 16.16%.

P14.30 (*a*) Neither makes 12%, but *A* is better with $EUAC_A$ = \$-6855; (*b*) neither makes 24%, but *A* is better with $EUAC_A$ = \$-22,956; (*c*) same answer.

P14.31 57.9%.

CHAPTER 15

P15.1

Year	0	1-7	8
Cash flow, \$	-350,000	17,460	406,660

P15.2

Year	0	1-4	5
CFAT, \$	-131,251	39,167	53,167

Recaptured depreciation involved.

P15.3

Year	0	1-4	5
CFAT, \$	-72,917	29,785	41,585

P15.4 (*a*)

Year	0	1(1981)	2(1982)	3–5	6(1986)
CFAT, $	–10,000	3320	3656	3608	4200

Total CFAT = $12,000.
(*b*) No total difference, only in timing.

(*c*)

Year	0	1(1981)–5(1985)	6(1986)
CFAT, $	–10,000	3154	6229

No difference in total, only in timing.

P15.5 (*a*)

Year	0	1	2	3	4	5
CFAT, $	–131,251	54,337	42,200	34,920	30,552	47,827

(*b*) Either method, same total CFAT; (*c*) either method; same total, different timing.

P15.6 (*a*)

Year	0	1(1981)	2(1982)	3–5	6(1986)
CAFT, $	-5000	2242	2578	2530	4200

Total CFAT = $11,610.
(*b*) No total difference, only in timing.

P15.7 Select machine *B* with $EUAC_B$ = $-18,756.
P15.8 Select used with $EUAC_{used}$ = $7159 (small margin in favor of used).
P15.9 Select machine *B* with $EUAC_B$ = $2124.
P15.10 (*a*) Select Roundee with P_R = $- 2043; (*b*) $EUAC_R$ = $-300.
P15.11 Yes, with P_R = $-3351 and $EUAC_R$ = $- 492.
P15.12 Select *A* with P_A = $11,550.
P15.13 Select *B* with P_B = $9841.
P15.14 Select plan *B* with P_B = $31,491.
P15.15 i^* = 6.12% (ROIDS).
P15.16 (*a*) i^* = 17.14%; (*b*) i^* = 28.49%.
P15.17 (*a*) 1981-ACRS i^* = 27.53%; (*b*) SL with no SV has i^* = 27.86%; (*c*) SL with SV has i^* = 25.50%.
P15.18 (*a*) i^* = 45.23%; (*b*) i^* = 46.18%, financing increases i^*.
P15.19 (*a*) i^* = 9.75%; (*b*, *c*, *d*) select plan *B*; (*e*) select plan *A*.
P15.20 i_A^* = 4.0%, i_B^* = 5.93%.
P15.21 i^* = 0.12% (ROIDS).
P15.22 (*a*) $7273; (*b*) $16,020.
P15.23 (*b*) i^* = 33.52% (ROIDS)
P15.24 i_A^* = 19.14%, i_B^* = 24.26%.
P15.25 (*a*) *P* = $42,666; (*b*) SV = $50,000; (*c*) *D* = $13,438.
P15.26 i^* = 1.75% (ROIDS).
P15.27 None.
P15.28 (*a*) Select defender with $EUAC_D$ = $1362; (*b*) select defender with $EUAC_D$ = $673.

P15.29 Select plan III with EUAC = \$3087.
P15.30 (*a*) i^* = 10.46%; (*b*) RV = \$11,028.
P15.31 (*a*) Select challenger with $EUAC_C$ = \$2388; (*b*) no.
P15.32 Retain defender 2 years.

P15.33

Year	0	1	2	3	4	5
CFAT (× 1000), \$	−1750	−50	50.5	189.2	470.8	169.4

$i^* < 0\%$.
P15.34 i^*_A = 15.11%, i^*_B = 18.32%.

CHAPTER 16

P16.1 n' = 5.66 years.
P16.3 (*a*) n' = 15.55 years; (*b*) no.
P16.4 n' = 34.4 years.
P16.5 n' = 3.9 years.
P16.6 (*a*) n' = 2.7 years; (*b*) yes; (*c*) n' = 3.4 years, yes.
P16.7 (*a*) Select machine *B* with n'_B = 3.86 years; (*b*) yes, select *A* with income $EUAC_A$ = \$333.51.
P16.8 (*a*) Purchase n' = 5.9 months, leave n' = 8.65, 17.01, or 25.06 months (multiple answers); (*b*) select purchase with EUAC = \$155.50 per month.
P16.9 (*a*) Select machine 1 with n' = 5.37 years; (*b*) yes, select 2 with $EUAC_2$ = \$951.31.
P16.10 2486 hours per year.
P16.11 \$16,891.
P16.12 94 days per year.
P16.13 566,935 yards per year.
P16.14 (*a*) 14.52 years; (*b*) \$94,450.
P16.15 (*a*) 333 samples per year; (*b*) 202 samples per year; (*c*) select present condition.
P16.16 \$288,557.
P16.17 \$5100.
P16.18 \$850.79.
P16.19 \$49.33 per year.
P16.20 (*a*) 25.9 days per year; (*b*) \$34,744 for 52 weeks per year.
P16.21 \$6739.
P16.22 LCC for proposed system at *P* = \$686,205.

CHAPTER 17

P17.1 Select *A* and *C* with *P* = \$23,748.54.
P17.2 (*a*) Select *B* and *C* with *P* = \$4600.86; (*b*) 15.

P17.3 Select 1, 2, and 4 with $P = \$2240$.
P17.4 (*a*) Select 1 and 3 with $P = \$4773.92$; (*b*) select 1, 3, and 4 with $P = \$5018.92$.
P17.5 Select 3 with $P = \$4261$.
P17.7 Select *A* and *C* with incremental $i^* > 35\%$.
P17.8 (*a*) Select II with incremental $i^* > 22\%$; (*b*) $P = \$1679.58$.
P17.9 Select 1 and 3 with incremental $i^* > 10\%$.
P17.10 Select 3 with incremental $i^* > 15\%$.
P17.13 (*a*) $X_1 = X_3 = 1, X_2 = X_4 = 0$; (*b*) $X_1 = X_3 = X_4 = 1, X_2 = 0$.

CHAPTER 18

P18.1 (*a*) Debt; (*b*) equity; (*c*) debt; (*d*) equity; (*e*) debt.
P18.2 0.5%, no.
P18.3 (*a*) Raise; (*b*) lower.
P18.4 (*a*) \$2,083,333; (*b*) $i^* = 4.48\%$.
P18.5 Recommend bonds with $i^* = 2.9\%$.
P18.6 Yes.
P18.7 (*a*) $i^* = 9.4\%$; (*b*) $i^* = 5.42\%$.
P18.8 $i^* = 7.80\%$.
P18.9 (*a*) BTCC = 12.4%; (*b*) ATCC = 6.52%.
P18.10 Yes, $i^* = 12.2\%$.
P18.11 BTCC = 20.9%.
P18.12 \$−0.18 per share (take funds from retained earnings to pay dividend).
P18.13 Earnings = 3.8% = \$0.22 per share.
P18.14 ATCC = 21.3%, high.
P18.15 Scheme 2 with BTCC = 8.75%.
P18.16 Scheme 1, 12%; scheme 2, 10.7%.
P18.17 $i^* = 12.4\%$, no.
P18.18 100% equity, MARR = 10.2%, return $i^* = 8.5\%$, no: 60% debt, MARR = 5.95%, return $i^* = 15.9\%$, yes.
P18.19 Plan 1, MARR = 15.3%, $i^* = 3.6\%$, no; plan 2, MARR = 16.2%, $i^* = 19.9\%$, yes; plan 3, MARR = 17.4%, $i^* = 100\%$, yes.
P18.20 On a 6-month basis, (*a*) BTCC = 3.02%; (*b*) BTCC = 6.15%; (*c*) BTCC = 4.59%; use plan 1.
P18.21 (*a*) Average BTCC = 19.01%; (*b*) average ATCC = 9.9%.
P18.22 (*a*) $(D/E)_1 = 0.11, (D/E)_2 = 0.67, (D/E)_3 = 4.0$; (*b*) $(D/E)_1 = 0, (D/E)_2 = \infty, (D/E)_3 = 1.0$; (*c*) $D/E = 1.0$.

P18.24

Firm	*A*	*B*	*C*
Return, %	54	67	40

P18.25 (*a*) *A*: debt = \$714,286, equity = \$1,785,714; *B*: debt = \$1,142,857, equity = \$457,143; (*b*) $(D/E)_A = 0.51, (D/E)_B = 5.26$.
P18.26 $i_A = 25.2\%, i_B = 86.9\%$.

CHAPTER 19

P19.1

Tons	10	15	20	25	30
PW, $	-134.922	-137,589	-140,257	-142,924	-166,078

P19.2 Investment approved for income of $18,000 with P_C = $12,514.
P19.3 Same conclusion as P19.2 with P_C = $6428.
P19.4 (*a*) Select lease with P_{lease} = $-41,033.

(*b*)

Cost, $	50	60–90
Decision	Build	Lease

P19.5

G, $	300	400	500	600	700
*i**, %	42.04	41.5	40.95	40.4	39.8

P19.6 Select system 1 for all MARR values.

P19.7

n	4	6	8	10	12
$EUAC_1$	$3832	2811	2307		
$EUAC_2$	. . .	$4610	3746	3239	2907

P19.8 Breakeven is 3.35 days, select cabin for all values.
P19.9 (*a*) Yes; (*b*) breakeven in 9.4 days is sensitive.

P19.10

n	10	12	15	18	20
Annual nominal *i**, %	6.84	6.64	6.40	6.24	6.16

P19.11 $n = 6, P =$ $1935; $n = 10, P =$ $-956; decision is sensitive.
P19.12 (*a*) $P =$ $264,801; (*b*) $P =$ $301,584.

P19.13

G, $	60	80	100	120	140
*n**	16.3	14.1	12.6	11.5	10.7

P19.14

G, $	60	80	100	120	140
*n**	19	16	14	13	12

P19.15 Same decision as P6.7; used is less.

n	2	3	4
$EUAC_{used}$, $	-15,542	-15,251	-15,035

P19.16 No effect, select immersion with EUAC = $26,099.

P19.17

AOC or lease cost, $/year	EUAC			
	50 days per year		100 days per year	
	Plan 1	Plan 2	Plan 1	Plan 2
1800	...	$3800	...	$5800
2000	$3664			
2500	4164	4500	...	6500
3200	...	5200	...	7200
3750	5414			
4000	...	...	$5664	
5000	...	...	6664	
7500	...	...	9164	

(*a*) Plan 1; (*b*) plan 2.

P19.18

Interest rate or life	Present worth	
	Plan *M*	Plan *Q*
(*a*) Pessimistic (15%)	$ -6,111	$ +8,927
Reasonable (10%)	+27,704	+51,758
Optimistic (8%)	+47,272	+76,544
(*b*) Pessimistic (22 years)	+31,573	+56,659
Reasonable (20 years)	+27,704	+51,758
Optimistic (16 years)	+17,356	+38,650

P19.19

Interest rate or life	Plan *A*	Plan *B*
(*a*) Pessimistic (15%)	$-13,317	$-37,293
Reasonable (10%)	-14,867	-38,600
Optimistic (8%)	-15,916	-39,472
(*b*) Pessimistic (44, 22)	-14,908	-38,519
Reasonable (40, 20)	-14,867	-38,600
Optimistic (32, 10)	-14,716	-38,762

P19.21 $E[X] = 9.84$.
P19.22 $1900.
P19.23 E[PW] = $-2,463.
P19.24 E[EUAC] = $-1169.
P19.25 Yes with E[EUAC] = $79,696.
P19.26 Yes with E[EUAC] = $950.
P19.27 Build for 2.5 inches with EUAC = $2183.
P19.28 Build for 2.5 inches with EUAC = $1403.

P19.29

Investment	Savings	Stocks	Property
i^*, %	6.35	10.7	8.54

Stocks are best.

APPENDIX D

D.1 (*a*) Current assets = \$77,000, fixed assets = \$1,055,000 liabilities = \$603,000, equity = \$529,000; (*b*) \$1,132,000 = \$603,000 + \$529,000.

D.2 (*a*) Cost of materials = \$41,000, prime cost = \$91,000, cost of goods sold = \$141,000; (*b*) \$21,000 decrease.

D.3 (*a*) Net profit = \$45,000; (*c*) 9%.

D.4 (*a*) Current ratio = 1.79, acid-test ratio = 1.07, equity ratio = 0.47; (*b*) 72%.

D.5 Inventory turnover = 16.13; (*a*) income ratio = 9%; (*b*) 9%.

D.6 (*a*) 8.33 per hour (M101), \$25 per hour (M102), \$6.25 per hour (M103, M104); (*b*) \$7.81 per hour.

D.7 (*a*) Space (February), DLC (March, April), space (May), material (June).

D.8 (*a*) \$0.41 per square foot; (*b*) \$6.73 per hour; (*c*) \$1.89 per dollar.

D.9 (*a*) \$5831 (M101), \$8750 (M102), \$4063 (M103), \$8125 (M104); \$1769 under; (*b*) \$23,430; \$1570 over.

D.10 (*a*) \$12,500(1) \$23,750(2) \$29,000(6); (*b*) \$90.

D.11 Preparation \$3230, subassembly \$6730, final \$4649.

D.12 Select make at cost of \$65,910; (*b*) continue to make (new equipment cost is \$71,669 per month).

BIBLIOGRAPHY

1. American Telephone and Telegraph: *Engineering Economy,* 3d ed., McGraw-Hill, New York, 1980.
2. Barish, N. N., and S. Kaplan: *Economic Analysis for Engineering and Managerial Decision Making*, 2d ed., McGraw-Hill, New York, 1978.
3. Bussey, L. E.: *The Economic Analysis of Industrial Projects*, Prentice-Hall, Englewood Cliffs, NJ, 1978.
4. Canada, J. R., and J. A. White: *Capital Investment Decision Analysis for Management and Engineering*, Prentice-Hall, Englewood Cliffs, NJ, 1980.
5. DeGarmo, E. P., J. R. Canada, and W. G. Sullivan: *Engineering Economy*, 6th ed., Macmillan, New York, 1979.
6. Fabrycky, W. J., and G. J. Thuesen: *Economic Decision Analysis*, 2d ed., Prentice-Hall, Englewood Cliffs, NJ, 1980.
7. Grant, E. L., W. G. Ireson, and R. S. Leavenworth: *Principles of Engineering Economy,* 7th ed., Wiley, New York, 1982.
8. Kasner, E.: *Essentials of Engineering Economics*, McGraw-Hill, New York, 1979.
9. Mallik, A. K.: *Engineering Economy with Computer Applications*, Engineering Technology, Mahomet, IL, 1979.
10. Morris, W. T.: *Engineering Economic Analysis*, Reston, Reston, VA, 1976.
11. Newnan, D. G.: *Engineering Economic Analysis*, rev. ed., Engineering Press, San Jose, CA, 1980.
12. Reisman, A.: *Managerial and Engineering Economics*, Allyn and Bacon, Boston, 1971.
13. Riggs, J. L.: *Engineering Economics*, 2d ed., McGraw-Hill, New York, 1982.
14. Smith, G. W.: *Engineering Economy: Analysis of Capital Expenditures*, 2d ed., Iowa State University Press, Ames, 1973.
15. Stevens, G. T., Jr.: *Economic and Financial Analysis of Capital Investments*, Wiley, New York, 1979.
16. Taylor, G. A.: *Managerial and Engineering Economy: Economic Decision-Making*, 3d ed., Van Nostrand, New York, 1980.
17. Thuesen, H. G., W. J. Fabrycky, and G. J. Thuesen: *Engineering Economy*, 5th ed., Prentice-Hall, Englewood Cliffs, NJ, 1977.
18. White, J. A., M. H. Agee, and K. E. Case: *Principles of Engineering Economic Analysis*, Wiley, New York, 1977.

INDEX

For HP 41 From IET 424

```
LBL F/P
I=?
XEQ Prompt
1
+
N=?
Xeq Prompt
Y^X
09 END
```

```
LBL F/A
I=?
X Prompt
Enter↑
Enter
1
+
N=?
X Prompt
Y^X
1
-
X<>Y
/
15 End
```

```
LBL A/G
I=?
X Prompt
Enter
Enter
Enter
N=?
X Prompt
X<>Y
1
+
X<>Y
Y^X
X<>Y
RDN
LastX
X<>Y
1
-
20 1/X
*
CHS
X<>Y
1/X
+
26 END
```

```
LBL A/P
I=?
X Prompt
Enter
Enter
1
+
N=?
Prompt
Y^X
*
LastX
1
-
15 1/X
*
17 END
```

Note: For conviniency it might have been better to solve all for P ie P/G, ...

```
LBL P/G
I=?
X Prompt
Enter
05 Enter
STO 42
1
+
N=?
X Prompt
STO 43
Y^X
STO 44
1/X
15 RCL43
16 *
CHS
RCL 44
1
-
RCL 44
RCL 42   ← from IET 42
*
/
25 +
RCL42
1/X
*
29 END
```

ASN ___ __

F/P	21
F/A	22
A/G	23
A/P	24
P/G	25

ENTER INTEREST As Absolute #, Not %

ie 0.095 NOT 9.5